Supercritical Fluid Science and Technology

Volume 7

In situ Spectroscopic Techniques at High Pressure

Supercritical Fluid Science and Technology

Series Editor – Erdogan Kiran

Vol. 1 Supercritical Fluids and Organometallic Compounds: From Recovery of Trace Metals to Synthesis of Nanostructured Materials.
By Can Erkey

Vol. 2 High-Pressure Fluid Phase Equilibria: Phenomenology and Computation.
By Ulrich K. Deiters and Thomas Kraska

Vol. 3 Phase Equilibrium Engineering.
By Esteban Brignole and Selva Pereda

Vol. 4 Introduction to Supercritical Fluids: A Spreadsheet-based Approach.
By Richard Smith, Hiroshi Inomata, and Cor Peters

Vol. 5 Hydrothermal and Supercritical Water Processes.
By Gerd Brunner

Vol. 6 Particle Formation with Supercritical Fluids: Challenges and limitations.
By Michael Türk

Supercritical Fluid Science and Technology

Volume 7

In situ Spectroscopic Techniques at High Pressure

Andreas Braeuer
Erlangen Graduate School in Advanced Optical Technologies (SAOT) and Lehrstuhl für Technische Thermodynamik (LTT)
Friedrich-Alexander-Universität
Erlangen-Nürnberg, Germany

ELSEVIER

AMSTERDAM • BOSTON • HEIDELBERG • LONDON • NEW YORK • OXFORD
PARIS • SAN DIEGO • SAN FRANCISCO • SINGAPORE • SYDNEY • TOKYO

Elsevier
Radarweg 29, PO Box 211, 1000 AE Amsterdam, Netherlands
The Boulevard, Langford Lane, Kidlington, Oxford OX5 1GB, UK
225 Wyman Street, Waltham, MA 02451, USA

Notices
Knowledge and best practice in this field are constantly changing. As new research and experience broaden our understanding, changes in research methods, professional practices, or medical treatment may become necessary.

Practitioners and researchers must always rely on their own experience and knowledge in evaluating and using any information, methods, compounds, or experiments described herein. In using such information or methods they should be mindful of their own safety and the safety of others, including parties for whom they have a professional responsibility.

To the fullest extent of the law, neither the Publisher nor the authors, contributors, or editors, assume any liability for any injury and/or damage to persons or property as a matter of products liability, negligence or otherwise, or from any use or operation of any methods, products, instructions, or ideas contained in the material herein.

British Library Cataloguing-in-Publication Data
A catalogue record for this book is available from the British Library

Library of Congress Cataloging-in-Publication Data
A catalog record for this book is available from the Library of Congress

ISBN: 978-0-444-63422-1
ISSN: 2212-0505

Contents

Foreword

It is with great pleasure that I introduce Volume 7 in the Elsevier Book Series on Supercritical Fluid Science and Technology. This volume entitled, '*In situ* spectroscopic techniques at high pressure' has been written by an eminently qualified scholar, Dr Andreas Braeuer, who is at the Friedrich-Alexander-Universität Erlangen-Nürnberg where he is the Head of the Applied Raman Scattering Laboratory and the High Pressure Process Engineering Group, and the Director of Administration of SAOT, the Erlangen Graduate School in Advanced Optical Technologies. The book is unique not only in its content that focuses on spectroscopy at high pressure as its major theme, but also in its pedagogical and warm personal writing style which reflects his passion and his dedication to teaching. The book has been developed in a manner to make it relatively easy for a beginner to follow and learn from, while satisfying the rigor that more advanced practitioners in the field may be looking for. Each topical area is presented with a detailed account of the fundamentals. Each chapter is enriched with superbly prepared diagrams, end of the chapter questions and illustrations and practical case studies.

Chapter 1 starts with an overview of spectroscopic techniques and their value in high-pressure research and the challenges one would be presented with. The author questions whether high pressure is a 'friend' that facilitates or is an 'opponent' that hinders the utilisation of spectroscopic techniques, and in doing so provides clear presentations of the strengths of the techniques while underlining the challenges and limitations. The appendix to the chapter provides the basic background pertaining to supercritical fluids and their properties.

Chapter 2 is devoted to properties of electromagnetic radiation and photons, and their interaction with matter. A thorough review of the translational, rotational, vibrational and electronic energy of molecules and their relevance to spectroscopy are presented. Reflection, refraction, diffraction are among the topics covered on the interaction of electromagnetic radiation with bulk matter. Absorption and emission processes such as IR and UV absorbance, chemiluminescence, fluorescence, phosphorescence and scattering processes such as the Rayleigh (elastic light scattering) and Raman (inelastic light scattering) scattering from molecules are discussed in detail.

Chapter 3 is focused on Raman spectroscopy. Among the topics covered are sensor design, excitation modes that are employed, system configurations, Raman imaging and signal processing for purification of signals from interferences. The chapter provides many insights on the various components and their engineering. The chapter also provides a rich collection of targeted case studies, illustrating how Raman spectroscopy is used in investigations of vapour–liquid equilibria; gas hydrates; reactive systems; analysis of internal combustion engines; fuel mixtures; heat transfer; mass transfer in sorption or extraction processes; and supercritical fluid-based processes such as the supercritical antisolvent processing.

Chapter 4 is on shadowgraph and Schlieren techniques that rely on the refraction of light passing through a media that displays index of refraction gradients. Working principles are discussed in detail. This chapter is also supported with illustrative case studies that demonstrate how the technique is used in characterisation of sprays and jets; or in determination of interfacial tension by analysis of pendant or levitated drops under pressure; or in drying of aerogels with supercritical fluids; and in applications involving microfluidics at high pressures.

Chapter 5 is devoted to laser-induced fluorescence and phosphorescence techniques with a focus on liquid and also vapour-phase thermometry. This chapter is enhanced with case studies as well. They illustrate the use of the technique in exploring droplet dynamics near critical conditions or jet mixing near critical point; and in analysis of internal combustion engines.

Chapter 6 treats absorption spectroscopy. Fourier transform absorption spectroscopy is discussed in detail. Illustrative case studies include NIR absorption spectroscopy for in line monitoring of extraction processes; absorption spectroscopy for temperature sensing at elevated pressure; and *in situ* imaging of diffusion and dissolution phenomena at high pressures using attenuated total reflection (ATR)–Fourier transform infrared (FTIR) spectroscopy.

I sincerely hope that you will share my enthusiasm and find this volume to be an indispensable resource for learning, teaching and also in applying spectroscopic techniques as non-invasive tools in your own research.

Erdogan Kiran
Series Editor
Blacksburg, Virginia
October 2015

Preface

Isn't it fascinating that humans can use light to carry out one-sided communications with molecules or matter? This kind of communication we call optical spectroscopy. Optical spectroscopy can greatly help us in gaining new and valuable insights into mechanical, thermal and reactive processes, even if they take place at elevated pressure. And, as we use optical measurement techniques, we do not even invade into the processes we intent to analyse. Maybe this attracted your attention and you would want to read about how optical spectroscopy works and how it can be applied in high-pressure process technology. Then I hope that this book can satisfy your inquisitiveness.

Of course in the preface two main questions have to be addressed, which are: firstly, 'why did you write this book' and secondly, 'why did you write it in this way'.

Let me first answer why I wrote this book. In autumn 2013 Erdogan Kiran, the editor of the book series *Supercritical Fluid Science and Technology* in which this book appears, asked me whether or not I would consider writing a book on 'spectroscopic techniques at high pressure'. At this time I was also very close to the completion of my habilitation thesis, which I wrote on '*in situ* Raman spectroscopy for high pressure process technology'. So I suggested to him to replace 'spectroscopic techniques' in the book title he had in mind with '*in situ* Raman spectroscopy'. This would have given me the unique opportunity to publish my habilitation thesis on the great platform the book series provides and—as most of the thesis had already been written—the workload seemed to be manageable to me. Unfortunately for me, but for the good of this book project, I was not able to convince Erdogan of my 'great' idea and so we finally agreed on the book title '*in situ* spectroscopic techniques at high pressure'. I am very grateful to Erdogan that I got the chance to write this book, which from my point of view is the first one of its kind. There are books on high-pressure process technology, there are books on optical measurement techniques, but there has not been a book which covers the physical basics of several measurement techniques together with application oriented discussions relevant for high-pressure process technology. Consequently I was convinced that this book could fill a gap in the education of students, whose background is not optical metrology, but who want to apply optical measurement techniques in order to study *in situ* chemical engineering processes in general, or especially at elevated pressure. For me this circumstance also justified the workload that I thought would arise in the preparation of this book. Now, after the completion of this book, I have to admit that I completely underestimated two things. I underestimated the time I had to invest in order to write this book and to prepare the illustrations. At this point I want to acknowledge the support of my doctoral-researchers for having sketched some of the figures shown in this book and especially Jaypee Quino for his support in gathering the licenses for the figures I took from other publications. Additionally, I underestimated the value that writing this book would have on my own knowledge about this field. I really learned a lot, which I would not have learned if I had not written this book.

Coming back to the title of the book, it was very important for me that '*in situ*' was considered in the title of the book for two reasons. Firstly, without it, the title of the book was too general and then I would not even had a chance to cover all relevant aspects, not even at a rudimental level. Thus '*in situ*' confined essentially the applications of spectroscopic techniques at high pressure that I had to cover. And secondly, I did not want to write a book on spectroscopic devices, which nowadays are commercially available and can be operated by anyone without any specific knowledge about the techniques. Thus '*in situ*', which for me means that we have to measure in place—there where the relevant processes take place—excludes the blind *ex situ* usage of any optical 'black box' instruments.

Why I wrote this book this way? I am a chemical engineer by training and—as the study programme did not focus on this aspect—I did not have the privilege of getting well educated in spectroscopic techniques, and of course, also not in their *in situ* application. Later, during my time as a (doctoral) researcher I became so fascinated by the potential *in situ* optical spectroscopy bears for chemical engineering that I acquired the knowledge required for the realization of *in situ* spectroscopic experiments. Now it is time to acknowledge my early mentors Frank Beyrau and Jan Egermann, who both invested a lot of time educating and training me in theory and in the laboratory. I am convinced that doctoral researchers can save valuable time in the laboratory if their supervisor takes the time to comprehensively instruct, educate and to discuss the experiments with them. During discussions with my doctoral researchers I realize, again and again, that not much has changed with respect to the situation I had to face after the graduation. Chemical as well as mechanical engineers and even physicists—irrespectively of where they had been educated—often lack the knowledge required for setting up self-built *in situ* spectroscopic experiments.

This is why I tried to describe especially the physical fundamentals in a very understandable way. Whenever possible I tried to show the parallels between the basics that graduates know for sure from their education and how these things are considered in optical spectroscopy. One example is the heat capacity, a quantity everybody is familiar with. But when we use the heat capacity in order to make thermodynamic computations, we for sure are not always aware of the fact that this quantity reflects translational, rotational and vibrational energy of molecules, properties of molecules relevant for spectroscopy. I also put a lot of emphasis illustrating the meanings of equations or correlations by many sketches. Most probably I spent more time preparing the figures than writing the text of this book. I hope that the illustrative presentation helped complex facts appear more straightforward.

I also want to acknowledge the support of colleagues from around the world who provided, upon my request, unpublished material I then used here and there for the illustration of the case studies. It is mentioned in the captions of the figures which material I took over from colleagues. I am very grateful to Erdogan Kiran for not only inviting to write the book, but also for reading and commenting on each chapter of the book. Also I am grateful to my mentor Alfred Leipertz who had been accompanying my research activities from the year 2003, when I started as doctoral researcher at his department, to the completion of my habilitation in the year 2014.

For sure I can say that this book project bestowed a time full of deprivation to me and of course this also affected the lives of my loved ones and my relationship with them. Thus I am very grateful for their patience and their support.

I dedicate this book to my beloved parents Josefa and Siegfried, who had been supporting me essentially on my way to a higher education.

Andreas Braeuer

List of Abbreviations and Parameters

Abbreviations and Acronyms

Ac Acetone
DMSO Dimethyl sulphoxide
EtAc Ethyl acetate
EtOH Ethanol
HPV High-pressure vessel
MCP Mixture critical point
PCL Polycaprolactone
PCM Paracetamol (acetaminophen)
Pr Prandtl number
RMSE Root mean square error relative
Sc Schmidt number
YAc Yttrium acetate

Parameters in Latin letters

A	Amplitude of wave function	
$A(\lambda)$	Absorbance at wavelength λ	–
B	Rotational constant	m^{-1}
Bi_D	Biot number of mass transfer	–
c	Speed of propagation of wave	$m\ s^{-1}$
c_{solute}	Solute concentration in the solvent	$g\ L^{-1}$
c_p	Isobaric specific heat capacity	$J\ (mol\ K)^{-1}$
c_V	Isochoric specific heat capacity	$J\ (mol\ K)^{-1}$
c_{medium}	Speed of electromagnetic radiation in medium	$m\ s^{-1}$
c_F	Concentration of potentially fluorescing molecules	$mol\ m^{-3}$
c_Q	Concentration of potential fluorescence quenchers	$mol\ m^{-3}$
D	Centrifugal elongation constant	m^{-1}
el	Electronic quantum number	
d	Drop diameter	m
D	Diameter	m
D_{Diff}	Diffusivity	$m^2\ s^{-1}$
el	Electronic quantum number	
E_0	Amplitude of the electric field	$V\ m^{-1}$
$\vec{E}_0$	Amplitude of electric field vector	$V\ m^{-1}$
E_{Photon}	Energy of one photon	J
$E_{thermal}$	Thermal energy stored in gas	J
E_{kin}	Kinetic energy stored in gas	J
E_{Laser}	Irradiance of the laser beam	$W\ m^{-2}$
E_{rot}	Energy of rotation	J

$\vec{E}(x,t)$	Electric field vector	$V\,m^{-1}$
F	Cross sectional area	m^2
$F(S_{molec})$	Maxwell velocity distribution	
F_{rot}	Rotational energy	m^{-1}
f_{length}	Focal length of a lens	m
f	Degrees of freedom	–
I	Irradiance of point source/Point irradiance	$W\,m^{-2}$
I	Irradiance of point source/Point irradiance	$W\,(sr)^{-1}$
$I_{incident,\lambda}$	Spectral intensity of incident light	$W\,m^{-3}$
$I_{transmitted,\lambda}$	Spectral intensity of transmitted light	$W\,m^{-3}$
J	Rotational quantum number	–
k	Propagation constant $2\pi/\lambda$	m^{-1}
$k_{binding}$	Spring constant	$N\,m^{-1}$
k_{index}	Calibration constant (Experimental efficiency factor) for one compound	–
$K_{index/index}$	Proportionality factor between the signal intensities and the number densities of two different compounds	–
L	Local field	–
L	Length	m
L_{rot}	Angular momentum of rotation	$kg\,m^{-2}\,s^{-1}$
M	Molar mass	$g\,mol^{-1}$
m	Mass	kg
m_{molec}	Mass of single molecule	kg
m_{red}	Reduced mass	kg
N	Number of scatterers, particles or molecules	–
N_F	Number of molecules which potentially could fluoresce	–
NA_{index}	Numerical aperture of a device (component)	–
n_{refr}	Index of refraction	–
n_F	Amount of substance of potentially fluorescing molecules	mol
n_{index}	Amount of substance of compound	mol
OPD	Optical path difference	m
P	Pressure	MPa
P_{Laser}	Power of laser beam	W
p_{Photon}	Momentum of one photon	$N\,s^{-1}$
R	Distance between nuclei	m
R_e	Distance between nuclei at the point of rest (equilibrium)	m
R	Intensity ratio	–
R	Reflectivity or reflectance	–
R_p	Reflectivity of p-polarised light	–
R_s	Reflectivity of s-polarised light	–
r	Distance from the centre of a drop to any point	m
S	Supersaturation	–
S_F	Number of fluorescence photons detected per time	s^{-1}
S_{molec}	Translational velocity of single molecule	$m\,s^{-1}$
T	Temperature	K
T	Transmission	–
T_p	Transmission of p-polarised light	–
T_s	Transmission of s-polarised light	–
T_{el}	Electronic energy	m^{-1}

t	Time	s
$U_{internal}$	Internal energy of gas	J
$U_{potential}$	Potential energy stored in gas	J
U_{chem}	Chemical energy stored in gas	J
V	Volume	m^3
v	Vibrational quantum number	
$x_{compound}$	Molar (mole) fraction of a compound in a mixture	–
$\Delta x_{screen\ wall}$	Distance between diffractive wall and screen	m
Z	Distance	m
$\Delta z_{slit\ centres}$	Distance between slit centres	m
$z_{max,\ n}$	Distance of nth from 0th order diffraction maximum	m

Parameters in Greek letters

a'	Isotropic component of the derived polarisability of the scatterers	$A\ m^2\ V^{-1}\ J^{-0.5}$
γ'	Anisotropic component of the derived polarisability	$A\ m^2\ V^{-1}\ J^{-0.5}$
Φ_F	Fluorescence quantum yield	–
Φ	Radiant power	W
Ω	Solid angle	sr
α_{index}	Angle of incidence/angle of refraction	°
$\alpha'_{xyz,ij}$	Derived polarisability	$A\ m^2\ V^{-1}\ J^{-0.5}$
β	Mass transfer coefficient	$m\ s^{-1}$
ε	Phase difference between waves	–
$\varepsilon(\lambda)$	Molar absorption coefficient	m^2 mol
η	Dynamic viscosity	Pa s
η	Efficiency factor	–
λ	Wavelength (spatial period)	m
λ_L	Wavelength of excitation laser radiation	m
λ_R	Wavelength of Raman scattered radiation	m
λ_{vacuum}	Wavelength of electromagnetic radiation in vacuum	m
ν	Kinematic viscosity	$m^2\ s^{-1}$
$\bar{\nu}_R$	Raman shift	m^{-1}
$\bar{\nu}_e$	Vibrational constant	m^{-1}
$\bar{\nu}$	Wave number	cm^{-1}
ν	Frequency	Hz
ν_e	Eigen frequency	Hz
ρ_{index}	Density of a compound	$kg\ m^3$
σ	Relative standard deviation	–
τ	Temporal period of wave	s
τ_{index}	Characteristic time to the onset of an event	s
φ	Plane angle of detection	°
$\Psi(x,t)$	Wave function	–
ω	Angular frequency	s^{-1}
Θ	Moment of inertia	$kg\ m^{-2}$
θ_{in}	Angle of incidence	°
θ	Plane angle of signal detection	°
θ_{refr}	Angle of refraction	°
θ_{refl}	Angle of reflection	°

CONSTANTS

c_{vacuum}	Speed of light in vacuum	$\approx 3 \cdot 10^8$ m s^{-1}
h	Planck's constant	$6.62606957 \cdot 10^{-34}$ J s
k	Boltzmann's constant	$1.3806488 \cdot 10^{-23}$ J K^{-1}
ε_0	Permittivity of vacuum	$8.854187817.. \cdot 10^{-12}$ A s V^{-1} m^{-1}
R	Ideal gas constant	8.314. J mol^{-1} K^{-1}

INDICES

a	Compound a
b	Compound b
I	Energy level i
HPV	High pressure vessel
i (in context with τ)	Interfacial tension disappearance
J	Energy level j
jb (in context with τ)	Jet break-up onset
k	Energy level k
I	Energy level l
p	Polarisation of electric field is parallel
p (in context with τ)	Start of particle generation
s	Polarisation of electric field is perpendicular
S	Scatterer
x	x direction
y	y direction
z	z direction
xyz	xyz coordinate system

Chapter 1

High Pressure: Fellow and Opponent of Spectroscopic Techniques

In this chapter I will first show the importance of high-pressure technology and then will emphasise the value of using spectroscopic techniques as *in situ* measurement tools. I will specifically emphasise

- how to modify the high-pressure apparatuses/equipment as little as possible to incorporate the essential elements when spectroscopic access has to be provided and at the same time emphasise
- how to modify the spectroscopic technique as much as possible to the needs of the high-pressure process

so that optimum measurement results can be gathered. This sometimes means that standard measurement equipment has to be modified significantly. I will describe in which context high pressure can be considered a fellow or an opponent of spectroscopic techniques and derive why sometimes spectroscopic techniques themselves can be considered as an opponent of research attempts in high-pressure process technology. The advantages of *in situ* spectroscopic techniques compared to conventional analysis techniques will also be described. The purpose of the exercise section of this chapter is especially to make the reader familiar with the nomenclature used in spectroscopy, such as 'temporal resolution', 'sampling rate', 'spatial resolution' and 'dimensionality'. The appendix of this chapter provides some information about the high-pressure process technologies addressed in this first chapter, which, even though would be well known to readers from the high-pressure community, might be useful for non-experts.

1.1 COMPRESSIBLE FLUIDS IN HIGH-PRESSURE PROCESS TECHNOLOGY

Owing to the 'Green Chemistry' (GC) initiative [1], process chemists and engineers have been ambitious to replace environmentally less acceptable solvents with environmentally benign ones. Today supercritical (sc) scH_2O [2–4] and

Supercritical Fluid Science and Technology, Vol 7. http://dx.doi.org/10.1016/B978-0-444-63369-9.00001-X

$scCO_2$ [5,6] are frequently used as solvents in Green Chemistry. A short description about 'What is a supercritical fluid?' can be found in Appendix 1.7.1.1 and a more comprehensive one in an earlier book of this series [7]. From the point of environmentally friendliness, availability and non-flammability water and CO_2 both fulfil the requirements [1], which make chemistry green. Nevertheless it is worth mentioning that in contrast to $scCO_2$, scH_2O is rather corrosive. But, these green solvents also have to satisfy the requirements needed by the process itself, which might be a specific polarity, solvation power, density, viscosity or permittivity [4]. Fortunately, the physicochemical properties of one single fluid can be 'tuned' finely by adjusting its density, or its pressure and temperature. Fluids in a state near their critical parameters feature the highest compressibility. This means that, along the isotherms near the critical temperature, the density and along with it the solvent characteristics of a process fluid can be changed significantly by changing the pressure only marginally. Consequently, one single compressible fluid can act as an appropriate solvent for different solutes in different processes, as its solvent properties can be tuned to the desired ones by adjusting pressure and temperature [8]. These features are discussed further in Appendix 1.7.1.2.

The critical temperature and pressure of water and CO_2 are 647 K and 22.1 MPa and 304 K and 7.4 MPa, respectively. Therefore, to fully exploit the unique tunability of the solvent characteristics of these compressible fluids the respective processes must be carried out at elevated pressures.

Today super- and subcritical[1] process technology can be found in several different disciplines, for example, in material sciences [9–13], in biological engineering [14–16], in medical engineering [6,17], in environmental engineering [18–21], in energy engineering [22–24], in mechanical engineering [25–27], in chemical engineering [5,28–30] and in process technology [31,32]. Especially in the field of high-pressure combustion of fuels, which under ambient conditions are liquid, for example, in diesel-like internal combustion engines, the impact of the unique characteristics of super- or near-critical fluid mixtures on the mixture generation and the combustion that follows is being increasingly recognised [33–39].

Optimisation of the processes taking place at elevated pressures is possible, if the entire functioning chain taking place inside the high-pressure apparatus is well understood. Sometimes, especially when the processes are inhomogeneous, turbulent, dynamic, far from equilibrium or non-stationary, this functioning chain cannot be reflected reliably based on thermodynamic and/or kinetic data or models alone. Then two different approaches, which are both illustrated in Fig. 1.1, can help achieve the objectives.

1. While 'supercritical' refers to operational conditions above the critical point of the employed fluid, 'subcritical' refers to a highly compressed gas phase slightly below its critical point. 'Subcritical' does not exist in thermodynamics, but is commonly used and absolutely accepted in high-pressure process technology.

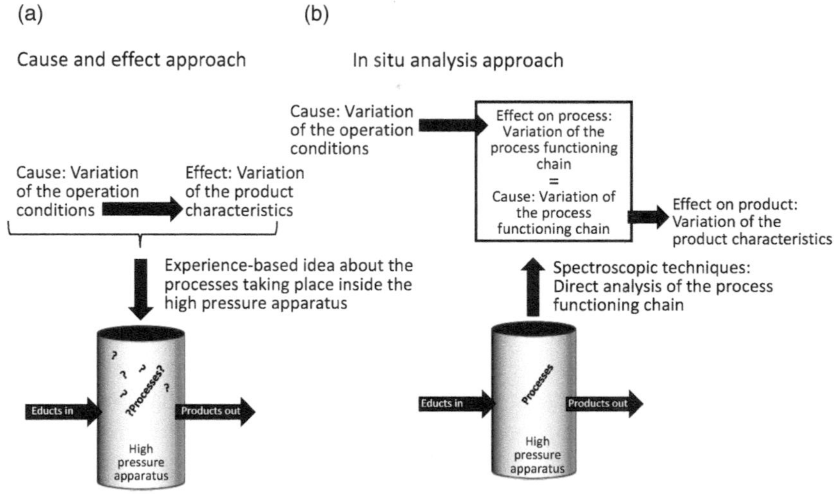

FIGURE 1.1 (a) Illustration of the 'cause and effect' approach and (b) the '*in situ* analysis' approach in high-pressure process technology.

- One approach (illustrated in Fig. 1.1a) is to perform many 'cause and effect' experiments. The variation of the product characteristics, which is the effect, can be analysed as a function of the variation of the operational conditions, which is the cause. Then the experiences can be summarised in an (semi-) empirical or physical model or can be fitted to numerical models, which reflect our current understanding about the processes taking place inside the high-pressure apparatus [40]. Often simplified models have to be considered, as a complete reflection of the real physical scenario can still not be coped with currently widely available computation powers. Nevertheless this 'cause and effect' approach with subsequent modelling can provide an experience-based knowledge-gain about the processes taking place inside the high-pressure apparatus.
- The second approach (illustrated in Fig. 1.1b) is to analyse the functioning chain taking place inside the high-pressure apparatus directly and experimentally by applying *in situ* measurement techniques. In this '*in situ* analysis' approach, the variation of the product characteristics, which still is the effect, can be directly correlated with the variation of the processes taking place inside the high-pressure apparatus, which is a cause with respect to the product characteristics. Simultaneously the variation of the processes taking place inside the high-pressure apparatus is also an effect with respect to the variation of the operational conditions, which is the original cause.

Each of these two approaches has its advantages and disadvantages. The first approach is time consuming and might result in a rather empirical familiarity with the process under investigation than in a physical understanding. Neverthe-less, if the experiments are planned carefully, even from a strategic selection of 'cause and effect' studies some physical understanding of the process can be derived. In contrast, the second approach can directly link the 'cause' with the 'effect', as the intermediate steps, which is the process functioning chain, taking place between the 'original cause', which is the variation of the operational conditions, and the 'final effect', which is the variation of the characteristics of the product, can be resolved with spectroscopic techniques. Thus, the *in situ* application of spectroscopic techniques in only few experiments can help develop a comprehensive physical understanding about high-pressure process technology.

Later I will underline the meaning of the 'cause and effect' approach and the '*in situ* analysis' approach giving one example. Today's knowledge about the precipitation of particles using the supercritical anti-solvent technology has been build up firstly following the 'cause and effect' and then later following the '*in situ* analysis' approach. Therefore I hold the opinion that this example is suitable to show the 'pros' and 'cons' of both approaches. A comprehensive description of the supercritical anti-solvent (SAS) technology is provided in Appendix 1.7.2.

In 1989 Gallagher et al. [41] showed for the first time the potential of supercritical fluids for the anti-solvent technology for the generation of small particles. Afterwards many papers treating this process were published by many different researchers and research groups. On 30th October 2014, a search for 'supercritical' and 'anti-solvent' and 'particles' in the WEB OF SCIENCE™ resulted in 677 counts. Some research groups analysed the influence of pressure, while others analysed the influence of temperature, or the overall mixture composition, or the flow conditions, or the solute concentration, or the apparatus geometry or many more parameters [42–45]. Unfortunately, the findings reported in literature were sometimes controversial, but only at first glance. Later I will describe that when looked more closely the published results have not been as controversial as they seemed to be at first glance. For example, while one research group reported decreasing particle sizes with increasing pressure, another group reported the opposite. Therefore at this stage of the research activities on the SAS-process, it seemed to be impossible to develop—based on the controversial reports on 'cause and effect' experiments—neither an empirical nor a physical understanding of the involved processes. Later in the year 2003 Reverchon et al. succeeded in reducing the number of parameters relevant for the SAS process [46]. They were able to summarise the effect of pressure, temperature and overall mixture composition onto the properties of the produced particles by considering the location of the SAS operation point in the pressure composition (*Px*)-diagram of the binary system composed of solvent and anti-solvent. This has been the decisive step towards a more strategic selection of the operational conditions to be analysed. Reverchon and co-workers were then able to correlate—for different SAS-systems—the location of the operation condition in the *Px*-diagram with the morphology of the produced particles [47–49] and with this significantly developed further our thermodynamic understanding of the SAS-process. It has to be mentioned here that Reverchon and co-workers did not develop their thermodynamic model solely based on their own 'cause and effect' experiments, but also additionally considered 'cause and effect' experiments and very few optical experiments reported by other research groups [44,50,51]. Summarising, the thermodynamic model was able to reflect many 'cause and effect' studies carried out by different research groups. Herewith an essential step towards the comprehensive understanding of the SAS-process was taken. Nevertheless the thermodynamic model was not able to describe all SAS-systems. Furthermore it was not able to reflect the non-thermodynamic influences, such as the flow conditions, or the presence of already precipitated particles and their impact on the mixing behaviour of flows. Therefore, after the 'cause and effect' approach, *in situ* optical measurement techniques, such as shadowgraphy or elastic and inelastic light scattering experiments, were applied to directly visualise the processes taking place inside the SAS-high-pressure-precipitation chamber [31,52–61]. The application of the *in situ* spectroscopic techniques brought two main insights.

- Firstly, the experience based thermodynamic understanding of the SAS-process, which has been based on the 'cause and effect' experiments could be confirmed.
- Secondly, the understanding of the many processes taking place within the functioning chain of the SAS-process could be developed further based on direct physical insights.

Today we know why certain SAS-systems form crystalline and why other SAS-systems form amorphous particles [62]. We know how and why the morphology of the precipitated particles can be tuned from non-spherical nanoparticles to spherical micro-particles or to spherical hollow structures. We know how and why the polymorphism can be influenced between different crystalline structures and how and why formulations or dispersions of different solutes can be formed [63]. Summarising, the combination of the 'cause and effect' and the '*in situ* analysis' approaches resulted in a comprehensive understanding of one example of high-pressure process, the SAS-process. I am convinced that if the '*in situ* analysis' approach had not been applied, our understanding of the SAS process would still remain behind the one we have today, thanks to the '*in situ* analysis' experiments.

1.2 SPECTROSCOPIC TECHNIQUES BRING LIGHT INTO THE DARKNESS OF HIGH-PRESSURE PROCESSES

Compressed or pressurised fluids have to be contained in pressure-resistant containments. Pressure-resistant vessels, pipes and fittings are usually made of steel. Consequently the access to the interior of the pressurised system for light is blocked by the steel walls. This is why we usually cannot see the processes taking place inside the pressurised fluid systems and why, in the heading of this section, I wrote '... the darkness of high-pressure processes'. If we can grant spectroscopic access to the high-pressure system we want to analyse, the spectroscopic techniques offer some advantages compared to conventional techniques. These are as follows.

- Their non-invasive—sometimes referred to as non-intrusive—measurement principle (assuming that the modification of the high-pressure equipment required to provide spectroscopic access does not influence the processes taking place inside the high-pressure apparatuses).
- Their high temporal resolution in combination with high repetition rates of measurements (sampling rates) up to several tens of kHz.
- Their high spatial resolution from the zero-dimensional (0D) to the three-dimensional (3D) domain.
- The potential of sampling from many different locations simultaneously, similar to a digital photograph, which samples information from as many points as pixels (today several mega-pixels) on the camera chip in a rather short exposure time in the order of milliseconds.

The detailed specifications of the different spectroscopic techniques described in this book will be presented in the subsequent main chapters about the individual techniques themselves. In Section 1.5 the aforementioned advantages of the *in situ* spectroscopic techniques are treated from a general point of view giving specific examples. Some exercises in Section 1.6 are supposed to make the reader familiar with the expressions 'temporal resolution', 'spatial resolution', 'sampling rate', etc. I highly recommend reading Section 1.5 before trying to solve the exercises in Section 1.6.

1.3 WHY HIGH PRESSURE IS AN OPPONENT OF SPECTROSCOPIC TECHNIQUES?

At high-pressure conditions, the adoption of *in situ* measurement techniques is much more challenging than at atmospheric pressure. High-pressure equipment is usually made of stainless steel. For the provision of spectroscopic access the high-pressure equipment has to be modified accordingly. With respect to nuclear magnetic resonance spectroscopy (NMR) the materials used inside the spectrometer should not interact with magnetic fields. Therefore micro-reactors or micro-capillaries made of glass or polymer are used, which can stand the pressure and are compatible with NMR. With respect to spectroscopic techniques working in the ultraviolet (UV), the visible (VIS), the near infrared (NIR) or the infrared (IR) spectral region, the high-pressure apparatuses have to be equipped with transparent windows in order to guarantee optical access. Then the access path to the interior of the high-pressure vessel has to be sealed and the windows have to be configured to withstand the pressure-induced stress. Strategies to provide spectroscopic access to high-pressure processes are not discussed here, but can be found in literature for rather small internal volumes but for high temperatures and pressures up to 200 MPa [2,64–66].

With respect to the title of this chapter 'High pressure: Fellow and opponent of spectroscopic techniques', it is the necessity to modify the high-pressure equipment for the adoption of spectroscopic techniques, which makes high pressure an opponent of spectroscopic techniques. To be more precise, it is the required modification which makes high pressure an **opponent of the reliability** of the results we gained having applied spectroscopic techniques. Each modification of the high-pressure equipment goes in hand with a modification of the processes taking place inside the high-pressure plant. For example, the installation of windows will influence the temperature field inside the plant and therefore will influence the processes. A miniaturisation of an apparatus, maybe required to bring it to an optical microscope, will change the length scales of the plant and thus change the processes. Therefore we have to pay attention that the modifications we install will not too much influence the processes themselves. Otherwise we measure in modified processes, which are very different from the real processes taking place in non-modified stainless steel apparatuses, and the results, which we produce cannot be directly utilised to improve our understanding of high-pressure process technology. This is why I recommend modifying the high-pressure apparatus/equipment as little as possible, but instead modifying the spectroscopic technique as much as possible in order to optimise it for the specific measurement purpose. This of course requires a profound understanding of the working principle of the spectroscopic technique as well as of the individual components the spectroscopic measurement technique/sensor is composed of. Therefore the interaction mechanisms between electro-magnetic radiation and matter will be described with respect to optical techniques in Chapter 2.

1.4 WHY HIGH PRESSURE IS A FELLOW OF SPECTROSCOPIC TECHNIQUES?

Still the title of this chapter makes us expect that high pressure also can be a fellow of spectroscopic techniques. High pressure is a fellow mainly due to the high density of fluids, expressed in number of molecules per volume, inherent to most of the high-pressure processes. Fluids at elevated pressure are also referred to as compressed fluids. Consequently, the probability of interaction of electro-magnetic radiation with matter in compressed fluid systems is high compared to low-pressure systems. With respect to, for example, Raman and IR spectroscopy this brings some advantages. The probability of interaction of the excitation light with matter according to the Raman effect is rather low. Therefore in non-compressed fluid systems, in which the number of molecules per volume is low, long signal integration times or very high excitation energies are required for the detection of reliable Raman signal levels. In contrast, in compressed systems reliable Raman signal levels can be achieved with significantly shorter integration times or significantly lower excitation energies. This opens possibilities of building up rather affordable Raman sensors for analysis of processes taking place in compressed systems. Chapter 3 therefore contains instructions how to configure and build up a low-budget Raman sensor and shows its performance with respect to specific measurement applications. Some species feature a low probability of interaction with electro-magnetic radiation due to the absorption of infrared or near infrared light. To compensate for this, multi-pass measurement chambers in different configurations [67–69], such as the White or Herriott cells, have been developed for measurement purposes in non-compressed systems. In contrast, in compressed systems, the number of molecules per volume is high enough to enable absorption experiments with only one pass through the object under investigation.

1.5 ADVANTAGES OF SPECTROSCOPIC TECHNIQUES

1.5.1 Non-invasive Measurement Principle of *In Situ* Spectroscopic Techniques

The most important advantage of *in situ* spectroscopic techniques is their non-invasive or non-intrusive measurement principle. Please keep in mind that especially at elevated pressure, the statement '*in situ* spectroscopic techniques are non-invasive' is only true, if the modification of the high-pressure apparatus, which maybe had to be realised to grant spectroscopic access to the interior of the high-pressure apparatus, did also not interfere with the processes. For the demonstration of the advantages accompanied with the usage of a non-invasive measurement technique relative to an invasive one, I chose one example, which is not related to high pressure, but to ambient pressure. It is the measurement of temperature at one location within the reaction zone of a Bunsen flame. Figure 1.2 shows a Bunsen flame for temperature measurements using either a thermocouple or a laser-spectroscopic technique (not further specified here).

FIGURE 1.2 Photographs of a Bunsen flame during temperature measurement in the reaction zone using a laser spectroscopic technique (a) or a thermocouple (c). The original flame without the application of any measurement technique is shown in the middle (b) for comparison. The flame operational conditions were kept constant while the three photographs were taken. (The laser [green in the web version] is visible in (a), as the ambient air was traced with smoke.)

Additionally a photograph of the original Bunsen flame is provided. It can clearly be seen that the presence of the thermocouple significantly influences the development of the flame. The thermocouple is an obstacle in the original flow of the gas and thus is invasive or intrusive with respect to the measurement system. The gas has to flow around the obstacle, which causes regions where the flow stagnates and others in which the flow is accelerated. A change in the flow field goes in hand with a change in the mixing process between fuel and oxidiser and thus changes the reaction of the fuel combustion. Furthermore, due to the presence of the thermocouple and the herewith-occurring manipulation of the flow and the combustion chemistry, also the heat transport processes are manipulated. Summarising, the presence of the thermocouple manipulates the flow, the heat and mass transport mechanisms and the reaction. This implies that the temperature measured with the thermocouple has not much to do with the temperature of the gas at the same location in the original flame without the thermocouple present.

In Fig. 1.2a a laser beam focused into the flame is shown. The focus of the laser beam forms the probe volume of this hypothetical laser-spectroscopic measurement technique and is located at the same position at which the tip of the thermocouple is positioned in Fig. 1.2c. Obviously the original flame (Fig. 1.2b) and the flame analysed using the laser-spectroscopic technique are identical. Therefore, we can conclude that the laser beam focused into the flame is non-invasive or non-intrusive to the measurement object. Consequently spectroscopic techniques, such as the laser spectroscopic techniques, bring the qualification to measure under un-altered conditions. If the spectroscopic techniques feature a high accuracy and high precision, the temperature within the flame can also be measured with a high accuracy and a high precision and the results will represent the reality.

What has been demonstrated earlier for the temperature measurement in the Bunsen flame, can be transferred to combustion processes taking place at elevated pressures, such as in internal combustion engines, rocket engines or pressurised gas turbines, to spray processes at elevated pressures, such as particles-from-sprays processes, or to solvation and extraction processes, such as the solvation of a compressed gas into a liquid for gas purification or the extraction of natural products. In general it also can be transferred to high-pressure processes, which are not dominated by the flow. For example, the analysis of high-pressure phase equilibria via conventional techniques relies on sampling of physical probes and their subsequent analysis. As the action of sampling is always accompanied by a disturbance to the system, the system under investigation is driven away from thermodynamic equilibrium just at that instant when the sample is taken. This is why Peper and Dohrn [70] emphasise in their review paper 'Sampling from fluid mixtures under high pressure: Review, case study and evaluation' under which circumstances the major error source in the analysis procedure, which is sampling, can be minimised. Remote analysis techniques, such as spectroscopic techniques, do not rely on sampling physical probes from the fluid system under investigation and therefore do not suffer from the errors inherent to the action of taking physical probes.

1.5.2 Temporal Resolution and Sampling Rates of *In Situ* Spectroscopic Techniques

In the context of this book, the temporal resolution is referred to the duration of one measurement, whose signal can be evaluated to a reliable result. It is not the rate at which measurements can be made, which is usually referred to as sampling rate. It has to be underlined that sometimes the temporal resolution and the sampling rate correlate, but sometimes they are independent of each other. I want to make this clear giving two examples.

Example I

Let us assume a spectroscopic experiment in which we excite the molecules in the high-pressure system continuously as it is represented by the light grey horizontal line in Fig. 1.3.

Due to the continuous excitation, the signal response of the system is supposed to be continuous, too, and is represented in Fig. 1.3 by the solid horizontal (red) line. During the signal acquisition times, represented by the (green) dashed line, the signal can be integrated on the detector. In the example given in Fig. 1.3 we integrate the signal over 50 ms, before it is read out each time. The integration time should be selected long enough to assure that the integrated signal is good enough to enable a reliable evaluation of the detected signal. In this example the **temporal resolution** of the spectroscopic technique is 50 ms.

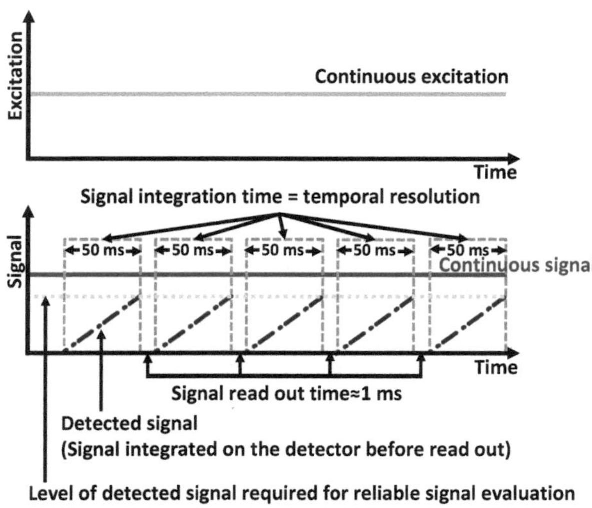

FIGURE 1.3 **Example of excitation–detection time scheme for the continuous excitation.**

But what is the **sampling rate**? We assume the time required to read out the integrated signal from the detector to be only 1 ms. Therefore, the read-out time (1 ms) is much shorter than the signal acquisition time (50 ms) and can be neglected, hence. This means that one event of measurement (signal acquisition plus read-out) takes approximately ~50 ms. Or in other words, as one measurement event takes 50 ms, 20 measurements can be realised per second which means that in our example the **sampling rate** is 20 Hz. Summarising, in the example shown in Fig. 1.3, the temporal resolution and the sampling rate are directly correlated, as the later one is the reciprocal of the first one. It must be underlined here that this correlation does not exist in general, which will be described by the second example.

Example II

Let us assume a spectroscopic experiment in which we excite the molecules in the high-pressure system pulse-wise as it is represented by the grey vertical lines in Fig. 1.4. One excitation pulse lasts 100 ns, which can be the laser pulse of an excimer laser. The repetition rate of the excitation pulses is 10 Hz, meaning that there are ~100 ms of 'no-excitation' between two subsequent excitation pulses. We assume that the temporal response behaviour of the signal directly follows the temporal shape of the excitation pulse. Therefore the signal pulses, which are represented by the red solid vertical lines, can be found at the same instant and with the same shape the excitation pulses can be found. For the complete detection of one signal pulse, which is 100 ns, the integration time of the detector has to be set to minimum 100 ns. In this case, the integration time additionally has to be synchronised perfectly to the signal (or excitation) pulse.

In practice, the realisation of short-gating times is challenging and the more challenging, the shorter the integration should be. To give an example, the shortest integration time realisable with a standard charge-coupled device (CCD) camera is in the order of 500 μs. But, of course, for the complete detection of the signal

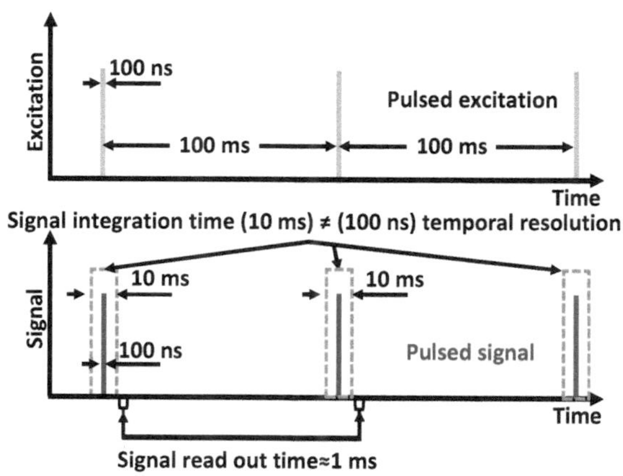

FIGURE 1.4 **Example of excitation–detection time scheme for the pulsed excitation.**

pulse, we can set the integration time of the detector also longer than the signal pulse duration. In Fig. 1.4 the signal integration time is set 10 ms, which is five orders of magnitude more than the signal pulse duration of 100 ns. Nevertheless with this configuration we can detect the entire signal of one pulse, as long as the signal pulse appears within the integration time window, as it does in Fig. 1.4. The signal integration time is followed by reading out the detected signal. In our example, the signal is read out after each signal integration time. This, in principle can also be modified to other scenarios, in which several signal integration windows are accumulated on the detector before the accumulated signal is read out in one event. If the signal integration time was set longer than 100 ms, more than one signal pulse could be accumulated on the detector before the read-out of the signal.

In contrast to the continuous excitation (Example I), here in the pulsed excitation example, it is not the signal integration time, but the signal pulse duration, which determines the temporal resolution of the spectroscopic technique. As the signal pulse duration is equal to the excitation pulse duration, one also can say that the temporal width of the excitation pulse determines the temporal resolution of the spectroscopic set-up. Though the signal integration time of the detector is set to 10 ms, the signal in fact is integrated only during the 100 ns the signal exists. So the **temporal resolution** of the spectroscopic technique, whose excitation–detection scheme is sketched in Fig. 1.4, is 100 ns. Also, the sampling rate is determined by the excitation pulse repetition rate. As there are 10 excitation pulses per second, there are 10 signal pulses per second. Consequently 10 single-shot measurements can be made per second, which results in a **sampling rate** of 10 Hz. It should be underlined that in contrast to the continuous excitation in Fig. 1.3, in the case of the pulsed excitation in Fig. 1.4, the sampling rate is not equal to the reciprocal of the temporal resolution.

Whether the spectroscopic technique employed is required to feature a high or a low temporal resolution depends on the field of application and on the information one intents to extract from the measurement result. For example, for the analysis of phase equilibria of fluid mixtures in high-pressure variable volume view cells, there is no need to measure with a high temporal resolution. The system analysed is in thermodynamic equilibrium, meaning that independent of the temporal resolution of the spectroscopic technique, the obtained results have to be the same. This is very much different if systems have to be analysed, which are not in equilibrium and or feature turbulent flows of different fluids. To make this very clear, Fig. 1.5 shows in the left column single-shot images of the CO_2 molar fraction (top) and of the distribution of phase boundaries (bottom) in the mixing zone of the SAS-process, when a solution of yttrium acetate dissolved in dimethyl sulphoxide was injected into supercritical CO_2. It will not be described in this chapter, but in a chapter that follows how and with which spectroscopic technique the images were acquired. For now it is important to realise the differences between the single-shot images in the left column and the mean-images in the right column. The turbulent structures in the mixing zone are only resolved in the single-shot images (left column) as here the measurements are made with a temporal resolution shorter than the characteristic time

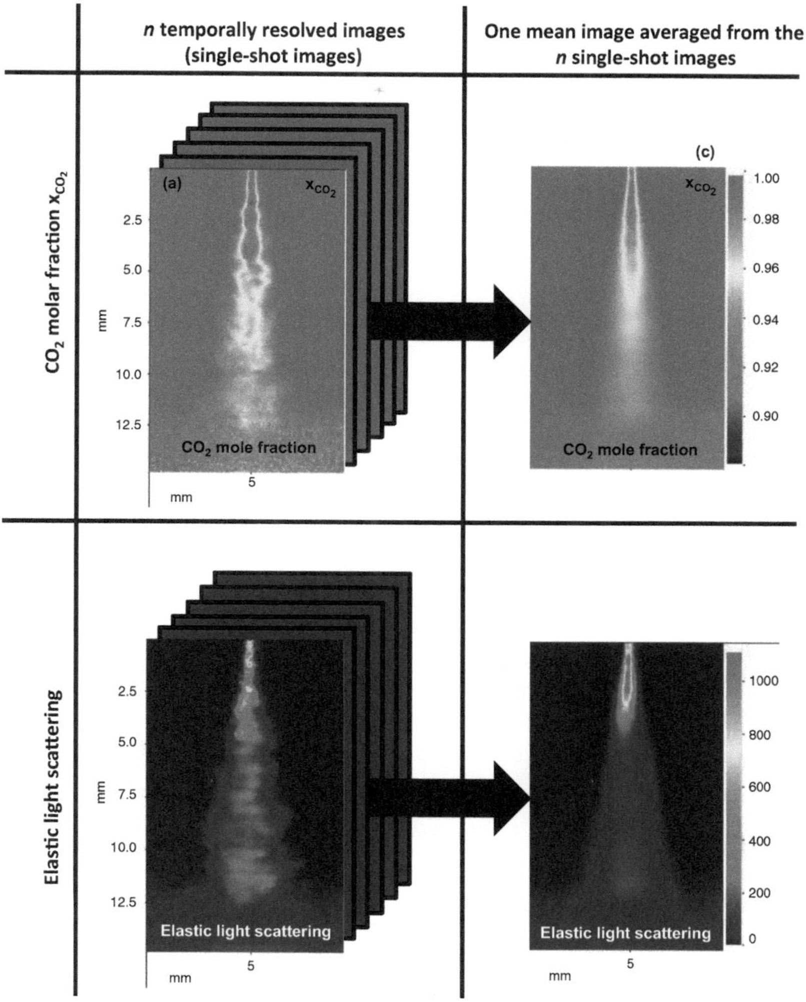

FIGURE 1.5 Single-shot (left) and mean (right) images of the molar fraction of CO_2 (top) and the distribution of phase boundaries (bottom) when a solution of yttrium acetate in dimethyl sulphoxide is injected into supercritical CO_2 according to the SAS-process.

of the turbulent motion of the flow. In other words, the turbulent motion is frozen in time in the single-shot images, which we therefore also call snapshots. Consequently, from several single-shot images one can deduce information about the fluctuations inside the mixing zone due to turbulence. If the temporal resolution of the spectroscopic technique was longer than the characteristic time of the turbulent motion, the acquired image would look smeared and no information about flow fluctuations was extractable.

As we did not carry out measurements with a temporal resolution longer than the characteristic time of the turbulent motion, I have to simulate this scenario by averaging several of the single-shot images. The resulting mean-images, which are shown in Fig. 1.5 in the right column, then look like measurements carried out with a temporal resolution longer than the characteristic turbulent time-scale.

The temporal resolution achievable with spectroscopic techniques spans the range from few nanoseconds to hours. Usually, very high temporal resolution in the nanosecond range can only be realised with short-pulse excitation strategies. Some exercises about 'temporal resolution' and 'sampling rate' are provided in Section 1.6.

1.5.3 Spatial Resolution of *In Situ* Spectroscopic Techniques

In the context of this book the spatial resolution is the smallest volume or the smallest area the detected signal can be assigned to. It is a volume, if the signals are taken from the bulk of the object or fluid, but it is an area, if the signals originate solely from the surface of the object or fluid. With respect to the application of spectroscopic techniques at high pressures, the exact determination of the spatial resolution is rather challenging. This is mainly due to the mostly unknown index of refraction and the heterogeneous distribution of the index of refraction of the fluid mixtures inside the high-pressure processes. Consequently, often the **resolution of the acquired signal**, which is the best achievable resolution, is given instead of the real **spatial resolution** of the spectroscopic technique. This will be described in detail later.

Figure 1.6 shows how two point light sources are imaged onto a detector array and how the signals appear after read-out of the single elements of the array detector.

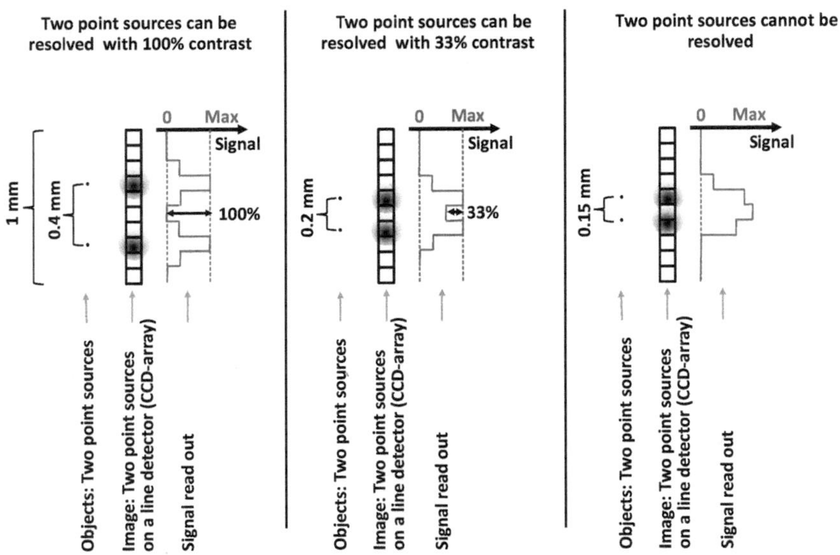

FIGURE 1.6 Illustrative sketch showing how two separate objects (point sources) are read out after their detection for different distances of 0.4, 0.2, and 0.15 mm between the objects.

Due to the modulation and transfer function of the detection system, the images of the originally infinitesimal small point light sources are bigger than the original point light sources. Therefore the images of the point light sources can exceed the size of one element of the array detector, though the original point light source is smaller than one element of the array detector. From left to right the distance between the two point light sources shrinks from 0.4 mm over 0.2 mm to 0.15 mm. For the distance of 0.4 mm the two point light sources can be identified based on their signals as two separate objects with a contrast of 100%. Nevertheless each of the images of the point light sources is spread over three elements on the detector's array.

For the distance of 0.2 mm the contrast goes down to 33%, but still the two point light sources can be identified as two separate objects. For the distance of 0.15 mm between the point sources, they no longer can be identified as two separate objects. Therefore the **spatial resolution** we achieve in Fig. 1.6 is somewhere between 0.2 mm and 0.15 mm. If the images of the objects on the detector were also infinitely small or at least significantly smaller than the dimension of one element of the detector array, than the optimum theoretically achievable spatial resolution would be 0.1 mm, which is the entire length of the object plane imaged onto the detector array (1 mm in Fig. 1.6) divided by the number of elements of the array detector (10 in Fig. 1.6). Summarising, the really achieved spatial resolution (here between 0.15 mm and 0.2 mm) and the resolution of the detected signal (here 0.1 mm), which is the best spatial resolution one can possibly achieve, are two different things. It is very simple to predict the later one, but it is rather challenging to experimentally measure the first one at high-pressure conditions. This is why in spectroscopy at high pressures often the later one is provided, solely. Some exercises are provided in Section 1.6 to make the reader more familiar with the spatial resolution.

1.5.4 Dimensionality of *In Situ* Spectroscopic Techniques

Within one measurement event spectroscopic techniques can simultaneously provide signals from more than one location spatially resolved.

1.5.4.1 Zero-Dimensional Spectroscopy (Point Measurements)

If during one measurement event the signal is detected from one surface or volume—spatially not resolved—the measurement technique is zero-dimensional (0D). A typical non-spectroscopic example is the measurement of the density of a fluid by weighing a certain volume of this fluid. The measured mass is assigned to the entire fluid volume. It cannot be resolved, whether or not the fluid at the bottom of the volume is denser than at the top. Therefore this is called a zero-dimensional measurement. A typical 0D spectroscopic technique is a set-up with signal detection in back-scattering direction, such as can be found in confocal microscopes. An example of 0D spectroscopic back-scattering set-up is integrated in Fig. 1.7. The excitation laser is focused by a lens along

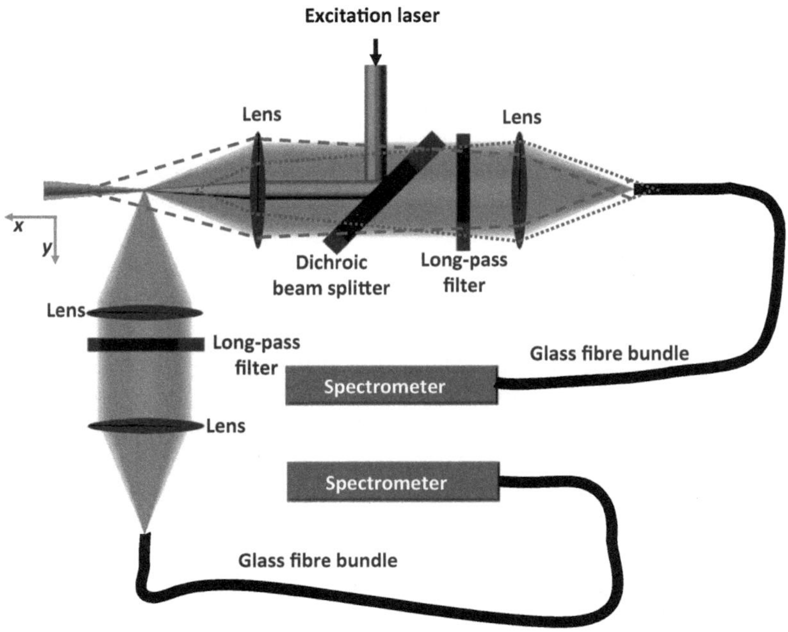

FIGURE 1.7 Sketch of two versions of a zero-dimensional spectroscopic set-up combined in one figure. The signal can either be detected in back scattering configuration (reverse *x*-direction) or in rectangular direction (*y*-direction).

the *x*-direction. The red-shifted signals are then imaged onto the entrance interface of the glass fibre and guided through the glass fibre to the spectrometer. The dashed and the dotted signal paths indicate that signals are not only collected from the excitation laser focus, but also from a certain depth of focus around the excitation laser beam focus. Consequently the signals launched into the glass fibre originate from a volume, which can be assumed cylinder-like. The diameter of this cylinder can be assumed to be the diameter of the focal beam waist and the length of this cylinder can be assumed to be the depth of focus of the set-up. This cylinder-like volume cannot be spatially resolved anymore, meaning that the entire volume determines the spatial resolution of the back-scattering set-up.

In addition to the back-scattering configuration, Fig. 1.7 also contains one 0D detection set-up in right-angle direction. Here the signals originating from the laser beam focus are imaged via the *y*-direction onto the entrance interface of the glass fibre. Now the depth of focus of the excitation path and the signal optical path show a much smaller overlap, as only their foci overlap. Consequently the spatial resolution of the 0D spectroscopic set-up with signal detection in right-angle configuration features a finer spatial resolution than the one with signal detection in back-scattering direction. In contrast, the back-scattering configuration collects the signals from a larger volume and therefore results in higher detected signal levels. The spatial resolution of the back-scattering

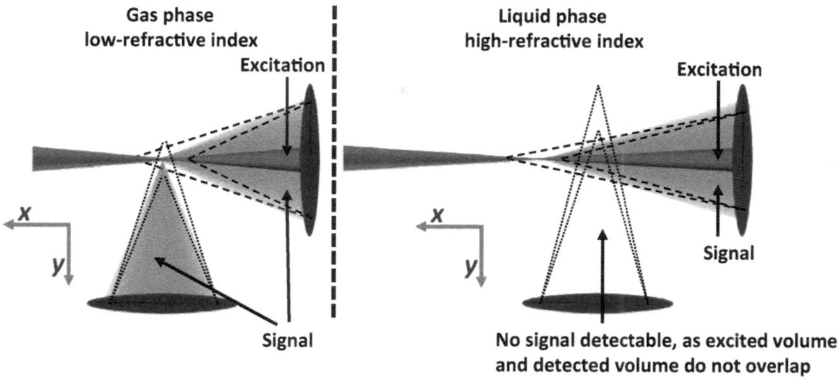

FIGURE 1.8 Sketch of the excitation and the detection paths for a back-scattering signal detection (along the negative *x*-axis) and a right-angle signal detection (along *y*-axis) configuration. Both configurations are aligned for the low refractive index (right) of the gas phase (left). The beam paths change with the variation of the refractive index but get misaligned only for right-angle signal detection.

configuration can be improved further by inserting a spatial filter (pinhole) into the detection path. Confocal microscopic set-ups contain such a spatial filter. Next to the advantage that with the back-scattering configuration higher signal levels can be detected—for the sake of the spatial resolution—the most important advantage of the back-scattering configuration is that the excitation and the detection focal point overlap after an initial alignment of the set-up, even if the index of refraction changes from experiment to experiment at high pressures. The influence of the index of refraction on the effective focal length of lenses is described in more detail in Section 2.3.2 of this book. Figure 1.8 illustrates the shift of the foci of the excitation laser beam and the detection path accompanied with a variation of the index of refraction inside a high-pressure apparatus for the right-angle-scattering configuration and the back-scattering configuration. As we move further, I will call the excitation laser beam focus the 'excited volume' and the focus of the detection path the 'detected volume'. Once the two detection scenarios are perfectly aligned, which is shown in Fig. 1.8 left, any change of the index of refraction shifts the excited and the detected volume, which is shown for the case of a gain of the index of refraction in Fig. 1.8 right. For the right-angle detection scenario the excitation and the detection volume are driven away from each other. In other words the excitation and the detection path get misaligned. This implies a significant decrease of the detected signal. To assure the detection of high signal levels the excitation and the detection paths have to be realigned again. In contrast, for the back-scattering scenario the excitation and the detection volume are both shifted in nearly the same way, meaning that even after the shift they still overlap according to Fig. 1.8. If one wants to bring back the measurement volume to the centre of the high-pressure chamber, either the chamber itself or the detection and the excitation set-up can be moved.

1.5.4.2 One-Dimensional Spectroscopy

If during one measurement event the signal is detected from one surface or volume, spatially resolved along a line, the measurement technique is one-dimensional (1D). An example of a 1D spectroscopic technique is the measurement of composition line profiles through a drop [71] within only one measurement event. Figure 1.9 shows how an excitation laser is focused through the drop, which is the measurement object. Inside the drop the focused excitation laser beam profile can be assumed like a long cylinder with the diameter equal to the beam waist of the laser and a length equal to the diameter of the drop.

The length of this cylinder-like excited volume is spatially resolved by the set-up. Therefore, in Fig. 1.9 on the right side the measurement results (here CO_2 concentration) can be given as a function of the radial position inside the drop. Figure 1.10 shows the approximation of the excitation laser beam waist as a long cylinder and how the spatial information is conserved during dispersion of the locally resolved signal wavelengths.

The spatial resolution along the long cylinder is represented by the slices the long cylinder stack is composed of. Three example slices are labelled with 'a', 'b', and 'c'. The section excited by the excitation laser, which in Fig. 1.10 is the '3D waist along the laser beam', is imaged onto the horizontally oriented entrance slit of an imaging spectrometer. Along the entrance slit of the spectrometer the spatial information along the probed line is conserved. When the light passes through the spectrometer, the spatial information is still conserved, while

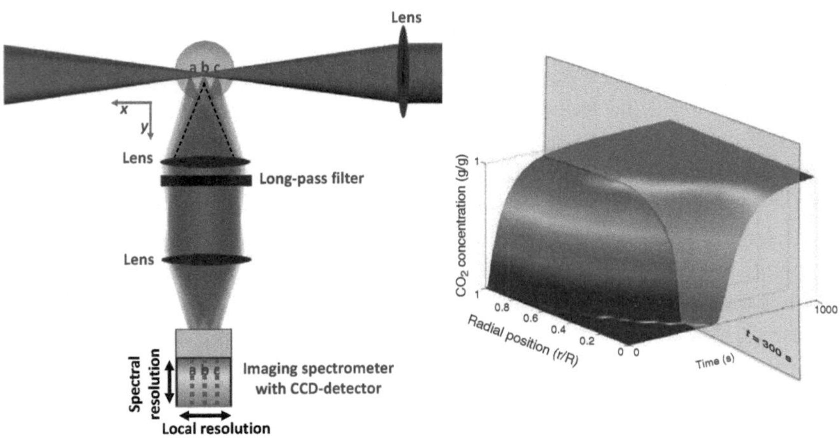

FIGURE 1.9 Sketch of a one-dimensionally (1D) resolving spectroscopic technique with signal detection in right angle direction. The signals originating from the line a, b, c through a drop of molten polymer are detected spatially (locally) and spectrally resolved on the charge-coupled device (CCD) chip of the camera simultaneously (left). As the diffusion of the compressed CO_2 from outside the drop into the drop is slow, and if the temporal resolution of the 1D spectroscopic technique is much shorter than the diffusion time, the evolution of the composition profile inside the drop can be determined, as it is shown on the right side. *(Source: [72].)*

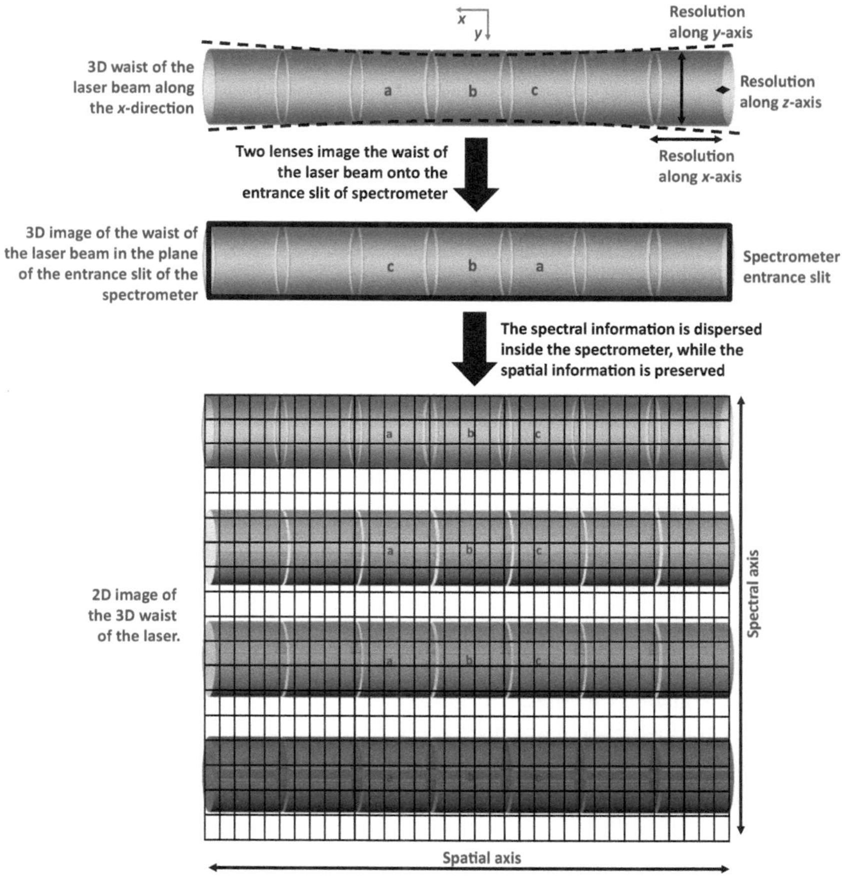

FIGURE 1.10 Sketch of how the 3D waist of the laser beam is transferred into one spatial and one spectral dimension. Locations a, b, and c correspond to locations a, b and c in Fig. 1.9.

the spectral information is dispersed along a second dimension of the detector chip. This is why one can see in Fig. 1.10 images of the waist of the laser at different vertical positions in different colours. The different colours are assumed to be contained in the spectrum of the detected signals. Consequently the one dimension of the camera chip mounted onto the exit of the imaging spectrometer is required to capture the spatial information while the other dimension is required to capture the spectral information.

If the measurement object behaves stationary, several 0D measurements can be scanned at different locations along a line (1D), along a plane (2D) or even throughout a volume (3D). The scanned results can be combined then in a post-processing step. In confocal microscopy images or 3D models are usually computed from scanned 0D measurements acquired at many different locations.

1.5.4.3 Two-Dimensional Spectroscopy

If during one measurement event the signal is detected from one surface or volume spatially resolved in one plane, the measurement technique is two-dimensional (2D). A typical example is making photos. Please keep in mind that photos of three-dimensional objects or volumes are reduced to the two dimensions of the detector chip or film. The detector chips of modern cameras—even if integrated into mobile phones—comprise several million detector elements (mega-pixels). Therefore the camera detector detects signal on several million detector elements each time a photo is made. These several million data points result in the 2D photo once the signals are read form the detector. In principle the photography with a simple colour-camera can be termed two-dimensional spectroscopy, as the detectors of the camera have to be sensitive to the spectrum of the light. Otherwise it would not be possible to acquire the photos in colour-mode.

For the computation of the CO_2 molar fraction distribution in the jet-mixing zone of the supercritical anti-solvent technology—the respective distributions are shown in Fig. 1.5—it was necessary to detect the distribution of the solvent (dimethyl sulphoxide) molecules and the anti-solvent (CO_2) molecules simultaneously in the mixing zone. This can be accomplished with a set-up shown in Fig. 1.11. A light sheet is formed of the excitation laser, which is guided through the mixing zone. Only molecules within the excitation light sheet are excited and provide signals. The signals originating from different kinds of molecules appear at different wavelengths and therefore can be separated spectrally by inserting dichroic mirrors into the detection path, which direct the signals specific for one species to the respective camera. In Fig. 1.11 the first dichroic mirror was utilised to detect the distribution of phase boundaries (droplets or particles) in addition to the distribution of the solvent and anti-solvent molecules. On the detection chip of the two other cameras the distribution of the molecular species in the excited plane—either solvent or anti-solvent—is imaged simultaneously. In a post-processing step the CO_2 molar fraction can be computed from two separately but simultaneously acquired images, one showing the distribution of solvent and the other one showing the distribution of anti-solvent. A detailed explanation of this set-up will follow in Chapter 3 on Raman techniques.

1.6 EXERCISES CORRESPONDING TO CHAPTER 1

Exercise 1.1: Temporal Resolution and Sampling Rate

Figure 1.12 shows an excitation detection time scheme for a pulsed excitation. The duration of one excitation pulse is 10 ns. The repetition rate of the laser pulses is 10 Hz. Due to the pulsed excitation the generated signal is also pulsed. The temporal response of the signal starts promptly with the laser excitation pulse but then decays slowly with a decay time of 1 ms. The decay time is also called the lifetime of the signal. The signal is detected with a detector, which

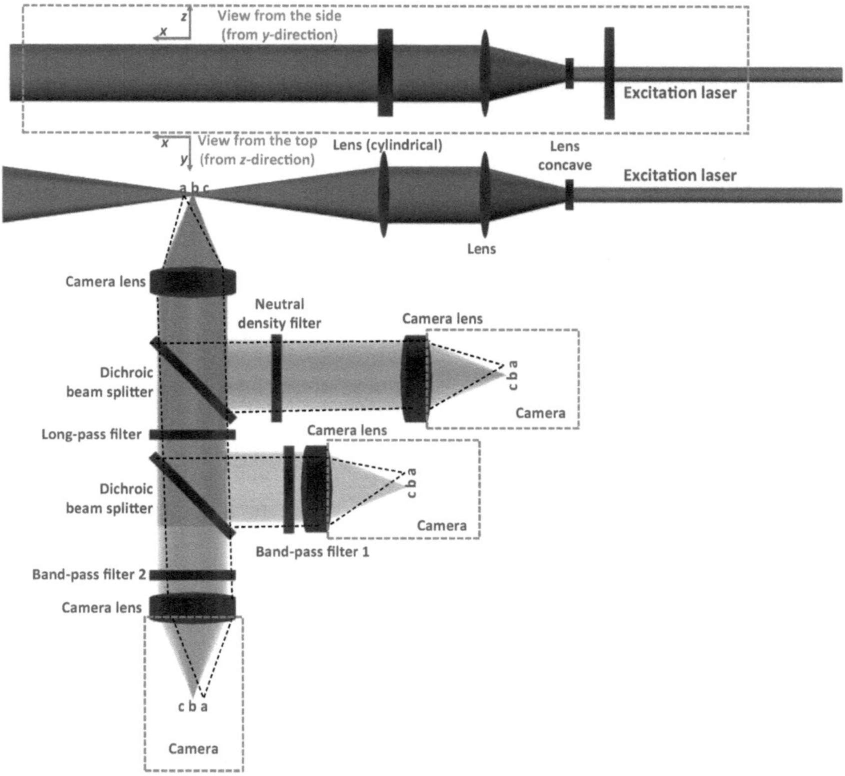

FIGURE 1.11 Sketch of a two-dimensional (2D) spectroscopic sensor with signal detection in right angle direction.

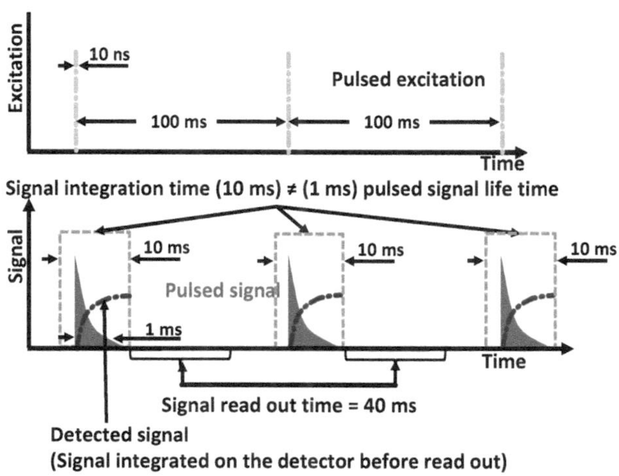

FIGURE 1.12 Excitation–detection time scheme for a pulsed excitation.

features a minimum signal integration time of 10 ms and a minimum signal read-out time of 40 ms.

Tasks: Exercise 1.1

1. What is the temporal resolution of the excitation detection scheme shown in Fig. 1.12?
2. What is the maximum sampling rate achievable with respect to solely the detector?
3. What is the actual sampling rate according to Fig. 1.12?
4. Which measures with respect to the excitation detection scheme in Fig. 1.12 can be taken to increase the intensity of the detected signal, if the signal intensities detected during one signal integration time window is not good enough to evaluate the signal reliably?
5. What are the consequences of averaging or accumulating signals with respect to the temporal resolution of the measurement technique?
6. Which measures can be taken in order to improve the temporal resolution of the excitation detection scheme shown in Fig. 1.12.

Answers: Exercise 1.1

1. The temporal resolution is the time during which the signal is detected. As the signal life time (1 ms) is shorter than the signal integration time window (10 ms), the temporal resolution is determined by the signal life time. The temporal resolution is 1 ms.
2. For the acquisition of one image, first the detector has to integrate the signal and then the detected signal has to be read from the detector. Therefore, the acquisition time is the summation of the signal integration time (10 ms) and the read-out time (40 ms), which sums up to 50 ms. Once the signal is read out, the next image can be acquired for which again 50 ms are required, 10 ms for the integration of the signal and 40 ms to read the detected signals from the detector. This means that the signals can be sampled 50 ms each which corresponds to a sampling rate of 20 Hz, which is the reciprocal of 50 ms.
3. The actual sampling rate in Fig. 1.12 is determined by the repetition rate of the excitation laser, which is 10 Hz. The actual sampling rate of 10 Hz is less than the maximum sampling rate achievable with the detector used.
4. With respect to the excitation detection time scheme there are two options:
   a. One can read the signal from the detector after each signal integration window. The signal is then referred to as single-shot signal. As the signal intensity of the single-shot images is not good enough to ensure a reliable signal evaluation, one can accumulate several single-shot signals. So accumulation of single-shot signals is one option.
   b. One can increase the signal integration time significantly in order to detect several signal pulses within one long lasting signal integration time window. If the signal integration time window is set to 1 s, 10 signal

pulses are accumulated on the detector within the signal integration time window, before the detected signal is read out. So signal accumulation on the detector before read-out is another option.

There are few detector types, which can realise the accumulation of several signal integration time windows, before reading the signal out. This is different to the example given earlier. We chose the signal integration time window to be 1 s aiming to accumulate 10 signal pulses on the detector. We also could accumulate 10 signal pulses on the detector, if we set the signal integration time windows as shown in Fig. 1.12, but reading-out the detected signal only each tenth signal pulse. This means that the signal pulses are also accumulated on the detector, before the accumulated signal is read out. The advantage of the latter scenario is that background interferences can be reduced if the overall signal integration time is short.

If one does not need the single-shot signal information, signal accumulation on the detector before read-out usually is better than the accumulation of single-shot signals, which already have been read out. At first glance both strategies might seem to bring the same effect. But at second glance accumulation on the detector is better than accumulation of already read out single-shot signals just because of the so called 'read-out noise'. This means that when accumulating single-shot signals, each read out single-shot image suffers from the read-out noise and thus also the read-out noise is accumulated. But in the strategy of signal accumulation on the detector before read-out the accumulated signal is only read out once! This means that the accumulated signal is only spoiled once with the read-out noise. Therefore the overall signal-to-noise ratio of signal accumulated on the detector before read-out is always better than the signal-to-noise ratio of a signal accumulated from already read-out single-shot signals. Nevertheless it has to be mentioned that next to read-out noise thermal noise also can influence the quality of detected signals. Thermal noise is proportional to the integration time. Thus with increasing signal integration time also the thermal noise increases. Fortunately with respect to silica-based CCD-technology the thermal noise can be reduced to nearly zero when cooling the detector's chip to temperatures as cool as 190 K and further. Some camera types are not able to cool their chip to such deep temperatures and thus provide—in the case of long integration times—signals spoiled by thermal noise.

5. The consequences of signal averaging or signal accumulation are as follows:
   a. No information about the single-shot signal's fluctuations can be extracted.
   b. The temporal resolution worsens and is often lost completely.
6. For the improvement of the temporal resolution the signal integration window can be shifted temporally with respect to the pulsed signal, as it is shown in Fig. 1.13. Now the signal is only detected during the time when the signal integration window and the pulsed signal overlap. Therefore the temporal resolution is improved, but, of course, the signal levels detected are lower.

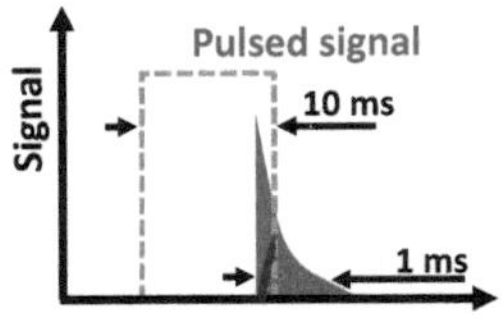

FIGURE 1.13 Signal integration time window (dashed [green in the web version]) shifted with respect to the signal pulse (dark grey [red in the web version]).

Exercise 1.2: Spatial Resolution

Figure 1.14 shows an image of a jet. The scales at the left and the bottom border of the image show that the image represents an area of 15 mm times 10 mm of the real jet mixing zone. The number of pixels the image is composed of vertically and horizontally is provided on the right and top border of the image. The area of 1 single pixel is 10 μm times 10 μm.

Tasks: Exercise 1.2

1. We already know that the image represents an area of 15 mm times 10 mm from the mixing zone of the jet. What is the area of the image of the jet?
2. By which number the image of the jet is magnified/miniaturised compared to the real jet?
3. What is the best spatial resolution achievable with the detector used here?

Answers: Exercise 1.2

1. The image is composed of 200 times 300 pixels. Each pixel features an area of 10 μm times 10 μm. Therefore the horizontal dimension of the image is 200 pixels times 10 μm/pixel, which is 2 mm. The vertical dimension of the

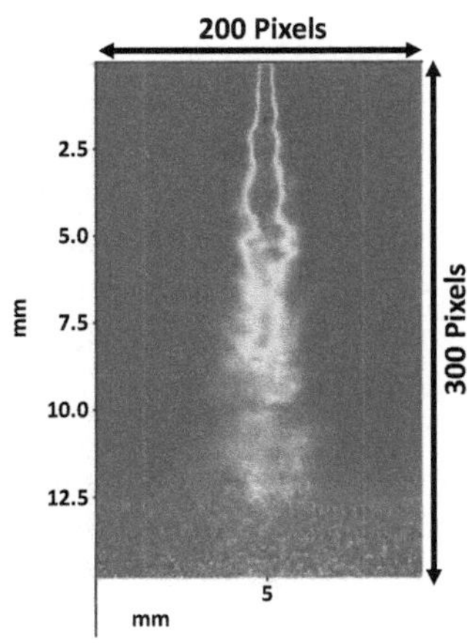

FIGURE 1.14 Image of a jet.

image is 300 pixels times 10 μm/pixel, which is 3 mm. Thus the area of the image of the jet is 2 mm times 3 mm.

2. The magnification/miniaturisation with respect to area is the area of the image over the area of the object. The object is the section of the mixing zone imaged, whose area is 10 mm times 15 mm (denominator). The area of the image is 2 mm times 3 mm (nominator). The magnification with respect to the area herewith is 6/150 = 1/25 = 0.04. So the area of the image is only 4% of the area of the imaged mixing zone.
3. The best spatial resolution achievable is the length of one dimension of the imaged mixing zone (15 mm) over the number of pixels resolving this (300 pixels). Therefore the best achievable spatial resolution in this case is 15 mm over 300 pixels, which is 0.05 mm/pixel. In other words, signals originating from two points, which are only 0.05 mm apart from each other, can be—in the ideal case—identified as two signals with separate origin. So the two dots ideally could be resolved as two dots, if the distance between them exceeds 0.05 mm.

Exercise 1.3: Spatial Resolution

Figure 1.15 shows a section of an image of a 1951 USAF resolution test chart acquired with a detector we want to determine the spatial resolution with. The USAF resolution test chart contains line patterns with a well-defined periodicity. The numbers on the test chart image define the number of line pairs per mm. One line pair is composed of one black and one white line. The line pair number of 125, for example, means that there are 125 line pairs per mm. The reciprocal value provides the width of one line pair, respectively the distance between two lines. Consequently the line pair number of 125 corresponds to a distance of 0.008 mm between two black or two white lines.

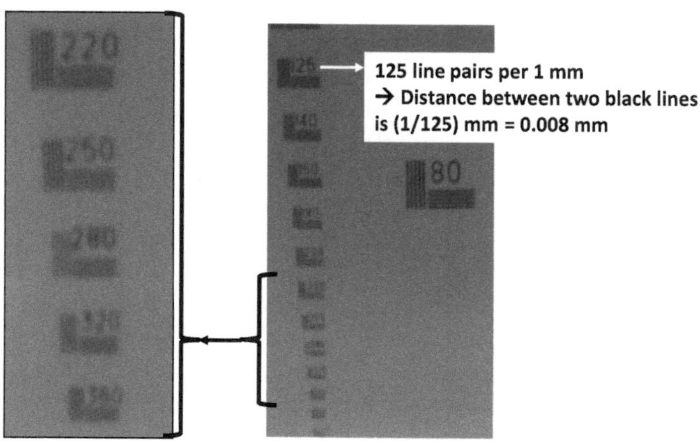

FIGURE 1.15 Image of a 1951 USAF resolution test chart.

Tasks: Exercise 1.3

1. What is the distance between two black or two white lines for a line pair number of 250?
2. What, according to your visual observation, is the vertical and the horizontal spatial resolution of the detector we used for the acquisition of the image of the USAF resolution test chart?
3. The area of one pixel is assumed to be 10 μm times 10 μm. Can you estimate, whether the USAF resolution test chart is magnified or miniaturised when imaged?

Answers: Exercise 1.3

1. The distance between two black or two white lines in mm is the reciprocal value of the 'lines per mm' number. The reciprocal of 250/mm is 0.004 mm.
2. According to my visual observation I can resolve a periodic line pattern vertically and horizontally for the 'lines per mm' number of 280, but not for 320. Therefore 280 lines per mm can still be resolved, which corresponds to approximately 0.0036 mm. Consequently the actual spatial resolution is 0.0036 mm.
3. The actual spatial resolution, which is 0.0036 mm, cannot be better than the ideal spatial resolution. Therefore we can assume the ideal spatial resolution to be in the worst case as good as the actual spatial resolution. The ideal spatial resolution is the length of the imaged object over the number of pixels on the length scale. Consequently at least 0.0036 mm of the object have to be imaged onto one pixel. As one pixel alone is 10 μm times 10 μm, this is only possible if the image exceeds the size of the object. Therefore the image is a magnification of the object.

1.7 APPENDIX—CHAPTER 1

1.7.1 Supercritical Fluids

1.7.1.1 What is a Supercritical Fluid?

A fluid is in its **supercritical state**, once its temperature T exceeds its critical temperature T_c and once at the same time its pressure P exceeds its critical pressure P_c. The pressure–temperature (PT) diagram in Fig. 1.16 classifies states, at which a fluid is solid, liquid, gaseous or supercritical.

Note that there is no phase transition, when a fluid is compressed isothermally from the gas phase to the supercritical state or when a fluid is heated isobaric from the liquid phase to the supercritical state.

The critical pressure P_c and the critical temperature T_c define the endpoint of the vapour–liquid equilibrium (vaporisation curve). The starting point of the vaporisation line is the triple point, at which a substance co-exists as a gas, a liquid and a solid. The vaporisation curve, therefore, differentiates the gaseous from the liquid regime and according to Gibb's rule defines the couples of pressure and temperature at which a liquid and a gaseous phase of a substance co-exist.

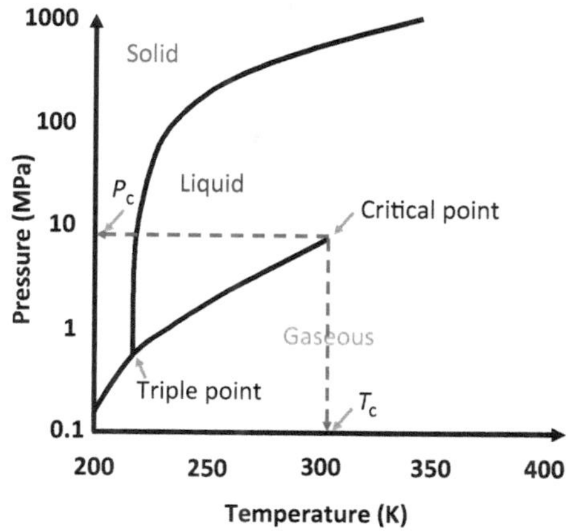

FIGURE 1.16 **Pressure–temperature (*PT*) diagram of a pure substance.**

In order to understand the properties at the critical point, we will move along the vaporisation curve from the triple point towards the critical point, assuming a closed chamber filled with a pure substance. At the triple point there will be a solid phase, a liquid phase and a gas phase coexisting inside the chamber. Moving away from the triple point towards the critical point, the solid phase will disappear. The remaining liquid and gas phase will be split, which is visible by the meniscus, which forms between the two phases. The liquid phase is much denser than the gas phase and will accumulate at the bottom of the chamber, while the light gas phase will fill the remaining volume of the chamber on top of the dense liquid phase. Moving nearer towards the critical point goes along with a density increase of the gas phase, while the density of the liquid remains relatively unaffected. At the same time, the meniscus differentiating the dense liquid from the lighter gas phase will get weaker and weaker. This is just because of the fact that the properties (including the density) of the liquid and the gas phase converge as we move towards the critical point. Exactly at the critical point (at T_c and P_c) the 'liquid' and the 'gas' phase are identical. Hence the meniscus completely disappears as the distinction between liquid and vapour disappears and the whole chamber is filled with a fluid. Above T_c and P_c the fluid is termed a supercritical fluid.

1.7.1.2 What Makes a Supercritical Fluid Attractive for Process Engineers?

Supercritical fluids combine some favourable properties of liquids and gases at once.

1.7.1.2.1 Supercritical Fluids are Compressible

Supercritical fluids near their critical temperature feature a liquid-like density. In contrast to liquids, see Fig. 1.17, they are still compressible.

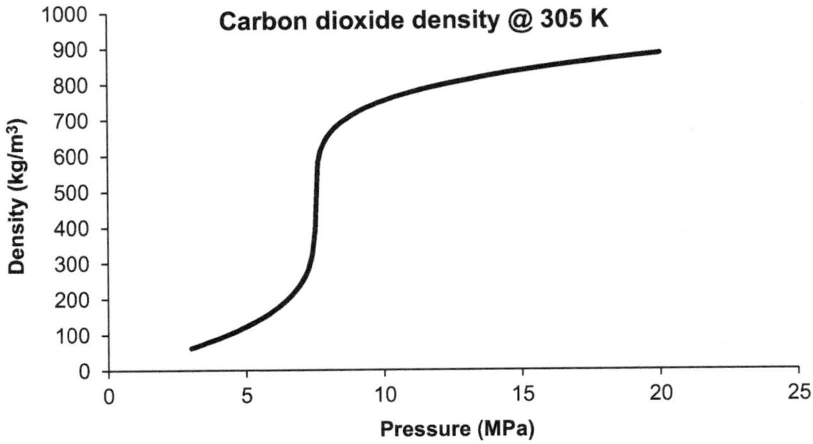

FIGURE 1.17 Density of CO_2 as a function of pressure at 305 K. Critical pressure and temperature of CO_2 are 7.38 MPa and 304 K, respectively. *(Source: Data from NIST.)*

Hence, one can manipulate the density of the supercritical fluid by adjusting the pressure. The solubility of solutes is proportional to the density of the solvent. Thus, whenever the supercritical fluids act as solvents, their solvation power can easily be varied by adjusting the pressure. The process 'rapid expansion of supercritical solutions' (RESS) makes use of this circumstance [73].

1.7.1.2.2 Supercritical Fluids Feature a Low Viscosity

Supercritical fluids near their critical point feature gas-like viscosities. The kinematic viscosity in square meter per second (m^2/s)—see Fig. 1.18—shows

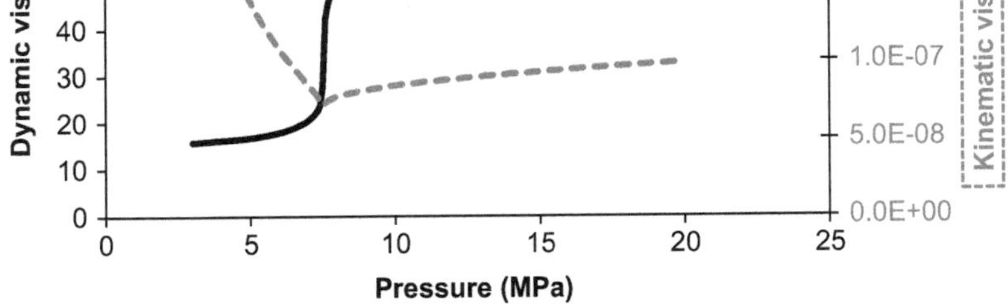

FIGURE 1.18 Viscosity of CO_2 as a function of the pressure at 305 K. Critical pressure and temperature of CO_2 are 7.38 MPa and 304 K, respectively. *(Source: Data from NIST.)*

a minimum near the critical point. This circumstance allows fast macro- and meso-mixing down to the Kolmogorov length scale.[2]

The dimensionless Schmidt (Sc) number

$$\mathrm{Sc} = \frac{\text{kinematic viscosity in m}^2/\text{s}}{\text{mass diffusivity in m}^2/\text{s}}, \tag{1.1}$$

is defined as the ratio of momentum diffusivity (kinematic viscosity) and mass diffusivity. It physically puts into relation the thickness of a hydrodynamic mixing layer and the thickness of the diffusive mass transfer mixing layer.

- Mixing in liquids: Sc >> 1
- Mixing in gases: Sc = 1
- Mixing in sc fluids: Sc ~ 10

For mixing processes in gases the thickness of the hydrodynamic mixing layer and the mass diffusivity mixing layer are approximately equal (Sc = 1). This means that composition inhomogeneities on the Kolmogorov scale are degraded once the inhomogeneity has reached the Kolmogorov scale. Therefore the Kolmogorov structures are not observable in the composition field of the mixing process, but still may be observable in the respective velocity field.

For mixing processes in liquids the hydrodynamic layer is much thicker than the mass diffusivity layer (Sc >> 1). This means that composition inhomogeneities at the Kolmogorov scale prevail. Mass diffusivity is too slow to completely degrade the inhomogeneities of the Kolmogorov scale. Consequently composition inhomogeneities are long living. One example is adding a coloured transparent liquid-A to a non-coloured transparent liquid-B. Inhomogeneities will remain very long. Only the macroscopic inhomogeneities can be degraded faster by stirring with a spoon.

For mixing processes in supercritical fluids the hydrodynamic mixing layer is approximately by a factor of 10 thicker than the mass diffusivity layer (Sc ~ 10). Hence the Sc number for supercritical fluid mixing is in between the mixing processes in liquids or in gases. Compared to liquids, inhomogeneities are degraded much faster in supercritical fluids. This is beneficial, as supercritical fluids feature liquid like densities but nevertheless—due to the gas-like viscosities—also exhibit an improved mixing behaviour. This is meant when stating that:

'Supercritical fluids combine the advantageous properties of liquids and gases at once'.

---

2. If fluids are to be mixed in a turbulent flow, the mixing energy (aerodynamics or convection) is initially provided in large-scale structures, which can be, for example, large eddies caused by stirring or jetting. In the turbulent flow these large-scale structures disintegrate into smaller-scale structures, which is known as the vortex or energy cascade. The smallest-scale structure existing in a turbulent flow—before this smallest structure completely dissipates—is called Kolmogorov length scale.

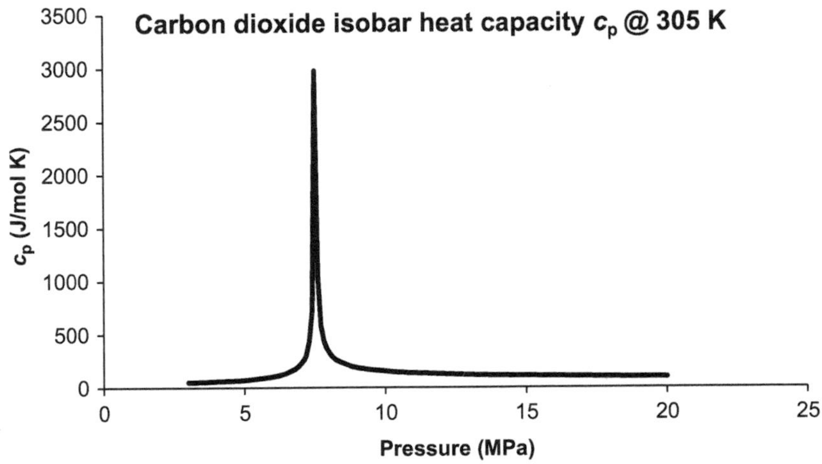

FIGURE 1.19 **Isobaric heat capacity of CO_2 as a function of pressure at 305 K.** Critical pressure and temperature of CO_2 are 7.38 MPa and 304 K, respectively. *(Source: Data from NIST.)*

1.7.1.2.3 Conductivity and Capacity of Heat of Supercritical Fluids

In addition to the compressibility and the viscosity supercritical fluids also exhibit few other properties worth to be mentioned. This is the heat conductivity and the heat capacity (isobaric).

Regarding the heat transfer phenomena in supercritical mixing processes, the Sc-number has to be replaced by the Prandtl **(Pr)-number**. The Pr-number puts into relation the momentum diffusivity (kinematic viscosity) and the thermal diffusivity.

$$\text{Pr} = \frac{\text{kinematic viscosity in m}^2/\text{s}}{\text{thermal diffusivity in m}^2/\text{s}}, \tag{1.2}$$

- Mixing in liquids: Pr >> 1
- Mixing in gases: Pr = 1
- Mixing in sc fluids: Pr ~ 10

What has been reported on the meaning of the Sc-number regarding the composition of a mixture can be transferred on a one-to-one basis onto the Pr-number regarding the temperature field of a mixing process.

The maximum of the isobaric heat capacity—see Fig. 1.19—near the critical point implicates that mixing processes in supercritical fluids can lead to temperature inhomogeneities.

If the isobaric heat capacity is changed during the mixing process, temperature changes have to occur. These temperature inhomogeneities are degraded rather fast due to the Pr-number of approximately 10.

The thermal conductivity of CO_2 as a function of the pressure at 305 K is given in Fig. 1.20

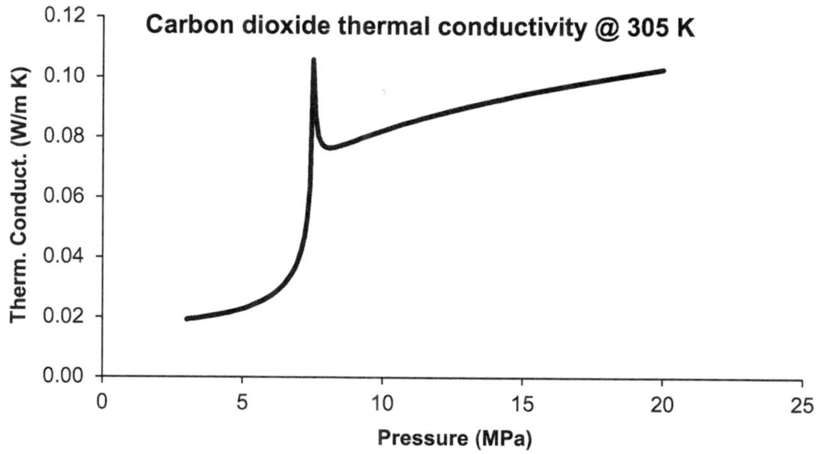

FIGURE 1.20 Thermal conductivity of CO_2 as a function of pressure at 305 K. Critical pressure and temperature of CO_2 are 7.38 MPa and 304 K, respectively. *(Source: Data from NIST.)*

1.7.1.3 What is a Supercritical Mixture?

1.7.1.3.1 Pressure-Composition (*Px*) Diagram of Binary Mixtures

The state of a substance composed of only one compound is defined by two state variables. This is explained in Section 1.7.1.1. In the *PT* diagram of one compound the state variables pressure and temperature define whether the compound is gaseous, liquid, supercritical or solid (in the two-phase region the state of phase or the specific volume is required additionally for a definition of the state of a fluid).

In process engineering often mixtures instead of single compound fluids are treated. This makes things more complex as the mixture composition has to be considered from now on additionally. For binary mixtures the mixture composition is defined as the mole fraction (or molar fraction) $x_{\text{constituent}}$ of one constituent. The molar fraction of component A (x_A)

$$x_A = \frac{n_A}{n_A + n_B}, \tag{1.3}$$

$$x_A + x_B = 1 \tag{1.4}$$

is thus defined as the amount of substance A (n_A) over the amount of substance A (n_A) and B (n_B).

The vapour–liquid phase behaviour of a binary mixture is usually given in a pressure-composition (*Px*) diagram. As an example, a *Px* diagram for a mixture of carbon dioxide (CO_2) and an organic solvent is given in Fig. 1.21. Please note that *Px* diagrams for many binary systems composed of CO_2 and an organic solvent look similar. The mixture critical pressure at 313 K is usually between

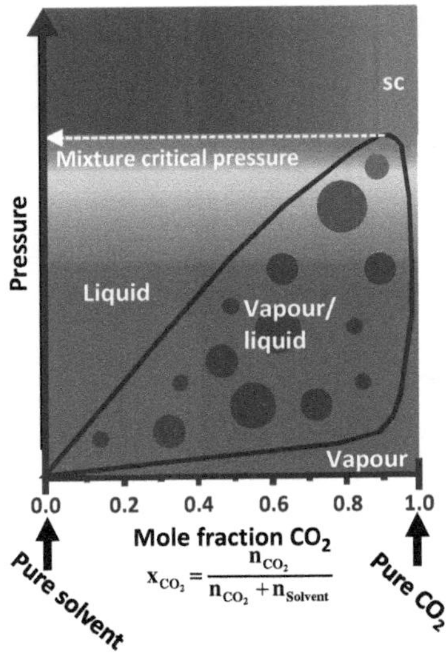

FIGURE 1.21 Qualitative sketch of a pressure-composition (*Px*) diagram of a binary mixture composed of CO_2 and an organic solvent at 313 K. The MCP (pressure) in this case is typically in the range between 7.8 MPa and 9.5 MPa.

the critical pressure of the pure CO_2 and 9.5 MPa. Systems not showing this behaviour are contained within this book series in the book *High-Pressure Fluid Phase Equilibria: Phenomenology and Computation* authored by Deiters and Kraska [74].

The *Px* diagram in Fig. 1.21 describes, at which couples of pressure and composition the mixture exists in which state.

For couples of pressure and composition located inside the envelope of the vapour/liquid two-phase region, a vapour and a gas phase coexist simultaneously. It has to be pointed out that the shape of the envelope of the vapour/liquid two-phase region is a function of temperature. Hence the *Px* diagram shows isothermal data. At high pressure and 'solvent-rich' compositions the mixture is usually liquid. At CO_2 compositions near one, the mixture is usually in the vapour phase. If the pressure is above the mixture critical point (MCP) and if additionally the composition of the mixture is richer in CO_2 than the composition at the MCP, the mixture then is in the supercritical state.

1.7.1.3.2 What is the Mixture Critical Point?

In order to understand the meaning of the MCP we will follow a compression of a mixture. We assume that the composition of the mixture is exactly the composition of the MCP. Additionally the pressure of the mixture is significantly below the MCP. Consequently the mixture is in the two-phase region. This means that the liquid phase of the mixture will accumulate on the bottom of the chamber, while the lighter gas phase mixture will fill the rest of the chamber volume on top of the liquid phase. The intersection of the respective isobar with the vapour/liquid envelope characterises the composition of the liquid and the vapour phase mixture, respectively. The composition of the liquid phase mixture is defined by the intersection of the isobar and the saturated liquid line left of the MCP. The composition of the vapour phase mixture is defined by the intersection of the isobar and the saturated vapour line right of the MCP.

Obviously at low pressures, the liquid phase mixture is rich in the solvent, while the vapour phase mixture is rich in CO_2. With increasing pressure the composition of the liquid phase mixture is getting richer in CO_2. On the contrary, the composition of the gas phase mixture remains relatively unaffected by the compression and keeps a quasi-constant composition. The closer the pressure increases towards the pressure of the MCP, the more similar the composition of the liquid and the vapour phase get. At the MCP the compositions of the liquid and the vapour phase finally are identical. Hence they are the same and cannot be differentiated any more. This is why the region above the mixture critical pressure and above the mixture critical composition is termed supercritical region of the mixture.

Fortunately many supercritical fluid based processes in chemical engineering can be investigated by considering binary mixtures. Even if the mixtures in the process carry more than two compounds, a theoretical reduction to binary mixtures is sometimes reasonable.

1.7.2 Supercritical Anti-solvent (SAS) Process

In Section 1.1 the 'cause and effect' approach as well as the '*in situ* analysis' approach are described considering the evolution of the understanding we developed over the years about the supercritical anti-solvent technology. Therefore, I here provide a brief description of the supercritical anti-solvent technology.

According to its name 'supercritical anti-solvent' the SAS technology applies the supercritical fluid as an anti-solvent. Hence the solute to be micronised has to be quasi non-soluble in the supercritical fluid. In a SAS experiment a solution composed of a solute and a solvent is injected into the anti-solvent. While the solvent and the anti-solvent are miscible the solute is quasi non-soluble in the mixture of the solvent and the anti-solvent. Consequently the mixture is supersaturated, which is compensated by the formation of solute particles.

Usually CO_2 is the supercritical anti-solvent. Figure 1.22 shows a flow sheet of a typical SAS experiment.

1. First the SAS vessel has to be pressurised with the anti-solvent CO_2 above its critical pressure. The vessel is also conditioned. Usually the experiment is carried out between 308 K and 335 K and between 8 MPa and 20 MPa. For optimum process conditions the CO_2 should be continuously purged through the SAS vessel.
2. The organic solvent is then injected into the pressurised SAS vessel and not the solution but the solvent is sprayed. Hence no particles are generated so

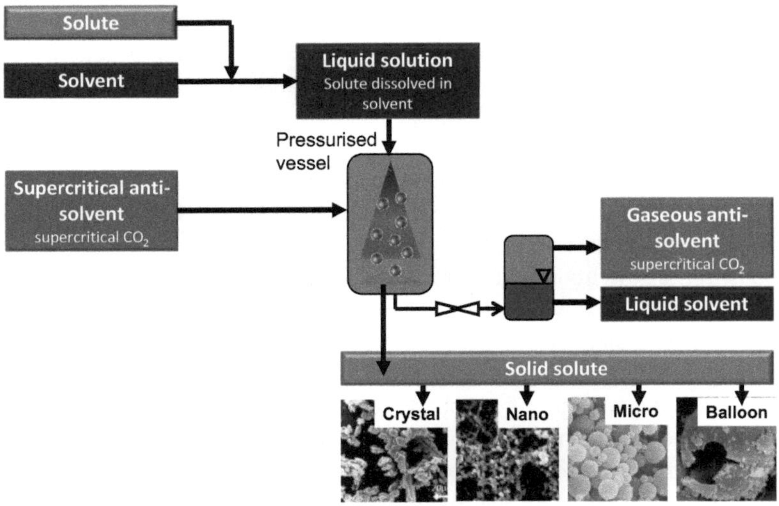

FIGURE 1.22 **Flow sheet of a typical SAS experiment.**

far. After a while of continuously purging the vessel with CO_2 and injecting the solvent, stationary conditions should be adjusted inside the vessel. This means that the composition, the pressure and the temperature inside the vessel are constant. The composition is usually adjusted to CO_2 molar fractions between 0.97 and 0.99.

3. Then the solution is injected instead of the solvent, while the chamber is still purged with fresh CO_2. Otherwise the composition would be changed during the injection of the solution. Once enough powder is produced the injection of the solution can be stopped.
4. The vessel is still purged with fresh CO_2 in order to clean the particles and the vessel from the residual solvent. This takes rather long.
5. Finally the vessel can slowly be depressurised and the powder can be taken from the vessel.

During the entire experiment, the mixture leaving the vessel is depressurised downstream of the vessel and experiences a phase change. Thus the liquid can be separated from the gas. Both fluids can be recycled.

By tuning the operational conditions;

- pressure,
- temperature,
- initial solute concentration or,
- composition of solvent mixture [75,76]

one can manipulate the morphology of the micronised particles from crystals, to amorphous nanoparticles, to spherical amorphous micro-particles and to hollow expanded micro-particles.

Several particle morphologies have been recurrently generated by SAS, comprising nanoparticles with mean diameters in the 30–200 nm range [47], micro-particles in the 0.2–10 μm range [48], expanded hollow micro-particles (balloons) with diameters between 10 μm and 200 μm and various crystal morphologies [49,77]. Nanoparticles have generally been obtained at fully developed supercritical conditions, at pressures significantly above the MCP of the binary system of the solvent and the sc anti-solvent. Micro-particles have been produced for many compounds, using various liquid solvents and usually operating at near critical conditions [78].

In the literature it is found that the fluid dynamic behaviour of the solution, which is injected into the pressurised anti-solvent, can be characterised by the jet break-up followed by droplet formation or by a gas-like jet mixing. Lengsfeld et al. [50] indicated that gas mixing is characteristic for completely miscible conditions due to the fast disappearance of the liquid/fluid interface. Dukhin et al. [51] introduced two characteristic times: a liquid/fluid interface disappearance time τ_i and a jet break-up time τ_{jb}.

The liquid/fluid interface disappearance time τ_i is defined by the time an interface exists between an injected fluid element and the bulk fluid phase. The bulk fluid phase may be either liquid or sc, which depends on the pressure and the mixture composition. The jet break-up time τ_{jb} is defined by the mean residence time of a fluid element between the moment when it leaves the nozzle exit and the moment when it passes the location of jet break up.

For $\tau_{jb} < \tau_i$ an interface exists between the injected liquid and the bulk fluid phase during jet break-up, so that the jet break-up transforms the liquid into droplets [79–82]. For $\tau_i < \tau_{jb}$, gas-like jet mixing is obtained without forming any interface. Dukhin et al. [51] discussed the existence of an additional dynamic interfacial tension mechanism, which is not driven by composition gradients but by temperature gradients. Temperature gradients come along with composition gradients during the mixing process and can be assigned to mixing enthalpies [83]. Systematic surface tension measurements of dynamic systems have been proposed by Sun and Shekunov [84] for the binary system ethanol/CO_2 showing that the surface tension at dynamic conditions is larger than the equilibrium surface tension and depends on pressure and temperature. Some researchers also studied the mechanisms that control the production of precipitates during SAS processing, assuming that droplets are formed after jet break-up and then result in micro-particles by solvent elimination [42,45,85]. If the liquid/fluid interface disappears before the jet break-up, gas mixing is obtained and nanoparticles are produced [47].

Some attempts to study the jet break-up were executed by optical measurement techniques such as shadowgraphy or light scattering techniques [43,50,80,86–88]. The advantages of optical measurement techniques are their non-invasive measurement principle (they do not alter the investigated processes), their high temporal (down to ns) and spatial resolution (down to μm) and their high sampling rates (up to kHz). In a recent work on the injection of pure solvent into $scCO_2$, Reverchon et al. [58] were able to identify a transition from the two-phase to the one-phase mixing regime by applying a particular elastic light scattering technique. They showed that, when the elastically scattered signal is detected in backward direction and under a particular angle, the evolution of the liquid/fluid interface could be monitored. By performing several experiments at different operational conditions, they concluded that spherical micro-particles result from precipitation from droplets when the liquid/fluid interface persists at the moment of precipitation. Nanoparticles result from precipitation from a supercritical phase when the liquid/fluid interface does not persist at the moment of precipitation. Therefore, the competition between the time of the jet break-up, the time of degradation of the liquid/fluid interface between the injected liquid solution and the supercritical bulk, and the precipitation time controls the formation of micro-particles or nanoparticles.

REFERENCES

[1] Furlan M, Kluge J, Mazzotti M, Lattuada M. Preparation of biocompatible magnetite-PLGA composite nanoparticles using supercritical fluid extraction of emulsions. J Supercrit Fluids 2010;54:348–56.

[2] Frantz JD, Dubessy J, Mysen B. An optical cell for Raman spectroscopic studies of supercritical fluids and its application to the study of water to 500°C and 2000 bar. Chem Geol 1993;106:9–26.

[3] Masten DA, Foy BR, Harradine DM, Dyer RB. *In situ* Raman spectroscopy of reactions in supercritical water. J Phys Chem 1993;97:8557–9.

[4] Venardou E, Garcia-Verdugo E, Barlow SJ, Gorbaty YE, Poliakoff M. On-line monitoring of the hydrolysis of acetonitrile in near-critical water using Raman spectroscopy. Vibration Spectrosc 2004;35:103–9.

[5] Licence P, Ke J, Sokolova M, Ross SK, Poliakoff M. Chemical reactions in supercritical carbon dioxide: from laboratory to commercial plant. Green Chem 2003;5:99–104.

[6] Reverchon E, Adami R, Cardea S, Porta GD. Supercritical fluids processing of polymers for pharmaceutical and medical applications. J Supercrit Fluids 2009;47:484–92.

[7] Smith R, Inomata H, Peters C. Introduction to supercritical fluids. Oxford, UK, Amsterdam, Netherlands: Elsevier; 2013.

[8] York P, Kompella UB, Shekunov BY. Supercritical fluid technology for drug product development. New York: Marcel Dekker; 2004.

[9] Kiran E. Modification of biomedical polymers in dense fluids. Miscibility and foaming of poly(p-dioxanone) in carbon dioxide + acetone fluid mixtures. J Supercrit Fluids 2012;66:372–9.

[10] Pastore Carbone MG, Di Maio E, Scherillo G, Mensitieri G, Iannace S. Solubility, mutual diffusivity, specific volume and interfacial tension of molten PCL/CO2 solutions by a fully experimental procedure: effect of pressure and temperature. J Supercrit Fluids 2012;67:131–8.

[11] Alnaief M, Smirnova I. *In situ* production of spherical aerogel microparticles. J Supercrit Fluids 2011;55:1118–23.

[12] Türk M, Lietzow R. Formation and stabilization of submicron particles via rapid expansion processes. J Supercrit Fluids 2008;45:346–55.

[13] Wong B, Yoda S, Howdle SM. The preparation of gold nanoparticle composites using supercritical carbon dioxide. J Supercrit Fluids 2007;42:282–7.
[14] Abrahamsson V, Rodriguez-Meizoso I, Turner C. Determination of carotenoids in microalgae using supercritical fluid extraction and chromatography. J Chromatogr A 2012;1250:63–8.
[15] Ferrentino G, Spilimbergo S. High pressure carbon dioxide pasteurization of solid foods: current knowledge and future outlooks. Trends Food Sci Technol 2011;22:427–41.
[16] Della Porta G, Castaldo F, Scognamiglio M, Paciello L, Parascandola P, Reverchon E. Bacteria microencapsulation in PLGA microdevices by supercritical emulsion extraction. J Supercrit Fluids 2012;63:1–7.
[17] Cinquemani C, Boyle C, Bach E, Schollmeyer E. Inactivation of microbes using compressed carbon dioxide; an environmentally sound disinfection process for medical fabrics. J Supercrit Fluids 2007;42:392–7.
[18] Steeper RR, Rice SF, Brown MS, Johnston SC. Methane and methanol diffusion flames in supercritical water. J Supercrit Fluids 1992;5:262–8.
[19] Shin H-Y, Lee S-H, Ryu J-H, Bae S-Y. Biodiesel production from waste lard using supercritical methanol. J Supercrit Fluids 2012;61:134–8.
[20] Schmieder H, Abeln J, Boukis N, Dinjus E, Kruse A, Kluth M, Petrich G, Sadri E, Schacht M. Hydrothermal gasification of biomass and organic wastes. J Supercrit Fluids 2000;17:145–53.
[21] Vezzú K, Sforza E, Sudiro M, Benedetti P, Losasso C, Bertucco A. Production of bioethanol under high pressure of CO_2: The effect of process conditions. J Supercrit Fluids 2009;51:67–73.
[22] Bozbag SE, Erkey C. Supercritical fluids in fuel cell research and development. J Supercrit Fluids 2012;62:1–31.
[23] Piqueras CM, Garcia-Serna J, Cocero MJ. Estimation of lower flammability limits in high-pressure systems. Application to the direct synthesis of hydrogen peroxide using supercritical and near-critical CO_2 and air as diluents. J Supercrit Fluids 2012;56:33–40.
[24] Wellig B, Weber M, Lieball K, Prikopsky K, von Rohr PR. Hydrothermal methanol diffusion flame as internal heat source in a SCWO reactor. J Supercrit Fluids 2009;49:59–70.
[25] Machmudah S, Goto M, Wahyudiono W, Kuwahara Y, Sasaki M. Gold nanoparticles fabricated by pulsed laser ablation in supercritical CO_2. Res Chem Intermediate 2011;37:515–22.
[26] Saitow K-i, Yamamura T, Minami T. Gold nanospheres and nanonecklaces generated by laser ablation in supercritical fluid. J Phys Chem C 2008;112:18340–9.
[27] Engelmeier L, Pollak S, Kilzer A, Weidner E. Liquid carbon dioxide jets for cutting applications. J Supercrit Fluids 2012;69:29–33.
[28] Ingram T, Rogalinski T, Bockemühl V, Antranikian G, Brunner G. Semi-continuous liquid hot water pretreatment of rye straw. J Supercrit Fluids 2009;48:238–46.
[29] Brunner G. Near and supercritical water. Part II: oxidative processes. J Supercrit Fluids 2009;47:382–90.
[30] Brunner G. Near critical and supercritical water. Part I. Hydrolytic and hydrothermal processes. J Supercrit Fluids 2009;47:373–81.
[31] Braeuer A, Dowy S, Torino E, Rossmann M, Luther SK, Schluecker E, Leipertz A, Reverchon E. Analysis of the supercritical antisolvent mechanisms governing particles precipitation and morphology by *in situ* laser scattering techniques. Chem Eng J 2011;173:258–66.
[32] Marre S, Roig Y, Aymonier C. Supercritical microfluidics: opportunities in flow-through chemistry and materials science. J Supercrit Fluids 2012;66:251–64.
[33] Bellan J. Supercritical (and subcritical) fluid behavior and modeling: drops, streams, shear and mixing layers, jets and sprays. Prog Energy Combust Sci 2000;26:329–66.
[34] Oschwald M, Smith JJ, Branam R, Hussong J, Schik A, Chehroudi B, Talley D. Injection of fluids into supercritical environments. Combust Sci Technol 2006;178:49–100.

[35] Falgout Z, Rahm M, Wang Z, Linne M. Evidence for supercritical mixing layers in the ECN Spray A. Proceedings of the Combustion Institut; 2015, 35 (2), 1579–1586.

[36] Manin J, Bardi M, Pickett LM, Dahms RN, Oefelein JC. Microscopic investigation of the atomization and mixing processes of diesel sprays injected into high pressure and temperature environments. Fuel 2014;134:531–43.

[37] Lacaze G, Misdariis A, Ruiz A, Oefelein JC. Analysis of high-pressure diesel fuel injection processes using LES with real-fluid thermodynamics and transport. Proceedings of the Combustion Institute; 2014.

[38] Qiu L, Reitz RD. Simulation of supercritical fuel injection with condensation. Int J Heat Mass Transf 2014;79:1070–86.

[39] Hofbauer P, Laimboeck F, Garrard T. Supercritical-state fuel injection system and method. US 8,511,287 B2, EcoMotors International; 2013.

[40] Gradl J, Schwarzer H-C, Schwertfirm F, Manhart M, Peukert W. Precipitation of nanoparticles in a T-mixer: Coupling the particle population dynamics with hydrodynamics through direct numerical simulation. Chem Eng Processing Process Intensificat 2006;45:908–16.

[41] Gallagher PM, Coffey MP, Krukonis VJ, Klasutis N. Gas antisolvent recrystallization: new process to recrystallize compounds insoluble in supercritical fluids. In: Johnston KP, Penninger JML, editors. Supercritical Fluid Science and Technology. American Chemical Society; 1989.

[42] Randolph TW, Randolph AD, Mebes M, Yeung S. Sub-micrometer-sized biodegradable particles of poly(l-lactic acid) via the gas antisolvent spray precipitation process. Biotechnol Prog 1993;9:429–35.

[43] Carretier E, Badens E, Guichardon P, Boutin O, Charbit G. Hydrodynamics of supercritical antisolvent precipitation: characterization and influence on particle morphology. Ind Eng Chem Res 2003;42:331–8.

[44] Shekunov BY, Baldyga J, York P. Particle formation by mixing with supercritical antisolvent at high Reynolds numbers. Chem Eng Sci 2001;56:2421–33.

[45] Rantakylä M, Jäntti M, Aaltonen O, Hurme M. The effect of initial drop size on particle size in the supercritical antisolvent precipitation (SAS) technique. J Supercrit Fluids 2002;24:251–63.

[46] Reverchon E, Caputo G, De Marco I. Role of phase behavior and atomization in the supercritical antisolvent precipitation. Ind Eng Chem Res 2003;42:6406–14.

[47] Reverchon E, De Marco I, Torino E. Nanoparticles production by supercritical antisolvent precipitation: a general interpretation. J Supercrit Fluids 2007;43:126–38.

[48] Reverchon E, Adami R, Caputo G, De Marco I. Spherical microparticles production by supercritical antisolvent precipitation: interpretation of results. J Supercrit Fluids 2008;47:70–84.

[49] Reverchon E, De Marco I, Adami R, Caputo G. Expanded micro-particles by supercritical antisolvent precipitation: interpretation of results. J Supercrit Fluids 2008;44:98–108.

[50] Lengsfeld CS, Delplanque JP, Barocas VH, Randolph TW. Mechanism governing microparticle morphology during precipitation by a compressed antisolvent: atomization vs. nucleation and growth. J Phys Chem B 2000;104:2725–35.

[51] Dukhin SS, Zhu C, Dave R, Pfeffer R, Luo JJ, Chavez F, Shen Y. Dynamic interfacial tension near critical point of a solvent-antisolvent mixture and laminar jet stabilization. Colloids Surf A Physicochem Eng Asp 2003;229:181–99.

[52] Braeuer A, Dowy S, Schatz R, Schluecker E, Leipertz A. Antisolvent precipitation: interaction of mixing, phase behaviour and particle formation. In: Ninth International Symposium on Supercritical Fluids, ISASF, Arachon; 2009.

[53] Dowy S, Braeuer A, Reinhold-Lopez K, Leipertz A. Laser analyses of mixture formation and the influence of solute on particle precipitation in the SAS process. J Supercrit Fluids 2009;50:265–75.

[54] Dowy S, Braeuer A, Schatz R, Schluecker E, Leipertz A. CO_2 partial density distribution during high-pressure mixing with ethanol in the supercritical antisolvent process. J Supercrit Fluids 2009;48:195–202.

[55] Braeuer A, Dowy S, Leipertz A. Simultaneous Raman and elastic light scattering imaging for particle formation investigation. Optics Lett 2010;35:2553–5.

[56] Braeuer A, Dowy S, Schatz R, Rossmann M, Schluecker E, Leipertz A. Supercritical antisolvent particle precipitation: *in situ* optical investigations. Chem Eng Technol 2010;33: 35–8.

[57] Dowy S, Braeuer A, Reinhold-López K, Leipertz A. *In situ* optical monitoring of the solution concentration influence on supercritical particle precipitation. J Supercrit Fluids 2010;55:282–91.

[58] Reverchon E, Torino E, Dowy S, Braeuer A, Leipertz A. Interactions of phase equilibria, jet fluid dynamics and mass transfer during supercritical antisolvent micronization. Chem Eng J 2010;156:446–58.

[59] Braeuer A, Adami R, Dowy S, Rossmann M, Leipertz A. Observation of liquid solution volume expansion during particle precipitation in the supercritical CO_2 antisolvent process. J Supercrit Fluids 2011;56:121–4.

[60] Dowy S, Torino E, Luther SK, Rossmann M, Braeuer A. Imaging the supersaturation in high-pressure systems for particle generation. Chem Eng J 2011;168:896–902.

[61] De Marco I, Knauer O, Cice F, Braeuer A, Reverchon E. Interactions of phase equilibria, jet fluid dynamics and mass transfer during supercritical antisolvent micronization: the influence of solvents. Chem Eng J 2012;203:71–80.

[62] Rossmann M, Braeuer A, Dowy S, Gallinger TG, Leipertz A, Schluecker E. Solute solubility as criterion for the appearance of amorphous particle precipitation or crystallization in the supercritical antisolvent (SAS) process. J Supercrit Fluids 2012;66:350–8.

[63] Rossmann M, Braeuer A, Leipertz A, Schluecker E. Manipulating the size, the morphology and the polymorphism of acetaminophen using supercritical antisolvent (SAS) precipitation. J Supercrit Fluids 2013;82:230–7.

[64] Spohn PD, Brill TB. A Raman spectroscopy cell for aqueous solutions to 500°C and 34.5 MPa. Appl Spectrosc 1987;41:1152–6.

[65] Gorbaty YE, Bondarenko GV, Venardou E, Garcia-Verdugo E, Sokolova M, Ke J, Poliakoff M. High-pressure high-temperature Raman spectroscopy of liquid and supercritical fluids. Appl Spectrosc 2003;57:1300–3.

[66] Gorbaty YE, Bondarenko GV. High-pressure high-temperature Raman cell for corrosive liquids. Rev Sci Instrum 1995;66:4347–9.

[67] Mohamed T, Zhu F, Chen S, Strohaber J, Kolomenskii AA, Bengali AA, Schuessler HA. Multipass cell based on confocal mirrors for sensitive broadband laser spectroscopy in the near infrared. Appl Optics 2013;52:7145–51.

[68] Tuzson B, Mangold M, Looser H, Manninen A, Emmenegger L. Compact multipass optical cell for laser spectroscopy. Optics Lett 2013;38:257–9.

[69] Tarsitano CG, Webster CR. Multilaser Herriott cell for planetary tunable laser spectrometers. Appl Optics 2007;46:6923–35.

[70] Peper S, Dohrn R. Sampling from fluid mixtures under high pressure: review, case study and evaluation. J Supercrit Fluids 2012;66:2–15.

[71] Braeuer A, Knauer OS, Quino J, Leipertz A. Quantification of the mass transport in a two phase binary system at elevated pressures applying Raman spectroscopy: pendant liquid solvent drop in a supercritical carbon dioxide environment. Int J Heat Mass Tran 2013;62: 729–40.

[72] Knauer OS, Pastore Carbone MG, Braeuer A, Di Maio E, Leipertz A. Investigation of CO_2 sorption in molten polymers at high pressures using Raman line imaging. Polymer 2013;54:812–8.

[73] Türk M, Formation of small organic particles by RESS: experimental and theoretical investigations. J Supercrit Fluids 1999;15:79–89.

[74] Deiters UK, Kraska T. High-pressure fluid phase equilibria: phenomenology and computation. Oxford, UK, Amsterdam, Netherlands: Elsevier; 2012.

[75] De Marco I, Rossmann M, Prosapio V, Reverchon E, Braeuer A. Control of particle size, at micrometric and nanometric range, using supercritical antisolvent precipitation from solvent mixtures: application to PVP. Chem Eng J 2015;273:344–52.

[76] Rossmann M, Braeuer A, Schluecker E. Supercritical antisolvent micronization of PVP and ibuprofen sodium towards tailored solid dispersions. J Supercrit Fluids 2014;89:16–27.

[77] Shen C-T, Chen P-Y, Wu J-J, Lee T-M, Hsu S-L, Chang C-MJ, Young C-C, Shieh C-J. Purification of algal anti-tyrosinase zeaxanthin from *Nannochloropsis oculata* using supercritical anti-solvent precipitation. J Supercrit Fluids 2011;55:955–62.

[78] Beckman EJ. Supercritical and near-critical CO_2 in green chemical synthesis and processing. J Supercrit Fluids 2004;28:121–91.

[79] Dukhin SS, Shen Y, Dave R, Pfeffer R. Droplet mass transfer, intradroplet nucleation and submicron particle production in two-phase flow of solvent-supercritical antisolvent emulsion. Colloids Surf A Physicochem Eng Asp 2005;261:163–76.

[80] Gokhale A, Khusid B, Dave RN, Pfeffer R. Effect of solvent strength and operating pressure on the formation of submicrometer polymer particles in supercritical microjets. J Supercrit Fluids 2007;43:341–56.

[81] Werling JO, Debenedetti PG. Numerical modeling of mass transfer in the supercritical antisolvent process. J Supercrit Fluids 1999;16:167–81.

[82] Werling JO, Debenedetti PG. Numerical modeling of mass transfer in the supercritical antisolvent process: miscible conditions. J Supercrit Fluids 2000;18:11–24.

[83] Dávila MJ, Cabañas A, Pando C. Excess molar enthalpies for binary mixtures related to supercritical antisolvent precipitation: carbon dioxide + N-methyl-2-pyrrolidone. J Supercrit Fluids 2007;42:172–9.

[84] Sun Y, Shekunov BY. Surface tension of ethanol in supercritical CO_2. J Supercrit Fluids 2003;23:73–83.

[85] Reverchon E. Supercritical antisolvent precipitation of micro- and nano-particles. J Supercrit Fluids 1999;15:1–21.

[86] Badens E, Boutin O, Charbit G. Laminar jet dispersion and jet atomization in pressurized carbon dioxide. J Supercrit Fluids 2005;36:81–90.

[87] Obrzut DL, Bell PW, Roberts CB, Duke SR. Effect of process conditions on the spray characteristics of a PLA plus methylene chloride solution in the supercritical antisolvent precipitation process. J Supercrit Fluids 2007;42:299–309.

[88] Sarkari M, Darrat I, Knutson BL. Generation of microparticles using CO_2 and CO_2-philic antisolvents. AIChE J 2000;46:1850–9.

Chapter 2

Interaction of Matter and Electromagnetic Radiation

Chapter 2 provides the fundamentals of interaction mechanisms of electromagnetic radiation and matter. I am aware of the fact that a comprehensive treatment of the different light–matter interaction mechanisms alone would require several books as they—in deed—can be found for each different light–matter interaction mechanism in literature. Nevertheless the focus of this book is on the application of *in situ* spectroscopic techniques at high pressure. The reader is supposed to see in which way the research activities in the field of high-pressure process technology in the past profited and in the future might profit from the *in situ* application of spectroscopic techniques. Of course, high-pressure-researchers have to understand the basics of light–matter interactions, if they intend to bring *in situ* spectroscopic techniques to ever new applications and to develop them further aiming to provide ever new insights into high-pressure process technology and science. Therefore I decided to provide in the sections that follow an extract of the basics of the interaction mechanisms between electromagnetic radiation and light, which—from my point of view—is essential for being able to understand the working principles of the different spectroscopic techniques and to transfer and adopt the spectroscopic techniques to new fields of high-pressure process technology and science. A reader who already has profound background knowledge about the properties of electromagnetic radiation and the quantum-physical properties of molecules can skip Sections 2.1 and 2.2. The subsequent sections of this Chapter 2 are structured according to the classification of different interaction mechanisms of matter and electromagnetic radiation shown in Fig. 2.1.

2.1 PROPERTIES OF ELECTROMAGNETIC RADIATION AND PHOTONS

Here the basics about electromagnetic radiation are provided, which are relevant for understanding the interaction mechanisms of matter and electromagnetic radiation addressed in this book. In detail I describe for a harmonic wave, which features a sinusoidal profile, the wavelength, the phase, the frequency,

Supercritical Fluid Science and Technology, Vol 7. http://dx.doi.org/10.1016/B978-0-444-63422-1.00002-X

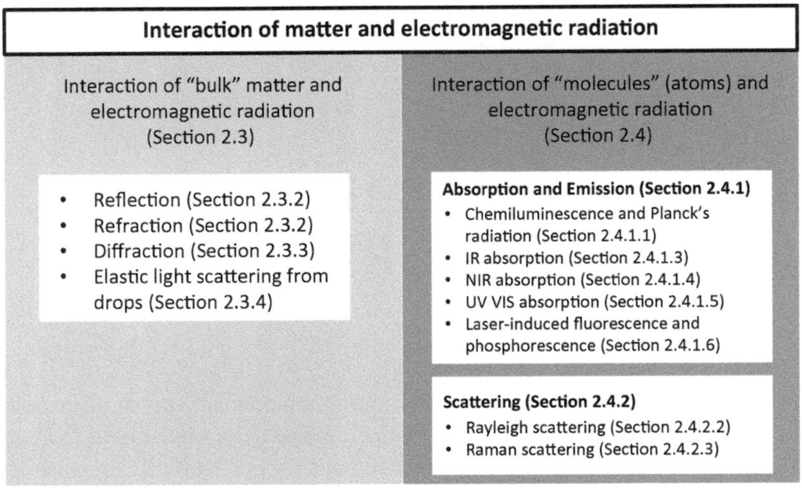

FIGURE 2.1 Classification of different interaction mechanisms of matter and electromagnetic radiation.

the wavenumber and the polarisation. We will see in the chapters that follow that under certain circumstances the interpretation of electromagnetic radiation in terms of photons is more elegant than in terms of harmonic waves. Therefore Section 2.1.4 provides the energy and the momentum of photons.

2.1.1 Equation of a Harmonic Wave

The wave function

$$\Psi(x,t) = A \cdot \sin(k(x \pm ct)) = A \cdot \sin(kx \pm \omega t) \qquad (2.1)$$

of a wave with a sinusoidal profile propagating with time t along the x-direction with a speed of propagation c is shown in Fig. 2.2. A is the amplitude of the wave, which is the maximum value of the magnitude $\Psi(x,t)$. Both, the electric and the magnetic fields oscillate perpendicular to the direction of their propagation. Therefore this kind of waves are said to be transverse. Furthermore, the oscillations of the electric and the magnetic fields are perpendicular to each other. As the argument of the sine function has to be dimensionless and as the wave function repeats itself

$$\Psi(x,t) = \Psi(x \pm \lambda, t) = \Psi(x, t \pm \tau) \qquad (2.2)$$

after one wavelength (spatial period) λ or one temporal period $\tau = \lambda / c$, the propagation constant must be given by $k = 2\pi / \lambda$.

Figure 2.2 shows only a length of two spatial periods (wavelengths) of the harmonic waves with sinusoid profile at three different instants.

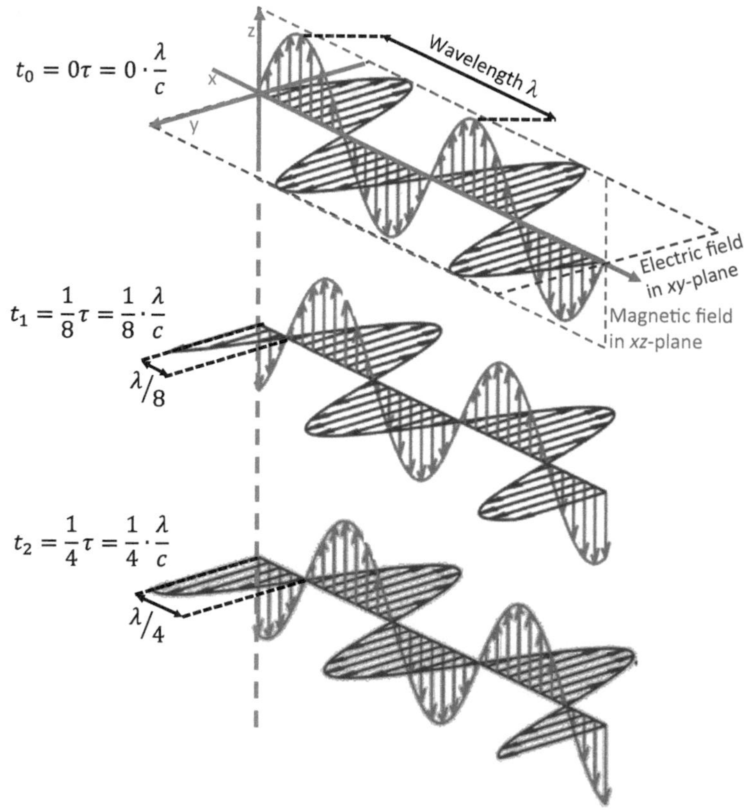

FIGURE 2.2 Sketch of an electric field oscillating in the ***xy*****-plane and propagating along the** ***x*****-axis and of the magnetic field oscillating in the** ***xz*****-plane and propagating along the** ***x*****-axis.** Both fields are sketched for three different instants t. *(Source: Sketch prepared by Medhanie Gebrekidan from the Erlangen Graduate School in Advanced Optical Technologies of the Friedrich-Alexander-Universität Erlangen-Nürnberg, Germany.)*

The time difference between the three instants is one-eighth of a temporal period each. Therefore, during one-eighth of a temporal period the waves propagate one-eighth of a spatial period. Due to the fact that the electric and the magnetic fields are coupled to each other, the properties of the magnetic field are changed automatically, if the properties of the electric field are changed due to an interaction with matter, even if the magnetic field did not interact with the matter. Likewise, the properties of the electric field would change, if the properties of the magnetic field changed due to an interaction with matter. For the interaction mechanisms between matter and electromagnetic radiation treated in this book, only the interaction between matter and the electric field is relevant. The interaction between matter and the magnetic field can be neglected. Therefore from now on the consideration of solely the electric field is sufficient. This directly brings us to the polarisation of the electric field.

2.1.2 Polarisation of the Electric Field

As mentioned before we—from now on—solely consider the electric field of the electromagnetic radiation. Therefore we also can neglect the magnetic field in the following figures and focus on the electric field. On this purpose we specify the wave function in Eqn (2.1) to

$$\vec{E}(x,t) = \overrightarrow{E_0} \cdot \sin(kx \pm \omega t) \tag{2.3}$$

and sketch few spatial periods for the instant $t = 0$ in Fig. 2.3 for three different plane states of polarisation. $\overrightarrow{E_0}$ is the amplitude of the electric field, which is the maximum disturbance of the electric wave propagating into the x-direction. One can see that the electric field vector $\vec{E}(x,t)$ for all scenarios shown in Fig. 2.3 oscillates in only one plane.

$$\overrightarrow{E_z}(x,t) = \overrightarrow{E_{0,z}} \cdot \sin(kx \pm \omega t) \tag{2.4}$$

$$\overrightarrow{E_y}(x,t) = \overrightarrow{E_{0,y}} \cdot \sin(kx \pm \omega t) \tag{2.5}$$

Therefore this kind of waves are said to be linearly or plane polarised. The plane of polarisation can—for plane waves only—be arbitrarily oriented between the cases of vertical and horizontal polarisation. In Fig. 2.3 the electric field vector $\vec{E}(x,t)$, for the vertically and the horizontally polarised waves, oscillates in the xz- and xy-plane, respectively. Consequently, for the proper description of an electric field, Eqn (2.4) has to be extended by indices z and y for vertically and horizontally polarised plane fields, respectively. Unfortunately by these two indices z and y we cannot describe a plane wave whose electric field

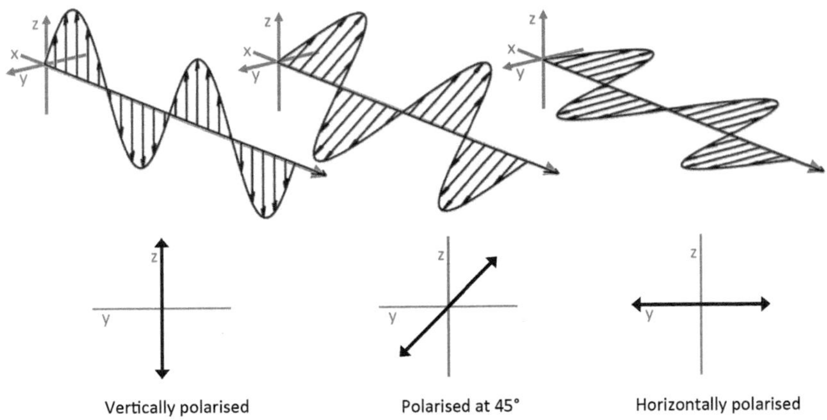

FIGURE 2.3 Sketches of linear polarised electric fields; (left) vertically polarised, (middle) polarised under 45°, (right) horizontally polarised. *(Source: Sketch prepared by Medhanie Gebrekidan from the Erlangen Graduate School in Advanced Optical Technologies of the Friedrich-Alexander-Universität Erlangen-Nürnberg, Germany.)*

vector oscillates in one plane oriented between vertical and horizontal polarisation, such as the plane wave sketched in the middle of Fig. 2.3. Furthermore we cannot describe electric fields, which do not reside in one plane (examples are given further). Therefore Fig. 2.4 shows that any state of polarisation of an electric field can be represented as a superposition of two

$$\vec{E}(x,t) = \overrightarrow{E_{0,z}} \cdot \sin(kx \pm \omega t) + \overrightarrow{E_{0,y}} \cdot \sin(kx \pm \omega t + \varepsilon) \tag{2.6}$$

orthogonal waves of the same wavelength (spatial period) and a phase difference ε between these waves.

- If there is no phase difference $\varepsilon = n \cdot 2\pi\ (n = 0, 1, 2, 3, 4, \ldots)$ between the two superimposing orthogonal waves of the same frequency, the resultant wave features a plane polarisation, too. Then the magnitudes of the amplitudes of the electric field vectors determine the final orientation of the plane of polarisation of the resultant field.
- If there is a phase difference $\varepsilon = n \cdot 2\ \pi - \pi\ (n = 1, 2, 3, 4, \ldots)$ of half of the wave's spatial period between the two superimposing orthogonal waves of

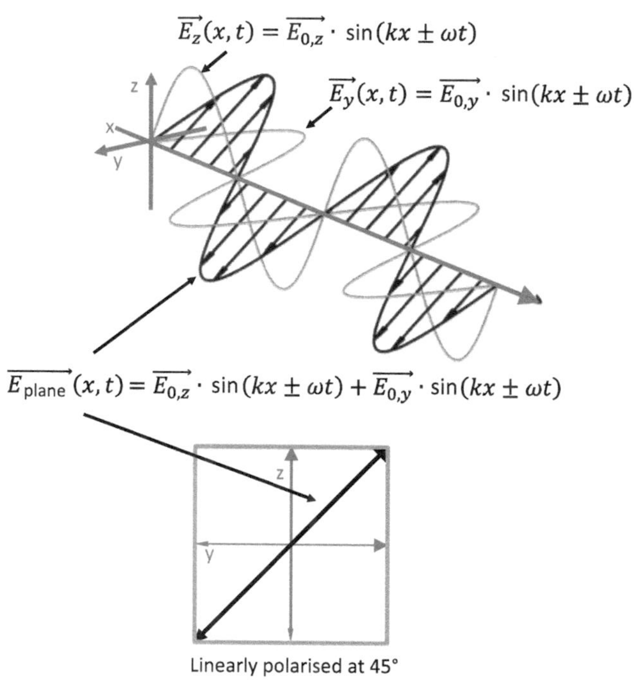

FIGURE 2.4 Representation of the polarisation of an electric field via the summation of two orthogonally oscillating electric fields. *(Source: Sketch prepared by Medhanie Gebrekidan from the Erlangen Graduate School in Advanced Optical Technologies of the Friedrich-Alexander-Universität Erlangen-Nürnberg, Germany.)*

the same frequency, the resultant wave still features a plane polarisation, which is oriented orthogonal to the case of no phase difference.

- If the phase difference $\varepsilon \neq n \cdot \pi\,(n = 0,1,2,3,4,\ldots)$ is not an integer value of half of the wave's period, the resultant electric field does not reside in a fixed plane, as it is also shown in Fig. 2.5. One special case is the phase difference $\varepsilon = n \cdot \pi + (1/2)\pi\,(n = 0,1,2,3,4,\ldots)$ of exactly one quarter of the wave's period of two orthogonal waves of the same wavelength and the same amplitude. In this case the magnitude of the resultant electric field vector $\vec{E}(x,t)$ is constant and equal to the amplitude of one of the two orthogonal electric fields $E_{0,z}$ or $E_{0,y}$ but rotates in time. This state of polarisation is said to be circularly polarised. If the amplitudes of the two orthogonal waves, which are shifted in phase by one quarter of the wave's period, are not the same, the electric field vector still rotates in time, but also its magnitude is a function of time and space. This state of polarisation is said to be elliptically polarised. Elliptically polarised electric fields also result from the superposition of two orthogonal waves of the same amplitude and wavelength whose phase difference $\varepsilon \neq n \cdot (1/2)\pi\,(n = 0,1,2,3,4,\ldots)$ is not equal to an integer value of one quarter of a wave's period.

For the description of the oscillation of a harmonic sinusoidal wave as well as for the different states of polarisation of electric fields we—so far—did not have to consider the magnitude of either the spatial period (wavelength) or

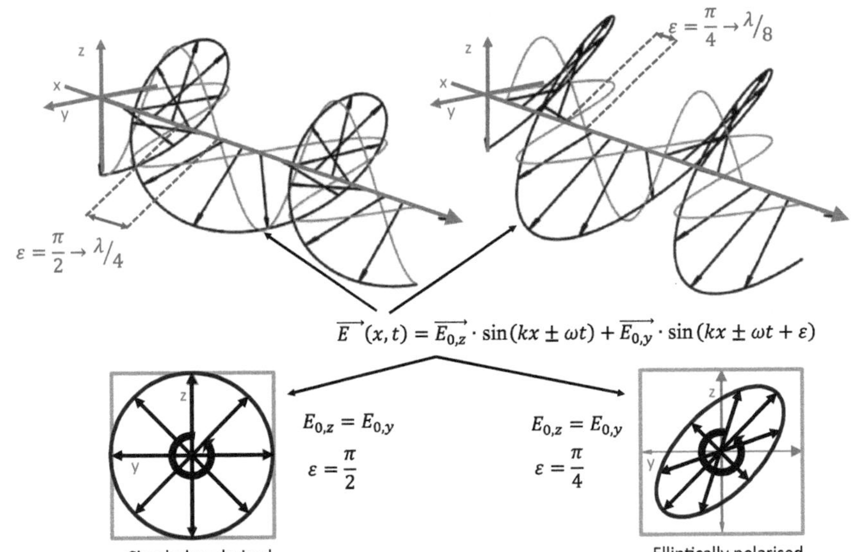

FIGURE 2.5 Illustration of circular and elliptical polarised electric fields. The resulting electric field is the summation of the two orthogonal oscillating fields with different phase shifts. *(Source: Sketch prepared by Medhanie Gebrekidan from the Erlangen Graduate School in Advanced Optical Technologies of the Friedrich-Alexander-Universität Erlangen-Nürnberg, Germany.)*

the temporal period. In the next section we will see that the periods of electromagnetic radiation cover a wide spectrum and that for different interaction mechanisms between electromagnetic radiation and matter electric fields with different periods are required.

2.1.3 Spectrum of Electromagnetic Radiation

Figure 2.6 aims at illustrating the spectrum of electromagnetic radiation between the radio- and gamma ray waves. For the different types of waves, which are radio, micro-, infrared-, visible-, ultraviolet-, X-ray- and gamma-ray waves the corresponding wavelength regimes (spatial periods) are also provided. To get a better impression of what a certain magnitude of a wavelength means, some objects are assigned to specific wavelength regimes. For example, the height of the editor of this book series, Erdogan Kiran, integrated in Fig. 2.6 as 'people' is in the order of the magnitude of metres, while the characteristic size of atoms and molecules is assigned to the nanometre (10^{-9} m) and sub-nanometre range. For the generation of or for the emission of electromagnetic radiation of different wavelengths also different sources of electromagnetic radiation are required. A microwave oven, for example, provides electromagnetic radiation with characteristic wavelengths in the range between 1 mm and 300 mm. The characteristic wavelength λ of a harmonic wave can be transferred into a wavenumber

$$\bar{\nu} = \frac{1}{\lambda(\text{in cm})}, \tag{2.7}$$

which is usually given in cm^{-1}, or into a frequency,

$$\nu = \frac{c}{\lambda} \tag{2.8}$$

which is usually given in Hz, where c is the velocity of the propagating electromagnetic wave. The meanings of the wavenumber $\bar{\nu}$ and the frequency ν are rather similar. While the wavenumber $\bar{\nu}$ tells how often the spatial period of the wave fits into 1 cm, the frequency ν tells how often its temporal period fits into 1 s. A wavenumber of 1000 cm^{-1}, for example, means that the wavelength of this wave fits 1000 times into 1 cm. Therefore the wavelength of this wave must be 10 μm. If this wave propagates in vacuum, the velocity of propagation can be approximated to $3\cdot10^{8}$ m/s and the frequency then according to Eqn (2.8) is $3\cdot10^{13}$ Hz. Therefore the temporal period of this wave fits $3\cdot10^{13}$ times into 1 s or, in other words, the temporal period of this wave is $3.33\cdot10^{-14}$ s.

We will see later that in spectroscopy waves are usually characterised by providing information about their wavenumber or their frequency. This is due to the fact that the wavenumbers and frequencies scale linearly with the energy of one wave, while the energy of one wave and the wavelength follow a nonlinear behaviour. Nevertheless it has to be mentioned—and we will see it in

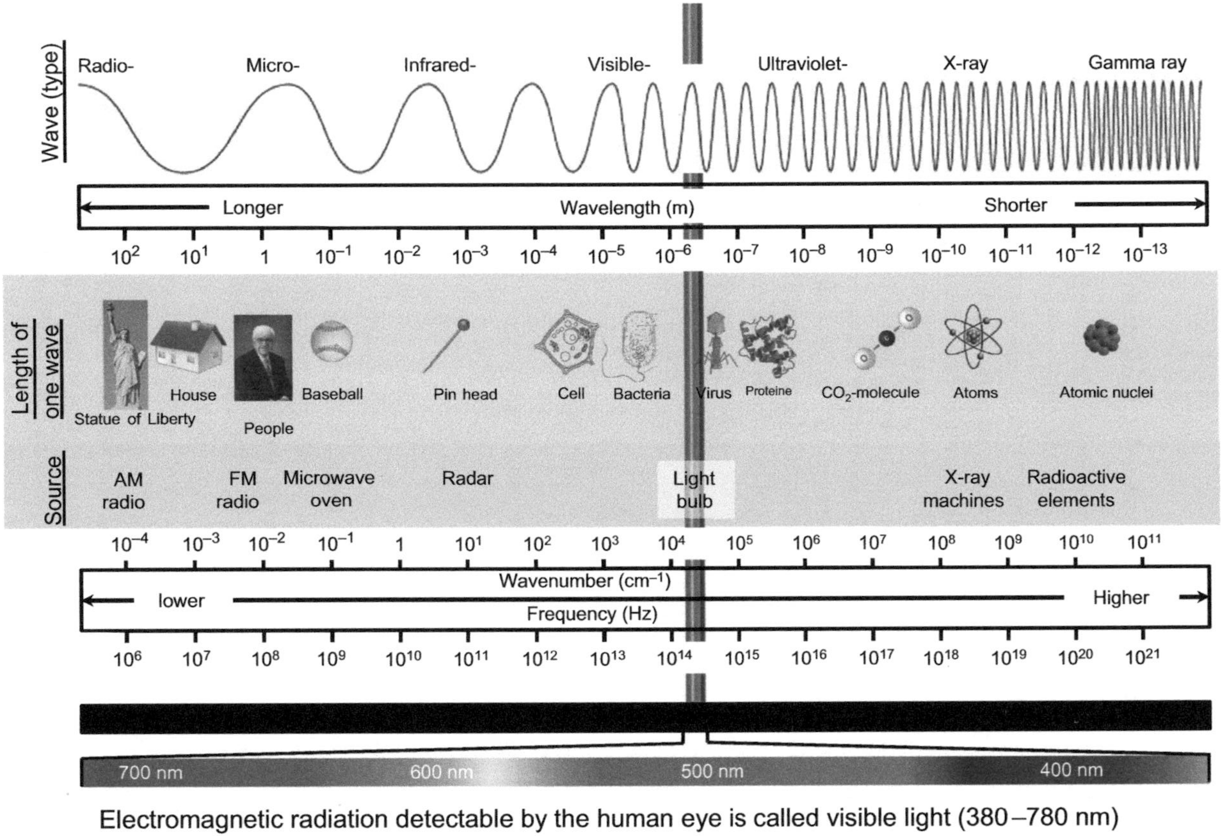

FIGURE 2.6 Schematic representation of the spectrum of electromagnetic radiation on the wavelength-, the wavenumber-, and the frequency scale. The spectrum of the visible radiation is provided as insert. *(Source: Schematic prepared by Sebastian Luther from the Erlangen Graduate School in Advanced Optical Technologies of the Friedrich-Alexander-Universität Erlangen-Nürnberg, Germany.)*

the subsequent section—that the most accurate way of characterising a wave is providing its frequency, as the frequency of the oscillation remains constant when the electric field propagates through different media of different indices of refraction, while the wavelength and with this the wavenumber of the wave would change. Nevertheless, in spectroscopy wavenumbers and wavelengths are more frequently used than frequencies.

2.1.4 Energy and Momentum of a Photon

Some interaction mechanisms between matter and electromagnetic radiation are much easier to describe and to understand if the 'electromagnetic radiation' is replaced by 'photons'. It is well known that some interaction mechanisms between matter and light can be described best by considering light to be a wave and others by considering light to be composed of photons. This duality between the wave and the photon properties of light is termed wave–particle duality. Both theories are required to explain all phenomena of light–matter interaction. The diffraction of light on a slit, for example, can be described considering light to be a wave. The fact that light carries momentum can be described by considering light to be composed of photons. Even for the computation of the energy of one photon

$$E_{\text{Photon}} = h \cdot \nu = h \cdot \frac{c}{\lambda} \tag{2.9}$$

properties characteristic for a wave, such as its frequency ν or its wavelength λ are required. Here h is the Planck's constant. Eqn (2.9) shows that the energy of one photon does linearly follow the frequency ν or the wavenumber $\bar{\nu} = 1/\lambda$ of the corresponding wave but that there is a non-linear $1/\lambda$-dependency between the photon energy and the wavelength of the corresponding wave. This dependency explains why in spectroscopy waves are characterised by either considering frequencies or wavenumbers.

Also for the computation of the momentum of one photon

$$p_{\text{Photon}} = \frac{E_{\text{Photon}}}{c} = \frac{h}{\lambda} \tag{2.10}$$

properties characteristic of a wave are required.

Figure 2.7 shows the normalised energy of photons emitted by common lasers available between the ultraviolet (UV) over the visible (VIS) and the infrared (IR) spectral region as a function of the wavelengths of the corresponding waves. The energy of the photons E_{Photon} is normalised to the energy of one photon emitted by an *excimer* (excited dimer), laser filled with a gas mixture of krypton, fluorine and helium. Therefore this laser is called KrF-excimer laser. One clearly can see that the photon energy E_{Photon} shows a $1/\lambda$ dependency. I want to emphasise here that the energy of a photon (photon energy) and the energy emitted by a laser (laser energy) are two different things. While the photon

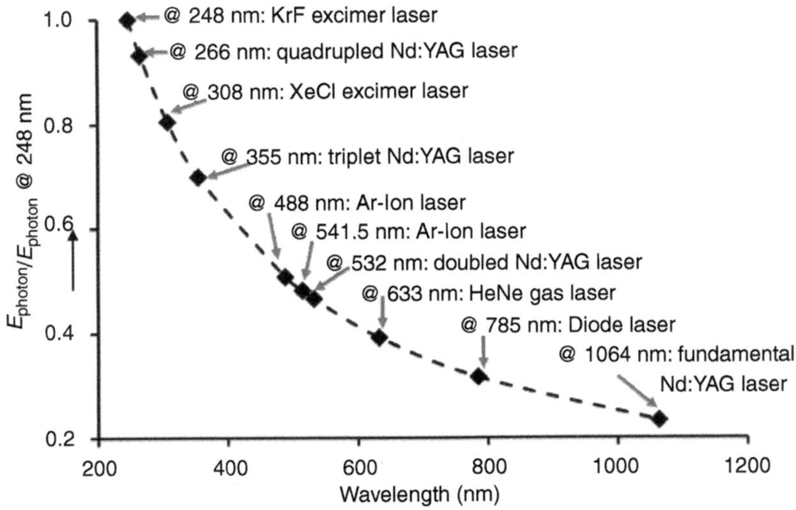

FIGURE 2.7 **The normalised photon energy as a function of the wavelength of the corresponding wave for common laser sources from the ultraviolet (UV) to the infrared (IR) spectral region.**

energy describes the energy of one single photon, the laser energy is the summation of the energies of all the photons contained in the emitted laser beam. The exercises relevant for this section aim at making clear this difference. They also aim at demonstrating the ultra-high peak intensities of pulsed lasers at rather low single-pulse energies.

2.1.5 Exercises Corresponding to Section 2.1

Exercise 2.1: Wave Function I

This is the wave function (in SI units)

$$\vec{E}(x,t)=1024\frac{V}{m}\cdot\sin\,\pi(2\cdot10^6x+6\cdot10^{14}t) \tag{2.11}$$

of an electromagnetic field.

Task: Exercise 2.1

Compute (1) the speed, (2) the wavelength, (3) the frequency, (4) the temporal period and (5) the amplitude of the electric field.

Answer: Exercise 2.1

One can rewrite Eqn (2.11) to

$$\vec{E}(x,t)=1024\frac{V}{m}\cdot\sin(2\cdot10^6(x\pm3\cdot10^8t)) \tag{2.12}$$

Now—when we compare Eqn (2.12) with Eqn (2.1) it becomes obvious that

1. the speed is $3 \cdot 10^8 \, m \, s^{-1}$
2. the wavelength is

$$\lambda = \frac{2\pi}{k} = \frac{2\pi}{2 \cdot 10^6 \, \text{m}^{-1}} = 3.142 \, \mu\text{m} \tag{2.13}$$

3. the frequency is

$$\nu = \frac{c}{\lambda} = \frac{3 \cdot 10^8 \, \text{m/s}}{3.142 \cdot 10^{-6} \, \text{m}} = 9.55 \cdot 10^{13} \, \text{Hz} \tag{2.14}$$

4. the temporal period is

$$\tau = \frac{1}{\nu} = \frac{\lambda}{c} = \frac{3.142 \cdot 10^{-6} \, \text{m}}{3 \cdot 10^8 \, \text{m/s}} = 1.05 \cdot 10^{-14} \, \text{s} \tag{2.15}$$

5. and the amplitude of the electric field is $1024 \, \frac{V}{m}$.

Exercise 2.2: Wave Function II

A plane electromagnetic harmonic wave with the frequency $8 \cdot 10^{14}$ Hz spreading out in the positive x-direction in vacuum, has an electric field amplitude of $70.71 \, \frac{V}{m}$. The wave is linearly polarised such that the plane of vibration of the electric field is at 45° to the xz-plane.

Task: Exercise 2.2

Determine the wave equation in the form of Eqn (2.1) so that the summation of $\vec{E_z}(x,t)$. and $\vec{E_y}(x,t)$ provides the earlier described wave.

Answer: Exercise 2.2

For the polarisation of 45° it holds that the amplitudes $E_{0,z} = E_{0,y}$ of the waves $\vec{E_z}(x,t)$ and $\vec{E_y}(x,t)$ are identical. As the amplitude of the resulting wave is known to be $70.71\frac{V}{m}$, the amplitudes $E_{0,y}$ and $E_{0,z}$ can be computed according to $70.71 \, \frac{V}{m} = (E_{0,z}^2 + E_{0,y}^2)^{1/2}$. With $E_{0,z} = E_{0,y}$ it follows that $E_{0,z} = E_{0,y} = 50\frac{V}{m}$.

Therefore the wave equation can be written as

$$\begin{aligned}\overrightarrow{E_z}(x,t) &= \overrightarrow{E_y}(x,t) = \overrightarrow{E_{0,z}} \cdot \sin(kx \pm \omega t)\\ &= \overrightarrow{E_{0,z}} \cdot \sin\left(\frac{2\pi \cdot \upsilon}{c} x \pm 2\pi \cdot \nu t\right)\\ &= 50\,\mathrm{V/m} \cdot \sin\left(\frac{2\pi \cdot 8 \cdot 10^{14}\,\mathrm{Hz}}{3 \cdot 10^{8}\,\mathrm{m/s}} x \pm 2\pi \cdot 8 \cdot 10^{14}\,\mathrm{Hz} \cdot t\right)\\ &= 50\,\mathrm{V/m} \cdot \sin\left(2\pi \cdot 8 \cdot 10^{14}\,\mathrm{Hz}\left(\frac{x}{3 \cdot 10^{8}\,\mathrm{m/s}} \pm t\right)\right)\end{aligned} \tag{2.16}$$

Exercise 2.3: Wave Function III

Task: Exercise 2.3

Write an expression for the wave equation of a harmonic wave $\Psi(x,t) = A \cdot \sin(kx \pm \omega t + \varepsilon)$ moving in the +*x*-direction for the instant $t = 0$ such that the following conditions hold:

1. $x = 0,\ \psi = 20$
2. $x = \frac{\lambda}{6},\ \psi = 20$
3. $x = \frac{5\lambda}{6},\ \psi = 0$

Answer: Exercise 2.3

To solve the task we consider three equations with three unknowns

$$\begin{aligned}&\Psi(x=0, t=0) = A \cdot \sin(\varepsilon) = 20\\ &\Psi\left(x = \frac{\lambda}{6}, t=0\right) = A \cdot \sin\left(\frac{\pi}{3} + \varepsilon\right) = 20\\ &\Psi\left(x = \frac{5\lambda}{6}, t=0\right) = A \cdot \sin\left(\frac{5\pi}{3} + \varepsilon\right) = 0\end{aligned} \tag{2.17}$$

From the third equation it follows from $\sin((5\pi/3) + \varepsilon) = 0$ that the phase is $\varepsilon = 1\pi/3$. If $\varepsilon = 1\pi/3$ is inserted in the first equation, the amplitude is $A = \frac{20}{\sin(\varepsilon)} = \frac{20}{\sin(\pi/3)} = \frac{20}{\sqrt{3}/2} = \frac{40}{\sqrt{3}}$. Finally we can insert the amplitude and the phase in the second equation $A \cdot \sin\left(\frac{\pi}{3} + \varepsilon\right) = 20 = \frac{40}{\sqrt{3}} \cdot \sin\left(\frac{\pi}{3} + \frac{\pi}{3}\right) = \frac{40}{\sqrt{3}} \cdot \frac{\sqrt{3}}{2} = 20$ to check our computation.

Exercise 2.4: Photons and Continuous-Wave Laser

Two laser pointers with different emission wavelengths have both an output power of 1 mW. The emission wavelengths are 532 nm for the ‘green’ and 785 nm for the ‘reddish’ laser pointer.

Task: Exercise 2.4

How many photons are emitted within one second by the 'green' and the 'reddish' laser pointer?

Answer: Exercise 2.4

The power (energy over time) of one laser pointer can be computed as the energy of all emitted photons per time t

$$P_{\text{laser}} = \frac{n_{\text{Photons}} \cdot E_{\text{Photon}}}{t}, \tag{2.18}$$

where n_{Photons} is the number of all photons and E_{Photon} is the energy of one single photon. Therefore considering Eqn (2.9) for the energy of a single photon, Eqn (2.65) can be rewritten as

$$n_{\text{Photons}} = \frac{P_{\text{laser}} \cdot t}{E_{\text{Photon}}} = \frac{P_{\text{laser}} \cdot t \cdot \lambda}{h \cdot c}. \tag{2.19}$$

Inserting the numbers, the 'green' laser emits per second

$$n_{\text{Photons,green}} = \frac{P_{\text{laser}} \cdot t}{E_{\text{Photon}}} = \frac{1 \cdot 10^{-3}\ \text{W} \cdot 1\,\text{s} \cdot 532 \cdot 10^{-9}\ \text{m}}{6.626 \cdot 10^{-34}\ \text{Js} \cdot 3 \cdot 10^{8}\ \text{m/s}} = 2.67 \cdot 10^{15} \tag{2.20}$$

photons, while the 'reddish' laser emits per second

$$n_{\text{Photons,reddish}} = \frac{P_{\text{laser}} \cdot t}{E_{\text{Photon}}} = \frac{1 \cdot 10^{-3}\ \text{W} \cdot 1\,\text{s} \cdot 785 \cdot 10^{-9}\ \text{m}}{6.626 \cdot 10^{-34}\ \text{Js} \cdot 3 \cdot 10^{8}\ \text{ms}^{-1}} = 3.95 \cdot 10^{15} \tag{2.21}$$

photons. Summarising, the number of photons contained in a laser beam of a certain power is a function of the wavelength of the emitted radiation. The longer the wavelength is, the more photons are required for the transportation of the same energy, as per photon the photon energy decreases with increasing wavelength of the corresponding wave.

Exercise 2.5: Peak Power and Average Power of Pulsed Laser

In the context of Figs 1.4 and 1.12 we already heard about pulsed lasers. Pulsed lasers are designed to store a huge amount of energy in the laser medium before the stored energy is suddenly realised within a very short time. This results in very high peak powers.

Task: Exercise 2.5

Please compute the peak power and the temporally averaged power of a pulsed laser, which emits with a frequency of 10 Hz laser pulses with a pulse energy of 3 J and a pulse duration of 5 ns.

Answer: Exercise 2.5

The peak power is

$$P_{\text{pulse}} = \frac{E_{\text{pulse}}}{t_{\text{pulse}}} = \frac{3\,\text{J}}{5 \cdot 10^{-9}\,\text{s}} = 600\,\text{MW}, \tag{2.22}$$

while the temporally averaged power is

$$P_{\text{averaged}} = \frac{10 \cdot E_{\text{pulse}}}{1\,\text{s}} = \frac{30\,\text{J}}{1\,\text{s}} = 30\,\text{W}. \tag{2.23}$$

For the temporally averaged power we consider 10 single laser pulses, which are fired during 1 s (pulse repetition rate is 10 Hz). For the computation of the peak power we only consider one single laser pulse. The computed peak power is very near to the power of a nuclear power plant (1 GW). In this task we considered—with 3 J—a rather high pulse energy. Nevertheless similarly high peak powers can be realised at much lower pulse energies, if not only the pulse energy but also the pulse duration is decreased accordingly. Today femtosecond lasers, for example, provide laser pulses with pulse energies in the order of 10 nanojoules (nJ) but pulse durations in the order of 10 femtoseconds (fs). Therefore they also reach peak intensities in the order of MW.

2.2 PROPERTIES OF MOLECULES

When I was a student in the chemical engineering program, I received—next to other subjects—a comprehensive education in 'engineering thermodynamics'. In 'engineering thermodynamics' one considers properties of macro systems, such as pressure, volume, temperature, composition and others, rather than quantum chemistry or the physics of single molecules, which is more apparently assignable to 'physical chemistry'. Especially for the understanding of the interaction of electromagnetic radiation with single molecules a profound background about the physics of single molecules is essential, which had not been available to me after my graduation from the chemical engineering program and which most probably is also not available to many researchers in the field of high-pressure process technology. Fortunately, properties which we use in thermodynamics for the characterisation of macro system are interconnected with the physics of all the single molecules forming this macro system. Therefore I decided to first show in the section that follows how the properties of single molecules and the properties of macro systems are interconnected and then—when it became clear that molecules have a kinetic, a rotational, a vibrational and an electronic energy—will describe the different forms of molecular energy.

2.2.1 Specific Heat Capacity of a Gas

In this section we consider diatomic molecules only, meaning molecules, which contain two nuclei. What will be derived for diatomic molecules could also be derived for multi-atomic molecules, but would be more complex and would exceed the scope of this book. I think that the consideration of the simplest case, which is the diatomic molecule, is sufficient to show how the molecular properties, such as the mass of the nuclei or the binding force between the nuclei affect the energy states of molecules. What I describe in the following for diatomic molecules can later be transferred to more complex molecules for the understanding of the interaction mechanisms between single—even complex—molecules and electromagnetic radiation.

In this section we consider a certain number of substance n_{subst} of a gas contained in a container. When we add a certain amount of heat Q to the gas, while we keep the volume of the container constant and while we assure that no other heat can be transferred to or from the container, the temperature of the gas inside the container will rise. In other words the gas stores the heat that we add and as a consequence the temperature, meaning its internal energy, increases. The magnitude of the temperature rise

$$\Delta T = \frac{Q}{n_{\text{subst}} \cdot c_{\text{V}}} \tag{2.24}$$

is not only a function of the added (stored) heat Q and the amount of substance in the system n_{subst}, but also a function of the specific isochoric heat capacity of the gas c_{V}. The specific isochoric heat capacity c_{V} is dependent on the gas species and a function of temperature and is tabulated in books or in the web databank of the NIST (National Institute of Standards and Technology) for many different gases and for a wide range of temperature. According to its unit J/(mol K), the specific isochoric heat capacity c_{V} tells us how much heat has to be added or removed to or from one mole of a substance in order to increase or decrease, respectively, its temperature by 1 K under isochoric conditions.

To make c_{V} dimensionless we divide it by the ideal gas constant $R = 8.3145$ J/(mol K) and plot this ratio—here computed for the diatomic hydrogen being the gas—in Fig. 2.8 as a function of the temperature.

It becomes obvious that c_{V}/R and with this also c_{V}, is a function of temperature, showing three plateaus, which in Fig. 2.8 are already assigned to—anticipating the following sections—translation, rotation and vibration of the gas molecules. Thus, the temperature dependency of the specific isochoric heat capacity, which is a macroscopic property of the entire amount of substance of the gas, is a result of the temperature-dependent population of different energy states of the gas molecules, which we will see later, is a molecular property. In the sections that follow we will see that when we add heat to a gas composed of diatomic molecules, the added heat is stored in form of translational energy of the molecules, rotational energy of the molecules and vibrational energy of the molecules. In general this is also true for gases composed of molecules with more than two nuclei. But, be aware that for the storage of energy in the forms of rotation (therefore, a moment of inertia is required) or vibration (therefore, a chemical bond is required) a molecule at minimum has to be diatomic.

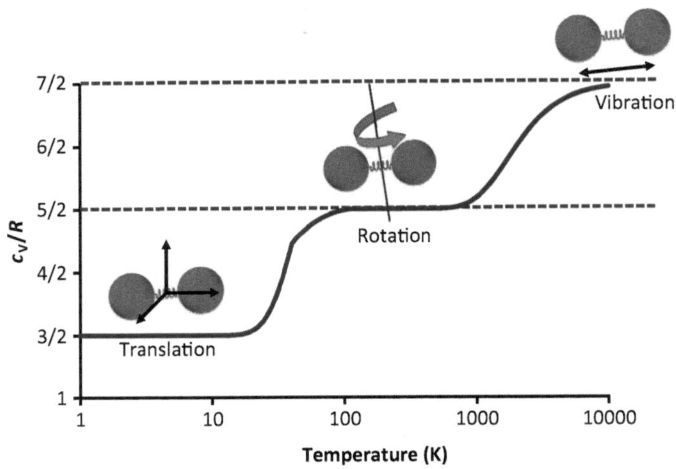

FIGURE 2.8 Specific isochoric heat capacity c_v over the ideal gas constant R as a function of the temperature. Curve computed for hydrogen. *(Source: Graphic prepared by Julian Schuster from the Lehrstuhl für Technische Thermodynamik of the Friedrich-Alexander-Universität Erlangen-Nürnberg, Germany.)*

Therefore, atomic gases, such as argon or helium, cannot store added heat in rotation or vibration.

The first plateau in Fig. 2.8 shows that at cold temperatures, or when the internal energy of the gas is small, the 'diatomic gases' store the added heat by increasing the translational energy of the molecules. The second plateau shows that at moderate temperatures the gas can store the energy by increasing both the translational and the rotational energy of the molecules. At high temperatures the gas can store the added heat by increasing the translational, the rotational and the vibrational energy of the molecules. Therefore, the specific isochoric heat capacity increases with increasing temperature. In other words, if we add the same amount of heat Q to three gas volumes, each containing the identical amount of substance n_{subst} of the same species, and one having a low, one having a medium and one having a high temperature with respect to Fig. 2.8, the temperature increase $\Delta T = Q/(n_{\text{subst}} \cdot c_v)$ accompanied with the storage of the heat will be highest for the initially cold gas (small c_v) and the lowest for the initially hot gas (high c_v).

The reason why in Fig. 2.8 the plateaus of c_v/R for translation, rotation and vibration of a diatomic molecule are at 3/2, 5/2 and 7/2, respectively, will be des cribed in the following section.

2.2.2 Translational Energy

Again we consider a certain number of diatomic molecules N_{molec} of a gas contained in a volume V at a temperature T. The temperature T—with respect to Fig. 2.8—is as low as the specific isochoric heat capacity c_v is dominated by the translational energy of the molecules. Then rotation and vibration can be neglected. Furthermore we assume that

- there is no contribution of chemical energy $U_{\text{chem}} = 0$ to the internal energy U_{internal}, as the gas molecules cannot react in the volume, and that
- there are neither attractive nor repulsive forces between the molecules, implying that also the potential energy $U_{\text{potential}} = 0$ does not contribute to the internal energy U_{internal}.

Under these circumstances $U_{\text{chem}} = U_{\text{potential}} = 0$ the internal energy U_{internal} of the gas containing volume,

$$U_{\text{internal}} = \sum_{N_{\text{molec}}} (E_{\text{kin, molec}}) = \sum_{N_{\text{molec}}} \left(\frac{1}{2} m_{\text{molec}} \cdot s_{\text{molec}}^2 \right) \tag{2.25}$$

is equal to the summation $\sum_{N_{\text{molec}}}$ of the kinetic energies $E_{\text{Kin, molec}}$ of all the molecules N_{molec} contained inside the volume, where m_{molec} and s_{molec} are the mass and the translational velocity of one single gas molecule, respectively. It is described in the appendix that at a certain temperature not all molecules move with the same velocity. The velocity distribution follows the Maxwell distribution. This is why I want to make clear that in Eqn (2.25) s_{molec} does not represent the average velocity of the molecules, but the velocity of one single molecule.

Under the circumstances defined ($U_{\text{chem}} = U_{\text{potential}} = 0$) the kinetic E_{kin} and thermal energy E_{thermal} of the gas—now we consider the summation of all the molecules—can be set equal

$$E_{\text{kin}} = \sum_{N_{\text{molec}}} \left(\frac{1}{2} m_{\text{molec}} \cdot s_{\text{molec}}^2 \right) = N_{\text{molec}} \cdot \frac{f}{2} \cdot k \cdot T = E_{\text{thermal}}. \tag{2.26}$$

where f represents the number of degrees of freedom of translational movement and $k = 1.380 \cdot 10^{-23}$ J/K the Boltzmann constant. The appendix (Eqn (2.147)) shows that Eqn (2.26) can also be derived from the consideration of the kinetic energy of the gas molecules and the pressure equation.

Eqns (2.25) and (2.26) can be combined to

$$U_{\text{internal}} = N_{\text{molec}} \cdot \frac{f}{2} \cdot k \cdot T = E_{\text{thermal}} = n_{\text{subst}} \cdot \frac{f}{2} \cdot R \cdot T \tag{2.27}$$

with the amount of substance $n_{\text{subst}} = N_{\text{molec}} / N_{\text{Avgadro}}$, the ideal gas constant $R = k \cdot N_{\text{Avgadro}}$ and Avogadro's constant $N_{\text{Avgadro}} \approx 6.022 \cdot 10^{23}$ mol^{-1}. In a next step we can see that if we consider the internal energy normalised to the amount of substance $U_{\text{internal}} / n_{\text{subst}} = (f/2) \cdot R \cdot T$ and then differentiate with respect to the temperature T

$$\frac{\mathrm{d}(U_{\text{internal}} / n_{\text{subst}})}{\mathrm{d}T} = \frac{f}{2} \cdot R = c_{\text{v}} \tag{2.28}$$

we derived the specific isochoric heat capacity c_V, which tells how much heat has to be added to or removed from a certain amount of substance in order to increase or decrease, respectively, the temperature of the amount of substance by 1 K at isochoric conditions.

Combining Eqn (2.28) and Fig. 2.8 we now understand why the plateaus of translation, rotation and vibration for the diatomic molecule are at 3/2, 5/2 and 7/2. Translation features three degrees of freedom of movement, in *x*-, *y*- and *z*-direction. Therefore at cold temperatures at which the diatomic molecules do neither rotate nor vibrate, $\frac{c_V}{R} = \frac{(f_{\text{translation}} = 3)}{2}$. At higher temperatures the molecules start to rotate. Rotation of a diatomic molecule in general has three degrees of freedom. Two of them are normal to the molecular axis and one is around the molecular axis. As the moment of inertia for rotation around the molecular axis is negligibly small with respect to the other rotations, we only consider two degrees of freedom for rotation of a diatomic molecule. Therefore at higher temperatures, $\frac{c_V}{R} = \frac{(f_{\text{translation}} = 3) + (f_{\text{rotation}} = 2)}{2} = \frac{5}{2}$. At the highest temperatures the diatomic molecules can also store the added heat in the form of vibrational energy. It is not obvious why vibration of a diatomic molecule should have two degrees of freedom, as during vibration the molecule elongates or shrinks only along one direction, which is the molecular axis. The two degrees of freedom of the vibration of a diatomic molecule are due to the fact that the vibrational energy is composed of a kinetic energy (velocity of the movement of the nuclei towards or away from each other) and a potential energy (displacement of the distance of the nuclei from the distance between them without vibration).Therefore, when vibrational energy also can store the added heat (at very high temperatures), $\frac{c_V}{R} = \frac{(f_{\text{translation}} = 3) + (f_{\text{rotation}} = 2) + (f_{\text{vibration}} = 2)}{2} = \frac{7}{2}$.

2.2.3 Rotational Energy of a Diatomic Molecule

Figure 2.9 shows a sketch of a diatomic molecule rotating with the angular frequency ω around one rotational axis.

The rotational axis is sketched as grey broken line passing the balance point of the molecule and being normal to the molecular axis. The molecule is composed of two nuclei of the masses m_{N1} and m_{N2}. The distances between the cores of the nuclei and the distances between the cores of the nuclei and the balance point of the molecules are labelled with *R*. With respect to the distance *R* the subscripts 'r' and 'nr' represent the 'rigid' and the **n**on-**r**igid rotator models, respectively. In the left column of Fig. 2.9, which shows the rigid rotator model, the distance between the two nuclei R_r is not affected by the increase of the angular frequency (top, slow rotation; bottom, fast rotation). Therefore, in the case of the rigid rotator model we assume that the centrifugal force, which acts to drive the nuclei away from the balance point, cannot increase R_r, as the

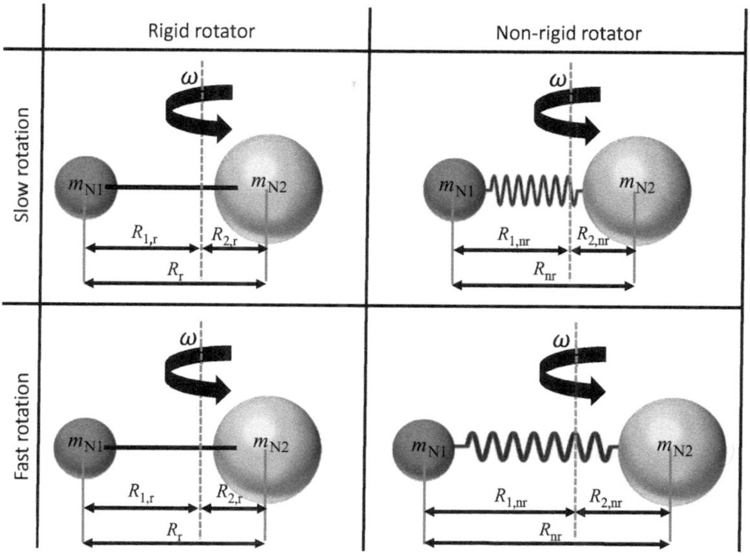

FIGURE 2.9 Sketch of a diatomic molecule rotating around the rotational axis with the angular frequency ω. In the lower part (fast rotation) the angular frequency is high compared to the upper part (slow rotation). The left and right columns show the diatomic molecule according to the rigid and the non-rigid rotator models, respectively.

binding between the two nuclei is rigid. In this section we will mainly consider the rigid rotator model in order to derive the energy of rotational energy levels. At the end of this section we will see then how the more realistic model, which is the non-rigid rotator model, impacts the findings we derived for the rigid rotator model. The non-rigid rotator model is shown in the right column of Fig. 2.9. Here the binding between the two nuclei is non-rigid and therefore represented by a spring. As the angular frequency increases, the centrifugal force drives the nuclei away from the balance point of the molecule. Consequently the distance between the two nuclei R_{nr} increases with increasing angular frequency of rotation. We will see soon that the distance between the nuclei and the balance point influences the momentum of inertia of the rotating molecule and with this also the rotational energy of molecules.

According to conventional mechanics, the energy of rotation (unit Joule)

$$E_{rot} = \frac{1}{2}\Theta\omega^2 = \frac{1}{2} \cdot \frac{L_{rot}^2}{\Theta} \tag{2.29}$$

of the molecules sketched in Fig. 2.9 can be computed from the angular momentum (unit kg m^2/s)

$$L_{rot} = \Theta\omega \tag{2.30}$$

normal to the molecular axis, which is the product of the momentum of inertia (unit kg m^2)

$$\Theta = m_{\text{red}} \cdot R^2 \tag{2.31}$$

and the angular frequency ω, with m_{red} being the reduced mass (unit kg)

$$m_{\text{red}} = \frac{m_{\text{N}1} \cdot m_{\text{N}2}}{m_{\text{N}1} + m_{\text{N}2}}. \tag{2.32}$$

Already now we can see that the momentum of inertia Θ, in the case of the rigid rotator model is not affected by the angular frequency ω, as $R_{\text{r}} \neq f(\omega)$ and $R_{\text{r}} = \text{const}$. On the contrary, in the case of the non-rigid rotator model, the momentum of inertia Θ increases with increasing angular frequency ω, as the centrifugal force increases the distance between the nuclei $R_{\text{nr}} = f(\omega)$.

According to quantum mechanics the angular momentum

$$L_{\text{rot}} = \frac{h}{2\pi} \cdot \sqrt{J(J+1)} \tag{2.33}$$

follows the quantum condition with $J = 0, 1, 2, 3, \ldots$ being the rotational quantum number and h being Planck's constant.

Combining conventional and quantum mechanics, meaning combining Eqns (2.29) and (2.33) and combining Eqns (2.30) and (2.33), yields firstly the quantisised energy of rotation

$$E_{\text{rot}}(J) = \frac{h^2}{8\pi \cdot \Theta} \cdot J(J+1) \tag{2.34}$$

and secondly the quantisised angular frequency of the rotation

$$\omega = \frac{h}{2\pi \cdot \Theta} \cdot \sqrt{J(J+1)}. \tag{2.35}$$

In molecular spectroscopy the energy of molecules is usually not given in 'Joules' (J), but in wavenumbers (cm^{-1}). This brings the advantage that we then can directly compare the energy of molecules with the energy of electromagnetic radion (or photons), which according to Eqn (2.7) is—at least in molecular spectroscopy—also quantified in wavenumbers.

For the conversion of the energy of rotation E_{rot} (J) in 'Joules' into the rotational energy F_{rot} (J) in 'wavenumbers'

$$F_{\text{rot}}(J) = \frac{E_{\text{rot}}(J)}{hc} = B \cdot J(J+1) \tag{2.36}$$

the energy of rotation E_{rot} (J) has to be divided by hc, with

$$B = \frac{h}{8\pi^2 \cdot \Theta \cdot c} \tag{2.37}$$

being the rotational constant. The rotational constant B is species specific, as it inversely depends on the momentum of inertia Θ, which according to Eqn (2.31) incorporates the molecular-specific reduced mass m_{red} and the molecular-specific distance R between the nuclei. Giving some examples, the rotational constant of the light hydrogen molecule H_2, of the heavy hydrogen molecule $^2H_2$, of the nitrogen molecule N_2 and of the carbon monoxide isotopes ^{12}CO and ^{13}CO are $B_{H_2} = 60.81\ \text{cm}^{-1}$, $B_{^2H_2} = 30.43\ \text{cm}^{-1}$, $B_{N_2} \approx 2\ \text{cm}^{-1}$, $B_{^{12}CO} = 1.921\ \text{cm}^{-1}$ and $B_{^{13}CO} = 1.8365\ \text{cm}^{-1}$, respectively.

B is the rotational constant and therefore cannot consider that, in the realistic case of the non-rigid rotator, the momentum of inertia Θ is not a constant, but increases with increasing angular frequency ω, due to the elongation of the molecule initiated by the centrifugal force. Therefore, for the non-rigid rotator the rotational energy (here only given for the first non-linear correction term of the series expansion)

$$F_{rot}(J) = \frac{E_{rot}(J)}{hc} = B \cdot J(J+1) - D \cdot J^2(J+1)^2 \tag{2.38}$$

is also a function of the centrifugal elongation constant (Eqn (2.38) only true for a harmonic force field)

$$D = \frac{h^3}{32\pi^4 \cdot k_{binding} \cdot \Theta^2 \cdot R_e^2 \cdot c}, \tag{2.39}$$

where $k_{binding}$ is the spring constant of the elastic binding between the two nuclei and R_e is the distance between the nuclei for an angular frequency $\omega = 0$. D is significantly smaller than the rotational constant B. Consequently the infuence of D onto the rotational energy F_{rot} (J) gets more pronounced with increasing rotational quantum number J and can be neglected for small rotational quantum numbers J.

Figure 2.10 compares the rotational energies F_{rot} (J) (black) as a function of the rotational quantum number J for the rigid (left) and the non-rigid (right) rotator model. The red dashed lines indicate that the rotational energy F_{rot} (J) of the non-rigid rotator is less than the one of the rigid rotator, and that the difference increases with increasing rotational quantum number J.

In grey colour the energy differences between neigbouring rotational energy levels

$$\Delta F_{rot}(J) = F_{rot}(J+1) - F_{rot}(J) \tag{2.40}$$

are provided, also for the rigid and the non-rigid rotator. While for the rigid rotator the energy difference between neighbouring rotational energy levels increases in increments of two times the rotational constant B, for the non-rigid rotator the energy difference between neighbouring energy levels ΔF_{rot} (J) also increases with increasing rotational quantum number J, but in increments getting continuously smaller than two times the rotaional constant B.

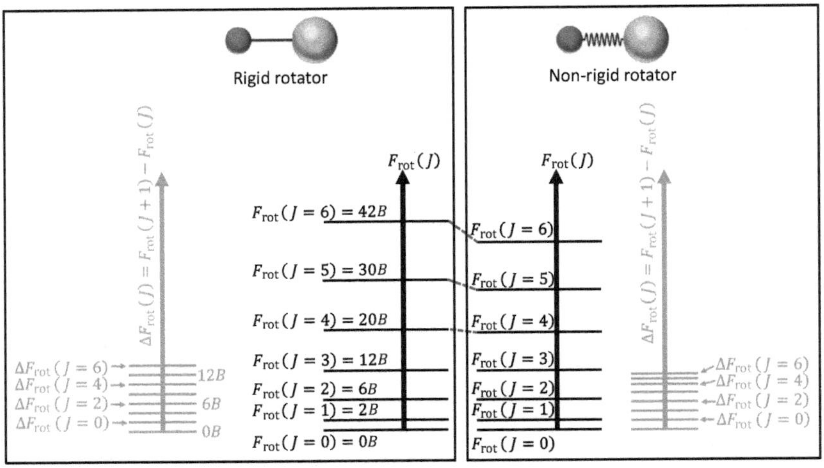

FIGURE 2.10 Energy level diagram (term scheme) of the rotational energy $F_{rot}(J)$ for rotational levels J for the rigid (left) and the non-rigid (right) rotator model. $\Delta F_{rot}(J)=F_{rot}(J+1)-F_{rot}(J)$ in grey represents the energy difference between two neighbouring energy levels J.

2.2.4 Vibrational Energy of a Diatomic Molecule

Similarly to the previous section on rotational energy, the vibrational energy will be derived for the simplest case, which is the diatomic molecule. I believe that the conceptual understanding of the diatomic oscillator is sufficient for being able to understand the working principles and the data evaluation procedures of the measurement techniques presented later. In this section we will also neglect that the diatomic molecules rotate while they vibrate and therefore only consider the vibrational energy.

Figure 2.11 shows three snapshots of two diatomic molecules during the molecular stretch vibration. The spheres represent the nuclei of the molecules and the spring represents the bond between the nuclei. While the left molecule is composed of two identical nuclei with the mass $m_{N1}=m_{N2}$, the right molecule is composed of a light and a heavy nucleus with the masses m_{N1} and $m_{N2}\gg m_{N1}$, respectively. The snapshot in the middle shows the two nuclei in the distance $R=R_e$, where R_e is the equilibrium distance between the nuclei at the—in reality not existing—point of rest. The upper and the lower snapshots show compressed $R<R_e$ and stretched $R>R_e$ molecules, respectively. It is also shown in Fig. 2.11 that in the case of the homo-nuclear molecule (left) the two nuclei move towards and away from each other, while in the case of the hetero-nuclear molecule (right) only the light nucleus moves with respect to the heavy one. Therefore the mass of the oscillating system is given by the reduced mass m_{red} (included in Fig. 2.11 and see Eqn (2.32)).

The energy stored in an oscillating system

$$E_{vib}=E_{vib,pot}+E_{vib,kin} \tag{2.41}$$

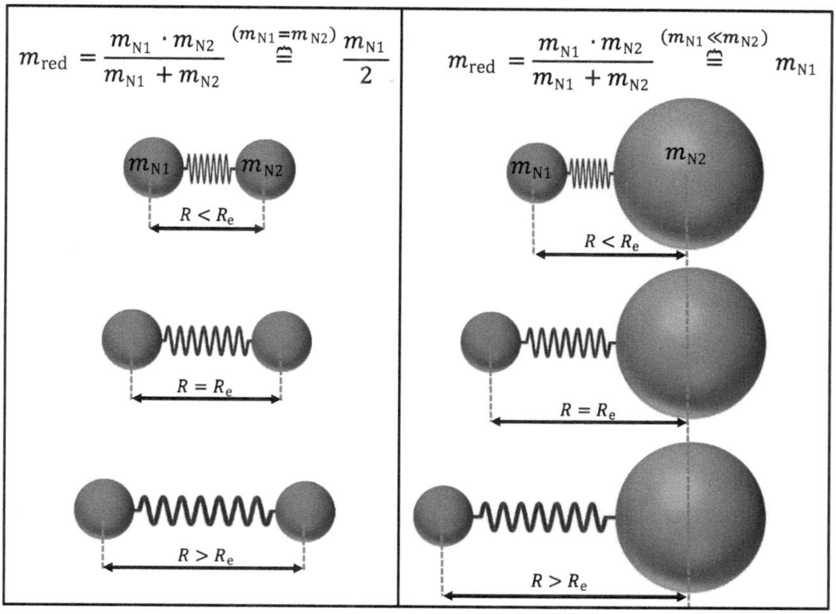

FIGURE 2.11 **Three snapshots of the stretch vibration of a diatomic molecule.** Left: Homo-nuclear molecule; right: hetero-nuclear molecule. Snapshots show a compressed molecule (top), a stretched molecule (bottom) and a molecule with the equilibrium extension.

is composed of the potential $E_{\text{vib, pot}}$ and the kinetic portion $E_{\text{vib,kin}}$. According to conventional mechanics the potential energy of a harmonic oscillating system (mass hanging on a spring and oscillating)

$$E_{\text{vib, pot}}(R) = \frac{k_{\text{binding}}}{2}(R - R_e)^2 \tag{2.42}$$

is parabolic (see Fig. 2.12 grey parabola) and a function of the spring constant k_{binding} ('binding force' between the nuclei; 'binding force' is written in quotation marks as the spring constant is not a force but a material property) and the square of the displacement $(R - R_e)^2$ of the distance R between the nuclei with respect to the distance at the point of rest R_e. Here the spring constant k_{binding} is closely related to the 'binding force' between the nuclei, which is set in quotation marks as the spring constant is not a force but a material property. Due to the conservation of energy (Eqn (2.41)), at the point $R - R_e = 0$, where the potential energy $E_{\text{vib, pot}} = 0$, all the energy of the oscillating system is stored in kinetic energy $E_{\text{vib}} = E_{\text{vib,kin}}$. On the contrary, at the reversal points of the oscillation $(R - R_e)^2 = \max$, where the displacement velocity of the two nuclei is zero and therefore the kinetic energy $E_{\text{vib,kin}} = 0$ too, the energy of the oscillating system is solely stored as potential energy

$$E_{\text{vib}} = E_{\text{vib, pot}} \tag{2.43}$$

According to quantum mechanics the vibrational energy

$$E_{\text{vib}}(\text{v}) = \frac{h}{2\pi}\omega\left(\text{v}+\frac{1}{2}\right) = h\nu_{\text{e}}\left(\text{v}+\frac{1}{2}\right) \tag{2.44}$$

follows the quantum condition with $\text{v} = 0, 1, 2, 3,\ldots$ being the vibrational quantum number, h being Planck's constant and ω being the angular frequency of oscillation

$$\omega = 2\pi \cdot \nu_{\text{e}}. \tag{2.45}$$

Here

$$\nu_{\text{e}} = \frac{1}{2\pi} \cdot \sqrt{\frac{k_{\text{binding}}}{m_{\text{red}}}} \tag{2.46}$$

is the eingenfrequency of the oscillating system, which according to conventional mechanics is a function of the spring constant k_{binding} (here the 'binding force') and the reduced mass m_{red}. Figure 2.12 shows as black horizontal lines

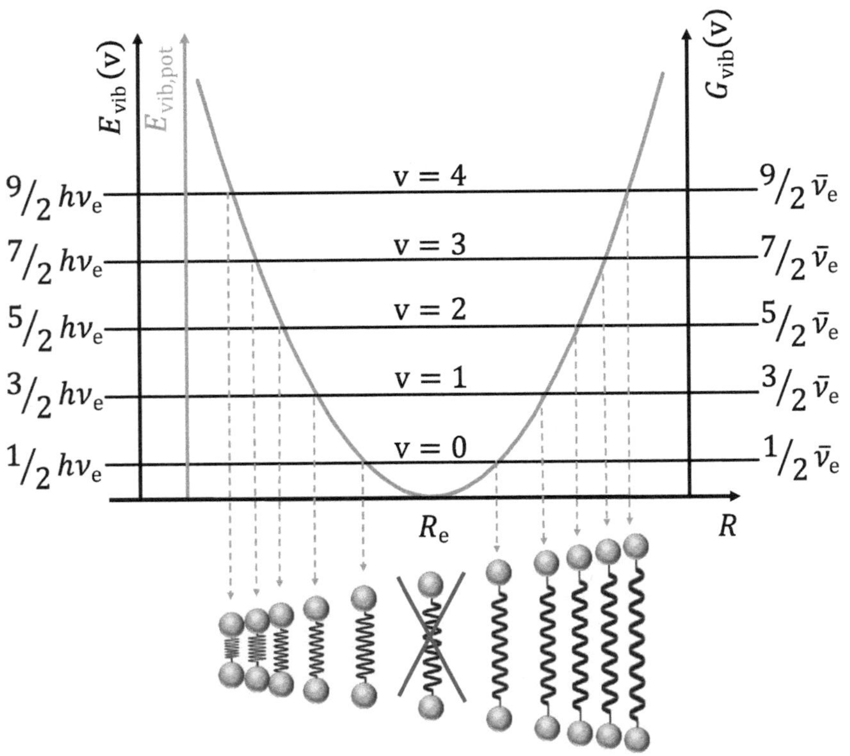

FIGURE 2.12 **Vibrational energy E_{vib} (v) as a function of the vibrational quantum number v and the potential portion of the vibrational energy $E_{\text{vib, pot}}$ as a function of the distance R between the nuclei in a 'harmonic oscillator' diatomic molecule.**

the quantised vibrational energy levels for the vibrational quantum numbers v = 0, 1, 2, 3, and 4. We will see later that in the electronic ground state the vibrational quantum numbers are often given as v″ instead of v.

Eqn (2.43) showed that at the reversal point of oscillation $E_{vib} = E_{vib,\,pot}$. Therefore in Fig. 2.12 the intersections of the horizontal lines representing $E_{vib}(\mathrm{v})$ and of the parabola representing $E_{vib,\,pot}$ indicate the distances R between the nuclei at the reversal points. Snapshots of the molecules at the reversal points are also provided for the vibrational quantum numbers v = 0, 1, 2, 3, and 4. Eqn (2.44) shows that the vibration energy $E_{vib}(\mathrm{v}=0)$ for the vibrational quantum number $\mathrm{v}=0$ is $(1/2)h\nu_e$. Accordingly, molecules in the vibrational ground state $\mathrm{v}=0$ do not rest at the point of rest R_e but oscillate even at a vibrational quantum number $\mathrm{v}=0$. Therefore I crossed the sketch of the molecule at the position R_e. The molecules in the vibrational ground state but at the reversal point of oscillation are represented for the compressed molecule as left and for the stretched molecule as right neighbours of the crossed molecule.

As it has been mentioned already in the previous section on the rotational energy, in spectroscopy the energy of molecules is quantified in wavenumbers. Therefore the vibrational energy has to be divided by $(h \cdot c)$

$$G_{vib}(\mathrm{v}) = \frac{E_{vib}(\mathrm{v})}{h \cdot c} = \frac{h}{2\pi \cdot h \cdot c}\omega\left(\mathrm{v}+\frac{1}{2}\right) = \bar{\nu}_e\left(\mathrm{v}+\frac{1}{2}\right) \tag{2.47}$$

which yields the vibrational constant

$$\bar{\nu}_e = \frac{h \cdot \omega}{2\pi \cdot h \cdot c} = \frac{\nu_e}{c}. \tag{2.48}$$

It has to be underlined here that ν_e is the eigenfrequency of the molecule in Hz and $\bar{\nu}_e$ is the vibrational constant of the molecule in wavenumbers (cm^{-1}). As ν_e is a function of the spring constant $k_{binding}$ (binding force) and the reduced mass m_{red} of the oscillating system (compare Eqn (2.46)) also the vibrational constant $\bar{\nu}_e$ is species specific. Giving some examples, the vibrational constant of the light hydrogen molecule H_2, of the heavy hydrogen molecule $^2H_2$, of the nitrogen molecule N_2 and of carbon monoxide CO and of oxygen O_2 are $\bar{\nu}_{e,H_2} = 4159.2\ cm^{-1}$, $\bar{\nu}_{e,^2H_2} = 2990.3\ cm^{-1}$, $\bar{\nu}_{e,N_2} = 2330.7\ cm^{-1}$, $\bar{\nu}_{e,CO} = 2143.3\ cm^{-1}$ and $\bar{\nu}_{e,O_2} = 1556.3\ cm^{-1}$, respectively. Comparing, for example, for nitrogen the rotational constant $B_{N_2} \approx 2\ cm^{-1}$ (see Eqn (2.37)) and the vibrational constant $\bar{\nu}_{e,N_2} = 2330.7\ cm^{-1}$ we can see that the vibration of molecules can store approximately three orders of magnitude more energy than the rotation and also that the energy difference between neighbouring vibrational energy levels is approximately three orders of magnitude higher than the energy difference between neighbouring rotational energy levels.

In reality the potential energy of vibration does not follow the parabolic curve shown in Fig. 2.12. Figure 2.13 compares the potential curves for the harmonic and the anharmonic oscillator. With a reduction of the binding length R

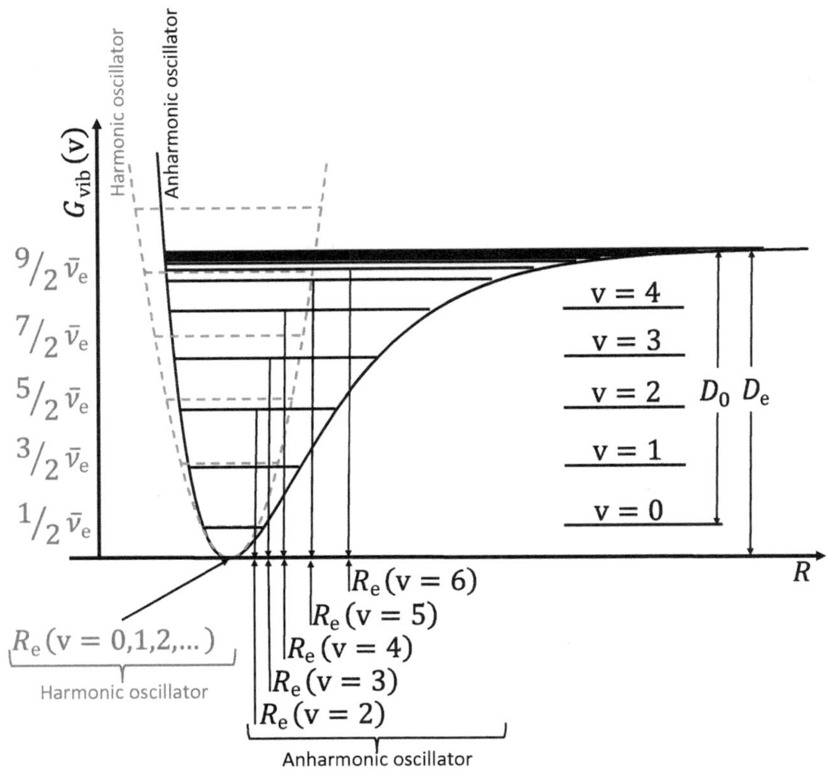

FIGURE 2.13 Comparison between the vibrational energy levels and the potential curve for the harmonic (grey and dashed lines) and the anharmonic (black solid lines) oscillator. Horizontal lines represent the quantity of the vibrational energy G_{vib} (v) as a function of the vibrational quantum number v and potential curves represent the potential portion of the vibrational energy $G_{vib,pot}$ as a function of the distance R between the nuclei in the diatomic molecule.

the repulsive forces become more and more dominant and make a fusion of the two nuclei impossible. Therefore the potential curve for the anharmonic oscillator for binding lengths $R < R_e$ is steeper than the potential curve of the harmonic oscillator. On the other hand an increase of the binding length R loosens the bond and finally can result in the dissociation of the molecule in two separate atoms. Therefore the potential curve of the anharmonic oscillator for $R > R_e$ is flatter than that one of the harmonic oscillator.

The potential curve of the anharmonic oscillator can be approximated by the Morse potential

$$G_{vib,pot}(R) = \frac{E_{vib,pot}}{h \cdot c} \underbrace{\approx}_{\text{Morse potential}} D_e(1 - e^{-a(R-R_e)})^2 \tag{2.49}$$

where D_e is the dissociation energy and

$$a = 2\pi \cdot \nu_e \left(\frac{m_{red}}{2D_e}\right)^{1/2} \tag{2.50}$$

represents specific properties of the molecule, such as its eigenfrequency ν_e (Eqn (2.46)), its reduced mass m_{red} (Eqn (2.32)) and its dissociation energy D_e. The energy of the vibrational levels v can be approximated by (giving only the first non-linear term of the series expansion)

$$G_{vib}(\mathrm{v}) = \bar{\nu}_e\left(\mathrm{v} + \frac{1}{2}\right) - x_e\bar{\nu}_e\left(\mathrm{v} + \frac{1}{2}\right)^2 \tag{2.51}$$

which in Fig. 2.13 are sketched as horizontal black lines. Higher order correction terms can be considered for high vibrational quantum numbers v.

$$x_e = \frac{h \cdot \nu_e}{4 \cdot D_e} \tag{2.52}$$

is the anharmonicity constant, which is the ratio of the vibrational energy $h \cdot \nu_e$ and four times the dissociation energy D_e. As x_e is typically in the order of 0.01, the correction term $x_e\bar{\nu}_e(\mathrm{v}+(1/2))^2$ becomes increasingly prominent with increasing vibrational quantum number v. This behaviour is illustrated in Fig. 2.13 showing an increasing difference between the vibrational energy levels of the harmonic (dashed grey horizontal) and the anharmonic (solid black horizontal) oscillator. For the vibrational quantum number of $\mathrm{v} = 0$ the energy difference between the anharmonic and the harmonic oscillator is least.

I want to point out here that Eqn (2.51), which we use for the description of the vibrational energy of the anharmonic oscillator, contains the vibrational constant $\bar{\nu}_e$, which we derived in the context of the harmonic oscillator. Therefore, the anharmonicity is not accounted for by correcting the vibrational constant directly, but by considering the vibrational constant of the harmonic oscillator $\bar{\nu}_e$ and then correcting the vibrational energy (compare Eqn (2.51)). Using the approach presented further—next to the fact that the vibrational energy of the anharmonic oscillator is less than the one of the harmonic oscillator—we also can demonstrate that the eigenfrequency ν (Eqn (2.46)) or the vibrational constant $\bar{\nu}$ of the anharmonic oscillator decreases with increasing vibrational quantum number v. In contrast to the eigenfrequency ν_e or the vibrational constant $\bar{\nu}_e$ of the harmonic oscillator we use ν and $\bar{\nu}$ without index 'e' for the eigenfrequency and the vibrational constant of the anharmonic oscillator, respectively. Instead of correcting the vibrational energy using the correction term as shown in Eqn (2.51), we now correct for the anharmonicity

$$G_{vib}(\mathrm{v}) = \overbrace{\bar{\nu}_e}^{\text{harmonic}} \cdot \left(\mathrm{v} + \frac{1}{2}\right) - x_e \cdot \overbrace{\bar{\nu}_e}^{\text{harmonic}} \cdot \left(\mathrm{v} + \frac{1}{2}\right)^2 = \overbrace{\bar{\nu}}^{\text{anharmonic}} \cdot \left(\mathrm{v} + \frac{1}{2}\right) \tag{2.53}$$

by considering the eigenfrequency ν or the vibrational constant $\bar{\nu}$ of the anharmonic oscillator with $\bar{\nu} = \nu / c$. The transformation of Eqn (2.54) to

$$\overbrace{\bar{\nu}}^{\text{anharmonic}} = \overbrace{\bar{\nu}_e}^{\text{harmonic}} \left(1 - x_e \left(v + \frac{1}{2}\right)\right) \tag{2.54}$$

shows that the vibrational constant $\bar{\nu}$ of the anharmonic oscillator is always less than the vibrational constant $\bar{\nu}_e$ of the harmonic oscillator and that the vibrational constant $\bar{\nu}$ of the anharmonic oscillator decreases with increasing vibrational quantum number v. Due to the dependency $\bar{\nu} = \nu / c$, also the eigenfrequency ν of the anharmonic oscillator is always smaller than the eigenfrequency ν_e of the harmonic oscillator and decreases with increasing vibrational quantum number v. Therefore the eigenfrequency ν(v) and the vibrational constant $\bar{\nu}$(v) of the anharmonic oscillator are a function of the vibrational quantum number v.

If the eigenfrequency ν_e(v) decreases with increasing vibrational quantum number v, Eqn (2.46) means that the spring constant k_{binding} of the oscillating system ('binding force' between the nuclei) decreases, too, and prognoses the dissociation of the molecule into two separate atoms.

Finally, it should be mentioned that due to the asymmetric potential curve of the anharmonic oscillator, the mid-point of the horizontal black solid lines shift with increasing vibrational quantum numbers v to higher mean distances R_e between the nuclei. For a clear presentation of this behaviour in Fig. 2.13, the mean distances R_e of the anharmonic oscillator are only given for the vibrational quantum numbers v = 2, 3, 4, 5,and 6. It should be highlighted here that even for the vibrational quantum number v = 0, the mean distance R_e between the nuclei for the anharmonic oscillator exceeds that one for the harmonic oscillator.

2.2.5 Electronic Energy

For a detailed description of the electronic energy of molecules a profound background about the configuration of electrons over the orbitals of the molecules was required, which also would demand a familiarity with the quantum-physical notations of these configurations. As this is out of the scope of this book, we here—similarly to the previous sections on rotational and vibrational energy—consider the electronic energy levels of a simple molecule, the hydrogen molecule. The simple concept demonstrated in the following for the hydrogen molecule can then be transferred in general to more complex molecules.

The left part in Fig. 2.14 shows that depending on the configuration of the electrons the molecule can store different amounts of energy. The difference of the minima of the two vibrational potential curves for hydrogen is approximately 100,000 cm^{-1}. This difference is simply due to the fact that in the molecule corresponding to the higher vibrational potential curve, one electron occupies a higher molecular orbital, compared to the molecule in the electronic ground state.

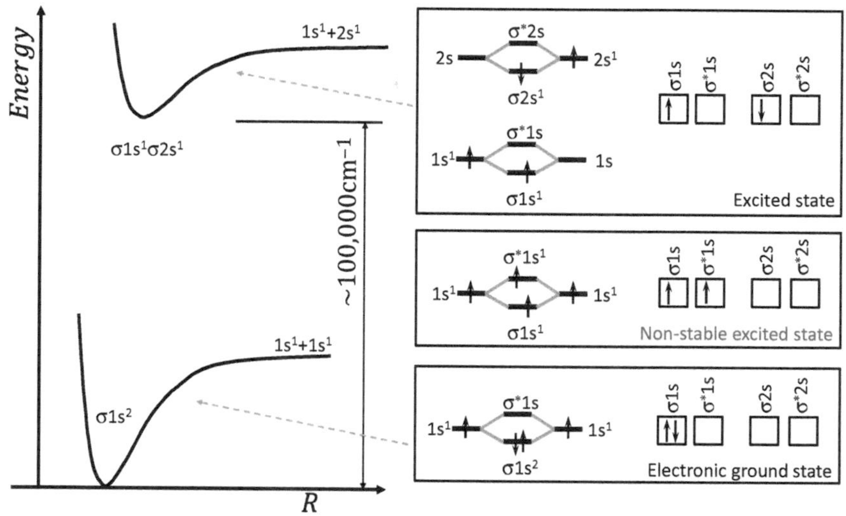

FIGURE 2.14 (left) Qualitative sketch of the potential curves of the hydrogen molecule in the electronic ground state and another hydrogen molecule in an excited electronic state. (right) Electron configuration for the hydrogen molecule occupying the electronic ground state (bottom) and the excited electronic state (top). The electronic configuration for a non-stable excited state is also given (middle).

For a more comprehensive description we first consider two separate hydrogen *atoms*. Each of these hydrogen *atoms* is composed of one proton and one electron. In the ground state the electron in the hydrogen *atom* occupies the s-orbital of the first shell, which is indicated in the right part of Fig. 2.14 by the little arrow (the arrow represents the electron) crossing the horizontal bar (horizontal bar represents the orbital), labelled as $1s^1$. Before the two hydrogen *atoms* form a hydrogen *molecule*, the electronic configurations of each of these atoms is $1s^1$ and each of these electrons is in the sphere of action of only one proton. When the hydrogen molecule is formed, each of the electrons is now in the sphere of action of two protons, which can bring the molecule to a state of lower energy. Consequently, the electronic configuration of the hydrogen *molecule* stores less energy than the summation of the electronic energy of two separate hydrogen *atoms*. In Fig. 2.14 in the right box at the bottom we can see that in the electronic ground state, the electrons which before the binding are characterised by $1s^1$, after the binding both occupy with anti-parallel spin the binding orbital $\sigma 1s$ of the *molecule* resulting from the two 1s orbitals of the *atoms*. As two electrons occupy this binding orbital $\sigma 1s$, the electronic configuration is labelled $\sigma 1s^2$. In Fig. 2.14 the right upper box (excited state) shows the electronic configuration of the hydrogen molecule in one excited state. Here one of the electrons moved from the binding $\sigma 1s$ molecular orbital to the binding $\sigma 2s$ molecular orbital. The $\sigma 2s$ binding molecular orbital results from the σ-binding between the two s-orbitals of the second shell of the hydrogen *atoms*.

As the binding molecular orbital σ2s contains one electron only, we write $\sigma 2s^1$. Nevertheless for the complete description of the electronic configuration of the excited hydrogen molecule we write $\sigma 1s^1\sigma 2s^1$ (see minimum of the potential curve of the excited molecule). The energy difference between the electronic configurations of the ground state hydrogen molecule $\sigma 1s^2$ and the excited hydrogen molecule $\sigma 1s^1\sigma 2s^1$ is approximately 100,000,cm^{-1}. The comparison between the rotational F_{rot} (J) and the vibrational G_{vib} (v) energy already yielded that between two and three orders of magnitude more energy can be stores in the vibration than in rotation. Now, the comparison between the vibrational G_{vib} (v) and the electronic T_{el} (el) energy yields that again between two and three orders of magnitude more energy can be stored in the electronic configuration. Later we will therefore see that very different photon-energies or radiation of very different wavelengths are assignable to the light–matter interactions based on either a change of the electronic, the vibrational or the rotational energy state. T_{el} (el) denotes the energy stored in the electronic configuration, where 'el' characterises the electronic quantum number, similarly to the way v and J quantified the vibrational and the rotational quantum numbers, respectively. Unfortunately, the dependency of the electronic energy T_{el} (el) on the electronic configuration 'el' cannot be easily derived based on mechanical models, as it has been the case for the rotational and the vibrational energy. Thus, I here skip the complex derivation of how the electronic energy can be computed for different electronic configurations. For the further progress of this book it is sufficient to define that we use $el = X$ and $el = A$ for the characterisation of the electronic quantum number of the energy of the electronic ground state T_{el} (X) and the first excited electronic state T_{el} (A).

A close look at the potential curves of the ground state and the excited state hydrogen molecule also shows that the minima of the two potential curves can be found at different mean distances R_e between the nuclei. For the excited state molecule R_e is larger than for the ground state hydrogen molecule. This is due to the fact that in the excited state molecule the spring constant ('binding force' between the nuclei) is less than in the ground state molecule. The smaller spring constant is the result of the transfer of one electron from a strongly binding molecular orbital σ1s to a weaker binding molecular orbital σ2s. Be aware of the fact that what is shown here for hydrogen also can be vice versa for other molecules. If, for example, electrons, which occupy an anti-binding molecular orbital in the electronic ground state, are transferred to binding molecular orbitals in the excited state, the overall spring constant ('binding force' between the nuclei) will increase and consequently R_e of the excited state will be smaller than R_e of the ground state. If the spring constant is not affected by the change of the electronic configuration, also R_e does not change.

The middle box on the right side in Fig. 2.14 shows an electronic configuration of an excited but non-stable hydrogen molecule. In this configuration $\sigma 1s^1\sigma^* 1s^1$ the number of electrons in binding and anti-binding molecular orbitals is identical. Therefore, the energy of the electron configuration of the *molecule* would be equal to the summation of the energies of the two *atoms* before binding. Consequently, as the binding does not contribute to a lower overall energy, there is no net force attracting the nuclei and thus stable molecules do not form.

Finally it should be mentioned that even for the simple hydrogen molecule there are many more excited state electronic configurations known. Consequently for molecules with more complex electronic configurations, there is a huge diversity of different excited state electronic configurations. Haken and Wolf provide a deeper insight into the electronic configurations of more complex molecules [1].

2.2.6 Energy of Molecules Relevant for Spectroscopy

In the previous Sections 2.2.2, 2.2.3, 2.2.4 and 2.2.5 the translational energy, the rotational- F_{rot} (J), the vibrational- G_{vib} (v) and the electronic energy T_{el} (el) were introduced. One application in which the translational energy of the molecules plays a role is, for example, the filtered Rayleigh scattering technique [2–4]. But later we will see that the translational energy of the molecules is not affected by most of the light–matter interactions on which the spectroscopic techniques provided in this book are based on. Therefore we neglect the contribution of the kinetic energy (translational energy—Eqn (2.25) for only one molecule) to the total energy of one molecule and express the total energy of the molecule —in wavenumbers—

$$T(\text{el}, \text{v}, J) = T_{\text{el}}(\text{el}) + G_{\text{vib}}(\text{v}) + F_{\text{rot}}(J) \tag{2.55}$$

as a function of the electronic- el, the vibrational- v, and the rotational quantum numbers J by the summation of the contributions of the electronic- T_{el} (el), the vibrational- G_{vib} (v) and the rotational energy F_{rot} (J).

Figure 2.15 shows an energy level scheme of various energy levels one molecule can occupy.

The quantum numbers el (here X for the electronic ground state and A for the first excited electronic state), v and J are not labelled for all the sketched energy levels. But in general it can be seen that each energy level is characterised by three quantum numbers el, v, and J and that each combination of these quantum numbers only exists once. In Fig. 2.15 the energy levels are sketched for a variation of the electronic quantum number from X to A, for vibrational quantum numbers $\text{v} = 0,1,2,3$ and for rotational quantum numbers $J = 0,1,2,3,4,5$. The black vertical arrow $T(\text{el}, \text{v}, J)$ is sketched broken in order to present the energy levels of different rotational- J, vibrational- v and electronic el quantum numbers in one diagram, whose energy in reality would differ by orders of magnitude.

With respect to Eqn (2.55) the term energy of one molecule occupying the energy level X, v_0, J_0 can be computed as

$$T(X, \text{v}_0, J_0) = T_{\text{el}}(\text{el}) + G_{\text{vib}}(\text{v}_0) + F_{\text{rot}}(J_0) = T_{\text{el}}(X) + G_{vib}(\text{v} = 0) + F_{\text{rot}}(J = 0). \tag{2.56}$$

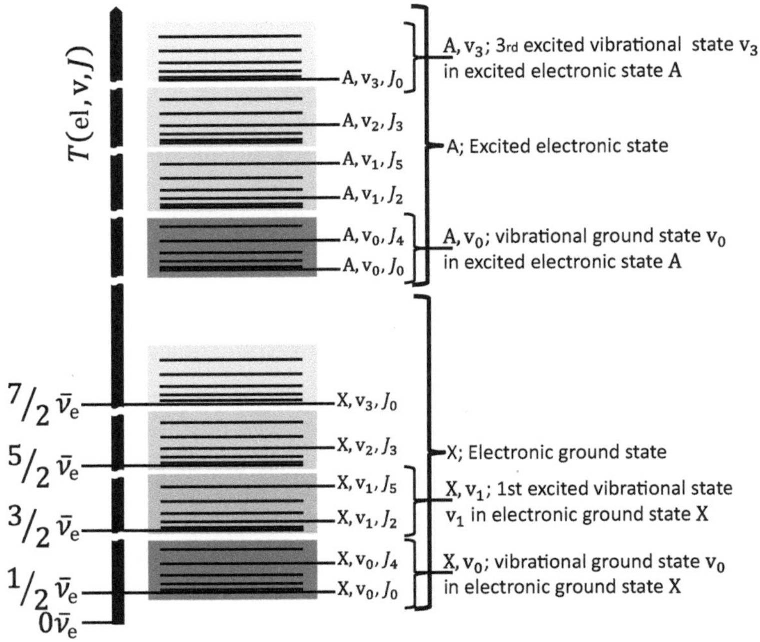

FIGURE 2.15 Scheme of the energy level diagram of a diatomic molecule, which is considered to behave as a rigid rotator and a harmonic oscillator. Horizontal lines indicate the energy of a molecule as a function of the quantum numbers of the electronic configuration (*X*: electronic ground state; *A*: first excited electronic state), the vibrational quantum number v and the rotational quantum number *J*.

Assuming that this molecule is composed of two nuclei we can use Eqns (2.38) and (2.51) for the computation of the term energy of the molecule

$$T(X,\mathrm{v}_0,J_0)=T_{\mathrm{el}}(X)+G_{\mathrm{vib}}(\mathrm{v}=0)+F_{\mathrm{rot}}(J=0)=\underbrace{0}_{T_{\mathrm{el}}(X)}+\underbrace{\bar{\nu}_{\mathrm{e}}\left(\mathrm{v}+\frac{1}{2}\right)-x_{\mathrm{e}}\bar{\nu}_{\mathrm{e}}\left(\mathrm{v}+\frac{1}{2}\right)^2}_{G_{\mathrm{vib}}(\mathrm{v}=0)}+\underbrace{B\cdot J(J+1)-D\cdot J^2(J+1)^2}_{F_{\mathrm{rot}}(J=0)}. \tag{2.57}$$

Assuming furthermore that the diatomic molecule behaves like a rigid rotator and a harmonic oscillator, Eqns (2.56) and (2.57) simplify to

$$T(X,\mathrm{v}_0,J_0)=\bar{\nu}_{\mathrm{e}}\left(\mathrm{v}+\frac{1}{2}\right)-\overbrace{x_{\mathrm{e}}\bar{\nu}_{\mathrm{e}}\left(\mathrm{v}+\frac{1}{2}\right)^2}^{0}+B\cdot J(J+1)-\overbrace{D\cdot J^2(J+1)^2}^{0}=\bar{\nu}_{\mathrm{e}}\left(\mathrm{v}+\frac{1}{2}\right)+B\cdot J(J+1) \tag{2.58}$$

and for $v = 0$ and $J = 0$ result in the term energy of

$$T(X, v_0, J_0) = \frac{1}{2}\bar{\nu}_e. \tag{2.59}$$

Eqn (2.58) does also provide—again assuming a diatomic rigid rotator and harmonic oscillator molecule—the term energies $T(\text{el},v,J)$ of the other energy levels labelled in Fig. 2.15. These term $T(\text{el},v,J)$ energies are summarised in Table 2.1.

TABLE 2.1 Term Energy of One Molecule When Occupying One Energy Level Characterised by the Quantum Numbers of the Electronic- el the Vibrational- v and the Rotational- *J* State

Energy level	T_{el} (el)	G_{vib} (v)	F_{rot} (*J*)	*T* (el, v, *J*)
X, v_0, J_0	$T_{el}(X)=0$	$\bar{\nu}_e\left(0+\frac{1}{2}\right)$	$B\cdot 0(0+1)$	$\frac{1}{2}\bar{\nu}_e$
X, v_0, J_4	$T_{el}(X)=0$	$\bar{\nu}_e\left(0+\frac{1}{2}\right)$	$B\cdot 4(4+1)$	$\frac{1}{2}\bar{\nu}_e+20B$
X, v_1, J_2	$T_{el}(X)=0$	$\bar{\nu}_e\left(1+\frac{1}{2}\right)$	$B\cdot 2(2+1)$	$\frac{3}{2}\bar{\nu}_e+6B$
X, v_1, J_5	$T_{el}(X)=0$	$\bar{\nu}_e\left(1+\frac{1}{2}\right)$	$B\cdot 5(5+1)$	$\frac{3}{2}\bar{\nu}_e+30B$
X, v_2, J_3	$T_{el}(X)=0$	$\bar{\nu}_e\left(2+\frac{1}{2}\right)$	$B\cdot 3(3+1)$	$\frac{5}{2}\bar{\nu}_e+12B$
X, v_3, J_0	$T_{el}(X)=0$	$\bar{\nu}_e\left(3+\frac{1}{2}\right)$	$B\cdot 0(0+1)$	$\frac{7}{2}\bar{\nu}_e$
A, v_0, J_0	$T_{el}(A)>0$	$\bar{\nu}_e\left(0+\frac{1}{2}\right)$	$B\cdot 0(0+1)$	$T_{el}(A)+\frac{1}{2}\bar{\nu}_e$
A, v_0, J_4	$T_{el}(A)>0$	$\bar{\nu}_e\left(0+\frac{1}{2}\right)$	$B\cdot 4(4+1)$	$T_{el}(A)+\frac{1}{2}\bar{\nu}_e+20B$
A, v_1, J_2	$T_{el}(A)>0$	$\bar{\nu}_e\left(1+\frac{1}{2}\right)$	$B\cdot 2(2+1)$	$T_{el}(A)+\frac{3}{2}\bar{\nu}_e+6B$
A, v_1, J_5	$T_{el}(A)>0$	$\bar{\nu}_e\left(1+\frac{1}{2}\right)$	$B\cdot 5(5+1)$	$T_{el}(A)+\frac{3}{2}\bar{\nu}_e+30B$
A, v_2, J_3	$T_{el}(A)>0$	$\bar{\nu}_e\left(2+\frac{1}{2}\right)$	$B\cdot 3(3+1)$	$T_{el}(A)+\frac{5}{2}\bar{\nu}_e+12B$
A, v_3, J_0	$T_{el}(A)>0$	$\bar{\nu}_e\left(3+\frac{1}{2}\right)$	$B\cdot 0(0+1)$	$T_{el}(A)+\frac{7}{2}\bar{\nu}_e$

2.2.7 Boltzmann's Distribution

In the previous section we saw that one molecule can occupy only one energy level, which is characterised by three quantum numbers, the electronic- el, the vibrational- v, and the rotational quantum number J and that based on these quantum numbers the term energy $T(\mathrm{el,v},J)$ of a single molecule can be computed as the summation of electronic- T_{el} (el), vibrational- G_{vib} (v) and rotational energy F_{rot} (J). Now we will see that at a certain temperature T many molecules of the same species are distributed over different molecular energy levels and that this distribution can be described by the 'Boltzmann distribution'. In general, the fact that at a certain temperature T not all the molecules occupy the same energy level is very similar to the fact that also at a certain temperature T not all molecules feature the same kinetic energy or the same translational velocity. In this context the Maxwell *velocity* distribution has been mentioned already before in Appendix 2.5.1.3 relevant for one of the previous Sections 2.2.2.

According to the Boltzmann distribution

$$
\begin{aligned}
\frac{N_{\mathrm{el,v},J}}{N_{\mathrm{total}}} &= \frac{(2J+1)\cdot e^{\left(-\frac{T(\mathrm{el,v},J)\cdot h\cdot c}{k\cdot T}\right)}}{\sum_{\mathrm{el,v},J}\left[(2J+1)\cdot e^{\left(-\frac{T(\mathrm{el,v},J)\cdot h\cdot c}{k\cdot T}\right)}\right]} \\
&= \frac{(2J+1)\cdot e^{\left(-\frac{[T(\mathrm{el})+G_{\mathrm{vib}}(\mathrm{v})+F_{\mathrm{rot}}(J)]\cdot h\cdot c}{k\cdot T}\right)}}{\sum_{\mathrm{el,v},J}\left[(2J+1)\cdot e^{\left(-\frac{[T(\mathrm{el})+G_{\mathrm{vib}}(\mathrm{v})+F_{\mathrm{rot}}(J)]\cdot h\cdot c}{k\cdot T}\right)}\right]} \\
&= \frac{(2J+1)\cdot e^{\left(-\frac{T(el)\cdot h\cdot c}{k\cdot T}\right)}\cdot e^{\left(-\frac{G_{vib}(\mathrm{v})\cdot h\cdot c}{k\cdot T}\right)}\cdot e^{\left(-\frac{F_{rot}(J)\cdot h\cdot c}{k\cdot T}\right)}}{\sum_{el,\mathrm{v},J}\left[(2J+1)\cdot e^{\left(-\frac{T(el)\cdot h\cdot c}{k\cdot T}\right)}\cdot e^{\left(-\frac{G_{vib}(\mathrm{v})\cdot h\cdot c}{k\cdot T}\right)}\cdot e^{\left(-\frac{F_{rot}(J)\cdot h\cdot c}{k\cdot T}\right)}\right]}
\end{aligned}
\tag{2.60}
$$

the number of molecules $N_{\mathrm{el,v},J}$ occupying a specific energy level relative to the total number of molecules N_{total} is a function of the rotational quantum number J, the energy of the respective energy level $T(\mathrm{el,v},J)$ and of the temperature T. In the nominator the expressions $(2J+1)$ and $e^{\left(-\frac{T(\mathrm{el,v},J)\cdot h\cdot c}{k\cdot T}\right)}$ reflect the degeneracy of the rotational energy levels and the Boltzmann factor, respectively. The denominator $\sum_{\mathrm{el,v},J}\left[(2J+1)\cdot e^{\left(-\frac{T(\mathrm{el,v},J)\cdot h\cdot c}{k\cdot T}\right)}\right]$ is the summation over the population of all energy states and used here for the normalisation to the overall number of molecules. Due to $T(\mathrm{el,v},J)=T(\mathrm{el})+G_{\mathrm{vib}}(\mathrm{v})+F_{\mathrm{rot}}(J)$, the Boltzmann factor can be separated in three exponential factors, one for electronic-, one for vibrational- and one for rotational energy.

For the electronic ground state $T(\mathrm{el}) = T(X) = 0$, Fig. 2.16 shows the population distribution $N_{\mathrm{el,v},J}/N_{\mathrm{total}}$ of molecules over energy levels for temperatures of 300 K (black bars), 1000 K (dark grey bars) and 3000 K (light grey bars) computed according to Eqn (2.60) and considering a rigid rotator and a harmonic oscillating diatomic molecule with a rotational and a vibrational constant of $B = 2\ \mathrm{cm}^{-1}$ and $\overline{\nu}_e = 2330.7\ \mathrm{cm}^{-1}$, respectively ($B$ and $\overline{\nu}_e$ taken from the nitrogen molecule, but nuclear spin effects are neglected). On the ordinate the rotational quantum number J is varied, while the upper, the middle and the lower part of the figure correspond to the vibrational ground state $\mathrm{v} = 0$, the first $\mathrm{v} = 1$ and the second $\mathrm{v} = 2$ excited vibrational state, respectively, all in the electronic ground state X.

The upper part of Fig. 2.16 (vibrational ground state) shows that with increasing temperature the population distribution $N_{\mathrm{el,v},J}/N_{\mathrm{total}}$ is broadened and that the maximum of the distribution is shifted to higher rotational quantum

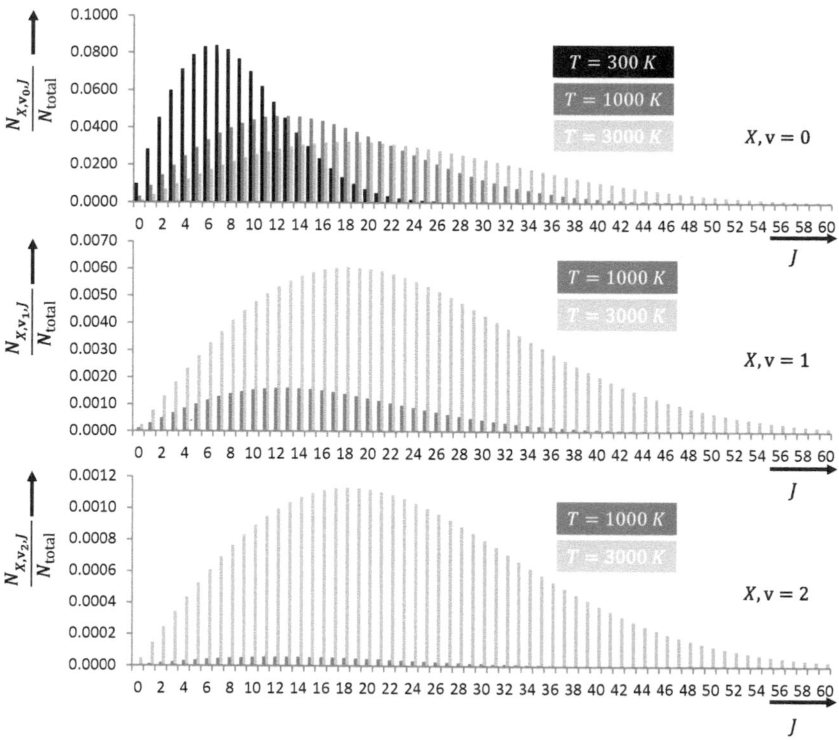

FIGURE 2.16 **Population distribution $N_{\mathrm{el,v},J}/N_{\mathrm{total}}$ of molecules ($B = 2\ \mathrm{cm}^{-1}$ and $\overline{\nu}_e = 2330.7\ \mathrm{cm}^{-1}$) as a function of the rotational quantum number J for the vibrational ground state $\mathrm{v} = 0$ in the electronic ground state X (top), for the first excited vibrational state $\mathrm{v} = 1$ in the electronic ground state X (middle) and for the second excited vibrational state $\mathrm{v} = 2$ in the electronic ground state X (bottom) for three different temperatures at 300 K (black bars), 1000 K (dark grey bars) and 3000 K (light grey bars).**

numbers J. While—according to our impression from Fig. 2.16—at 300 K the highest rotational energy level populated traceable corresponds to the quantum number $J = 26$, at 1000 K and 3000 K the highest rotational energy levels populated correspond to quantum numbers $J = 45$ and $J > 60$, respectively. Furthermore, the comparison of the upper $\mathrm{v} = 0$, the middle $\mathrm{v} = 1$ and the lower part $\mathrm{v} = 2$ of Fig. 2.16 shows that at a temperature of 300 K only the vibrational ground state is remarkably populated, as we cannot see any black bars in the first $\mathrm{v} = 1$ and second excited vibrational states $\mathrm{v} = 2$. At temperatures of 1000 and 3000 K also the first and the second vibrational ground states are populated remarkably. Nevertheless, I want to emphasise that the population distributions $N_{\mathrm{el,v},J}/N_{\mathrm{total}}$ are scaled not identically for the vibrational ground, the first excited and the second excited vibrational states. Considering, for example, for the temperature of 3000 K the maximum of the population distribution, we find that in the vibrational ground state $\mathrm{v} = 0$ approximately 4% of the molecules occupy the energy level corresponding to the rotational quantum number $J = 18$ while for the first excited vibrational quantum number only 0.6% and for the second excited vibrational quantum number only 0.1% occupy the energy levels corresponding to the rotational quantum number $J = 18$. This population behaviour over **vibrational** energy levels v is expressed in the most popular form of the Boltzmann distribution,

$$\frac{N_{X,\mathrm{v}=i}}{N_{X,\mathrm{v}=0}} = \frac{e^{\left(-\frac{G_{\mathrm{vib}}(\mathrm{v}=i)\cdot h\cdot c}{k\cdot T}\right)}}{e^{\left(-\frac{G_{\mathrm{vib}}(\mathrm{v}=0)\cdot h\cdot c}{k\cdot T}\right)}} = e^{\left(-\frac{[G_{\mathrm{vib}}(\mathrm{v}=i)-G_{\mathrm{vib}}(\mathrm{v}=0)]\cdot h\cdot c}{k\cdot T}\right)} \tag{2.61}$$

which is derived from Eqn (2.60) in Appendix 2.5.2. Here $N_{X,\mathrm{v}=i}$ and $N_{X,\mathrm{v}=0}$ are the summation of all molecules occupying an excited vibrational state $\mathrm{v} = i = 1,2,3,\ldots$ and the vibrational ground state $\mathrm{v} = 0$, respectively. Due to the facts that always

- $G_{\mathrm{vib}}(\mathrm{v} = i) > G_{\mathrm{vib}}(\mathrm{v} = 0)$,
- that the constants h, c and k are positive and
- that also the temperature $T > 0$ is positive, the ratio $N_{X,\mathrm{v}=i}/N_{X,\mathrm{v}=0}$ is always smaller than 1 and follows the behaviour

$$\frac{N_{X,\mathrm{v}=0}}{N_{X,\mathrm{v}=0}} = 1 > \frac{N_{X,\mathrm{v}=1}}{N_{X,\mathrm{v}=0}} > \frac{N_{X,\mathrm{v}=2}}{N_{X,\mathrm{v}=0}} > \frac{N_{X,\mathrm{v}=3}}{N_{X,\mathrm{v}=0}} > \ldots$$

One also can derive from Eqn (2.60) a simplified Boltzmann distribution

$$\frac{N_{X,\mathrm{v}=\mathrm{const},J=i}}{N_{X,\mathrm{v}=\mathrm{const},J=0}} = \frac{(2J+1)\cdot e^{\left(-\frac{F_{\mathrm{rot}}(J=i)\cdot h\cdot c}{k\cdot T}\right)}}{(2(J=0)+1)\cdot e^{\left(-\frac{[F_{\mathrm{rot}}(J=0)=0]\cdot h\cdot c}{k\cdot T}\right)}} = (2J+1)e^{\left(-\frac{F_{\mathrm{rot}}(J=i)\cdot h\cdot c}{k\cdot T}\right)} \tag{2.62}$$

for the description of the population behaviour of molecules over the **rotational** energy levels J in one vibrational state $\mathrm{v} = \mathrm{const}$. In this case the degeneracy $(2J+1)$ does not cancel out and has to be considered. It should be already underlined here that it also depends on the nuclear spin whether a rotational energy level J can in general be occupied by any molecule or not. This we have not considered until now but will consider it later in Section 2.2.7.2. Nevertheless Eqn (2.62) shows that the distribution $\frac{N_{X,\mathrm{v}=\mathrm{const},J=i}}{N_{X,\mathrm{v}=\mathrm{const},J=0}}$ over rotational energy levels J is not a function of the vibrational quantum number v. Therefore irrespectively of the vibrational quantum number v the molecules are distributed in the same way over the rotational energy levels. Of course—this is stated by Eqn (2.61)—the overall number of molecules occupying the vibrational ground state is always the highest and the vibrational population decays with increasing vibrational quantum number v. But within one vibrational quantum number v, the distribution $\frac{N_{X,\mathrm{v}=\mathrm{const},J=i}}{N_{X,\mathrm{v}=\mathrm{const},J=0}}$ over the rotational energy levels J is always the same for all vibrational levels.

2.2.7.1 Comparison of Different Molecules

The population distribution shown in Fig. 2.16 was computed for rigid rotating and harmonically oscillating molecules assuming a rotational and a vibrational constant of $B = 2\ \mathrm{cm}^{-1}$ and $\bar{\nu}_e = 2330.7\ \mathrm{cm}^{-1}$, respectively, which would correspond to the properties of a nitrogen molecule. Please be aware that we did not consider the nuclear spin effects in Fig. 2.16. Therefore the population distribution of real nitrogen—just due to the nuclear spin effect—would look different. We will see this later in Section 2.2.7.2. Here the influence of the molecular properties on the population distribution will be shown, again neglecting the influence of the nuclear spin. Therefore we now consider again rigid rotating and harmonically oscillating molecules, but now with a rotational and a vibrational constant of $B = 60.81\ \mathrm{cm}^{-1}$ and $\bar{\nu}_e = 4159.2\ \mathrm{cm}^{-1}$, respectively, which would correspond to a hydrogen molecule. Again I have to point out that the population distribution we will discuss here is not true for hydrogen, as we will not consider the nuclear spin effect. The population distributions shown in Fig. 2.17 are computed for virtual molecules featuring only the properties B and $\bar{\nu}_e$ of hydrogen (black bars) and nitrogen (grey bars). One significant difference between the populations shown in Fig. 2.17 is the widths of the distributions. This is due to the significant difference of the magnitude of the two rotational constants B we assumed for the computations of the population distributions. For the rigid rotator we can replace $F_{\mathrm{rot}}\ (J)$ with $\left[B \cdot J \cdot (J+1)\right]$ and simplify Eqn (2.62) to

$$\frac{N_{X,\mathrm{v}=\mathrm{const},J=i}}{N_{X,\mathrm{v}=\mathrm{const},J=0}} = (2J+1)e^{\left(-\frac{F_{\mathrm{rot}}(J)\cdot h\cdot c}{k\cdot T}\right)} = (2J+1)e^{\left(-\frac{B\cdot J\cdot(J+1)\cdot h\cdot c}{k\cdot T}\right)}, \quad (2.63)$$

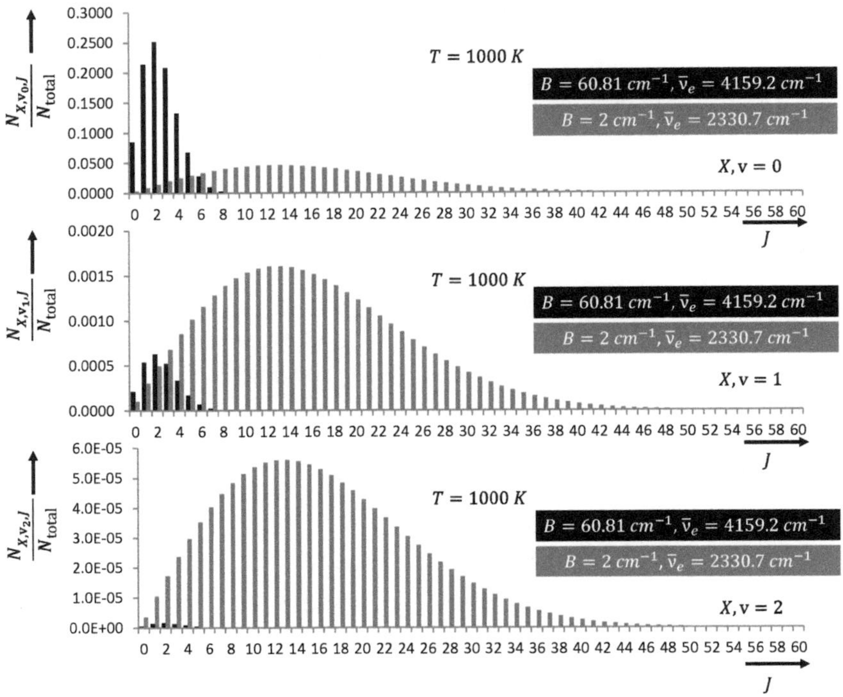

FIGURE 2.17 **Comparison of the population distribution $N_{el,v,J}/N_{total}$ of two different species with molecular properties of $B = 60.81\ cm^{-1}$ and $\bar{\nu}_e = 4159.2\ cm^{-1}$ (in black bars) and $B = 2\ cm^{-1}$ and $\bar{\nu}_e = 2330.7\ cm^{-1}$ (in grey bars) as a function of the rotational quantum number J for the vibrational ground state $v = 0$ in the electronic ground state X (top), for the first excited vibrational state $v = 1$ in the electronic ground state X (middle) and for the second excited vibrational state $v = 2$ in the electronic ground state X (bottom) for a temperature of 1000 K.**

which makes clear that the exponential function decays faster with increasing B. Therefore for the rotational constant $B = 60.81\ cm^{-1}$ fewer rotational energy levels are occupied.

Another significant difference is $N_{X,v=i}/N_{X,v=0}$, which according to Eqn (2.61) quantifies the ratio of the number of molecules occupying energy levels with the vibrational quantum numbers $v = i = 1,2,3,\ldots$ and $v = 0$. For the harmonic oscillator we can replace $G_{vib}(v)$ with $\left[\bar{\nu}_e \cdot \left(v + \frac{1}{2}\right)\right]$ and simplify Eqn (2.61) to

$$\frac{N_{X,v=1}}{N_{X,v=0}} = e^{\left(-\frac{[G_{vib}(v=1) - G_{vib}(v=0)]\cdot h\cdot c}{k\cdot T}\right)} = e^{\left(-\frac{\bar{\nu}_e\cdot h\cdot c}{k\cdot T}\right)}, \tag{2.64}$$

which makes clear that the exponential function decays faster with increasing $\bar{\nu}_e$. Therefore for the vibrational constant $\bar{\nu}_e = 4159.2\ cm^{-1}$ fewer molecules occupy the excited state vibrational quantum numbers.

2.2.7.2 Comparison for Ideal and Real Molecules

In this section the population distributions of ideal and real molecules will be compared. But in a first step we again neglect the nuclear spin effects and only compare the population distributions for ideal and non-ideal molecules. In this context *ideal* stands for the rigid rotating and harmonically oscillating and *non-ideal* for the non-rigid rotating and non-harmonically oscillating molecule. In Fig. 2.18 the black and the grey bars represent the population distributions for the non-ideal and the ideal molecules. For the non-ideal molecules the rotational F_{rot} (J) and the vibrational energy G_{vib} (v) was computed according to Eqns (2.38) and (2.51), assuming a centrifugal elongation constant

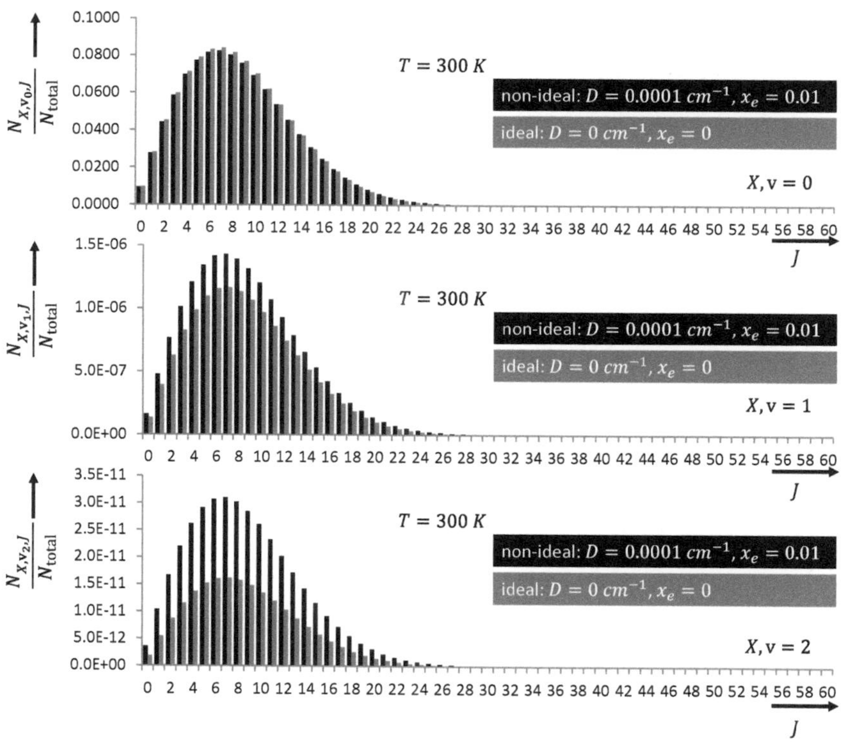

FIGURE 2.18 Comparison of the population distribution $N_{el,v,J}/N_{total}$ for species composed of realistic (non-rigid rotating and non-harmonically oscillating) molecules (black bars) and ideal (rigid rotating and harmonically oscillating) molecules (grey bars). For both species the rotational and the vibrational constants are $B = 2\ cm^{-1}$ and $\bar{\nu}_e = 2330.7\ cm^{-1}$, respectively. For the non-rigid rotating and non-harmonic oscillating molecules additionally the centrifugal elongation constant $D = 0.0001\ cm^{-1}$ (Eqn (2.39)) and the anharmonicity constant $x_e = 0.01$ (Eqn (2.52)) were assumed. Population distribution given as a function of the rotational quantum number J for the vibrational ground state $v = 0$ in the electronic ground state X (top), for the first excited vibrational state $v = 1$ in the electronic ground state X (middle) and for the second excited vibrational state $v = 2$ in the electronic ground state X (bottom) for a temperature of 300 K.

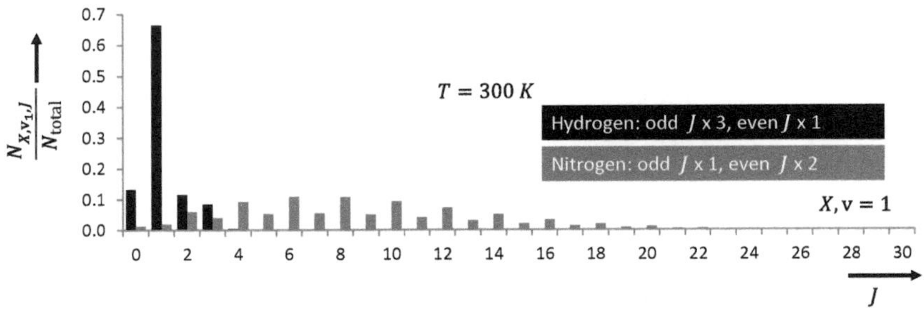

FIGURE 2.19 Population distribution of hydrogen and nitrogen at 300 K in the vibrational v = 0 and the electronic ground state *X* as a function of the rotational quantum number *J*.

and an anharmonicity constant $D = 0.0001\ \text{cm}^{-1}$ (Eqn (2.39)) and $x_e = 0.01$ (Eqn (2.52)), respectively.

As both, the rotational as well as the vibrational energy are smaller for the non-ideal molecules than for the ideal molecules, the exponential functions of Eqns (2.61) and (2.62) decay faster for the ideal molecules. Therefore, in one vibrational quantum number v = const, the non-ideal molecules occupy slightly more rotational quantum numbers J'', while fewer ideal molecules occupy the excited state vibrational energy levels.

Finally it has to be mentioned that in the case of homo-nuclear diatomic molecules as well as—in general—for molecules featuring an inversion centre, such as carbon dioxide, some rotational quantum numbers *J* are not occupied at all or that there is a statistical and alternating fluctuation between the odd and even rotational quantum numbers. This statistical fluctuation is due to the nuclear spin and depends on the spin quantum number of the nuclei the molecule is composed of. According to Fig. 2.19, in the case of the hydrogen molecule (nuclear spin of hydrogen nucleus ^{1}H is 1/2) the statistical weight of the population of odd and even rotational quantum numbers is odd/even = 3/1, while in the case of nitrogen (nuclear spin of nitrogen nucleus ^{14}N is 1) the statistical weight of the population of odd and even rotational quantum numbers is odd/even = 1/2. For the oxygen molecules (nuclear spin of oxygen nucleus ^{16}O is 0) the even rotational quantum numbers *J* are not occupied at all. For carbon dioxide, which has an inversion centre, the odd rotational quantum numbers *J* are not occupied at all.

In detail, the statistical fluctuations of the population of rotational energy levels of diatomic homo-nuclear molecules or of molecules, which have an inversion centre, depends on the parity of their electronic ground state. Further and deeper descriptions are beyond the scope of this book and are therefore skipped. More detailed descriptions can be read in *Molecular Physics And Elements Of Quantum Chemistry* by Haken and Wolf [1].

2.3 INTERACTION OF BULK MATTER AND ELECTROMAGNETIC RADIATION

This section contains the fundamentals of the interaction of bulk matter and electromagnetic radiation, which are required for the understanding of the working principles of measurement techniques such as the schlieren technique, shadowgraphy or elastic light scattering from sprays. As this interaction mechanism can be described mathematically considering the index of refraction, I first introduce the index of refraction and how it is interconnected with the properties of electromagnetic radiation. Then I describe how phenomena such as reflection and refraction can be described mathematically using the index of refraction. Then diffraction is described, which has nothing to do with the index of refraction. But we will see that the working principle of an optical spectrometer, which is an essential component in nearly each spectroscopic experimental set-up, is based on the combination of refractive and diffractive elements and that based on this combination radiation can be separated as a function of its wavelength. Also 'elastic light scattering from drops' is described in this section, as I consider drops to be bulk matter, due to the fact that many molecules are contained within one single drop. Finally an exercise section is provided.

2.3.1 Index of Refraction

For a transparent medium, meaning that the electromagnetic wave propagating through this medium is not attenuated, the index of refraction of this medium

$$n_{\text{refr}} = \frac{c_{\text{vacuum}}}{c_{\text{medium}}} \tag{2.65}$$

is the ratio of the speed of the electromagnetic wave in vacuum c_{vacuum} relative to the speed of that wave in the medium c_{medium}. The index of refraction of liquid water at 293 K is, for example, 1.333 for an electromagnetic wave with a wavelength of approximately 600 nm. This means that this wave propagates 1.333 times faster in vacuum than in water, but still oscillates with the same frequency. Therefore the wavelength

$$\lambda_{\text{medium}} = \frac{\lambda_{\text{vacuum}}}{n_{\text{refr}}} \tag{2.66}$$

of this electromagnetic wave is shorter when it passes through water than when it passes through vacuum. Consequently, the wavelength of a wave as well as the corresponding wavenumber (Eqn (2.7)) are a function of the index of refraction of the medium $n_{\text{refr,medium}}$ the wave is propagating through. In contrast, the frequency (Eqn (2.8)) is not a function of the index of refraction and therefore is the most accurate way of describing the oscillation of the electric field, as it has already been mentioned in the previous Section 2.1.3.

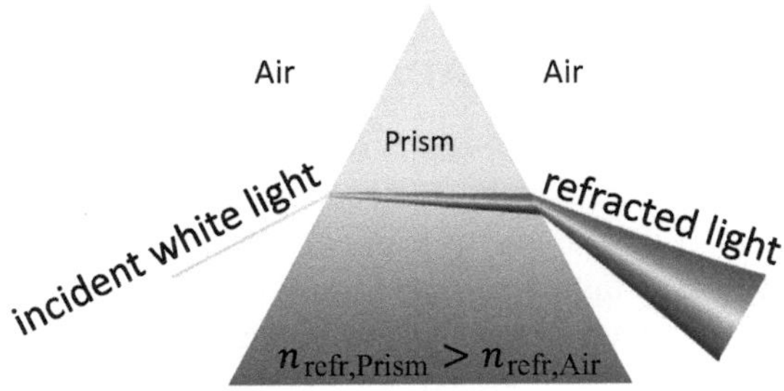

FIGURE 2.20 Sketch of the dispersion of the waves of different wavelengths when white light passes through a glass prism. The index of refraction of the prism exceeds that one of air $n_{refr,Prism} > n_{refr,Air}$. *(Source: Sketch prepared by Julian Schuster from the Lehrstuhl für Technische Thermodynamik of the Friedrich-Alexander-Universität Erlangen-Nürnberg, Germany.)*

Usually, the index of refraction of a medium $n_{refr,medium}$ increases with decreasing wavelength. This effect is called dispersion and is responsible for the fact that waves of different wavelengths split spatially when they are refracted, which is shown in the sketch in Fig. 2.20, or split temporally when they travel through a long glass fibre, which is shown in Fig. 2.21. We will see later in Chapter 4 that the working principles of shadowgraphy- and schlieren techniques are based on refraction.

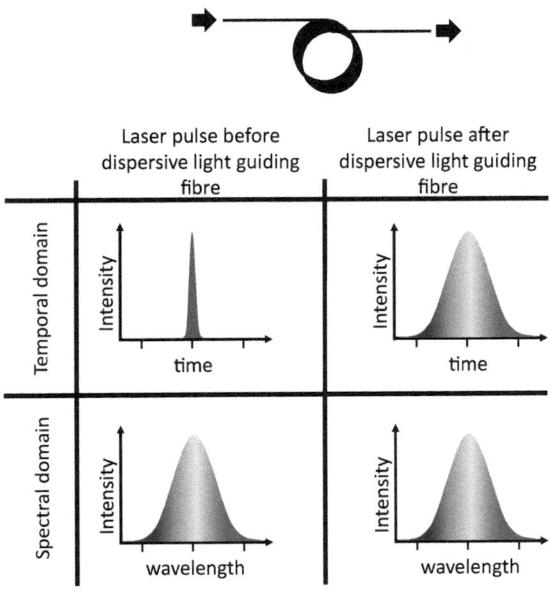

FIGURE 2.21 Dispersion of the wavelengths of a short spectrally broadband laser pulse in a very long dispersive light guiding fibre.

Figure 2.21 shows that if a temporally short light-pulse, which carries waves of different wavelengths, propagates through a long glass fibre, the dispersion delays the short-wavelength-waves more than the long-wavelength-waves. In other words, the long-wavelength-waves travel faster through the fibre, as the index of refraction is smaller for longer wavelengths. Therefore, at the exit of the very long glass fibre the long-wavelength-waves will appear first. The shorter the wavelength is, the more delayed the respective waves will appear at the fibre exit. We will see later in Section 6.1 that the supercontinuum time of flight absorption technique makes use of the dispersion of spectrally broadband light in very long glass fibres.

2.3.2 Reflection and Refraction of Electromagnetic Radiation

Figure 2.22 shows how an electromagnetic wave, incident to the air/glass interface at an angle of θ_{in} is reflected at an angle of $\theta_{refl,0} = \theta_{in}$ and refracted at an angle of $\theta_{refr,0}$. The refracted portion propagates into the glass and at the bottom glass/air interface is reflected and refracted again at the angles of $\theta_{refl,1}$ and $\theta_{refr,1}$, respectively. The portion reflected internally to the top glass/air interface is again internally reflected and refracted under the angles of $\theta_{refl,1} = \theta_{refl,2}$ and $\theta_{refr,2}$, and so on. While for *reflection* the angle of incidence is equal to the angle of reflection, the interrelation between the angle of the incident electric field and the angle of the *refracted* field is described by Snell's law

$$n_{refr,\,Air} \cdot \sin\theta_{in} = n_{refr,\,Glass} \cdot \sin\theta_{refr}. \tag{2.67}$$

What is sketched in Figs 2.20 and 2.22 for an electromagnetic wave passing *interfaces* between two media of different indices of refraction n_{refr} can

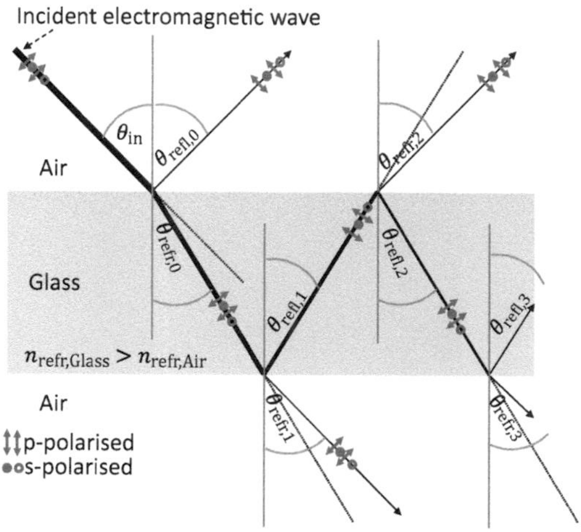

FIGURE 2.22 Illustration of reflection and refraction for electromagnetic radiation passing interfaces between media (air and glass) of different indices of refraction.

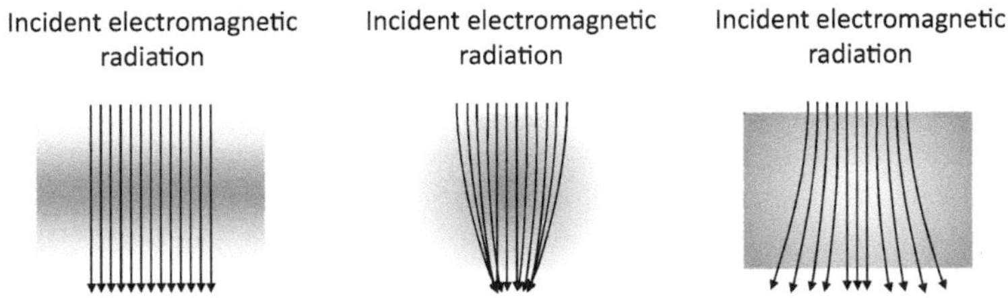

FIGURE 2.23 Illustration of the refraction of electromagnetic radiation (black arrows) in systems featuring a gradient (not an interface) of the index of refraction. Left: The directions of the gradient and the propagation of the radiation are identical (e.g., index of refraction gradient in layers of different composition due to diffusion); centre: the index of refraction increases to the centre (e.g., substance of high index of refraction surrounded by other substance of low index of refraction); right: the index of refraction decreases to the centre (e.g., hot flame surrounded by cool air).

also similarly be applied when an electromagnetic wave travels through a medium, which does not contain any interfaces, but which features *gradients* of the index of refraction n_{refr}, as it is sketched in Fig. 2.23 and shown for the hot off-gas stream of a crème brûlée torch in Fig. 2.24. One example could be a flame (right in Fig. 2.23). If we observe a flame in a backlight configuration, we all know that we see schlieren in the flame's shadow. They are caused by the refraction of the light-rays (electromagnetic radiation) when the direction of propagation of the light and the gradient of the index of refraction do not

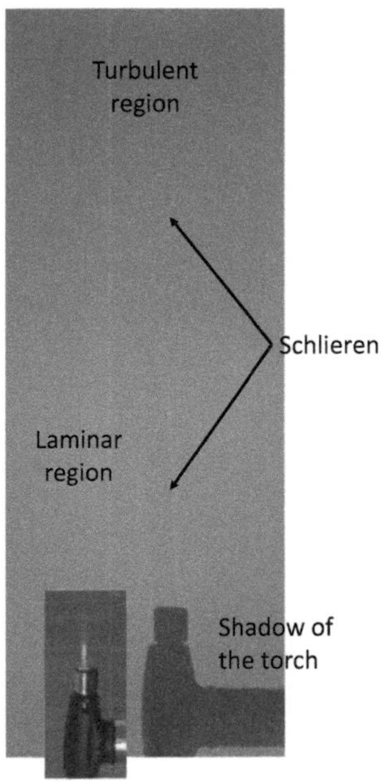

FIGURE 2.24 Shadowgraph of the hot off-gas flow of a crème brûlée torch.

point to the same direction. We will see later in the chapter on shadowgraph and schlieren techniques that not only temperature gradients, but also concentration and composition gradients, which are more likely to exist in high-pressure process technology, can cause gradients in the field of the index of refraction. This is the reason why we, for example, can observe the mixing of two substances via shadowgraph or schlieren techniques.

Finally it has to be mentioned that Snell's law is not a function of the polarisation of the electric field (light). This implies that p- and s-polarised light are refracted into the same directions. In Fig. 2.22 the red arrows and the combination of the red disc and circle represent parallel (p-) and perpendicular (s-) polarised electric fields, respectively. Snell's law does not describe which portions of the incident electric field are finally refracted or reflected. Nevertheless—this is what we will see later in the exercise section—Snell's law can describe that under certain circumstances 100% of the incident electromagnetic radiation is reflected, while 0% is refracted. This effect is said to be total internal reflection, an effect on which the working principle of light-guiding glass fibres is based on.

With respect to Fig. 2.22 and for the transition of a light ray from the air to the glass, the reflected portion of the incident light can be computed polarisation dependent from the Fresnel equations

$$R_s = \left| \frac{n_{\text{refr,Air}} \cdot \cos\theta_{\text{in}} - n_{\text{refr,Glass}} \cdot \cos\theta_{\text{refr,0}}}{n_{\text{refr,Air}} \cdot \cos\theta_{\text{in}} + n_{\text{refr,Glass}} \cdot \cos\theta_{\text{refr,0}}} \right|^2 = \left| \frac{n_{\text{refr,Air}} \cdot \cos\theta_{\text{in}} - n_{\text{refr,Glass}} \sqrt{1 - \left(\frac{n_{\text{refr,Air}}}{n_{\text{refr,Glass}}} \cdot \sin\theta_{\text{in}} \right)^2}}{n_{\text{refr,Air}} \cdot \cos\theta_{\text{in}} + n_{\text{refr,Glass}} \sqrt{1 - \left(\frac{n_{\text{refr,Air}}}{n_{\text{refr,Glass}}} \cdot \sin\theta_{\text{in}} \right)^2}} \right|^2 \quad (2.68)$$

$$R_p = \left| \frac{n_{\text{refr,Air}} \cdot \cos\theta_{\text{refr,0}} - n_{\text{refr,Glass}} \cdot \cos\theta_{\text{in}}}{n_{\text{refr,Air}} \cdot \cos\theta_{\text{refr,0}} + n_{\text{refr,Glass}} \cdot \cos\theta_{\text{in}}} \right|^2 = \left| \frac{n_{\text{refr,Air}} \sqrt{1 - \left(\frac{n_{\text{refr,Air}}}{n_{\text{refr,Glass}}} \cdot \sin\theta_{\text{in}} \right)^2} - n_{\text{refr,Glass}} \cdot \cos\theta_{\text{in}}}{n_{\text{refr,Air}} \sqrt{1 - \left(\frac{n_{\text{refr,Air}}}{n_{\text{refr,Glass}}} \cdot \sin\theta_{\text{in}} \right)^2} + n_{\text{refr,Glass}} \cdot \cos\theta_{\text{in}}} \right|^2, \quad (2.69)$$

where R_s and R_p represent the reflectivity or the reflectance (ratio of the intensity of reflected and incident light) of the s- and the p-polarised electric field. In Eqns (2.68) and (2.69) the angles of the incident and the refracted fields are labelled as they are labelled in Fig. 2.22. For the transition of a light ray from the glass to the air, the Fresnel equations would read as

$$R_s = \left| \frac{n_{\text{refr,Glass}} \cdot \cos\theta_{\text{refr},0} - n_{\text{refr,Glass}} \cdot \cos\theta_{\text{refr},1}}{n_{\text{refr,Glass}} \cdot \cos\theta_{\text{refr},0} + n_{\text{refr,Glass}} \cdot \cos\theta_{\text{refr},1}} \right|^2 \tag{2.70}$$

$$R_p = \left| \frac{n_{\text{refr,Glass}} \cdot \cos\theta_{\text{refr},1} - n_{\text{refr,Air}} \cdot \cos\theta_{\text{refr},0}}{n_{\text{refr,Glass}} \cdot \cos\theta_{\text{refr},1} + n_{\text{refr,Air}} \cdot \cos\theta_{\text{refr},0}} \right|^2, \tag{2.71}$$

where again the angles are labelled as they are labelled in Fig. 2.22. If the electric field of the incident light is unpolarised, the reflectivity is

$$R = \frac{R_p + R_s}{2}. \tag{2.72}$$

Figure 2.25 illustrates in the upper part which portion of the incident electromagnetic radiation (light) is reflected R and transmitted T due to refraction

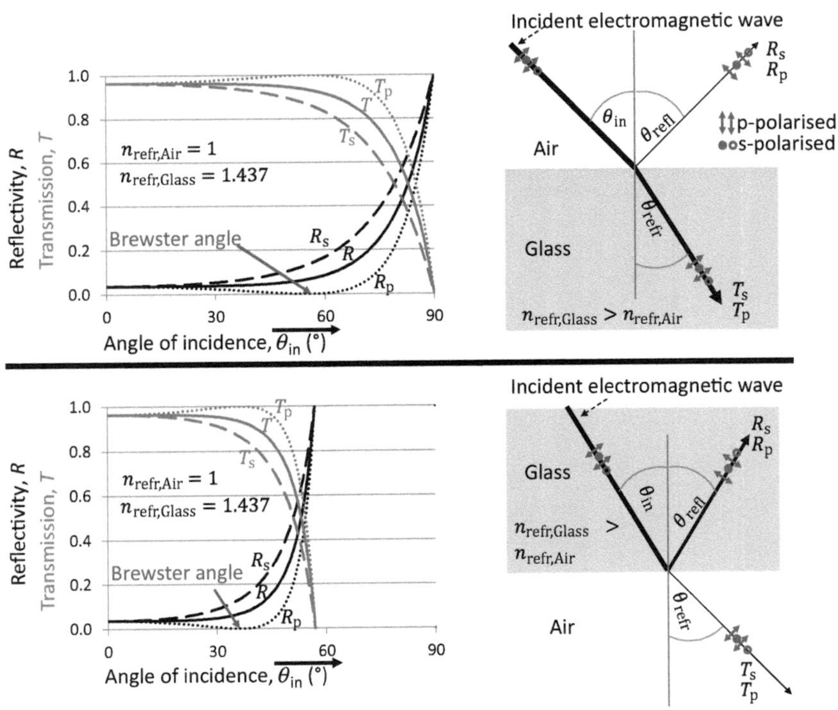

FIGURE 2.25 Reflectivity R and transmission T as a function of the angle of incidence for the transition of electromagnetic radiation from (top) a low index of refraction medium (air) to a higher index of refraction medium (glass) and from (bottom) a high index of refraction medium (glass) to a lower index of refraction medium (air). Reflectivity (black) and transmission (grey) curves are provided for s- (dashed), p- (dotted) and unpolarised (solid) electric fields.

as a function of the angle of the incident radiation for the transition from air to glass (compare upper right sketch). Alternatively, the lower part illustrates the portions of the reflected and the transmitted light for the transition of glass to air (compare bottom right sketch). For s-polarised light the reflectivity continuously increases with increasing angle of incidence until 100% of the incident light is reflected. For the transition of air to glass this 100% reflectivity is reached at $\theta_{in} = 90°$, while for the transition from glass to air this 100% reflectivity is already reached at $\theta_{in} = 56°$. For p-polarised light, the reflectivity first decreases with increasing angle of incidence θ_{in}, till it reaches $R_p = 0$ at the Brewster angle, from where it increases to $R_p = 1$ with increasing angle of incidence θ_{in}.

2.3.3 Diffraction of Electromagnetic Radiation

We will see later that measurement techniques such as the schlieren technique or shadowgraphy are based onto the *refraction* of light, when the incident light passes, for example, through a system that we want to analyse (compare Fig. 2.24). In contrast, none of the measurement techniques covered in this book is based on the *diffraction* of the incident light when it passes through the measurement object. A measurement technique that would rely on diffraction of light was, for example, the laser diffraction technique for the determination of particle sizes, which are computed from the diffraction pattern generated by the particle. Nevertheless, with respect to some of the techniques treated in this book, such as fluorescence spectroscopy, absorption in the visible spectral range, laser-induced fluorescence or Raman spectroscopy, the diffraction of light is exploited for the analysis of the detected signals and therefore plays a crucial role in the context of the equipment we use for the realisation of the measurement techniques. Within a common spectrometer—Fourier-transform spectrometers are mainly applied in (N)IR absorption spectroscopy and are described in Section 6.1.2 of this book—there is one grating which according to the diffraction of light spatially separates light of different wavelengths. Summarising, the diffraction of light is described here in order to demonstrate the working principle of spectrometers.

Figure 2.26 shows an electromagnetic wave incident to a wall with a double slit. Left of the wall the electric wave is considered to be a plane wave, which is indicated by the parallel (red) lines representing the maxima of the electric field at one instant of time. The plane waves would move from the left towards the wall. The two slits can be considered as the origin of two identical new waves, having the same wavelength as the incident wave and propagating into all directions of the room. Figure 2.26 shows that right of the wall there are some paths along which the two waves interfere positively, where the effective phase shift between the two waves is zero, while along other paths the two waves interfere destructively, where the effective phase shift is half of the spatial period of the wave $\lambda/2$. Therefore in the far field right of the double slit wall a diffraction pattern characteristic for a double slit experiment occurs. The maxima visible are called 0th, 1st, 2nd and so on order maxima.

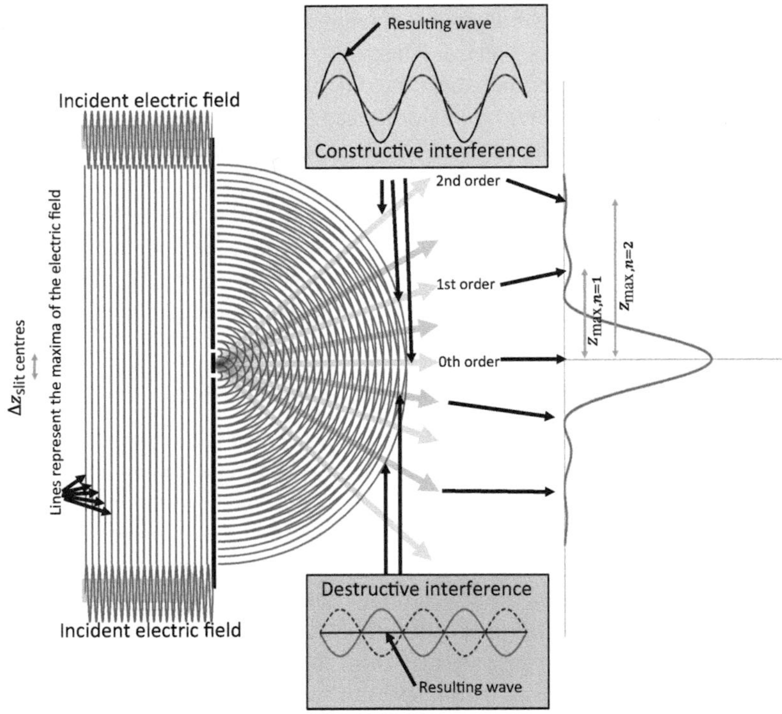

FIGURE 2.26 Sketch of a double slit experiment when a plane wave hits a double slit wall and as a consequence due to constructive and destructive interference a typical diffraction pattern is generated. *(Source: Sketch prepared by Jaypee Quino from the Erlangen Graduate School in Advanced Optical Technologies of the Friedrich-Alexander-Universität Erlangen-Nürnberg, Germany.)*

While the position of the 0th order maximum is always in the centre directly opponent to the double slit, the positions of the higher order maxima (for $\Delta x_{\text{screen-wall}} \gg z_{\max,n}$)

$$z_{\max,n} = n_{\text{order}} \cdot \lambda \cdot \frac{\Delta x_{\text{screen-wall}}}{\Delta z_{\text{slit centres}}}. \tag{2.73}$$

and minima

$$z_{\min,n} = (2n_{\text{order}} - 1) \cdot \frac{\lambda}{2} \cdot \frac{\Delta x_{\text{screen-wall}}}{\Delta z_{\text{slit centres}}}. \tag{2.74}$$

are a function of the wavelength λ, the respective order n_{order}, the distance between the wall and the screen $\Delta x_{\text{screen-wall}}$ and the distance between the centres of the two slits $\Delta z_{\text{slit centres}}$. Consequently, if the double slit wall is illuminated with collimated light (plane wave) covering a certain wavelength range, the higher order maxima of the different wavelengths would appear spatially separated in the diffraction pattern. The longer the wavelength is, the further away from the 0th order maximum the maxima of the higher orders would appear. The gratings we use inside the spectrometers can be considered as multi-slit walls. Typical numbers of slits per millimetre on the grating, which is specified in lines per millimetre are 300, 600, 1200 or 2400 lines mm^{-1}. With increasing number of lines per millimetre the distance between the maxima of different wavelengths increases. Therefore the higher the number of lines per millimetre is, the higher is the spectral resolution, but the smaller is the spectral range that still can be covered on the plane where the diffraction pattern is detected. If $\Delta x_{\text{screen-wall}} \gg z_{\max,n}$ cannot be assumed, the appearance of the maxima can be computed from a more general grating equation

$$\Delta z_{\text{slit centres}} \cdot \sin(\theta_n = \text{angle between } n = 0 \text{ and } n) = n_{\text{order}} \cdot \lambda. \tag{2.75}$$

Finally, in order to bring together the phenomena of refraction and diffraction, the working principle of a spectrometer is described. The general set-up of an optical spectrometer equipped with a reflective grating for diffraction and the paths of the light are sketched in Fig. 2.27. White light, here comprising the visible spectral range is focused through the entrance slit into the spectrometer. The first pupil, which is positioned exactly in the distance of its focal length f_{length} collimates the white light and guides it as a plane wave to the grating. This grating is not a transmission grating (diffraction was derived earlier for the double slit wall transmission experiment), but a reflection grating. Nevertheless, due to the same phenomenon of constructive and destructive interference the 1st order maxima of the different wavelengths are shifted by a certain angle α from the 0th order diffraction maximum and relative to each other. Therefore,

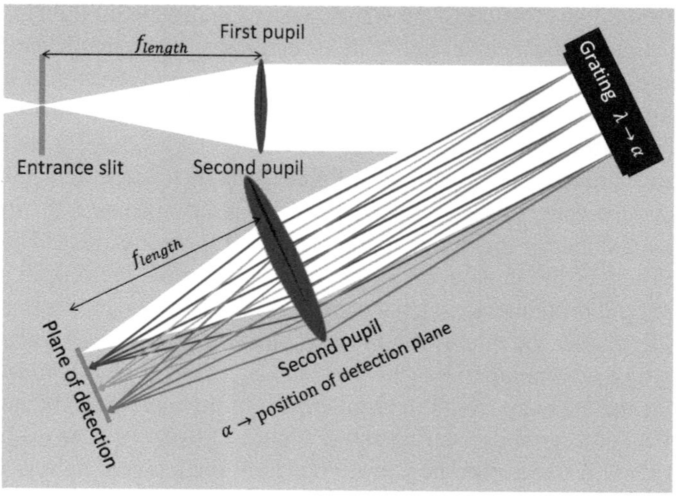

FIGURE 2.27 Sketch of the path of light inside an optical spectrometer, composed of an entrance slit, two pupils and one reflective diffraction grating. Path only sketched for the 0th and one 1st order maximum.

in Fig. 2.27 the 0th order diffraction maximum is still the white light, as the angle of the 0th order is the same for all wavelengths. According to Eqn (2.73) the diffraction maxima of longer wavelengths (here red) are more separated from the 0th order diffraction maximum than shorter wavelengths (here blue). Thus, the grating converts a wavelength information $\lambda \rightarrow \alpha$ into an angular information. The still plane waves of each colour reach the second pupil at different angles and therefore are focused spatially separated on the plane of detection. Thus, the second pupil converts the angular information α into a spatial information. Summarising, due to the combination of diffractive (grating) and refractive elements (lenses = pupil), the spectrometer can spatially separate different wavelengths, which is called spectral resolution.

2.3.4 Elastic Light Scattering From Drops

The final section we treat in the context of light↔bulk–matter interactions is about the interaction of light with drops. Of course, one can argue that a drop usually is any phase dispersed inside another bulk phase and thus should not be treated in the context of light↔bulk–matter interactions but more appropriately in the context of light↔dispersed–matter interactions. On the other hand, a drop is composed of a bulk of molecules. Therefore I decided to group the light↔drop interactions into the Section 2.3 relevant for light↔bulk–matter interactions, but treat it as bulk↔dispersed–matter interaction. We also here will consider only spherical drops. In reality, drops can also be non-spherical, for example, due to aerodynamic forces or any other force fields. Additionally we assume that the drop as well the bulk medium is completely transparent for the wavelength of the incident and the scattered light. We will see that in the case of elastic light scattering the wavelength of the scattered light is equal to the wavelength of the incident light. All the diagrams shown in this section were computed using the free software 'MiePlot' provided and developed by Philip Laven (http://www.philiplaven.com).

'Elastic' in the terminology 'Elastic light scattering' stands for the fact that there is no energy exchange during the light↔drop interaction. In Section 2.1.4 we saw that the energy of a photon is inherently interlinked with the frequency or the wavenumber or the wavelength of the corresponding wave. Thus, if the energy of neither the photon nor of the molecules inside the drop is changed during the light/drop interaction, the incident light before the interaction and the scattered light after the interaction must feature exactly the same frequency, or wavelength or wavenumber. But the scattered light, due to the light/drop interaction is scattered away from the original path of the incident light and can be observed under a certain scattering angle. Figure 2.28 shows as an example and qualitatively in a 'radar plot' how much power of the incident light is scattered elastically by the particle under which angle.

In this radar plot the numbers oriented around the radar plot indicate the scattering angle, while the distance between the centre of the radar plot and the

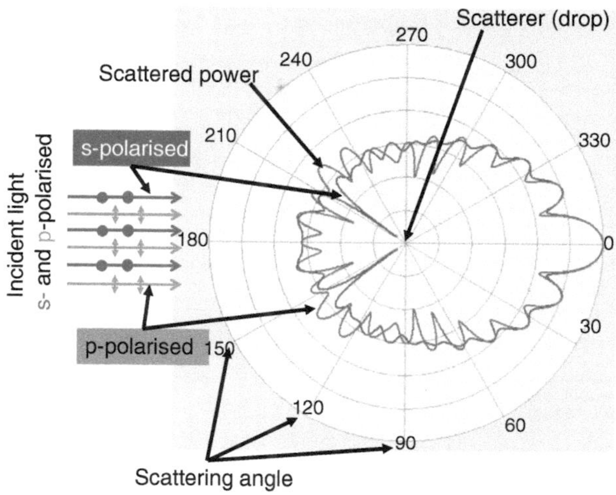

FIGURE 2.28 **Radar plot showing qualitatively how much power of the incident radiation is scattered under which angle from the particle for s- and p-polarised light.** *(Source: www.philiplaven.com)*

(green) lines represent the scattered power (here the scattered power is scaled logarithmic). We can see that under a scattering angle of 150°, the scattered power of the p-polarised light exceeds the scattered power of the s-polarised light, while for 0° or 180° the scattered powers are the same for s- and p-polarised light. This phenomenon of elastic light scattering from drops we experience when in a disco we see laser beams passing through foggy air, or when we are blinded by our own car's headlights scattered back from fog towards the eyes of the driver.

For the characterisation of different regimes of elastic light scattering a dimensionless size parameter is used

$$x = \frac{\pi d}{\lambda}, \tag{2.76}$$

where d is the diameter of the drop and λ the wavelength of the incident light. In Fig. 2.29 the power (in Watts) of the light scattered from a single drop, which is irradiated with an irradiance power of 1 W/m^2 is computed using the MiePlot software as a function of the size parameter x. The scattered power comprises that portion of the light scattered between 0° and 180° in the scattering plane, which is defined by the direction of the incident light and the direction of detection. The indices of refraction of the drop and the bulk medium are 1.4 and 1.13, respectively, which would correspond to a liquid solvent drop surrounded at elevated pressure by compressed carbon dioxide. The incident radiation has a wavelength of 532 nm. Figure 2.29 reveals three different regimes of light scattering. In the Rayleigh regime the circumference of the drop is significantly smaller than the wavelength of the light ($x \ll 1$) and the scattering power scales with the diameter of the drop to the power of six ($I_{\text{scatter}} \propto d^6$). For $x \gg 1$ the

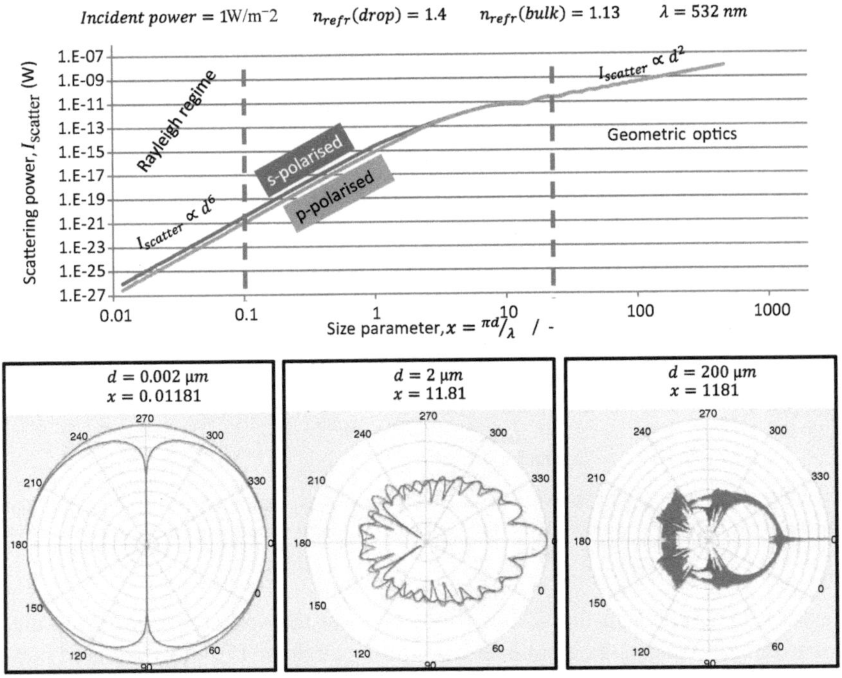

FIGURE 2.29 Scattered power for s- and p-polarised incident light of incident power of 1 W/m^2 as a function of the size parameter of a drop with index of refraction of 1.4 surrounded by a bulk medium of index of refraction of 1.13. Only elastic light scattering is considered. Bottom part shows typical scattering intensities as a function of the scattering angle. *(Source: www.philiplaven.com)*

circumference of the drop significantly exceeds the wavelength. Then the scattering power is proportional to the projection area of the drop ($I_{\text{scatter}} \propto d^2$). Within the regime of geometric optics one can use Snell's law and the Fresnel equations, both laws of geometric optics, for the computation of the light scattering profile (radar plot) and its intensity. In the bottom part of Fig. 2.29 radar plots of the scattered light power are given also on a logarithmic scale as a function of the scattering angle for three different size parameters or drop diameters, respectively. Again the dark (green) lines represent the s-polarised scattered light while the bright (green) lines represent the p-polarised scattered light.

The radar plots corresponding to the Rayleigh regime ($x \ll 1$) are point symmetric for the s- and the p-polarised light and cannot be computed using the laws of geometric optics. The point-symmetries are lost in the regime of geometric optics ($x \gg 1$) and in the regime ($0.1 < x < 10$) between the Rayleigh regime and the regime of geometric optics.

In the context of high-pressure process technology the density of any high-pressure system might change significantly, for example, during the starting or shut-down phase of a process. As the density correlates with the index of refraction, also the index of refraction of the bulk or the dispersed phase (or of both) might change. Therefore it might happen that at any instant of time the indices of refraction of the bulk phase and the dispersed drops become identical. Then a two-phase system existing within the high-pressure system would look like a single-phase system. Therefore, the knowledge of the behaviour of the index of refraction as a function of composition, temperature or pressure is beneficial, if elastic light scattering techniques are exploited for the investigations into high-pressure process technology.

2.3.5 Exercises Corresponding to Section 2.3

Exercise 2.6: Diffraction

Figure 2.30 describes the meaning of 'pitch' in the context of CDs, DVDs or blue ray discs. The pitch defines the distance between the lines on a reflecting substrate. Each line can be assumed to be a slit. Therefore such a disc can work as a reflective diffraction grating. And indeed most probably you have already noticed the appearance of a spectral separation of the colours when you looked at the disc.

Task 1: Exercise 2.6

A disc as it is shown in Fig. 2.31 is illuminated with a laser beam of wavelength 632 nm. Calculate the pitch (groove spacing) of the disc if the first order diffraction is found at distance of 0.4 m. Which type of disc the pitch corresponds to?

Answer 1: Exercise 2.6

The disc grooves act as grating. Thus we can use the grating equation

$$\Delta z_{\text{slit centres}} \cdot \sin(\theta_1 = \text{angle between } n = 0 \text{ and } n) = n_{\text{order}} \cdot \lambda. \tag{2.75}$$

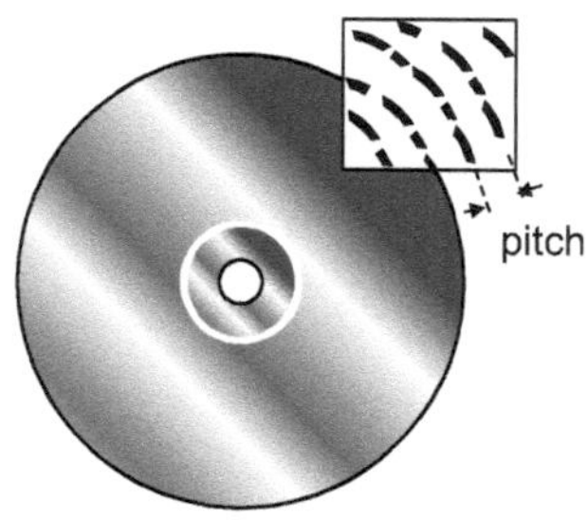

	Pitch (µm)	Laser, λ (µm)	Storage (GB)
CD	1.60 ± 0.10	0.780	0.70
DVD	0.74	0.650	4.70
Blue-Ray	0.32	0.405	25.00

FIGURE 2.30 **Sketch and table of the specifications of a CD, a DVD and a blue ray disc.** *(Source: Sketch prepared by Jaypee Quino from the Erlangen Graduate School in Advanced Optical Technologies of the Friedrich-Alexander Universität Erlangen-Nürnberg, Germany.)*

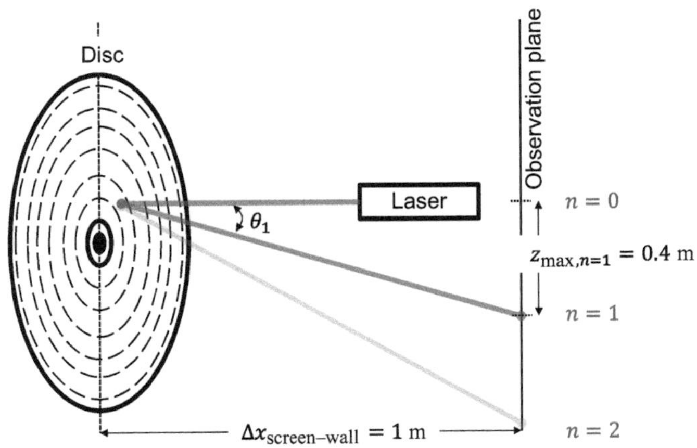

FIGURE 2.31 Sketch of the diffraction of a laser pointer directed to a data storage disc. *(Source: Sketch prepared by Jaypee Quino from the Erlangen Graduate School in Advanced Optical Technologies of the Friedrich-Alexander Universität Erlangen-Nürnberg, Germany.)*

The unknown 'angle between $n = 0$ and $n = 1$' can be computed as

$$\theta_1 = \tan^{-1}\left(\frac{z_{max,n=1}}{\Delta x_{screen\text{-}wall}}\right) = \tan^{-1}\left(\frac{0.4 \text{ m}}{1 \text{ m}}\right) = 21.80°. \tag{2.77}$$

Consequently

$$\Delta z_{slit\ centres} = \frac{n_{order} \cdot \lambda}{\sin(\theta_1)} = \frac{1 \cdot 632 \cdot 10^{-9} \text{ m}}{\sin(21.80)} = 1.7\,\mu\text{m} \tag{2.78}$$

Thus the disc is a CD.

Task 2: Exercise 2.6

If we use a DVD disc, how large will be the separation between the 0th and the 1st order diffraction maximum?

Answer 2: Exercise 2.6

For a DVD the diffracted angle at first maximum would be

$$\theta_1 = \sin^{-1}\left(\frac{n_{order} \cdot \lambda}{\Delta z_{slit\ centres}}\right) = \sin^{-1}\left(\frac{n_{order} \cdot \lambda}{\Delta z_{slit\ centers}}\right) = 58.7°. \tag{2.75}$$

Therefore

$$z_{max,n=1} = \Delta x_{screen\text{-}wall} \tan(\theta_1) = 1\,\text{m} \cdot \tan(58.7°) = 1.64\,\text{m} \tag{2.79}$$

the first order diffraction maximum would appear in a distance of 1.64 m of the 0th order diffraction maximum.

2.4 INTERACTION OF MOLECULES AND ELECTROMAGNETIC RADIATION

When I realised that light can be utilised for the communication with molecules I was more than fascinated and I am still today. With 'communication' I mean that we can send light to the molecules and get back from the molecules a signal, which can contain information about the species (structure, functional groups, bonding) of the molecule, about the energy of the molecules, which correlates with the temperature, and even about the molecules in the neighbourhood of the excited molecule. For the comprehensive interpretation of the obtained signals a profound knowledge of the properties of molecules as well as of the incident light is beneficial. This is why I organised the previous Sections 2.1 and 2.2 rather comprehensively.

Usually we feel much more familiar with the interactions of bulk matter with light, which we treated in the section before. Most probably, this is due to the fact that we personally have already observed many of these optical effects such as refraction, or rainbows, or schlieren or shadows. Indeed, we all have also already experienced many interaction mechanisms taking place between molecules and light, but we might not be aware of. Typical 'daily life' examples of interactions between molecules and light are the blue sky, the blue sea, the reddish sun during sunrise or sunset and the greenish colour of plants. The pigment chlorophyll is responsible for the greenish colour of the plants. It absorbs the bluish and the reddish part of the white light emitted from the sun and utilises the absorbed energy for photosynthesis. If the bluish and the reddish parts are absorbed, the remaining spectrum looks greenish, like the plants do. This is a typical example where the incident light is absorbed. We already can see here that obviously not all colours are absorbed with the same probability!

The phenomena responsible for the blue sky, the blue sea and the reddish sun are not based on an absorption effect, but on a light scattering effect. In Section 2.3.4 we have already been in contact with elastic light scattering, not from molecules, but from droplets and saw that—when the droplets are significantly smaller than the wavelength of the incident light—this interaction mechanism is called Rayleigh scattering. As single molecules usually are very small, elastic light scattering from molecules is also called Rayleigh scattering. I want to remind here again that 'elastic' means that there is no energy exchanged between the molecule (scatterer) and the scattered photon. Again, from our daily life observations '*blue* sky', '*blue* sea' and '*reddish* sun' we can speculate that the probability of Rayleigh scattering is also dependent on the wavelength of the incident light. Otherwise the daily life observations should not feature specific colours. The details about absorption and emission processes as well as about different kinds of scattering mechanisms will be described in the following two sections.

2.4.1 Absorption and Emission Processes

Figure 2.32 illustrates absorption and emission interactions between molecules and electromagnetic radiation, which are exploited when applying IR-absorption

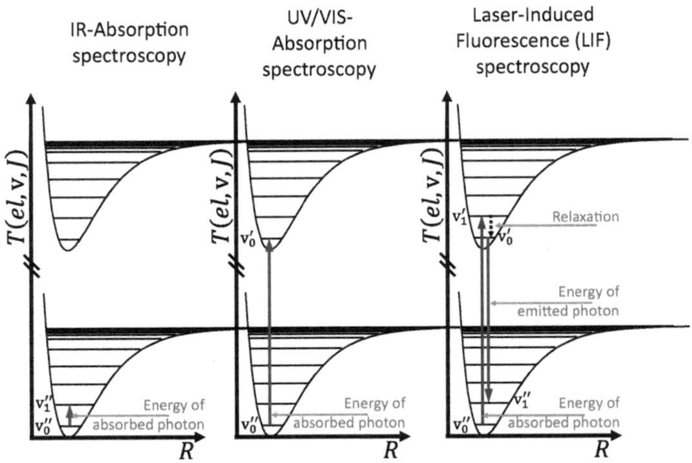

FIGURE 2.32 **Illustration of the interaction mechanisms between light and molecules relevant for the IR-absorption spectroscopy (left), the UV/VIS-absorption spectroscopy (centre) and the laser-induced-fluorescence spectroscopy (right) in energy level diagrams.** Rotational energy levels are not included in this figure. Vibrational quantum numbers in the electronic ground state and in the first excited electronic state are decorated with superscripts ″ and ′, respectively.

spectroscopy, UV/VIS-absorption spectroscopy and laser-induced fluorescence (LIF)-spectroscopy.

In Fig. 2.32 and in the corresponding descriptions the rotational energy levels as well as rotational quantum numbers are not regarded and we consider the potential curve as well as the vibrational levels as we derived them in Section 2.2. They all have in common that one photon is absorbed by one molecule. In the case of IR-absorption, the energy of the absorbed photon fits between two neighbouring vibrational energy levels. In Section 2.2.4 we already saw that the energy difference between two neighbouring vibrational energy levels is in the range of the energy of infrared radiation. Therefore, the corresponding spectroscopic technique, where the involved molecules—due to the absorption of a photon—remain in the electronic ground state but only change their vibrational quantum number v'' is called infrared (IR) absorption spectroscopy. It should be underlined here again that molecules only can occupy distinct energy levels (see Section 2.2) and that thus also the energy difference between neighbouring vibrational energy levels is species specific (fingerprint of a molecule). Thus, not arbitrary photons can be absorbed by a molecule but only those whose photon energy fits exactly to the energy difference between the energy levels of the transition the molecule undergoes. A more detailed description will follow.

In the case of UV/VIS-absorption spectroscopy the energy of the absorbed photon fits between two energy levels, which belong to two different electronic configurations. Therefore the molecule undergoes a transition from the electronic ground state to the first excited electronic state. Next to the electronic quantum number also the vibrational quantum number can change. In Fig. 2.32 the vibrational quantum number does not change. In Section 2.2.4 we already saw that the energy difference between two neighbouring electronic energy levels is in the range of the energy of ultraviolet radiation. For some molecules, such as dye molecules, it also can be in the range of the visible radiation. Therefore, the corresponding spectroscopic technique, where the involved molecules—due to the absorption of a photon—undergo a transition from one electronic state to another one is called ultraviolet/visible (UV/VIS) absorption spectroscopy. Again, only those photons can be absorbed whose photon energy fits between the energy levels the molecule occupies before and after the absorption process. Coming back to the chlorophyll, which is responsible for the green colour of the plants. Obviously chlorophyll can only absorb photons with photon energies corresponding to electromagnetic waves in the blue and red spectral range. The photon energy of the 'green' photons does not fit to the energy difference between two energy states, which can be occupied by chlorophyll, and thus green is not absorbed. A more detailed description will follow in Section 2.4.1.5.

So far we have not considered which processes might take place after the absorption of the photon. Let us assume that a fluid, which is composed of molecules, is in thermodynamic equilibrium at any temperature before some of the molecules absorb photons. We could use the Boltzmann distribution in order to compute which percentage of the number of molecules occupies which specific energy level before the absorption. Then we launch light of those wavelengths through the fluid, which can be absorbed by the molecules. Due to the absorption some of the molecules undergo a transition to a vibrationally (IR spectroscopy) and/or an electronically (UV/VIS spectroscopy) excited state. This of course disturbs the thermodynamic equilibrium and directly after the absorption process the real distribution over the energy states does no more follow Boltzmann's statistics. The system inherently tries to reach back a state of thermodynamic equilibrium. Thus, the excited molecules try either to distribute the energy they absorbed over the other molecules in the system, or to emit the absorbed energy (or parts of it) by the emission of a photon. In the first scenario where no photon is emitted (IR and UV/VIS absorption spectroscopy), the energy of the excited molecules is distributed among the entire fluid by non-radiative energy transfer mechanisms, such as collisions with other molecules. Consequently, after a while (it depends on the system) the fluid is again in thermodynamic equilibrium, but due to the absorbed energy at marginally (it depends on the amount of absorbed energy) higher temperature. The second scenario, where a photon is emitted, is illustrated in the right part of Fig. 2.32 and is called laser-induced fluorescence (LIF), as the absorbed photon is usually provided by a laser source. First, one photon is absorbed and as a consequence the molecule undergoes a transition from the electronic ground state to the first excited electronic state. In the sketch of Fig. 2.32 (right part) the excited molecule also occupies the first excited vibrational state. In order to make very clear the difference between *absorption–emission* processes and light *scattering* processes, which will follow later, I have to underline that the photon, which was absorbed, does no more exist at the time when the molecule is in the excited state. It has been quasi eaten up by the molecule. As already mentioned before, each system, which is composed of many molecules, intents to reach a state of thermodynamic equilibrium. Within the excited electronic state the excited molecule relaxes to the vibrational ground state. This again is a radiationless energy decay process, which, for example, via collisions with other molecules distributes the absorbed energy over the entire fluid. Parallel to the relaxation process the excited molecules also can emit a photon, whose photon energy again fits to the energy difference between two energy levels the molecule can occupy. In Fig. 2.32 the excited molecule first relaxes radiationless from v_1' to v_0' and then emits a photon to undergo a transition from v_0' to v_1''. It is not included in Fig. 2.32 that in order to reach thermodynamic equilibrium again, the molecule still has to relax radiationless from v_1'' to v_0''. A more detailed description will follow in Section 2.4.1.6. At this point it has to be mentioned that the vibrational energy levels of the electronic ground and first excited electronic state are decorated with superscripts $''$ or $'$, respectively. This brings the advantage that the electronic quantum number does not have to be mentioned separately.

2.4.1.1 Chemiluminescence and Planck's Radiation

In order to complete the emission processes I want to mention that there are also naturally occurring emission processes, which do not follow a previous absorption process, as it is the case for laser-induced fluorescence. Examples are flames or glowing metals. At high temperatures, such as in the reaction zone of flames, the molecules in thermodynamic equilibrium occupy electronically excited states. In order to depopulate back to the electronic ground state they can emit a photon. As this emission process is induced by a chemical reaction, it is said to be chemiluminescence. From a chemiluminescence spectrum, as it is shown in Fig. 2.33, one can read which species are present—even as intermediate products—in the reaction zone of flames or, if one analyses the population distribution of molecules over energy levels, one even can measure the temperature in the flame front.

I want to make clear that the majority of the chemiluminescence spectrum of hydrocarbon flames is in the UV spectral region. That part which we can see with the human eye (from ~380 nm to ~780 nm) is only the bluish border of the chemiluminescence between 380 nm and 500 nm.

Maybe some of us during our education in analytical chemistry have analysed the flame coloration for the determination of ions present in a salt. For this experiment we had to seed the salt to the flame. Then the energy of the photons emitted by the excited ions—they are excited due to the very high temperatures in the Bunsen burner flame—are characteristic for the respective ions, such as calcium (red), potassium (violet), cupper (green) or sodium (yellow) and others. Thus, from the coloration of the flame one can estimate which ions

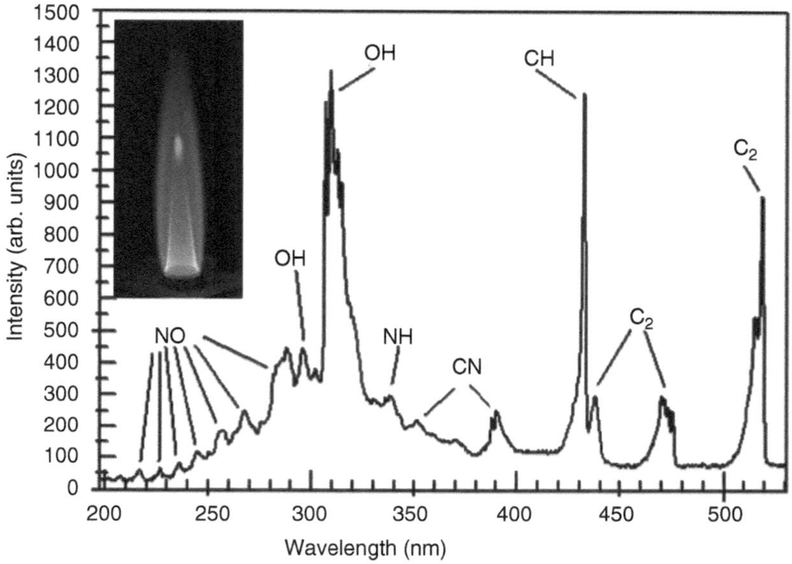

FIGURE 2.33 Example emission spectrum of a hydrocarbon flame doped with ammonia (from [5]). The electronic transitions accompanied with the emission of photons are labelled for NO, OH, NH, CN, CH and C_2. OH, for example, is an intermediate radical, which is formed in the reaction zone (flame front) and is present in the hot off-gas regions upstream of the flame front. The yellow region in the flame is due to Planck radiation of hot soot particles.

are present in the salt seeded to the flame. For an accurate determination of the ion, its emitted spectrum has to be analysed spectrally.

With respect to the flame of a candle the bluish part of the chemiluminescence spectrum is sparely visible below the yellow part of the flame, as it is shown in Fig. 2.34.

The yellow part has nothing to do with chemiluminescence. In the yellow part of the flame of a candle tiny soot particles are present. The temperature of these soot particles is as hot as they emit radiation in the visible spectrum due to a process we call Planck radiation or black-body-radiation. The spectrum emitted by a blackbody radiator according to Planck's equation and visualised

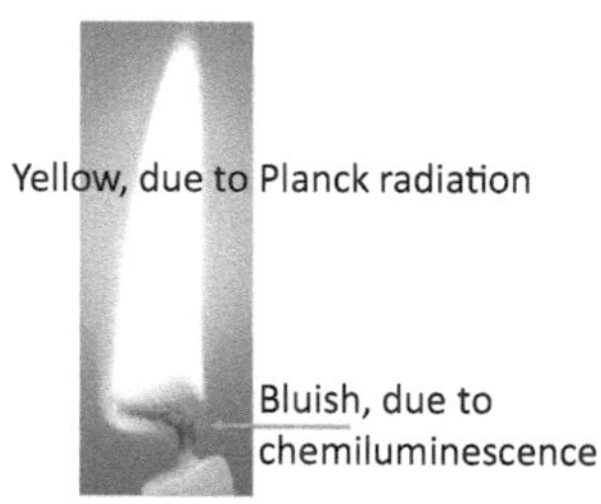

FIGURE 2.34 Photograph of the flame of the candle showing regions visible due to chemiluminescence and due to Planck radiation from hot soot particles.

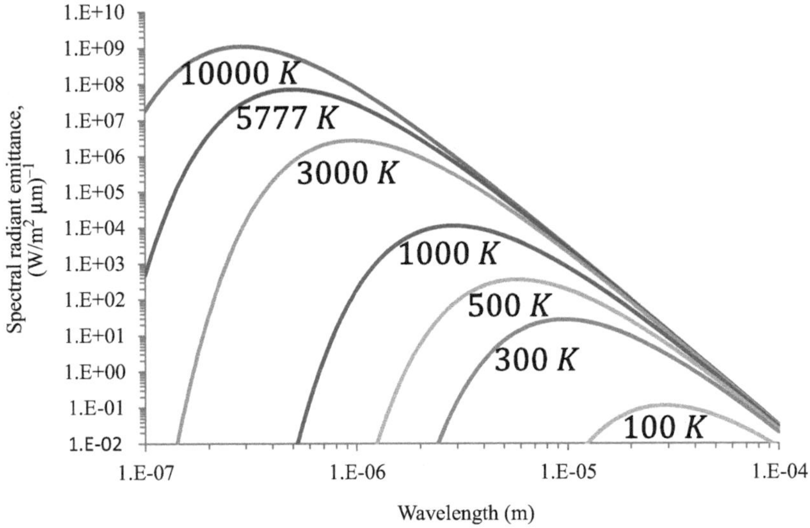

FIGURE 2.35 Spectral radiant emittance of a black body radiator as a function of the wavelength for various temperatures of the radiator. Curves are computed from Planck's equation. *(Source: Spectra computed by Jaypee Quino from the Erlangen Graduate School in Advanced Optical Technologies of the Friedrich-Alexander-Universität Erlangen-Nürnberg, Germany.)*

in Fig. 2.35 (double logarithmic scale) is a function of the temperature. Each body with a temperature above 0 K emits Planck radiation. With increasing temperature the emitted spectrum widens more towards shorter wavelengths or to higher photon energies. Thus, only if the black body radiators feature high temperatures we can see the emissions with the human eye, while the majority of the emitted Planck spectrum covers the infrared spectral range and longer wavelengths. A blacksmith, for example, estimates the temperature of the iron he wants to treat from the colorant at which the metal glows. A whitish glowing metal is hotter than a reddish glowing metal. Also the sun is a black body radiator. From the spectrum emitted from the sun one can compute the temperature of the sun's surface, which is approximately 5777 K.

As spectroscopic techniques based on 'chemiluminescence' or 'Planck radiation' have not frequently been exploited in the field of high-pressure process technology and science, they are not treated in detail later. Nevertheless it should be mentioned that they have been applied in order to analyse, for example, the influence of pressure onto the flame front in high-pressure combustion devices [6] or in order to measure the temperature in sooting flames [7].

In the following subsections of Section 2.4.1 the basic working principles of the respective spectroscopic techniques are provided. Examples of their applications at high pressure and fancy modifications of these techniques will be reported in Chapters 5 and 6. Before, the Section 2.4.1.2 introduces the quantitative nature of absorption spectroscopy.

2.4.1.2 Quantitative Nature of Absorption Spectroscopy

In absorption spectroscopy usually the absorbance A (on the ordinate) is given as a function of the wavenumbers $\bar{\nu}$ or of the wavelengths λ of the radiation (on the abscissa) send through the sample, as it is shown in Fig. 2.36 for pure $C^{12}O_2$ at 0.1 MPa and 320 K in a cuvette with 0.1 m path length. Due to the dependency $\bar{\nu} = 1/\lambda$ the two absorbance spectra appear non-linearly vertically flipped. From the wavenumber range and from the wavelength range on the abscissa we can see that this is an IR absorption spectrum. In Section 2.4.1.3 the physical meaning of the many lines the IR absorption spectrum is composed of will be described. The absorbance

$$A(\lambda) = -\log_{10}(T(\lambda)) = \log_{10}\left(\frac{I_{\text{incident},\lambda}}{I_{\text{transmitted},\lambda}}\right). \tag{2.80}$$

is the negative decadal logarithm of the transmission $T(\lambda)$ or the decadal logarithm of the ratio of the intensities (or powers) of the incident $I_{\text{incident},\lambda}$ and the transmitted $I_{\text{transmitted},\lambda}$ light. Figure 2.37 shows three scenarios where the intensity of the incident light is attenuated on its way through a cuvette.

In the upper part of Fig. 2.37 the droplets dispersed inside the bulk fluid scatter some portion of the incident light intensity away from its original path, while the bulk fluid inside the cuvette does not at all attenuate the light intensity. In this example the bluish part of the incident spectrum is scattered more intense than the reddish part.

Thus the transmitted spectrum is not homogeneously attenuated by a constant factor, but also spectrally non-symmetric with respect to the incident one. The bluish part of the spectrum is attenuated more than the reddish part. I chose this example, where no light is absorbed but only scattered in order to make clear that *absorbance* and *absorption* are two different things. Absorption is shown in the centre illustration of Fig. 2.37. Here, there are no drops or particles contained in the sample volume but this time the fluid inside the sample volume can absorb distinct wavelengths of the incident light. Consequently, certain wavelengths are attenuated more than others. The absorbance then is mainly due to absorption. I have to point out that light is not only scattered from particles, drops or bubbles but also from molecules. Thus, there is always a very small portion of the incident light, which is scattered away due to light scattering from molecules, which we neglect or can correct for experimentally. In the bottom illustration of Fig. 2.37 the absorbance is due to both absorption and scattering. Under this circumstance it is rather challenging to quantify which of both effects contributed how to the measured absorbance.

If the measured absorbance is solely due to absorption, the Beer–Lambert law

$$A(\lambda) = \varepsilon(\lambda) \cdot c_{\text{abs}} \cdot l \tag{2.81}$$

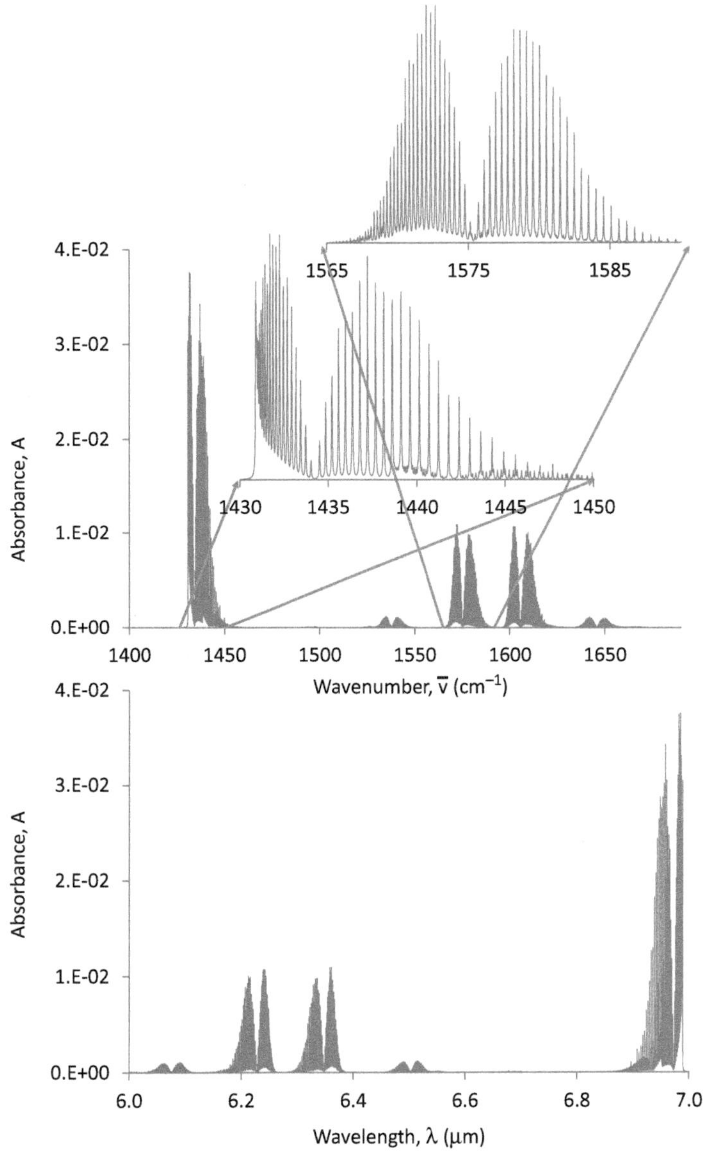

FIGURE 2.36 Absorption spectrum of $C^{12}O_2$ between 1400 cm^{-1} and 1700 cm^{-1} (top) or 6 μm and 7 μm (bottom) computed with the open software (databank) HITRAN (https://www.cfa.harvard.edu/hitran/) for 0.1 MPa, 320 K and an absorption path length of 0.1 m. *(Source: Spectra computed by Thomas Werblinski from the Lehrstuhl für Technische Thermodynamik of the Friedrich-Alexander-Universität Erlangen-Nürnberg, Germany.)*

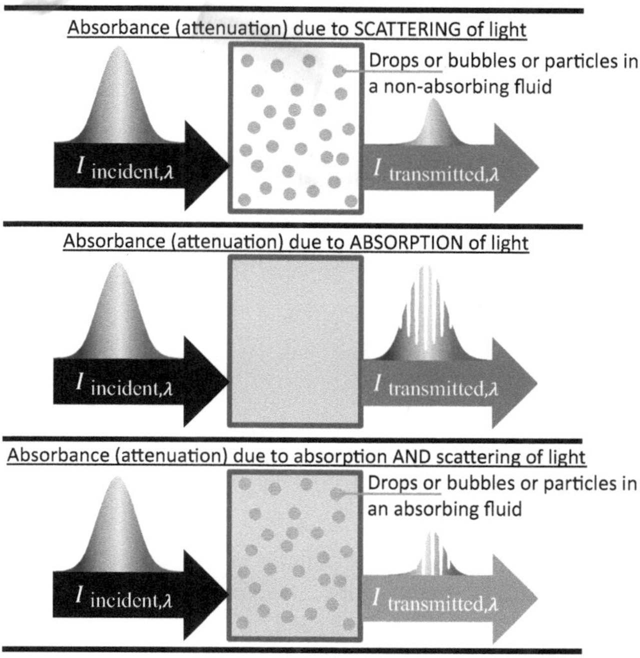

FIGURE 2.37 **Sketch for the illustration of absorbance due to elastic light scattering, absorption of light and both phenomena together.**

puts into relation the properties of the absorbing system and the measured absorbance $A(\lambda)$, where $\varepsilon(\lambda)$ is the molar attenuation (or absorption) coefficient in $m^2/mole$, c_{abs} is the molar concentration of the absorbing species in mol/m^3 and l is the path length the radiation has to pass through the sampling volume in m. The Beer–Lambert law is derived in Section 2.5.3. If c_{abs} is not given as molar concentration in mol/m^3 but as number density of absorbing molecules in m^{-3}, also $\varepsilon(\lambda)$ is given as absorption cross-section in m^2. It might be not obvious why $\varepsilon(\lambda)$ has the unit of a cross-section m^2, and what the physical meaning of such a cross-section is. Therefore I provide in the appendix a description, which is based on the computation of the probability that an incident photon is absorbed by a single molecule.

Based on the Beer–Lambert law we can compute the concentration

$$c_{abs} = \frac{\varepsilon(\lambda) \cdot l}{A(\lambda)} \qquad (2.82)$$

of the potentially absorbing molecules from the measured absorbance $A(\lambda)$, if $\varepsilon(\lambda)$ and l are known. The path length l is usually the thickness of the cuvette containing the probe to be analysed and therefore known. $\varepsilon(\lambda)$ is tabulated for many compounds and if not available also can be quantified before the

TABLE 2.2 Comparison of Transmission and Absorbance

Transmission T	Absorbance A
1	0
0.9	0.046
0.5	0.3
0.1	1
0.01	2
0.001	3
0.0001	4
0.00001	5

Dark-grey background values indicate the optimum-working regime of absorption spectrometers.

measurements in a calibration routine at defined and known concentrations c_{abs}. Summarising, absorption spectroscopy, irrespectively of whether IR, NIR or UV/VIS absorption spectroscopy is applied, is a straightforward technique for the quantification of the composition or the concentrations of mixtures (if scattering can be neglected). In the following absorption related sections we will see that absorption spectroscopy is also suitable for qualitative analysis.

Regarding Eqns (2.80) and (2.81) it becomes obvious that doubling c_{abs} or l also results in doubling the absorbance $A(\lambda)$, but that there is no linear interrelation between the absorbance $A(\lambda)$ and the transmission $T(\lambda)$. According to Table 2.2 an absorbance $A(\lambda)=1$ means that only 10% of the intensity of the incident light is transmitted through the sample.

Or, if only 10% of the incident radiation are absorbed on their way through the sample, meaning that 90% are transmitted, this would correspond to an absorbance of $A(\lambda)=0.046$. In this range $0.046 \leq A(\lambda) \leq 1$ absorption spectrometers show their best performance. Of course, it sounds logical that it is very challenging to quantify small differences between the incident and the transmitted intensity, which would correspond to an absorbance close to zero ($A(\lambda) \rightarrow 0$). On the other hand, it also sounds logical that it is challenging to quantitatively detect the intensity of spurious amounts of transmitted intensities, which would correspond to an absorbance significantly exceeding 1 ($A(\lambda) > 1$). Of course, there are also state-of-the-art absorption spectrometers available, making possible the reliable quantification of concentrations in a range of absorbance below or above the range $0.046 \leq A(\lambda) \leq 1$. I prepared one exercise which puts emphasise on the challenges inherent to absorption spectroscopy which might arise, when the composition of a mixture of two components has to be quantified, in which one component is much more concentrated than the other one.

2.4.1.3 Absorption of Infrared Radiation

In this section I describe mainly which molecules can absorb infrared (IR) and near infrared (NIR) radiation as well as the physical meaning of the many lines an absorption spectrum is composed of (compare Fig. 2.36). These two main aspects of this section will later make clear, why (N)IR absorption spectroscopy can be considered to be complementary to Raman spectroscopy and why the composition, the temperature and the pressure of a fluid can be computed from its absorption spectrum. Raman spectroscopy will be treated later in the section related to the light scattering techniques.

But before, I want to mention some tools or databanks available for the computation or generation of absorption spectra. These databanks do not provide ready to use absorption spectra as shown in Fig. 2.36, but they provide the centre wavenumbers of the absorption lines, their absorbance and the temperature and pressure broadening. From these parameters the user has to generate the absorption spectrum himself. Available databanks, without the claim of completeness, are HITRAN and HITEMP (https://www.cfa.harvard.edu/hitran/). While HITRAN contains many molecules, HITEMP is specialised for high temperatures and contains only CO_2, CO and H_2O. The Barber Tennyson databank (ftp://cdsarc.u-strasbg.fr/cats/VI/119/) contains only H_2O but for this molecule is the most detailed and comprehensive one. A third databank is the GEISA databank (http://ether.ipsl.jussieu.fr/etherTypo/?id=950&L=0), which is similar to the HITRAN databank. Unfortunately I cannot provide more information as I have not been working with either of them, except for the generation of Fig. 2.36.

Which properties a molecule must have, if it is supposed to absorb IR radiation? In order to answer this question we have to recall that the energy of IR radiation corresponds to the energy of a molecular transition, where the rotational J and vibrational v quantum numbers of a molecule may change, but the electronic quantum number el remains unaltered. Therefore IR absorption spectroscopy is also called rotational/vibrational absorption spectroscopy. I will mention later which transitions are allowed from the point of quantum mechanics, without deriving the reasons for this. Thus atoms—as they can neither vibrate nor rotate—do not have an IR absorption spectrum. But this does not mean that all molecules, which can rotate and or vibrate show an IR absorption spectrum. Only these molecules, whose rotation and/or vibration can couple to the IR radiation also can absorb it. The IR radiation is an oscillating electromagnetic field. From basic physics we know that electric fields can influence charged objects or couple to systems, which also feature an electric field. Thus only molecules, whose vibration and/or rotation feature an electric field, can absorb IR radiation. With respect to molecules, the electric field is termed dipole. Consequently, molecules, which can absorb IR radiation, must feature

- a vibrational and/or rotational transition whose energy corresponds to the energy of the incident IR radiation and
- a dipole momentum, which can be either a permanent dipole or an induced dipole.

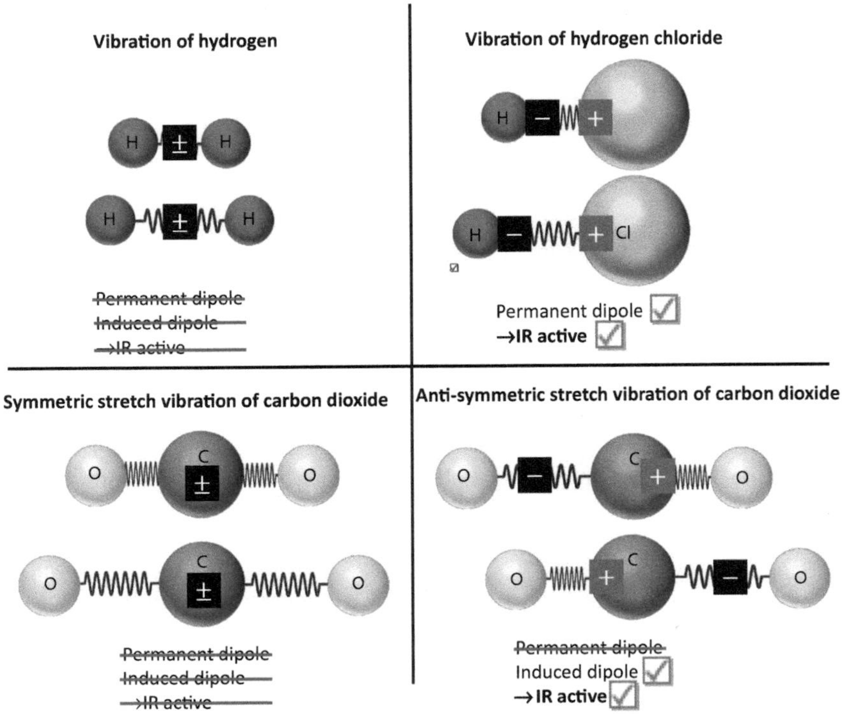

FIGURE 2.38 Sketch of some molecular vibrations, which can and cannot absorb IR radiation. Plus (+) and minus (−) indicate the centre points of the positive and negative charges. If both centre points are at different locations, the molecule is a dipole.

It should be mentioned here that rotation cannot induce a dipole. Thus only molecules featuring a permanent dipole can absorb radiation solely due to a pure rotational transition (Only *J* changes while the other quantum numbers v and el remain unaltered.). In contrast, some vibrations can induce a dipole to a molecule, which does not feature a permanent dipole. In this context Fig. 2.38 shows some molecular vibrations, which can and cannot absorb IR radiation.

Only these molecules, which feature a permanent dipole or a dipole, which is induced by the vibration, absorb IR radiation due to a vibrational transition. With respect to the examples provided in Fig. 2.38, hydrogen does neither feature a rotational nor a vibrational absorption spectrum and is thus said to be infrared (IR) inactive. Hydrogen chloride features a permanent dipole and thus is IR-active with respect to vibration and rotation. Carbon dioxide does not feature a permanent dipole, but the anti-symmetric stretch vibration in contrast to the symmetric stretch vibration shows an induced dipole. Therefore, the anti-symmetric stretch vibration is IR-active, but the symmetric one is IR inactive. **Exercise 2.8** in Section 2.4.4 is designed in order to show which motions/transitions of a molecule are active with respect to IR absorption spectroscopy

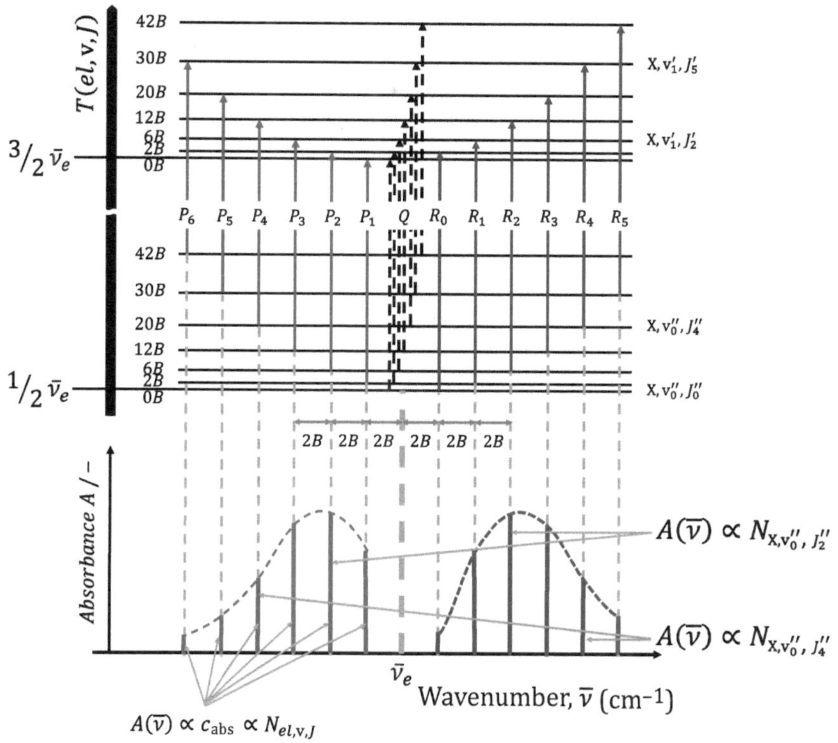

FIGURE 2.39 Rotational/vibrational transitions of a molecule due to the absorption of IR radiation (top) and the corresponding absorption lines (bottom). The vibrational quantum numbers are decorated with the superscripts ″ and ′ in order to indicate from which energy level the transitions starts and where it ends.

and/or Raman spectroscopy. Therefore it is recommended to read the introduction into Raman spectroscopy in Section 2.4.2.3 before solving this exercise.

Finally it has to be described why an IR absorption spectrum is composed of so many absorption lines and why these absorption lines are not homogeneously spread over the entire absorption spectrum but appear to be concentrated in groups at certain spectral regions. As already mentioned before, and as it is shown in Fig. 2.39, IR absorption spectroscopy is rotational/vibrational absorption spectroscopy, meaning that only the quantum numbers of rotation J and vibration v change, while the electronic quantum number el remains unaltered. (Pure rotational absorption spectroscopy is not discussed in the context of this book, but would correspond to molecular transitions where solely the rotational quantum number J changes, while v and el remain unaltered. Pure rotational transitions are also not included in Fig. 2.39.) According to quantum mechanics, not any arbitrary transition is possible. There are quantum mechanically 'allowed' and 'forbidden' transitions. The probability that a forbidden transition might take place is low, but is not always zero. Thus, sometimes even forbidden transitions are visible in an absorption spectrum. Figure 2.39 shows P-, Q- and R-transitions.

- P-transition: rotational $\Delta J = -1$ and vibrational $\Delta \text{v} = +1$ quantum numbers change
- Q-transition: only vibrational quantum number $\Delta \text{v} = +1$ changes; rotational quantum $\Delta J = 0$ number remains unaltered
- R-transition: rotational $\Delta J = +1$ and vibrational $\Delta \text{v} = +1$ quantum numbers change

For the computation of ΔJ or Δv we subtract the respective quantum number before the transition from the quantum number after the transition. For the P_4-transition this would be, for example, $\Delta J = 3 - 4 = -1$ and $\Delta \text{v} = 1 - 0 = +1$. I already want to mention here that the selection rules $\Delta J = \pm 1$ and $\Delta \text{v} = +1$ introduced here for rotational/vibrational *absorption* spectroscopy are different for rotational/vibrational Raman spectroscopy $\Delta J = \pm 2$ and $\Delta \text{v} = \pm 1$, which will be treated in a following section.

The length of the red (P-branch), black (dashed Q-branch) and blue (R-branch) arrows in Fig. 2.39 represent the energy difference between the states a molecule occupies after (v′) and before (v″) the transition. Therefore the lengths of these arrows also correspond to the energy (photon energy, wavelength, wavenumber, etc.) of the absorbed IR radiation, which in an absorption spectrum usually is given on the abscissa. The different Q-branch transitions (black dashed arrows) feature—in Fig. 2.39 for the case of a rigid rotating and harmonically oscillating dipole molecule—all the same energy (arrow length), which is the vibrational constant $\bar{\nu}_e$ (compare Eqn (2.48)). Therefore the Q-branch absorption transition would appear in the absorption spectrum on the abscissa at the position $\bar{\nu}_e$. In the absorption spectrum contained in Fig. 2.39 (bottom part) there is no absorption shown for the Q-branch transitions, as this transition is not possible for any IR-active molecular vibration, but only for few, whose angular momentum is in the axis of the molecule. The P-branch transitions appear at smaller wavenumbers, with respect to $\bar{\nu}_e$, as also the energy difference between the states involved into the P-branch transitions is smaller than the one corresponding to the Q-branch transitions. Also with increasing rotational quantum number *J*, the energy absorbed due to a P-branch transition decreases in increments of two times the rotational constant *B*. R-branch transitions absorb more energy than Q-branch transitions and therefore appear at wavenumbers higher than $\bar{\nu}_e$. The absorbed energy increases with increasing quantum number in increments of 2*B*. It has to be underlined that the line spacing of 2*B* is not found for any molecule. At the end of Section 2.2.7.2 it was described that molecules featuring an inversion centre, such as carbon dioxide, do not at all occupy energy levels of specific rotational quantum numbers *J*. This population behaviour is due to the nuclear spin and depends on the spin quantum number of the nuclei the molecule is composed of. CO_2

molecules, for example, only occupy rotational energy levels with even rotational quantum number. Consequently, the line spacing between the different P- and R-transitions would be $4B$ instead of usually $2B$.

Finally the question arises, which valuable information we can extract from an IR-absorption spectrum for high-pressure process technology or process technology in general. According to Eqn (2.81) the absorbance $A(\bar{\nu}) \propto c_{abs} \propto N_{el,v,J}$ is a function of the wavenumber (or wavelength) of the absorbed radiation and proportional to the concentration of the potentially absorbing molecules c_{abs}. Figure 2.39 shows that each transition starts at a certain energy level. This means that only these molecules can be involved in a specific molecular transition, which already before the transition occupy the respective energy level. For example, only the molecules occupying the energy level X, v_0'', J_3'' before the transition, can potentially undergo a P_3-transition to X, v_1', J_2', an R_3-transition to X, v_1', J_4' or a Q_3-transition to X, v_1', J_3'.

- *Derivation of temperature T*: In Section 2.2.7 the Boltzmann distribution is described, according to which the number of molecules $N_{el,v,J}$ occupying a specific energy level can be computed, with respect to the total number of molecules of the same species. Of course, we convert $N_{el,v,J}$ into a concentration (or a number density) c_{abs} of potential absorbers, meaning that the absorbances $A(\bar{\nu})$ at different wavenumber positions $\bar{\nu}$ measures indirectly the temperature sensitive Boltzmann distribution of this species. Also a change of the absorbance $A(\bar{\nu})$ would correlate with a change in the population $c_{abs} \propto N_{el,v,J}$ of the corresponding energy level. As the Boltzmann distribution according to Eqn (2.60) is a function of the temperature T, one can compute from the absorbances $A(\bar{\nu})$ of different absorption lines belonging to one species the temperature T. Figure 2.16 in detail showed the influence of temperature onto the population distribution of molecules. Another temperature evaluation strategy could be theoretically based on the fact that increasing temperature results in broadening of the absorption lines. According to my knowledge this approach has not been put in practice, but is mentioned here in order to recall the influence of the temperature onto the Maxwell velocity distribution of molecules. From the broadening of the absorption lines, with respect to a reference point, also the temperature could be extracted. The temperature induced (Gaussian) line broadening is due to the Doppler effect. Molecules moving within the probe volume counter the propagation direction of the incident light, effectively recognise a higher frequency of the incident electromagnetic radiation and therefore absorb radiation, whose energy is smaller than the energy of the molecular transition. Simultaneously, molecules, which move into the direction of propagation of the incident electromagnetic radiation, effectively recognise a smaller frequency of the incident radiation and thus absorb radiation, whose energy is higher than the energy of the molecular transition. The temperature sensitive velocity distribution of molecules within this book is described in the context of the Maxwell distribution in Section 2.5.1.3. The aforementioned web-data-pools provide parameters for line broadening due to temperature. (In **Exercise 2.10** the shapes of Gaussian-, Lorentzian and Voigt profile peaks are provided.)

- *Qualitative composition of a mixture*: Furthermore, the spectral locations of the absorption lines are a function of the vibration and rotation constants $\bar{\nu}_e$ and B, respectively, two properties, which are absolutely species specific (molecular fingerprint). Therefore, from the spectral locations at which the absorption lines appear, one can deduce information about which compounds—at least the IR-active ones—are contained in a mixture.
- *Quantitative composition of a mixture*: Once it is identified, which absorption lines correspond to which species, the composition of the mixture can be quantified based on the absorbances $A(\bar{\nu})$ of the different species.
- *Pressure*: In the context of temperature measurements, it has already been mentioned that the temperature leads to a Gaussian broadening of the absorption lines due to the Doppler effect. Also pressure leads to a Lorentzian broadening of the absorption lines of gases. With increasing pressure the number of collisions between molecules in a gas increases. In other words, with increasing pressure any molecule is more and more influenced by its environment, or by the molecules in its environment. At low pressure the molecules can be considered isolated from their environment. Collisions take place rarely and thus the molecules rotate and vibrate more or less all at the quantised energies derived in Sections 2.2.3 and 2.2.4. At high pressure collisions take place more frequently and thus the rotational and vibrational energies of the molecules get more distributed around the mean rotational and vibrational energy, which is still the one computed in Sections 2.2.3 and 2.2.4 for isolated molecules. With increasing pressure the distribution gets broader, which directly is observable in the pressure induced broadening effect of absorption lines. While Fig. 2.36 shows a computed absorption spectrum for CO_2 at 0.1 MPa, Fig. 2.40 shows the corresponding absorption spectra for the same temperature and absorption path length but for 1 MPa and 10 MPa. At a pressure of 0.1 MPa (see Fig. 2.36) the absorption lines are still separated. At 1 MPa and 10 MPa—due to pressure induced line broadening—the absorption lines cannot be resolved anymore. They merge to broad absorption bands. As the web-data-pools (HITRAN, HITEMP, Barber Tennyson, GEISA) mentioned earlier also provide parameters for the consideration of the pressure broadening, one in general can extract the pressure from the signature/appearance of an experimentally recorded absorption spectrum. (In **Exercise 2.10** the shapes of Gaussian-, Lorentzian and Voigt profile peaks are provided.)

We will see later that not only absorption lines but also emission lines and scattering lines are broadened due to temperature (Gaussian broadening) or pressure (Lorentzian broadening) effects.

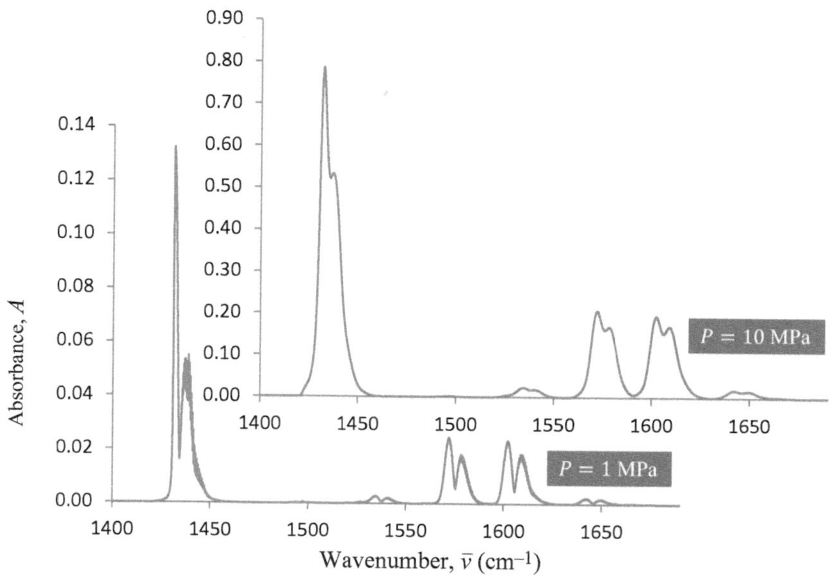

FIGURE 2.40 Absorption spectra of $C^{12}O_2$ between 1400 cm^{-1} and 1700 cm^{-1} computed with the open software (databank) HITRAN (https://www.cfa.harvard.edu/hitran/) for 320 K and an absorption path length of 0.1 m at pressures of 1 MPa and 10 MPa. Figure 2.36 shows the corresponding absorption spectrum at 0.1 MPa. *(Source: Spectra computed by Thomas Werblinski from the Lehrstuhl für Technische Thermodynamik of the Friedrich-Alexander-Universität Erlangen-Nürnberg, Germany.)*

2.4.1.4 Absorption of Near Infrared (NIR)-Radiation

With respect to absorption spectroscopy, the near infrared spectral range covers the wavelengths between 780 nm and 2500 nm. If we convert this wavelength range into wavenumbers, we would get the wavenumber range from 13,333.33 cm^{-1} to 4,000 cm^{-1}. We already saw in Section 2.2 that the vibrational and rotational constants, $\bar{\nu}_e$ and B, respectively, of molecules or of the vibrations of certain molecular bonds are smaller than 4000 cm^{-1}. Only the hydrogen molecule features a vibrational constant $\bar{\nu}_e$, which is above 4000 cm^{-1}. Therefore the photon-energies available in the spectral range between 780 nm and 2500 nm are too high to be absorbed by a transition of a fundamental vibration. In Fig. 2.41 the transitions corresponding to the fundamental stretch vibrations are allocated to the IR spectral range. In Fig. 2.41 one absorption band is represented by one black bar, but in reality would comprise the branches of the P-, Q- and R-transitions, as it is also indicated only for the fundamental stretch vibration. At this point I want to remind again that for symmetric molecules the Q-branch transitions are not always allowed and therefore in this example not visible. Combinations of these two fundamental vibrations as well as harmonics of these vibrations are also not allowed from a quantum mechanical point of view, but in reality occur with a much lower probability than fundamental

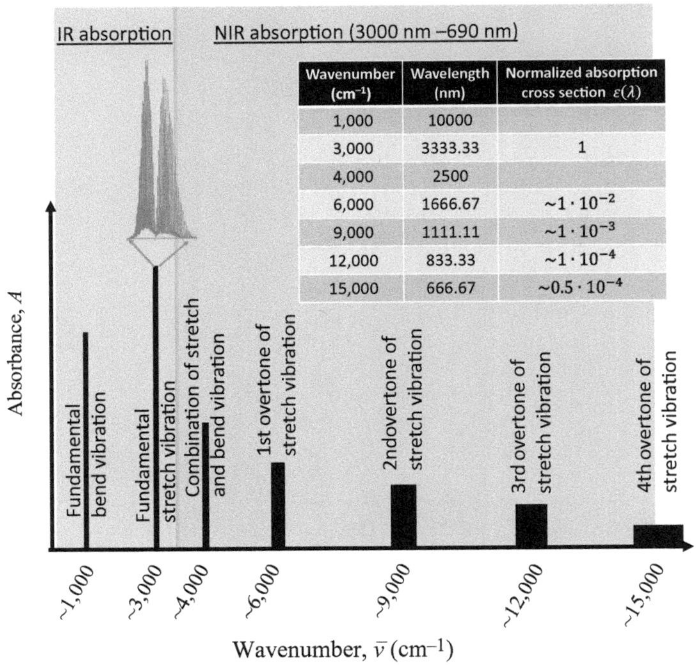

Wavenumber (cm^{-1})	Wavelength (nm)	Normalized absorption cross section $\varepsilon(\lambda)$
1,000	10000	
3,000	3333.33	1
4,000	2500	
6,000	1666.67	$\sim 1 \cdot 10^{-2}$
9,000	1111.11	$\sim 1 \cdot 10^{-3}$
12,000	833.33	$\sim 1 \cdot 10^{-4}$
15,000	666.67	$\sim 0.5 \cdot 10^{-4}$

FIGURE 2.41 Schematic illustration of an IR and NIR absorption spectrum of a molecule with two IR-active fundamental vibrations. The locations of the combinative and overtone absorptions are indicated. The inserted table provides the corresponding wavelengths and the magnitudes of the cross-sections normalised to the cross-section of the fundamental stretch vibration.

transitions. These combinative and harmonic vibrations can be observed in the NIR spectral range with a significantly smaller absorbance. The energy of the combinative vibration in wavenumbers is approximately the summation of the wavenumbers of the fundamental vibrations. Similarly, the harmonics of the stretch vibration, which are also called overtone absorption bands, appear at integer multiple wavenumbers of the fundamental vibration. The higher the order of the absorption band (first overtone is the second order absorption band, second overtone is the third order absorption band,…), the lower is the respective absorption cross-section ε (λ). This behaviour is reflected by the magnitudes of the normalised absorption cross-sections tabulated in the table inserted in Fig. 2.41. This table also shows that in the wavelength scale the overtone absorption bands get closer with increasing order.

While the wavelength difference between the first and the second overtone absorption band is approximately 555 nm, the wavelength difference between the third and the fourth overtone absorption band is approximately 167 nm. Therefore, the spectral resolution gets worse with increasing order of the absorption band. Consequently, especially for combinative vibrations, due to the interaction of different vibrations the many emerging absorption lines merge together to not spectrally resolvable broad absorption bands. The circumstance that the spectral resolution is worse in the NIR spectral rang is reflected in Fig. 2.41 by the increasing width of the black bars (absorption bands).

At first glance NIR absorption spectroscopy compared with IR absorption spectroscopy only has disadvantages. This is firstly the lower absorption cross-sections $\varepsilon(\lambda)$ of the combinative and the overtone absorption lines and secondly their worse spectral resolution. But at second glance both drawbacks can be turned into advantages and make NIR absorption spectroscopy a wide spread measurement technique in process technology.

Let us first consider the absorption cross-sections $\varepsilon(\lambda)$, which for combinative or overtone vibrations are significantly (orders of magnitude) smaller than those of the fundamental vibrations. Due to the smaller absorption cross-sections NIR radiation features a higher penetration depth into or even through samples. Thus NIR absorption spectroscopy can be applied in transmission and even in reflectance to objects directly, such as leaves (agriculture), tablets (pharmacy) or tissue (medicine), without the need of preparing the sample before the measurement. Another advantage of the decreasing magnitude of the absorption cross-sections with increasing order becomes clear in consideration of the problem mentioned in **Exercise 2.7**. In this exercise it is demonstrated that based on the Beer–Lambert law in combination with the dynamic range of the absorbance of an IR absorption spectrometer the composition of a binary mixture can only be measured reliably in a certain range of molar fractions. This is also true for NIR absorption spectroscopy, if one considers for two very differently concentrated compounds the same order of overtone vibration. But—what is not possible in the IR spectral range—in the NIR spectral range one could, for example, consider the second order overtone vibration for the compound being less concentrated in the mixture and the third or fourth order overtone for the compound being highly concentrated in the same mixture. As the absorption cross-section of the third or fourth overtone vibration is significantly smaller than that one of the second order overtone vibration, the absorbances of both compounds (one less and one highly concentrated) would reach similar values in the NIR absorption spectrum. Therefore, the range of mixture compositions measurable with a NIR absorption spectrometer is superior to that of an IR absorption spectrometer featuring the same dynamic range of absorbances. Also in this context it is advantageous that the resolution of the spectra, which at first glance seems to be a disadvantage, in the NIR spectral range is worse. The NIR spectral range covers the combinative as well as up to four overtones of the fundamental vibrations. Nevertheless the wavelength range of the NIR spectral range is rather narrow compared to the IR spectral range. Consequently, it is experimentally possible to simultaneously detect different overtones of one compound and also different overtones of several compounds of a mixture within one absorption spectrum. Thus, the 'worse' spectral resolution is the basis that provides to the experimenter the freedom of choosing between different combinative or overtone vibrations of different order.

Nevertheless, one disadvantage remains. Due to the mutual interaction of different vibrations (combinative vibrations) of one compound and the spectral overlap of the absorption bands of different compounds and the influence of the environment (solvent, temperature, pressure, mixture composition, etc.) onto the absorption bands, the NIR absorption spectra especially of mixtures are rather complex. Experimentally measured NIR absorption spectra of mixtures cannot easily be reproduced with theoretically computed spectra. Therefore the quantitative evaluation of NIR absorption spectra, with respect to, for example, temperature, or the mixture composition, usually requires a calibration and statistical computation routines.

2.4.1.5 Absorption of Ultraviolet and Visible Radiation

With respect to absorption spectroscopy the UV and VIS spectral ranges cover the wavelength range from 380 nm to 780 nm for the visible part (VIS) and wavelengths shorter than 380 nm down to 10 nm for the ultraviolet (UV) part. Wavelengths between 150 nm and 200 nm are absorbed by oxygen and wavelengths shorter than 150 nm are absorbed by nitrogen, which are the two major compounds the air in each laboratory is composed of. Consequently, wavelengths shorter than 200 nm are practically less relevant for absorption spectroscopic experiments. Therefore, in the following we consider the interaction of molecules and electromagnetic radiation in the spectral range between 200 nm and 780 nm. It should be mentioned here also that atoms can absorb radiation in this spectral range. Nevertheless we here consider only the absorption of UV and VIS radiation by molecules.

In the previous sections on (N)IR-absorption spectroscopy we saw that the vibrational v and/or rotational J quantum numbers of the molecule change due to the absorption of the (N)IR-radiation. The electronic quantum number el did not change. This is the reason why (N)IR absorption spectroscopy is called vibrational/rotational spectroscopy. In the case of UV/VIS-absorption spectroscopy the electronic quantum number el changes, while the quantum numbers of vibration v and rotation J can change, too. An example of electronic transition where solely the electronic quantum number is changed is sketched in Fig. 2.32.

In the following we will see that a molecule—depending on its functional groups—can undergo different electronic transitions. The electronic quantum number of a molecule changes whenever an electron of the molecule undergoes a transition from an orbital it originally occupies to another orbital, which features a different energy. Consequently, the energy, which is stored in the electronic configuration of the molecule, is changed. With respect to electronic transitions there are many selection rules, which we here will not consider further. Most of the potentially possible electronic transitions are forbidden, which does not mean that all forbidden transitions do not occur in reality, most of them just happen with a lower probability and are therefore less intense in the absorbance spectrum. Only few electronic transitions are allowed from a quantum mechanical point of view and are therefore much more pronounced

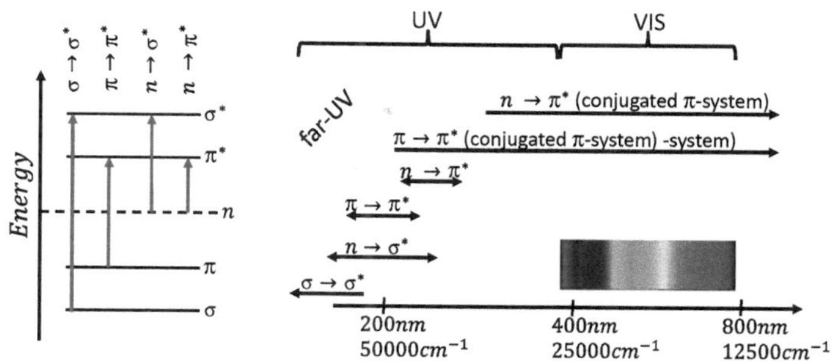

FIGURE 2.42 Electronic transitions between molecular binding orbitals σ and π and molecular anti-binding orbitals π* and σ*, as well as orbitals *n* not involved into the molecular bond (left). Spectral regions in which light is absorbed due to electronic transitions (right) if and if not molecules feature conjugated π-electron systems.

in the absorbance spectrum than the forbidden ones. Figure 2.42 shows on the left side electronic transitions between different binding, anti-binding (asterisk) and non-bonded (n) orbitals and on the right side the spectral regions in which these transitions can be observed in an UV/VIS-absorption spectrum. If the molecule features a conjugated π-electronic system, the transitions from, or to π-electronic states are shifted to longer wavelengths into the VIS part of the spectrum and could even be shifted to the NIR spectral range.

Summarising, a molecule such as acetone can absorb in the UV spectral range due to an

- $n \rightarrow \pi^*$ transition of the carbonyl group C=O, an
- $\sigma \rightarrow \sigma^*$ transition of the CH bondings and another
- $\sigma \rightarrow \sigma^*$ transition of the CC bondings.

For each occurring electronic transition there is one electronic absorption band. It is called absorption *band*, as in its substructure—if spectrally resolvable—one can find bands belonging to transitions for which the vibrational quantum numbers changed. Of course this vibrational band substructure features the substructure of the rotational/vibrational absorption *lines*. In reality the substructure of an electronic transition is usually not resolvable. I want to make this clear giving one example. Let us have a look at the $n \rightarrow \pi^*$ transition of the carbonyl group C=O of any ketone, which we assume has its electronic absorption maximum at approximately 280 nm. From the previous Sections 2.2.3 and 2.2.4 on the rotational and vibrational energy of molecules we know that the energy difference between neighbouring rotational and vibrational energy levels is few wavenumbers and few several hundreds of wavenumbers, respectively. Assuming arbitrarily an energy difference between rotational energy levels of 10 cm^{-1}, their rotational absorption lines would be separated at 280 nm by only 0.08 nm. Furthermore, at elevated pressures or in liquid systems the rotational–vibrational lines are usually broadened due to collisions. Therefore, due to the small wavelength difference between the rotational/vibrational absorption lines in combination with the broadening effects, the single rotational/vibrational lines are usually not resolvable in the UV/VIS absorption spectrum. Exceptions are UV/VIS-absorption spectra recorded with highly resolving equipment from the vapour of substances at low pressures, when the absorbing molecules can be considered as isolated molecules not interacting with their environment and not colliding with their neighbouring molecules, or if even at elevated pressures the molecules are so small that the energy difference between neighbouring rotational/vibrational energy levels is big enough to resolve them in the absorption spectrum.

Due to the usually not resolvable vibrational/rotational structure, which is the fingerprint of each species, UV/VIS-absorption spectra are not very specific with respect to the detailed structure of the absorbing molecule but can provide information about the functional groups of the molecule. Furthermore, as the absorption bands in the UV/VIS spectral range are rather broad (acetone, e.g., from 250 nm to 300 nm) the absorption bands of different compounds can easily overlap which makes the analysis of multi-component mixture compositions more challenging than with (N)IR-absorption spectroscopy.

In a previous section I already mentioned that chlorophyll absorbs in the bluish and reddish visible spectral range. This is due to the very extended conjugated π-electronic system of the chlorophyll molecule and the reason why plants appear in the green colour, which is not absorbed, but 'reflected'. Another example and very prominent example of a molecule with a very extended π-electronic system is the 'indigo–blue' molecule, which originally was used for dyeing the 'blue jeans'. Molecules, which do not feature a conjugated π-electronic system, absorb in the UV spectral range, if they undergo an electronic transition. The carbonyl functional group of acetone, for example, absorbs between 250 nm and 300 nm. The electronic absorption spectra of many compounds are tabulated in literature, including the information in which solvent and at which concentration the spectra were recorded. This information is essential as the solvent as well as the concentration influence the absorption behaviour due to interactions between the molecules and their environment.

It is important to mention here that the compounds or the samples the UV/VIS absorption spectrum should be analysed from, have to be prepared first. This becomes obvious when having a look at the extinction coefficient ε of electronic transitions, which can be between 100 and 20,000 cm^2/mol. We already saw that absorption spectrometers show a good performance if the absorbance is $0.046 \le A(\lambda) \le 1$. Thus, if we assume an absorbance of 1, an extinction coefficient of 1000 cm^2/mol and a path length l of the cuvette of 1 cm, the concentration of the absorbing species inside the cuvette according to the law (Eqn (2.81)) of Beer and Lambert has to be attenuated to

$$\frac{A(\lambda)}{\varepsilon(\lambda)\cdot l} = \frac{1}{1000\,\mathrm{cm}^2/\mathrm{mol}\cdot 1\,\mathrm{cm}} = 1\,\mathrm{mol/L}.$$

Consequently species featuring a high electronic extinction coefficient have to be diluted before an electronic absorption spectrum can be recorded. Ideally they are diluted in solvents whose absorption spectrum does not overlap with the absorption spectrum of the species under investigation. On this purpose perfluorinated alkanes, saturated hydrocarbons, water, acetonitrile and some alcohols are suitable solvents, depending on the UV-spectral range the substance diluted in the solvent absorbs. Of course, a measurement technique such as UV/VIS-absorption spectroscopy, for which a sample first has to be prepared for the measurement is not first choice with respect to *in situ* applications, where the measurements are supposed to be made directly inside the reactor or the process.

Nevertheless, UV/VIS-absorption spectroscopy is frequently applied in the context of high-pressure process technology, not directly at elevated pressure but in order to analyse the product generated via a high-pressure process. Particles containing an active pharmaceutical ingredient (API) can be generated using high-pressure process technologies. In order to analyse the release kinetics of the API and the concentration of the API inside the particles or tablets (the particles matrix can be a biodegradable polymer) the particles are first added to a solvent, which mimics the biological environment in which the API is supposed to be released. Then the solvent is continuously pumped through an UV/VIS-absorption spectrometer and the absorbance spectra are recorded as a function of time. In this example the API is usually concentrated low enough to make the measurement of the absorbance possible directly without any previous preparation step. We already saw in Section 2.4.1.2 how one can compute the concentration of the API dissolved in the solvent using the Beer–Lambert law.

2.4.1.6 Laser-Induced Fluorescence and Phosphorescence

The laser-induced fluorescence (LIF) and—phosphorescence (LIP) techniques are usually applied in order to analyse mass- and heat-transfer mechanisms during mixing processes or in order to detect species which are generated as intermediates or products during a reaction. LIF is, for example, frequently applied as *in situ* technique in order to study the mixing between the fuel and the air in internal combustion engines or turbines or for studying the chemistry of reactions.

The laser-induced fluorescence (LIF) as well as the laser-induced phosphorescence (LIP) belong to the absorption–emission processes. For the LIF such an absorption–emission event is sketched on the right side of Fig. 2.32 (for LIP a sketch will follow in Fig. 2.44). A photon, which has been emitted from a laser is absorbed by the molecule, which as a consequence of the absorption undergoes a transition to an excited electronic state. I want to underline here again that the absorbed photon after its absorption does no more exist. It has been quasi 'eaten up' by the molecule. The system containing the excited molecule(s) intents to reach again a state of thermodynamic equilibrium, at which electronically excited molecules usually do not exist (compare Boltzmann distribution in Section 2.2.7). Consequently the excited molecule(s) relaxes down to the electronic ground state. One path of relaxation is the emission of a photon, which can be either a fluorescence or a phosphorescence event. Once again I want to underline that the emitted photon and the absorbed photon are two different photons, without any relation between them. I put so much emphasis on this as we will see later when we give attention to light *scattering* mechanisms that in the case of light *scattering* the incident photon is never absorbed but instantaneously scattered! This is the main difference between *absorption–emission* processes and *scattering* processes, which is relevant when it comes to the point why—under certain circumstances—absorption–emission processes and why—at other circumstances—scattering processes are the better choice for *in situ* diagnostics.

In the following I will first describe the working equation of laser-induced fluorescence. Then I am going to briefly describe why different kinds of experimental equipment are required for laser-induced fluorescence experiments with 'small' or 'big' fluorescing molecules.

In the following we will assume that our measurement object, from which we intend to detect LIF signals is a cuvette filled with a mixture of water and acetone. For the excitation of the LIF signal we have a laser with an emission wavelength of 266 nm (frequency quadrupled Nd:YAG laser compare Fig. 2.7). In the previous section on UV/VIS absorption we saw that water is often used as not absorbing diluting agent in the wavelength range above 200 nm. From this we know that the water molecules will not absorb the laser radiation and consequently will not fluoresce. But the acetone molecules feature a carbonyl group, which due to the n → π* transition (compare Fig. 2.42) can absorb the laser radiation and consequently after the absorption can fluoresce. In the previous section on UV/VIS absorption we also saw that the UV/VIS-absorbing agent has to be very much diluted due to the high molar absorption coefficients. This is why we assume in the following that there are only few acetone molecules and many water molecules contained inside the cuvette and homogeneously mixed so that the excitation laser beam can pass through the cuvette without any recognisable absorbance $A(266\,\mathrm{nm}) \rightarrow 0$. In a certain time interval some of the acetone molecules will absorb one laser photon. The probability of absorption is proportional to the molar absorption coefficient $\varepsilon(266\,\mathrm{nm})$. Some of these excited molecules—not all of them—will fluoresce in a certain time interval, which is called fluorescence lifetime, after the absorption. We will see later that next to fluorescence there are other competing radiationless channels an excited molecule can chose in order to get rid of its excess (absorbed) energy. The probability that an excited molecule chooses the fluorescence channel in order to reach back into the electronic state is called fluorescence quantum yield, which will be defined later. We also will see later that the fluorescence signal usually features longer wavelengths than the absorbed radiation (emissions red-shifted with respect to the absorption). This assures that the fluorescence is not reabsorbed by acetone molecules in the neighbourhood of the emitting one.

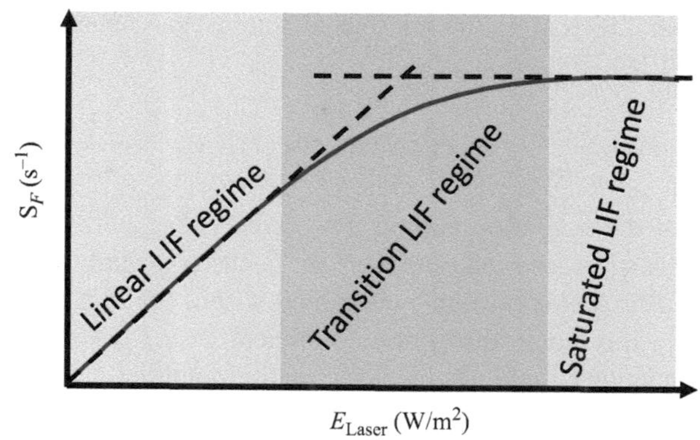

FIGURE 2.43 Detected fluorescence signal intensity S_F as a function of the excitation irradiance E_{Laser}.

Nevertheless, those molecules, which fluoresce, emit a photon in any direction of the room in order to transit back to the electronic ground state. The fluorescence events of different molecules happen at different time instants within the fluorescence lifetime and are not in any relation with each other. This means that one fluorescing molecule (stimulated emissions are an exception [8–12]) does not recognise the other fluorescing molecules. Consequently, the fluorescence radiation is not polarised. Each fluorescing molecule can emit only one fluorescence photon. As the fluorescence photons are emitted in any direction by different acetone molecules, we would be able to see the fluorescence in any polarisation from any direction with the same intensity; from the top, the bottom and from any side of the cuvette.

So far it sounds logical that when we increase the irradiance (W/m^2) of the excitation laser beam E_{Laser}, we would expect also the detected fluorescence signal intensity S_F (number of fluorescence photons detected per time interval (s^{-1})) to increase linearly, as it, according to Fig. 2.43, can be observed in the liner LIF regime.

For higher excitation irradiance the fluorescence signal intensity in the transition LIF region converges to a plateau and finally in the saturated LIF region remains constant irrespectively of the excitation irradiance. As the fluorescence signal intensity is proportional to the excited molecules, operating in the saturated LIF region simply means that already all molecules, which potentially can be excited, are excited. More than 100% of these potentially excitable molecules cannot be excited, even if the laser excitation irradiance is increased further. In contrast, in the linear LIF regime the number of excited molecules scales linearly with the excitation irradiance. This of course is only possible if a small portion of the potentially excitable molecules is excited at moderate excitation irradiances. Usually LIF experiments are operated in the linear LIF regime. Exceptions are known [13–16] but are not treated here.

In the linear LIF regime the number of fluorescence photons detected per time interval [17]

$$S_F = \frac{E_{Laser}}{h\nu} \cdot c_F \cdot \varepsilon(\lambda, T, P) \cdot \Phi_F(x, T, P) \cdot V \cdot \frac{\Omega}{4\pi} \cdot \eta \tag{2.83}$$

is a function of the excitation photon flow density $E_{Laser}/h\nu$ (number of photons per cross-section area and time) ($s^{-1}\ m^{-2}$), the concentration of potentially fluorescing molecules c_F (mol/m^3) contained within the excited and observed volume V (m^3), the molar absorption coefficient $\varepsilon(\lambda, T, P)$ (m^2/mol) and the fluorescence quantum yield $\Phi_F(x, T, P)$ of the potentially fluorescing molecules, the solid angle Ω with which the fluorescence photons are detected and an experimental efficiency coefficient η, which considers that the optical components used in the set-up do not operate ideally (signal intensity losses at mirrors and lenses, quantum efficiencies of detectors smaller than 100%). The molar absorption coefficient $\varepsilon(\lambda, T, P)$ is a function of the wavelength λ the temperature T and the pressure P. The fluorescence quantum yield $\Phi_F(x, T, P)$ is also a function of temperature T and pressure P and additionally also depends on the composition x of the mixture the fluorescing molecules are dissolved in (in our example it is the molar fraction of acetone in the cuvette filled with the acetone water mixture). E_{Laser} is the irradiance of the excitation laser beam in W/m^2.

2.4.1.6.1 Laser-Induced Fluorescence, a Quantitative Measurement Technique?

If we keep the experimental parameters, such as $E_{Laser}/h\nu$, V, $\Omega/4\pi$, λ and η, as well as the temperature and pressure constant, Eqn (2.83) simplifies to

$$S_F \propto c_F \cdot \Phi_F(x). \tag{2.84}$$

Under these circumstances the number of detected fluorescence photons per second S_F, which from now on I will call the fluorescence intensity, is proportional to the number of potentially fluorescing molecules and their fluorescence quantum yield. In order to understand the fluorescence quantum yield and its meaning for the fluorescence signal intensity, we have to consider all mechanisms which contribute to the laser-induced fluorescence event, and which are in competition to it. Figure 2.44 shows in an energy term scheme, which vibrational energy levels a molecule could potentially occupy in the electronic singlet[1] ground state, the first excited electronic singlet state and the first excited triplet[2] state.

- *Absorption*: Initially the molecule—before the absorption of UV radiation—occupies the vibrational ground state v_0'' in the electronic singlet ground state. In Section 2.2.7 we saw that at ambient temperature nearly 100% of

1. In a singlet state, the spins of the electrons are paired. The total spin is zero.
2. In a triplet state, the spins of the electrons are not paired, they are parallel. The total spin is one.

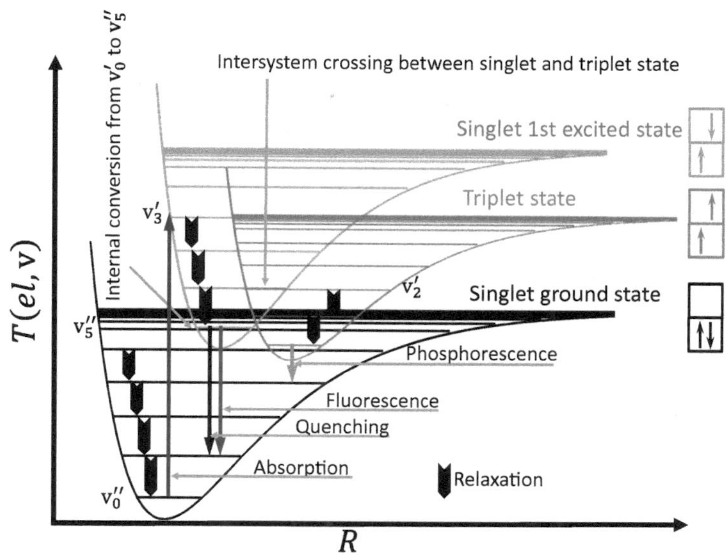

FIGURE 2.44 Term scheme of vibrational energy levels (rotational energy levels are not included) a molecule can occupy in the electronic singlet ground state, the first excited singlet state and the excited triplet state. The arrows indicate transitions, which are relevant for the laser-induced fluorescence and phosphorescence.

all molecules occupy the vibrational ground state. Due to the absorption of UV radiation the absorbing molecule undergoes then an electronic transition from the singlet ground state to the first excited electronic state. For molecules with conjugated π-electron systems also VIS radiation could be absorbed in order to achieve an electronic transition. If, due to the electronic transition, one electron is shifted from a binding to an anti-binding orbital, the binding between the nuclei is weakened and consequently the equilibrium distance R_e at the potential curve minimum increases, as it is shown in Fig. 2.42. If the binding force is not influenced by the electronic transition, the minima of the potential curves of the excited and the ground states occur at the same distance between the nuclei R_e. In detail, according to Fig. 2.42 the molecule directly after the absorption occupies the vibrational energy level v_3' in the 1st excited singlet state. If only one or very few molecules out of many molecules absorb the UV radiation we can assume that the temperature of the entire system composed of many molecules is the same before and after the absorption event. Thus, the fact that the absorbing molecule(s) occupy electronically and vibrationally excited states is not in agreement with the Boltzmann statistics. Consequently the excited molecules have to relax back to states of lower vibrational and electronic energy in order to thermodynamically equilibrate. In general there are different paths for the realisation of the relaxation.

- *Radiationless relaxation*: Usually the radiationless rotational/vibrational relaxation (rotational energy levels are not included in Fig. 2.42) down to the vibrational ground state v_0' in the first excited electronic state is a fast process. Due to intermolecular interactions (collisions) the excess energy of the few excited molecules is distributed over the many not excited molecules. As there are many more molecules not excited, the relaxation of the few excited ones does not cause a recognisable change of the energy distribution of the not excited ones.
- *Fluorescence*: The fluorescence is a radiative electronic transition from an excited electronic singlet state to a lower singlet electronic state. 'Radiative' means that a photon is emitted whose energy corresponds to the energy difference between the state the molecule occupies before and after the fluorescence event. As the radiationless rotational/vibrational relaxation is a fast process at elevated pressures (high collision frequency with other molecules), the fluorescence usually starts from the vibrational ground state of the excited electronic singlet state. Due to the rotational/vibrational relaxation in the excited electronic state and due to the fact that the fluorescence does not always bring the molecule directly back to the vibrational ground state v_0'' in the electronic singlet ground state, the fluorescence emission is usually red-shifted with respect to absorbed wavelength. 'Red-shifted' (shifted to less energy) in this context means that the emissions appear at wavelengths longer than the absorption. In contrast 'blue-shifted' (shifted to more energy) would mean that the emissions appear at wavelengths shorter than the absorption. If the fluorescence does not bring back the molecule directly from v_0' to v_0'', but, for example, as it is sketched in Fig. 2.42 to v_1'', radiationless relaxation will drive the molecule finally back to v_0''.
- *Quenching*: During the collision[3] of an electronically excited molecule with a so called quencher molecule, the excited molecule can transfer its excess energy to the quencher molecule so that the initially excited molecule makes a radiationless transition to the electronic ground state [18–21]. Consequently each excited molecule, which relaxed due to the radiationless quenching process, is no more available for a potentially occurring fluorescence or phosphorescence event. Therefore the radiationless energy decay from the excited to the ground electronic state is called quenching (it quenches the fluorescence emissions). The probability that electronically excited molecules are quenched depends on the chemical environment of the excited molecule, so to say the molecules the excited molecule potentially can collide with, and the collision frequency between the molecules (pressure and temperature). Furthermore, the lifetime of the excited electronic state plays an important role. The longer the lifetime of the excited state is, the higher is

3. There is also a quenching mechanism known for which the excited molecule and its quencher do not have to collide but have to approximate.

the probability of a collision with a quencher before the fluorescence emission. Aromates typically feature a rather long lifetime in the excited electronic state [22–25] and therefore their fluorescence is heavily quenched if potential quenchers are around. Ketones typically feature a rather short lifetime in the excited electronic state [26] and therefore their fluorescence signal intensity is quasi not sensitive to quenching.

- *Intersystem crossing*: In the sketch in Fig. 2.42 the intersystem crossing can take place between the energy level v_1' of the excited singlet state and the energy level v_2' of the excited triplet state, as here the energies of both states are approximately the same. The more similar the term energies of an excited singlet and an excited triplet state are, the more probable is intersystem crossing. In this case the spin of the 'excited' electron changes and as a result the singlet-excited state becomes a triplet excited state. Then again the fast radiationless relaxation drives the molecule to the v_0' energy level of the electronically excited triplet state.
- *Phosphorescence*: In thermodynamic equilibrium, the initially excited molecule has finally to reach the vibrational ground state v_0'' in the electronic ground state. Therefore also the molecules undergoing the intersystem crossing from the singlet to the triplet state have to make the reverse transition in order to come back to the singlet electronic configuration in the electronic ground state. In this context the phosphorescence is the radiative process bringing the electronically excited triplet molecules back to the singlet ground state [27–30]. Consequently phosphorescence can also be seen as a radiative version of intersystem crossing. Not only the fluorescence but also the phosphorescence can be quenched by collisions with other quencher molecules. Similarly to the fluorescence, the phosphorescence usually appears red-shifted with respect to the absorbed wavelength. As intersystem crossing is a forbidden transition, the lifetime of molecules in the triplet state is rather long. Consequently the phosphorescence emissions can be observed rather long after a short pulse excitation of the molecules, but only, if no quenchers are around. Due to the long lifetime of the triplet state its phosphorescence is quenched completely if quenchers are around.
- *Internal conversion*: During internal conversion a molecule in the excited singlet state converts into the electronic singlet ground state, while the term energy of the molecule undergoing this transition remains quasi constant. Consequently after the internal conversion the molecule occupies a highly excited vibrational state in the electronic ground state. In Fig. 2.42 the internal conversion takes place between the vibrational ground state level v_0'' of the first excited singlet state and the excited vibrational state v_5'' of the singlet ground state. Again the probability of internal conversion between an excited and a singlet ground state is the higher, the more similar the term energies of the involved states are. Finally the radiationless rotational/vibrational relaxation drives the molecule to the vibrational ground state v_0'' in the electronic ground state.

With respect to the fluorescence quantum yield contained in Eqn (2.84), Fig. 2.44 and the descriptions given in its context show that the fluorescence quantum-yield at constant pressure and temperature

$$\Phi_F(x) = \frac{k_F}{k_F + k_{ISC} + k_{IC} + \left[k_Q \cdot c_Q\right]} \tag{2.85}$$

can be defined as the fluorescence rate k_F over the summation of the rates of all depopulation processes, which are the fluorescence itself k_F, the intersystem crossing k_{ISC}, the internal conversion k_{IC} and the quenching $k_Q \cdot n_Q$. As usual, the rates k_F, k_{ISC}, and k_{IC} feature the unit 'events per second' (s^{-1}). These rates are not influenced by the chemical environment (molecular neighbours) of the excited molecule. This is different for quenching, which relies onto a collision of the excited molecule with a quencher molecule. Therefore the unit of k_Q is cubic metres per second and mole (m^3/s mol) and the unit of the concentration of quenchers c_Q is as usual moles per cubic metre (mol/m^3). Quenching is the only reason why the fluorescence quantum-yield $\Phi_F(x)$ is a function of the composition of the chemical environment of the excited molecule. We will see soon that oxygen, for example, is the most prominent quencher with respect to excited aromatic molecules [22–25], while excited ketones, such as acetone, are nearly not quenched by oxygen [26].

If we combine Eqns (2.84) and (2.85) and multiply the nominator and the denominator with $(k_F + k_{ISC} + k_{IC})^{-1}$

$$S_F \propto c_F \cdot \frac{\dfrac{k_F}{k_F + k_{ISC} + k_{IC}}}{\dfrac{k_F + k_{ISC} + k_{IC} + (k_Q \cdot c_Q)}{k_F + k_{ISC} + k_{IC}}} \tag{2.86}$$

$$S_F \propto c_F \cdot \frac{\overbrace{\dfrac{k_F}{k_F + k_{ISC} + k_{IC}}}^{b}}{1 + \underbrace{\dfrac{k_Q}{k_F + k_{ISC} + k_{IC}}}_{k_{SV}} \cdot c_Q}$$

we obtain the short form

$$S_F \propto \frac{c_F \cdot b}{1 + k_{SV} \cdot c_Q} \tag{2.87}$$

of the LIF working equation. The Stern–Volmer constant k_{SV} (it is not a rate!) with the unit of an inverse concentration (m^3/mol) was first introduced by Otto Stern and Max Volmer in 1919 [31]. The Stern–Volmer constant k_{SV} can be determined graphically from the slope of the Eqn (2.88) of a line. Here S_{F_0} is the

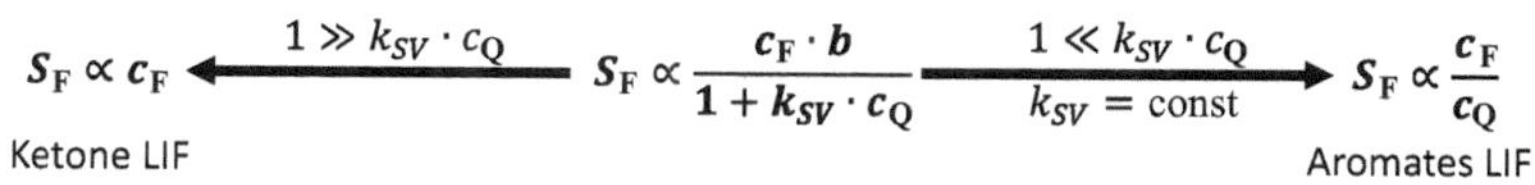

FIGURE 2.45 Simplification of Eqn (2.87).

fluorescence signal detected from a sample with a concentration c_F of potentially fluorescing molecules, in which no quenchers are contained $c_Q = 0$. S_F is the fluorescence signal detected from the same sample with the same concentration of potentially fluorescing molecules c_F, but now with quencher molecules present $c_Q \neq 0$. Due to quenching, S_F decreases with increasing concentration of quenchers c_Q.

$$\frac{S_{F_0}}{S_F} = \frac{(c_F \cdot b)_0 \cdot (1 + k_{SV} \cdot c_Q)}{\left(1 + \underbrace{k_{SV} \cdot c_Q}_{0}\right)_0 \cdot \underbrace{(c_F \cdot b)}_{=(c_F \cdot b)_0}} = 1 + k_{SV} \cdot c_Q \tag{2.88}$$

According to Fig. 2.45 Eqn (2.87) can further be simplified to

$$S_F \propto c_F \tag{2.89}$$

if $1 \gg k_{SV} \cdot c_Q$ or to

$$S_F \propto \frac{c_F}{c_Q} \tag{2.90}$$

if $1 \ll k_{SV} \cdot c_Q$ and $k_{SV} = \text{const}$.

2.4.1.6.2 Quantitative Concentration and Composition LIF

Concentration LIF: $1 \gg k_{SV} \cdot c_Q$

In the first case $S_F \propto c_F$ (Eqn (2.89)) where the LIF signal intensity scales linearly with the concentration of potentially fluorescing molecules, the LIF measuring technique can be considered to be a quantitative *concentration* measurement technique. Based on a calibration of S_F as a function of c_F or alternatively based on computations of Eqn (2.83) the *concentration* of potentially fluorescing molecules can be computed from the detected LIF signal intensity S_F. Usually, the fluorescence of carbonyl molecules is only marginally quenched by the quencher 'oxygen' ($1 \gg k_{SV} \cdot c_{O_2}$) except for high-oxygen concentrations c_{O_2}. Thus, under many circumstances, carbonyl molecules can be used as concentration indicating LIF molecules in, for example, combustion experiments, where the 'quencher' oxygen is present. Nevertheless the requirement $1 \gg k_{SV} \cdot c_{O_2}$ always has to be checked before detected LIF signals are interpreted according to the *concentration* LIF regime.

Composition LIF: $1 \ll k_{SV} \cdot c_Q$ and $k_{SV} = \text{const}$

In the latter case (Eqn (2.90)), where the LIF signal intensity S_F is directly proportional to the ratio of the concentration of potentially fluorescing and quenching molecules c_F / c_Q, the LIF measuring technique can be considered to be a quantitative *composition* measurement technique. Based on a calibration of S_F as a function of the ratio c_F / c_Q or alternatively based on careful computations of Eqn (2.83) the ratio of potentially fluorescing and quenching molecules can be determined from the detected LIF signal intensity S_F. Usually the fluorescence of aromatic molecules is quenched strongly by the quencher 'oxygen' $(1 \ll k_{SV} \cdot c_{O_2})$ except for small oxygen concentrations $c_{O_2} \rightarrow 0$. Furthermore it has to be checked whether k_{SV} is a constant or not. Under many circumstances aromatic molecules can be used as composition indicating LIF molecules in, for example, combustion experiments, where the quencher oxygen is present. Nevertheless the requirements $1 \ll k_{SV} \cdot c_{O_2}$ and $k_{SV}=\text{const}$ always have to be checked before detected LIF signal are interpreted according to the *composition* LIF regime. At this point I want to underline that during the calibration of S_F as a function of c_F / c_Q it is not sufficient to vary the ratio c_F / c_Q by varying solely the nominator c_F. It should be recalled that quenching is a process relying on collisions between excited molecules, which potentially can fluoresce, and quenchers. The probability that such an excited molecule collides with a quencher molecule is directly proportional to the concentration of quenchers c_Q. Thus, if solely c_F is varied in order to vary $\frac{(c_F)_{\text{varied}}}{(c_Q)_{\text{const}}}$, the probability of quenching is not changed at all. In this case S_F would always be directly proportional to $\frac{(c_F)_{\text{varied}}}{(c_Q)_{\text{const}}}$, irrespectively of which LIF indicator (ketonic or aromatic tracers) is used and irrespectively of the magnitude of $k_{SV} \cdot c_{O_2}$. It is just logic that the detected LIF signal intensity increases by a factor of two if c_F is doubled. The probability of quenching is only varied if c_Q is varied in the denominator of $\frac{(c_F)_{\text{const}}}{(c_Q)_{\text{varied}}}$. Then the probability that an excited molecule, which potentially could fluoresce, is quenched, is varied and only then it can be checked, whether a molecule can really act as composition indicating LIF molecule or not [18,32,33]. As I have the feeling that the quenching of 'composition' indicating LIF molecules is not always calibrated correctly, I additionally want to make this clear emphasising the importance of the correct calibration routine in the appendix.

In the following I want to make clear the difference between concentration and composition LIF. Let us assume that we need to measure the fuel/air-ratio inside the combustion chamber of an internal combustion engine equipped with fuel direct injection and variable valve operating mechanism. The combustion engine is operated with exhaust gas recirculation. While the recirculated exhaust flows into the combustion chamber through the 'out-valves', the fresh air flows into the combustion chamber through the 'in-valves'. Consequently inside the internal combustion engine the fresh air with an oxygen content of approximately 21% and the recirculated exhaust gas with an oxygen content of approximately 0% mix. Additionally, the fuel, which is injected as a liquid (spray) has to evaporate and to mix with the fresh air and the recirculated exhaust gas. Summarising, during the mixing process, after the complete evaporation of the injected fuel and before the homogenisation of the mixture, regions can exist inside the combustion chamber, which are dominated by fuel vapour, by recirculated exhaust gas or by fresh air.

In order to realise a quantitative LIF measurement technique we replace the common gasoline fuel, which is a more or less unknown and varying blend of several compounds with additionally unknown photo-physical properties, with a single component fuel substitute, which does not fluoresce at all. As an example, isooctane is frequently used as a non-fluorescing substitute for gasoline. Then a fluorescence tracer, whose photo-physical properties are well known, is added in small amounts to the isooctane. The amount of added fluorescence tracer has to be chosen appropriately so that the LIF signal is high but that also still the laser can propagate through the fuel plume without getting attenuated too much. As the concentration of the fluorescence tracer inside the non-fluorescing isooctane is known, one can compute from the LIF signal intensity either the fuel concentration (concentration LIF) or the fuel/air-ratio (composition LIF).

According to the task described, the application of a concentration LIF tracer (such as acetone, 3-pantanone and others) would not be meaningful. In fact the concentration LIF technique could provide information about the concentration of the tracer substance in a certain region of the combustion chamber, from which one can compute the concentration of the fuel. But it would not provide any information about the chemical composition of the neighbourhood of the tracer (fuel) molecules. Consequently we would obtain from two different regions inside the combustion chamber with the same concentration of fuel but different oxygen concentrations the same LIF signal intensity. This was not optimum as oxygen is required in the neighbourhood of the fuel molecules for their combustion. Therefore the application of a composition LIF tracer, such as aromates, is much more meaningful. In this case we would obtain from two different regions inside the combustion chamber with the same concentration of fuel but different oxygen concentrations different LIF signal intensities. From the LIF signal intensity of the composition indicating tracer, we are then able to derive the fuel/air-ratio, or more precisely the fuel/oxygen-ratio, which indicates whether the mixture of fuel, fresh gas and recirculated exhaust gas is well mixed or not, with respect to the following combustion. Summarising, for the measurements in an internal combustion engine with exhaust gas recirculation the composition LIF approach is more suitable than a concentration LIF approach.

Now let us assume a second similar scenario with the essential difference that we now operate the direct injecting internal combustion engine without exhaust gas recirculation. Then the entire combustion chamber can be assumed to be filled with fresh air when the liquid isooctane (fluorescence tracer dissolved in isooctane) is injected. Everywhere inside the combustion chamber the oxygen content would be approximately 21%. The presence of the evaporated fuel would not remarkably change this, as less than 1.7 vol.% of isooctane vapour in air would be required for a stoichiometric combustion. Consequently, having injected the fuel, the probability of quenching would also be the same everywhere inside the combustion chamber. Therefore we could use both, either a concentration or a composition LIF tracer, as the detected LIF signal—at circumstances where the quencher concentration is constant—

$$S_F \propto \frac{c_F}{\underbrace{c_Q}_{\text{const}}} \propto c_F \tag{2.91}$$

would provide information about the concentration of the fuel.

2.4.1.6.3 Laser-Induced Fluorescence of 'Small' and 'Big' Molecules

What I described so far in the context of LIF is true for 'small' as well as for 'big' molecules. Now I want to underline why and that different equipment is required depending on whether 'small' or 'big' molecules should be efficiently excited to fluoresce. We will see in this section that for the efficient excitation of 'small' molecules narrowband tuneable lasers (usually in the UV) are required. On the contrary, for the efficient excitation of 'big' molecules the excitation laser does neither have to be narrowband nor tuneable.

We already saw in Fig. 2.42 in Section 2.4.1.5 on UV/VIS absorption that molecules with extended and conjugated π-electron systems, such as chromophores or dye molecules, feature broad absorption bands in the UV and even in the VIS spectral region. Molecules with smaller conjugated π-electron systems or without any conjugated π-electron system feature absorption bands near or even in the far-UV spectral region. Thus, it depends on the molecule that has to be excited to fluoresce whether an excitation laser emitting in the UV or in the VIS spectral region is suitable. Furthermore, the absorption spectrum of the molecule that has to be excited to fluoresce has to be analysed in more detail. We saw in Section 2.2 that the energetic distance between neighbouring rotational/vibrational energy levels decreases with the mass and the complexity of the molecule. Consequently, 'small' molecules feature spectrally well resolvable lines in the UV absorption spectrum, while the spectral spacing between the absorption lines of 'big' molecules is too small to resolve them spectrally. A more detailed description on the resolvability of the absorption lines in an UV absorption spectrum is provided in Section 2.4.1.5.

In Fig. 2.46 the absorption spectra of a 'small' and a 'big' molecule are compared.

The 'small' molecule is represented by the OH-radical, which is formed during the combustion processes in the flame front and thus occurs with a certain

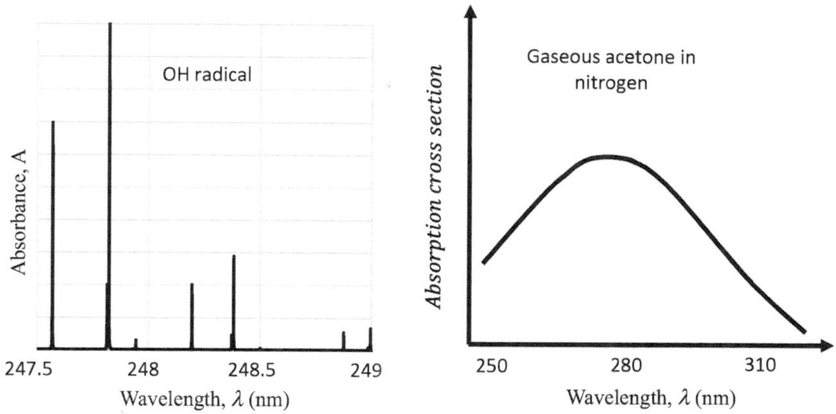

FIGURE 2.46 Absorption spectrum of the OH radical occurring as intermediate during the combustion (left) and of acetone vapour (right). *(Source: (Left) Spectrum computed by Tobias Klima from the Erlangen Graduate School in Advanced Optical Technologies of the Friedrich-Alexander-Universität Erlangen-Nürnberg, Germany.)*

lifetime in the exhaust gas region. This intermediate species is relevant for understanding the combustion chemistry [34–36], which will not be further described here. The 'big' molecule is represented by the acetone molecule, which is a common LIF tracer for mixture formation analysis in internal combustion engines. While the absorption spectrum of the OH radical is computed using the free LIFbase software (http://www.sri.com/contact/form/lifbase), the absorption spectrum of gaseous acetone in a nitrogen atmosphere was measured [37] experimentally and is here represented only qualitatively. Please take note of the very different widths of the spectral regions shown in Fig. 2.46. The absorption spectrum of the OH radical comprises less than 2 nm. The distinct absorption lines are well separated. The absorption spectrum of acetone comprises more than 50 nm and no fine structure can be resolved within the absorption band. Therefore, for the excitation of acetone one can use any UV laser, which emits at a wavelength the acetone absorbs. Of course for an efficient excitation the optimum wavelength would be at the maximum of the absorption band. This is different for the efficient excitation of a 'small' molecule, such as the OH radical. A standard UV laser, which is neither narrowband nor tuneable usually has a bandwidth exceeding approximately 0.5 nm. With respect to the absorption spectrum of the OH radical this would mean that the majority of the spectral width of the excitation laser would not overlap with the spectrally well separated and in comparison to the bandwidth of the excitation laser very narrow absorption lines. Only a minor portion of the energy of the excitation laser can be used efficiently for the excitation of the OH radicals. Consequently a narrowband UV excitation laser has to be used for the efficient excitation of 'small' molecules. Narrowband KrF excimer lasers can be as narrow as 1 pm.

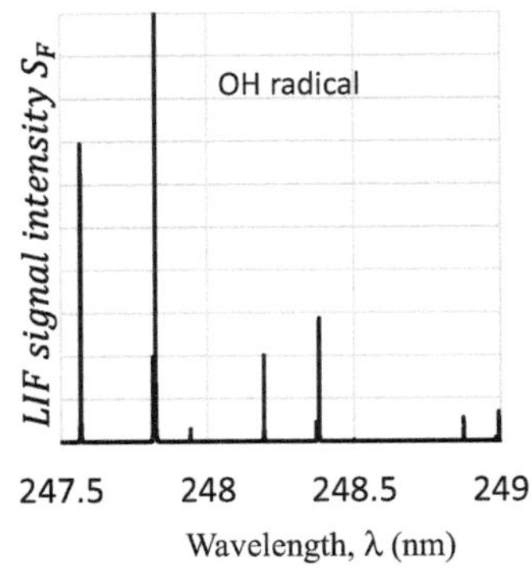

FIGURE 2.47 Excitation scan of the OH radical computed with the LIFbase software for exactly the same conditions considered for the computation of the absorption spectrum shown left in Fig. 2.46. *(Source: Spectrum computed by Tobias Klima from the Erlangen Graduate School in Advanced Optical Technologies of the Friedrich-Alexander-Universität-Erlangen-Nürnberg, Germany.)*

Furthermore, in order to assure that the narrow emission of the narrowband UV excitation laser really spectrally overlaps with one of the narrow absorption lines of the 'small' molecule also, the narrowband laser has additionally to be tuneable. Usually the tuning range is the bandwidth of the non-narrowband version of the same laser. Then the narrowband excitation laser emission can be tuned spectrally to overlap optimally with one of the absorption lines.

In general from the absorption spectrum of a compound we only can extract that the compound absorbs in certain spectral regions. An absorption is not always followed by an emission. Some species can absorb in the UV spectral region but do not emit as consequence of the absorption. Therefore with respect to LIF experiments it might sometimes be more reasonable to measure an excitation scan, which is shown in Fig. 2.47, instead of an absorption spectrum. For the acquisition of an excitation scan the fluorescence signal intensity (not the absorbance) of the fluorescing species is detected as a function of the excitation wavelength. Then a tuneable laser is required as excitation source.

2.4.2 Scattering Processes

Figure 2.48 illustrates scattering interactions between molecules and electromagnetic radiation, which are exploited when applying Rayleigh scattering, Raman Stokes spectroscopy and Raman anti-Stokes spectroscopy. In Fig. 2.48 and in the corresponding descriptions the rotational energy levels as well as rotational quantum numbers are not regarded and we consider the potential curve as well as the vibrational levels as we derived them in Section 2.2. The scattering processes summarised in Fig. 2.48 have in common that the incident photon is *not* absorbed.

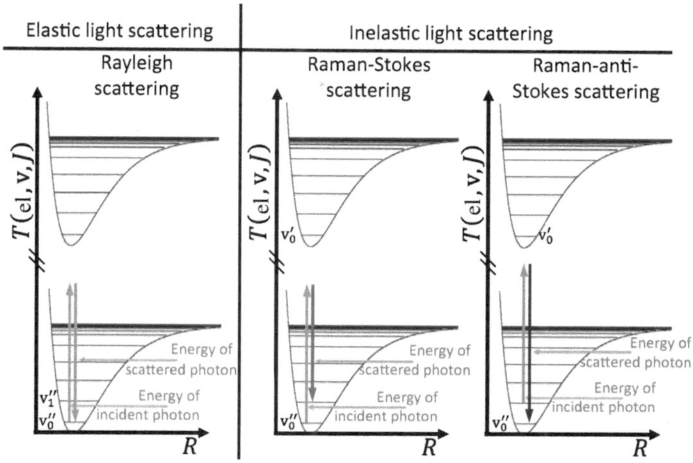

FIGURE 2.48 Illustration of the interaction mechanisms between light and molecules relevant for the Rayleigh scattering (left), the Raman Stokes scattering (centre) and the Raman anti-Stokes scattering in energy level diagrams. Rotational energy levels are not included in this figure.

This is the main difference between the *absorption/emission* processes treated in Section 2.4.1 and the spontaneous *scattering* processes treated here. As the incident photon is not absorbed, the molecule, which is involved into the scattering process at no instant (*never*), occupies an excited electronic state.

The scattering process can be regarded—if we consider the term scheme in Fig. 2.48—as a collision between the photon and the scatterer and, as collisions can be either elastic or inelastic, also the scattering process can be elastic or inelastic. Due to the collision, the photon can change its direction of propagation and is said to be scattered away from its original path. The collision of the photon and the scatterer takes place on a time scale shorter than the characteristic time scale of mutual collisions between molecules. Consequently, collisions between molecules are not relevant for the evaluation of scattering signals. Please remember that in the case of LIF, it is exactly the collisions between molecules, which make the evaluation of detected LIF signals very challenging, especially if one of the colliders also can act as fluorescence quencher. Thus—compared to LIF—the concentration of a species or the composition of a mixture can be extracted from scattering signals straightforward.

Rayleigh scattering is an *elastic* scattering process during which neither the scatterer nor the photons change their energy. Consequently, in Fig. 2.48 the photon before and after the collision (scattering process) is represented by the green arrows of identical length (energy). Furthermore, the upward directed arrow starts exactly on the energy level at which the downward directed arrow stops, which means that the energy of the scatterer before and after the collision did neither change.

Raman scattering is an *inelastic* scattering process. In this case energy is transferred between the scatterer and the photon, while the summation of the energy of the scatterer and the photon is conserved. If during the collision energy is transferred from the photon to the scatterer, the process is called Raman Stokes scattering. Then the Raman Stokes scattered radiation is red-shifted with respect to the incident radiation. If during the collision energy is transferred from the scatterer to the photon the process is called Raman anti-Stokes scattering. Then the Raman anti-Stokes scattered radiation is blue-shifted with respect to the incident radiation. From the sketch in Fig. 2.48 we can see that the energy of the photon involved into one of the Raman processes is changed exactly by the value of the energy difference between two neighbouring vibrational energy levels. We already know from Section 2.2 that the energy difference between neighbouring energy levels can be considered as fingerprint of a species. Thus, the analysis of the difference of the photon energy of the incident radiation before the scattering process and the Raman scattered radiation after the scattering process reveals the molecule that scattered the photon. A comprehensive description will follow in Section 2.4.2.3.

2.4.2.1 Rayleigh and Raman Scattering Derived Classically

What is described in the context of Fig. 2.48 for Rayleigh and Raman scattering can also be based on the theoretical computation of the interaction of an external oscillating electric field and a molecule or an atom. This theoretical approach is summarised here in short as an extract of the descriptions described by Long [38] in detail. The molecule or the atom is composed of negatively charged light electrons and positively charged heavy (with respect to the mass of the electrons) nuclei (only one nucleus for atoms), which tend to direct according to the external electric field

$$E = E_0 \cos(\omega_0 t) \tag{2.92}$$

oscillating with the angular frequency ω_0 (compare Section 2.1.1). From now on we will assume molecules only. Only the light electrons are able to follow the fast oscillation frequency of the external field. The heavy nuclei are too 'lethargic' to follow. Due to the shift of the electrons with respect to the nuclei the external electric field induces a polarisation to the molecule. In the linear regime the polarisation of the molecule

$$P = \alpha E \tag{2.93}$$

is directly proportional to the polarisability α of the molecule. Different molecules feature different polarisabilities, as their polarisability depends on the 'mobility' of the electrons. The stronger the electrons are attracted by the nuclei, the smaller is the polarisability of the molecule and vice versa. We already saw in Section 2.2 that molecules can vibrate and rotate with respect to a certain coordinate system. Considering a vibration of a diatomic molecule it becomes obvious that due to the stretching and the compression of the molecule during the vibration also the polarisability of the molecule changes. In the compressed state (minimum distance between nuclei) the average distance between the nuclei and the electrons is minimum resulting in a small polarisability, while in the stretched state (maximum distance between the nuclei) the average distance between the nuclei and the electrons is maximum, which results in a large polarisability. Therefore the polarisability more correctly

$$\alpha = \alpha_0 + \alpha'_{\text{vib}} Q_{\text{vib}} \cos(\omega_{\text{vib}} t + \delta_{\text{vib}}) \tag{2.94}$$

has to be described by a part α_0, which reflects the average polarisability for the distance R_e between the nuclei (compare Section 2.2), and a second part composed of the derived polarisability α'_{vib} of this vibration, multiplied with the harmonic vibration of the molecule $Q_{\text{vib}} \cos(\omega_{\text{vib}} t + \delta_{\text{vib}})$, where ω_{vib} is the angular frequency of the vibration, Q_{vib} can be considered to represent the amplitude of the molecule's vibration and δ_{vib} is the phase of the vibration. Eqn (2.94) thus describes the polarisability of the vibrating molecule as a function of time. Combining Eqns (2.94) and (2.92) in Eqn (2.93) results in

$$P = \alpha_0 E_0 \cos(\omega_0 t) + \alpha'_{\text{vib}} E_0 Q_{\text{vib}} \cos(\omega_0 t) \cos(\omega_{\text{vib}} t + \delta_{\text{vib}}) \tag{2.95}$$

the time dependent polarisation of the molecule. With the 'trick'

$$\begin{aligned} &\cos(\omega_0 t)\cos(\omega_{\text{vib}} t + \delta_{\text{vib}}) \\ &= \frac{1}{2}\left\{\cos((\omega_0 t) + (\omega_{\text{vib}} t + \delta_{\text{vib}})) + \cos((\omega_0 t) - (\omega_{\text{vib}} t + \delta_{\text{vib}}))\right\} \end{aligned} \tag{2.96}$$

Eqn (2.95) can be formulated as

$$\begin{aligned} P = {} & \alpha_0 E_0 \cos(\boldsymbol{\omega}_0 t) + \frac{1}{2} \alpha'_{\text{vib}} E_0 Q_{\text{vib}} \cdot \left[\cos(t(\boldsymbol{\omega}_0 + \boldsymbol{\omega}_{\textbf{vib}}) + \delta_{\text{vib}})\right] \\ & + \frac{1}{2} \alpha'_{\text{vib}} E_0 Q_{\text{vib}} \cdot \left[\cos(t(\boldsymbol{\omega}_0 - \boldsymbol{\omega}_{\textbf{vib}}) + \delta_{\text{vib}})\right]. \end{aligned} \tag{2.97}$$

Eqns (2.92) to (2.97) and their combinations are represented graphically in Fig. 2.49 for two cases of $\alpha'_{\text{vib}} = 0$, which would represent an atom instead of a molecule or a molecule whose polarisability does not change during the vibration, and $\alpha'_{\text{vib}} > 0$, which would represent a molecule whose polarisability changes during the vibration. Now we have to consider that charged particles produce electromagnetic radiation by themselves whenever they are accelerated. With respect to Eqn (2.97) the electrons are—according to the time-dependent polarisation of the molecule—accelerated with three different frequencies (the electron cloud oscillates with three different frequencies), which are ω_0, $(\omega_0 + \omega_{\text{vib}})$ and $(\omega_0 - \omega_{\text{vib}})$.

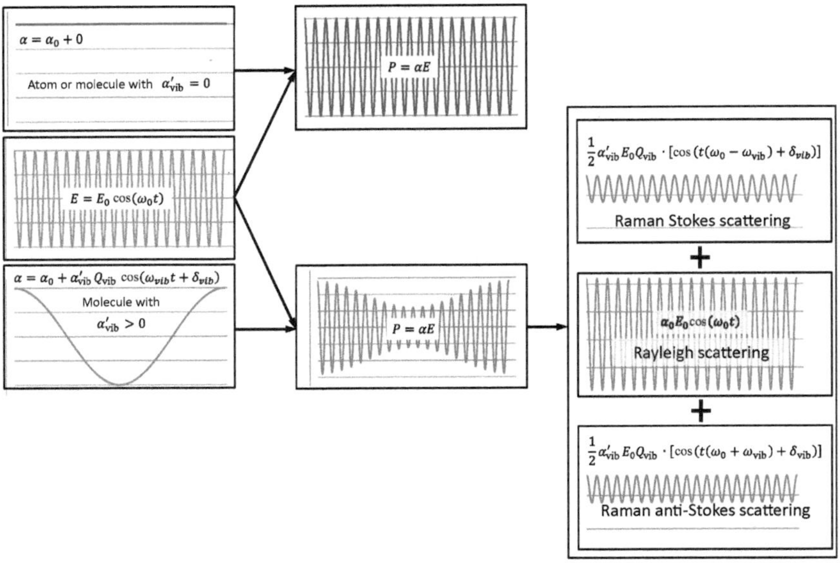

FIGURE 2.49 Graphical representation of the Eqns (2.92) to (2.97). *(Source: Graphs computed by Jaypee Quino from the Erlangen Graduate School in Advanced Optical Technologies of the Friedrich-Alexander-Universität Erlangen-Nürnberg, Germany.)*

In Fig. 2.48 we saw that for Rayleigh scattering the energy of the photon is identical before and after the scattering process. In Section 2.1.4 we saw that the energy of one photon correlates with the frequency of the electric wave representing this photon. Thus it is clear that the first part of Eqn (2.97) has to represent Rayleigh scattering. The second part shows an oscillation at a frequency $(\omega_0 + \omega_{\text{vib}})$ and thus has to represent the Raman anti-Stokes process, where the photon after the scattering process features more energy than before. The third part of Eqn (2.97) describes a polarisation oscillating at the frequency $(\omega_0 - \omega_{\text{vib}})$ and thus has to represent the Raman Stokes process, where the photon after the scattering process features less energy than before.

2.4.2.2 Rayleigh Scattering (Elastic Light Scattering From Molecules)

Rayleigh scattering has already been mentioned in the context of Section 2.3.4, in which the elastic scattering mechanism between drops and light is described. We can imagine that during the scattering process a photon collides with a scatterer (atom or molecule) and as a consequence is scattered away from its original path of propagation. 'Elastic' in the term 'elastic light scattering' stands for the circumstance that no energy is transferred between the scatterer and the scattered light. Consequently the scattered photon features the 'same' energy and its corresponding wave features the 'same' frequency or wavelength before and after the scattering process, but its propagation direction might have changed. This is true for light scattered elastically from drops and for Rayleigh scattering from molecules treated here. 'Same' is put in between quotation marks, as we will see that there might be a small energy shift due to the Doppler effect when a photon collides with a molecule or an atom [39,40]. We saw in the previous Section 2.4.2.1 that atoms as well as molecules scatter light due to the Rayleigh mechanism, but also that atoms cannot be Raman active.

Here mainly the Rayleigh scattering mechanism between visible light and molecules will be considered. In this case the size parameter $x = \pi d / \lambda$, which we defined in Eqn (2.76) as the ratio of the diameter d of the scatterer and the wavelength λ of the incident light, is significantly smaller than 1. Furthermore, as we assume that light is scattered elastically from molecules we can use term schemes of molecules—as sketched in Fig. 2.48 (left) —in order to illustrate the Rayleigh mechanism. Also according to this sketch the energy of the scattered photon before and after the scattering process remains the same. Therefore the scattered photon does not carry any information about the molecule it had been colliding with. Consequently, if we excite a mixture of different gas molecules with a laser, the Rayleigh signals scattered from the different compounds will all appear exactly at the same wavelength, which is the wavelength of the laser radiation. This already indicates that Rayleigh scattering is not the appropriate technique for the determination of the composition of a mixture, as the Rayleigh signals of the different mixture compounds cannot be separated spectrally. Furthermore, another drawback relevant for practical applications of Rayleigh scattering is the behaviour of the elastic scattering intensity as a function of the size parameter x, which is shown in Fig. 2.29. Be aware that the abscissa and the ordinate are both scaled logarithmic. This means that if there are some gas molecules—they are characterised by a small size parameter—contained in a cuvette together with very few aerosol particles (or dust)—they are characterised by a significantly larger size parameter –, the scattering intensity of the particles—even if they are with respect to their number significantly less than the molecules—by orders of magnitude overwhelms the Rayleigh scattering intensity of the molecules. This indicates that Rayleigh scattering from molecules can be applied only in very clean environments, as otherwise the elastically scattered signals are not dominated by the Rayleigh process but by elastic light scattering from drops or particles. With respect to high pressure applications for which the excitation laser light always has to pass through glass windows or through a glass fibre, also the light scattered from the surfaces of the high pressure vessel and windows is so much more intense than Rayleigh scattering from molecules, so that Rayleigh scattering for another reason under many circumstances is not a suitable measurement technique for high pressure applications. Because of these circumstances I here only describe briefly the Rayleigh process. A detail derivation of the Rayleigh working equation is derived by Long [38]. In the following emphasise is laid on the understanding which parameters influence

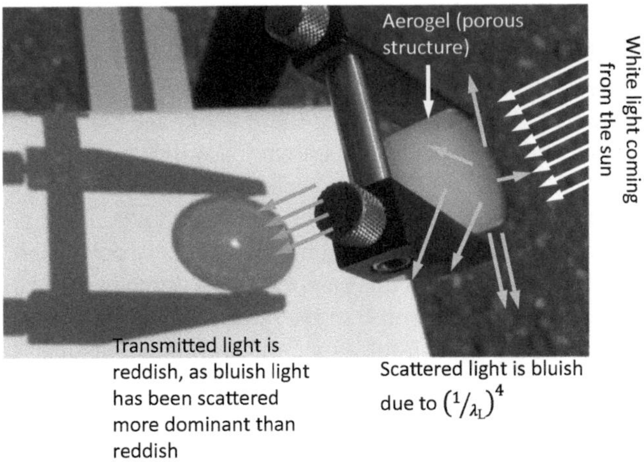

FIGURE 2.50 Photo of a silica-based aerogel exposed to sunlight. The aerogel is a nano-porous usually colourless solid structure, which is more or less transparent but turbid. Due to Rayleigh scattering it looks bluish while its shadow looks reddish. *(Source: Photo made by Jaypee Quino from the Erlangen graduate School in Advanced Optical Technologies of the Friedrich-Alexander-Universität Erlangen-Nürnberg, Germany.)*

the Rayleigh signal intensity. The point irradiance I (Rayleigh signal intensity) from molecules N undergoing a Rayleigh transition

$$I \propto \cdot E_{\text{Laser}} \cdot L_{ij} \cdot N \cdot \left(\frac{1}{\lambda_L}\right)^4 \cdot \alpha_0^2, \tag{2.98}$$

in $W(sr)^{-1}$ is proportional to the laser beam irradiance E_{Laser}, the local field L_{ij}, which is defined later in Eqn (2.108), the number of molecules scattering N, the inverse wavelength of the excitation light $(1/\lambda_L)^4$ to the fourth power and the squared polarisability α_0^2. Usually the polarisability of molecules scales with the molecular size. Thus, big molecules typically feature a larger polarisability than small molecules. The wavelength λ_L of the excitation laser radiation shows the most dominant influence onto the Rayleigh signal intensity I. The shorter the wavelength is, the higher is the scattering intensity. I want to illustrate this showing a photo in Fig. 2.50. Herein a silica-based aerogel, which is a nano-porous structure, is exposed to the sunlight. The aerogel usually is a transparent but turbid and colourless material. In Fig. 2.50 the aerogel looks bluish and its shadow looks reddish. Within the aerogel structure there are many small particles for which the size parameter can also be assumed to be smaller than 1.

As the scattering intensity scales with $(1/\lambda_L)^4$ the light of short wavelengths, which is the bluish part of the spectrum of the sun, is scattered significantly more intense than the long wavelength light, which is the reddish part of the spectrum of the sun. As the photo of the aerogel is taken from the side while it is illuminated with sunlight from a different direction, mainly the scattered portion of the sunlight is detected. Consequently the aerogel looks bluish if we look at it from the side. But the shadow of the aerogel looks reddish. This is due to the fact that due to scattering the bluish part of the spectrum is attenuated more intense than the reddish one. Consequently the transmitted light is poor in the bluish part of the spectrum and appears therefore reddish. What has been demonstrated with the aerogel can be experienced at any sunny day. When we look into the sky, but not directly into the sun, the sky appears bluish, as we then can see only the light, which is scattered from the molecules of the atmosphere into our eyes. At sunrise or sunset we can look directly into the sun. This is because of the long way the light has to pass through the atmosphere so that it is significantly attenuated due to scattering mechanisms. Again, the bluish part of the spectrum is scattered more intense than the reddish part. But during sunset or sunrise we observe the transmitted spectrum, which consequently has to appear reddish.

2.4.2.3 Raman Scattering (Inelastic Light Scattering From Molecules)

This chapter treats the fundamentals of the Raman scattering process, making it a species specific and—at certain circumstances—a quantitative measurement technique. It is also shown which vibrations of molecules or functional groups and which rotations of molecules are Raman active.

The influence of the index of refraction on the Raman signal intensity is discussed, as this is the parameter, excluding the possibility of quantifying *concentrations* in terms of 'number of molecules per volume' from Raman signals at changing pressures. This—as it will be shown later—is possible only at a constant index of refraction. Nevertheless, the quantification of *compositions* in terms of 'number of species "i" molecules per total number of molecules' is usually not excluded by an alteration of the index of refraction and is the proper and common way in order to quantify the composition of mixtures at elevated pressure.

The influence of the index of refraction on the geometrical optics of the installed Raman set-up and its consequences on the detection efficiency of the Raman experiment are also discussed briefly and have already been addressed in a more general way in the context of Fig. 1.8 in Section 1.5.4.1.

Finally a physical model is derived, which makes possible the determination of the composition of mixtures from Raman spectra. The advantage of the physical model, compared to empirical models is that only a very limited number of calibration measurements are required for the quantification of mixtures.

2.4.2.3.1 Raman Active or Not?

Not all molecular rotations and/or vibrations are Raman active. This is similar to the fact that also not all molecular vibrations or rotations are IR-active, what we have seen in Section 2.4.1.3. Fortunately, many molecular vibrations and/or rotations, which are not IR-active, are Raman active and vice versa. This is why Raman spectroscopy and IR absorption spectroscopy are said to be complementary measurement techniques. In Section 2.4.1.3 we saw that an IR-active molecule has to feature either a permanent or an induced dipole momentum. Only then the external excitation field can couple to the molecular vibration and/or rotation and can potentially be absorbed.

This is different in the case of Raman scattering. Only molecules, whose polarisability changes during the vibration or the rotation, are Raman active. We saw this already during the classical derivation of the Rayleigh and the Raman mechanism in Section 2.4.2.1. The polarisability describes the ability of a molecule to get polarised in an external electric field. The external field is the excitation field of the excitation laser beam and oscillates with a high frequency (see Section 2.1). The electrons (negatively charged) or the nuclei (positively charged) of the molecule realise the external electric field and as a consequence the negative electrons and the positive nuclei are shifted spatially with respect to each other. In fact, the heavy nuclei are too lethargic to be able to follow the excitation field with such a high oscillation frequency. Only the light and mobile electrons are deflected 'back and forth' or 'up and down' with the oscillation frequency of the external field and are therefore responsible for the oscillating displacement of the charge centres of the positively charged nuclei and the negatively charged electrons. This oscillating displacement is induced by the excitation radiation and has already been described in the context of Rayleigh scattering in Section 2.4.2.2. The easier the electrons can follow the external electric field, the less they are attracted by the nuclei. During the vibration of the hydrogen molecule, which is shown top left in Fig. 2.51 also the extension of the electron cloud changes. For the compressed hydrogen molecule the mean distance between the electrons and the nuclei is smaller than for the stretched hydrogen molecule. Consequently, the polarisability of the compressed hydrogen molecule is small compared to the stretched one and we so can say that the polarisability of the hydrogen vibration changes during its vibration. Thus, the vibration of the non-polar molecule 'hydrogen' is Raman active. Another example of a Raman active molecular vibration is the symmetric stretch vibration of carbon dioxide, which is also shown in Fig. 2.51. In contrast, the anti-symmetric stretch vibration, as well as the bend vibrations (in plane and out of plane), are not Raman active, as the polarisability of the molecule does not change during these molecular vibrations. Examples of Raman active and non-Raman active rotations are given in the bottom line of Fig. 2.51. The rotation of hydrogen is Raman active, as—assume the excitation electric field propagates from left to right and oscillates up and down—the polarisability perpendicular to its molecular axis is different than the polarisability in its molecular axis. The rotation of methane is not Raman active, as during rotation the polarisability remains constant.

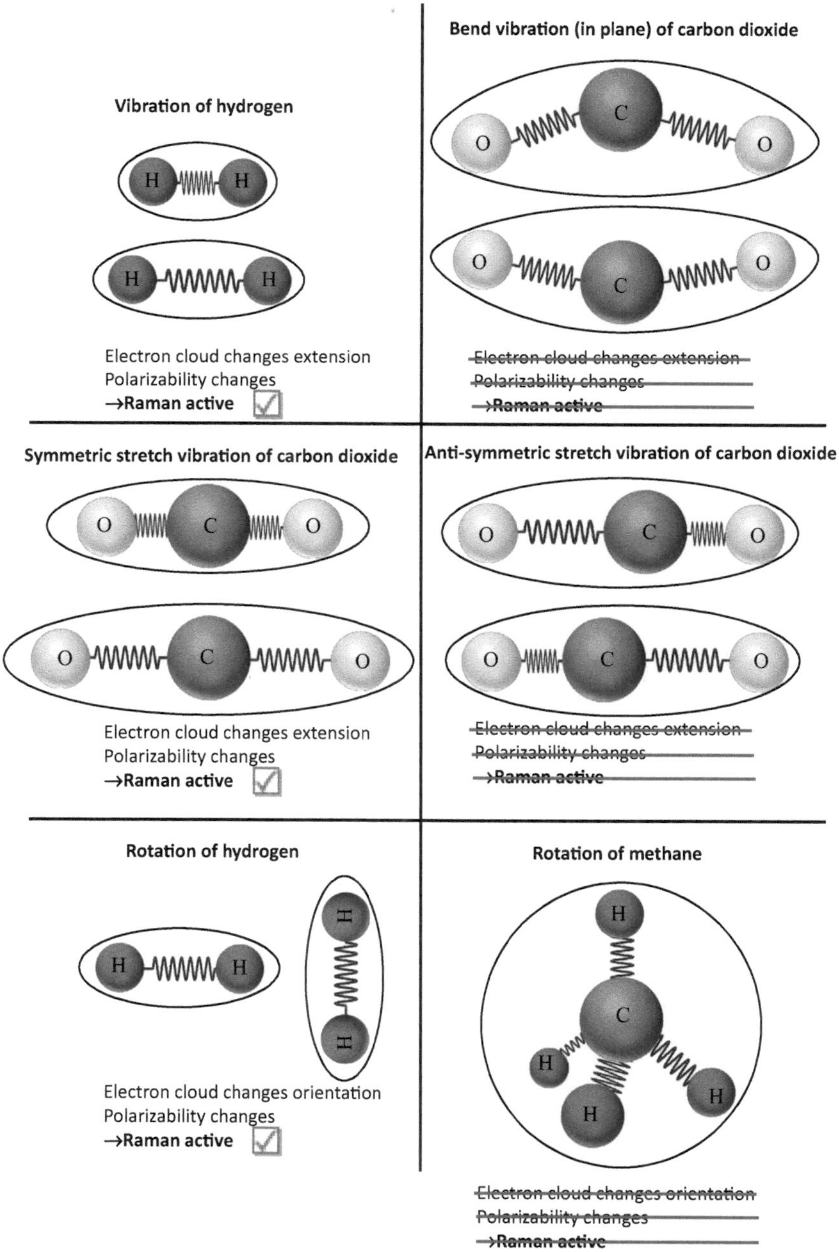

FIGURE 2.51 Sketch of some molecular vibrations and rotations, which are and are not Raman active. Black envelope shows qualitative representation of the polarisability ellipsoid ('form' of the electron cloud).

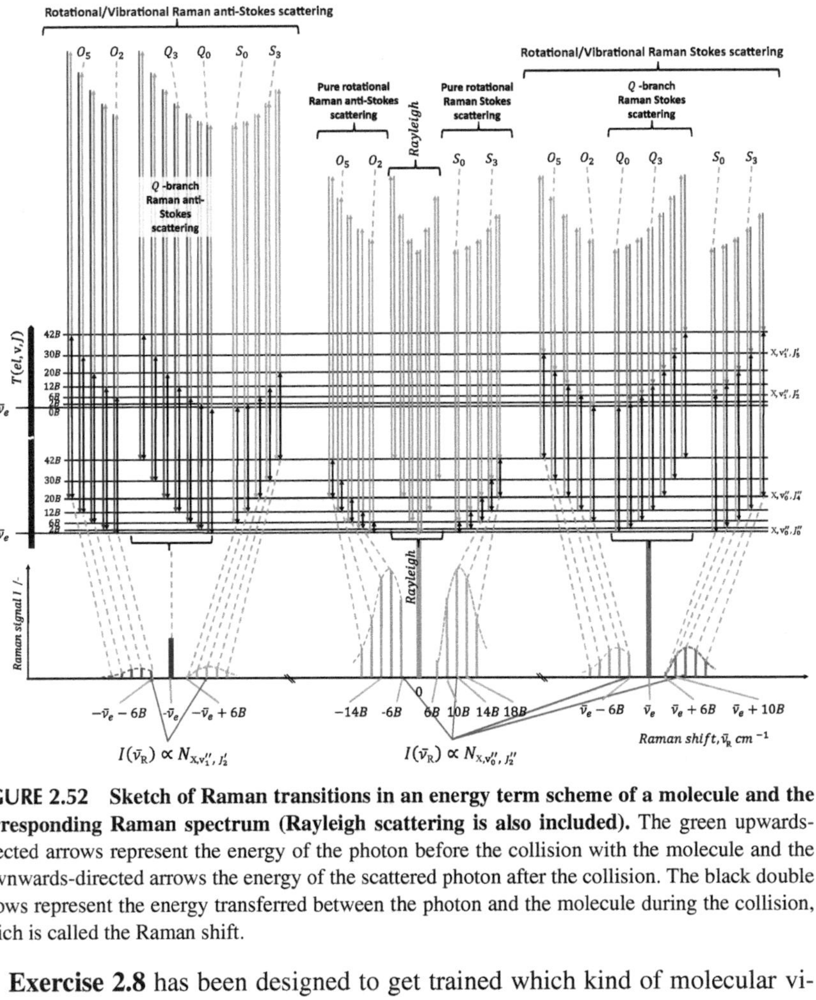

FIGURE 2.52 Sketch of Raman transitions in an energy term scheme of a molecule and the corresponding Raman spectrum (Rayleigh scattering is also included). The green upwards-directed arrows represent the energy of the photon before the collision with the molecule and the downwards-directed arrows the energy of the scattered photon after the collision. The black double arrows represent the energy transferred between the photon and the molecule during the collision, which is called the Raman shift.

Exercise 2.8 has been designed to get trained which kind of molecular vibrations and rotations and which kind of vibrations of functional groups are IR-active and/or Raman active.

2.4.2.3.2 Raman Transitions and the Species Specificity of Raman Signals

Raman scattering—according to Fig. 2.48—is an inelastic interaction mechanism (collision) between a photon and a molecule[4] (scatterer). Figure 2.52 in more detail shows Raman transitions in a term scheme of a molecule (rigid rotator and harmonic oscillator assumed), which can potentially take place when a molecule and a photon collide inelastically. The horizontal black lines indicate different rotational energy levels in the vibrational ground state v_0'' and the first

4. The inelastic light scattering process between incident radiation and lattice vibrations (phonon) of solids is also called Raman scattering.

excited vibrational state v_1'' in the electronic ground state X. Example energy levels are specified with respect to their quantum numbers el, v and J in detail on their right border and with respect to their overall energy $T(\text{el,v},J)$ on their left border. The green upwards directed arrows represent the energy of the photon before the scattering process. The green arrows start at different energy levels of the molecule, depending on the energy level the respective molecule occupies before the scattering process, respectively the collision with the photon.

The collision between the molecule and the photon can be elastic according to the Rayleigh mechanism (compare Section 2.4.2.2), meaning that the energy of both, the photon and the molecule remain unaffected by the scattering process. Thus, the downwards-directed arrow, which represents the energy of the photon after the scattering process must be exactly as long as the upwards-directed arrow representing the photon before the scattering process. Only then the energy of the molecule before and after the scattering process can remain the same. The Rayleigh transitions are sketched for molecules occupying the rotational energy levels from $J = 0$ to $J = 6$ in the vibrational ground state v_0''. As in the case of Rayleigh scattering the energy difference of the photon before and after the scattering process is zero, the Rayleigh signal appears in the Raman spectrum at a Raman shift of $\bar{\nu}_R = 0\,\text{cm}^{-1}$.

Directly left and right of the Rayleigh peak the pure rotational Raman signals appear. They are called pure rotational Raman signals, as the molecule involved into the inelastic collision changes solely its rotational quantum number J. The vibrational quantum number v, before and after the Raman transition, is identical. If energy is transferred from the photon to the molecule's rotation—then the scattered photon is said to be red-shifted—the collision is said to be a Raman Stokes transition. If energy is transferred from the molecule's rotation to the photon—then the scattered photon is said to be blue-shifted—the collision is said to be a Raman anti-Stokes transition. All transitions changing the rotational quantum number J follow the selection rule $\Delta J = \pm 2$. For the computation of ΔJ we subtract the respective quantum number before the transition from the quantum number after the transition. For the S_3-transition this would be, for example, $\Delta J = 5 - 3 = +2$. For S-branch transition $\Delta J = +2$ and for O-branch transitions $\Delta J = -2$. It should be recalled that the selection rules for IR absorption spectroscopy are $\Delta J = +1$ for R-transitions and $\Delta J = -1$ for P-transitions (compare Fig. 2.39), instead. Due to the different selection rules of 'Raman' and 'IR', also the distance between the Raman rotational lines ($4B$) is larger than the distance between the rotational lines ($2B$) in an IR absorption spectrum (compare Fig. 2.39).

Further left and right of the Rayleigh peak we can see in the Raman spectrum the assembles of O-, Q- and S- branch rotational/vibrational Raman transitions with energy transfer from the photon to the molecule (Stokes side) and with energy transfer from the molecule to the photon (anti-Stokes side). The Q-branch transitions are also called 'pure vibrational Raman transitions', as the molecules involved into this scattering process only change their vibrational quantum number $\Delta v = \pm 1$ while the rotational quantum number remains $\Delta J = 0$ unchanged.

Again, for the computation of Δv we subtract the respective quantum number before the transition from the quantum number after the transition. Therefore the Stokes sided and the anti-Stokes sided Q-branches correspond to $\Delta v = +1$ and $\Delta v = -1$, respectively. In the case of a rigid rotator and harmonic oscillator the energy exchanged between the molecule and the photon during a Q-branch transition is not a function of the rotational level *J* the molecule occupies before the scattering process, but is simply the vibrational constant of the molecule $\bar{\nu}_e$. Therefore all Stokes sided Q-branch transitions appear in the Raman spectrum at the same spectral position as one dominant peak. This is also true for the anti-Stokes sided Q-branch transitions. Of course, the anti-Stokes rotational/vibrational transitions can take place only, if molecules before the scattering process already occupy the first excited vibrational energy level v_1''. In Section 2.2.7 we saw that at moderate temperatures nearly 100% of all molecules occupy the vibrational ground state v_0'' and that high temperatures are required in order to obtain a 'realisable' population of the first excited vibrational state and that the population in the first excited vibrational state—even at infinite temperature—will never exceed the population of the vibrational ground state. Therefore the intensity of the anti-Stokes rotational/vibrational Raman signals is always smaller and at moderate temperatures even significantly smaller than the intensity of the Stokes rotational/vibrational Raman signals. **Exercise 2.11** in this context shows how the temperature-dependent population of energy levels can be utilised for temperature measurements.

Left and right of the Q-branch transitions in Fig. 2.52 one can again find the O-branches ($\Delta J = -2$) and the S-branches ($\Delta J = +2$) of the rotational/vibrational Raman transitions ($\Delta v = \pm 1$) on the Stokes ($\Delta v = +1$) and the anti-Stokes side ($\Delta v = -1$).

Having described in detail in the context of Fig. 2.52 the origin of the structure of a complete Raman spectrum covering the anti-Stokes and the Stokes side, it will now be described why the Raman spectrum is said to be species specific. On this purpose we will only consider Q-branch Stokes Raman transitions, though the species specificity could also be shown for rotational transitions. As we saw in Fig. 2.52 that all Q-branch transitions appear at the same spectral position $\bar{\nu}_e$, irrespectively of the rotational energy level *J* the scatterer occupies before the scattering process, we can neglect the rotational energy levels, whenever we only regard Q-branch transitions of rigid rotator and harmonic oscillator molecules. On this account Fig. 2.53 shows the Raman scattering process in an energy level diagram for two different diatomic symmetric molecules, nitrogen and hydrogen.

Please recall that diatomic symmetric molecules, such as hydrogen or nitrogen, are not infrared-active and therefore cannot be detected via (N)IR absorption spectroscopy. But, as shown in Section 2.4.2.3.1, they are Raman active.

In Fig. 2.53 again the rotational energy levels are skipped, but a 'virtual energy level' is introduced where the upwards-directed arrow ends and the downwards-directed arrow starts. I integrated this 'virtual energy level' here

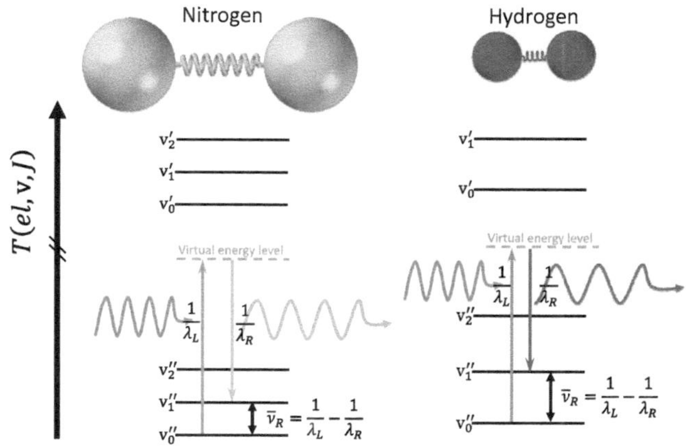

FIGURE 2.53 Sketch of pure vibrational Raman transitions in an energy level diagram (term scheme) of nitrogen and hydrogen. The position of the energy levels is qualitatively sketched.

just because in literature this 'virtual energy level' is usually employed. For me the introduction of this virtual energy level in the context of spontaneous Raman scattering does not make any sense and misleads students to imagine that the photon might be absorbed by the molecule. Therefore I did not integrate a 'virtual energy level' in Fig. 2.48 and will not consider the 'virtual energy level' for the description of the Raman process in the following. Due to energy conservation, the energy difference of the Raman scattered photon before and after the scattering process is equal to that of the scatterer before and after the inelastic scattering process. The energy of the rotational/vibrational levels, which can be occupied by a specific scatterer, is a function of the scatterer's molecular properties, such as the mass of the nuclei, the distance between the nuclei and the binding force between the nuclei [38]. This has been described comprehensively in Section 2.2 and makes the energy of rotational/vibrational levels and with this the energy difference between them species specific. The species-specific energy difference between the rotational/vibrational levels of a scatterer is also called Raman shift $\bar{\nu}_R$ and makes a Raman spectrum a 'fingerprint' of a molecule. Different approaches for the exact computation of the energy of levels occupied by a molecule are given by Herzberg [41] and Atkins [42]. In spectroscopy the Raman shift $\bar{\nu}_R$ is usually given in wavenumbers and is computed from the 'energy difference' (energy in wavenumbers) of the radiation before and after the scattering process

$$\bar{\nu}_R = \frac{1}{\lambda_L} - \frac{1}{\lambda_R}, \tag{2.99}$$

where λ_L and λ_R are the wavelengths of the incident laser light and the Raman scattered signal, respectively. Raman shifts of example molecules are listed in Table 2.3.

TABLE 2.3 Pure Vibrational Raman Shifts of Simple Molecules (Q-Branch Transitions)

Species	Raman shift $\bar{\nu}_R$, (cm^{-1})	Wavelength of the Raman Stokes scattered radiation (nm, wavelength of incident radiation is 532 nm)
Hydrogen	4155	683.03
Nitrogen	2331	607.34
Oxygen	1556	580.03

Figure 2.54 shows illustrative Raman spectra of different fluids and proofs that a Raman spectrum of a substance can be considered as a fingerprint of the molecules the substance is composed of. In more detail the so-called 'fingerprint region' of a Raman spectrum is the wavenumber region smaller than 2000 cm^{-1}. At higher wavenumbers the Raman spectrum of organic compounds is dominated by the strong CH vibration (approximately 2900 cm^{-1}) whose

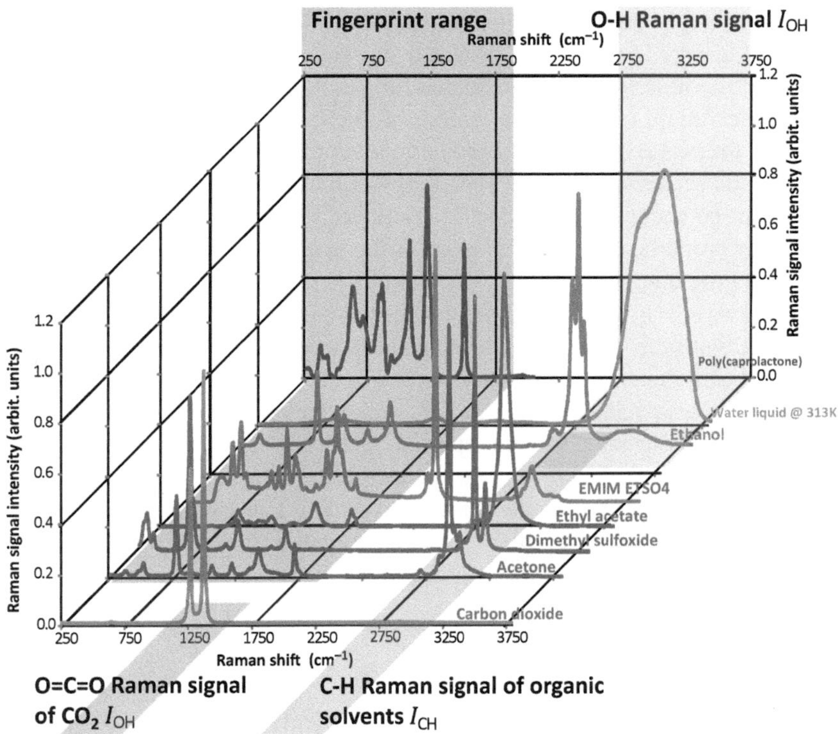

FIGURE 2.54 Raman spectra of various compounds, which can be found in the further progress of this book.

signature cannot be easily exploited in order to extract the constituents in a mixture of different hydrocarbons. The fingerprint region—on the contrary—comprises the various intra-molecular interactions, from which the constituents of a mixture can be extracted (qualitative Raman spectroscopy).

The Raman spectra in Fig. 2.54 are visually grouped in four classes:

The first class contains CO_2 only, which—with respect to high-pressure process technology—might take over the role of a solvent, anti-solvent, blowing agent or others. It is known that the central wavenumber position and the shape of the Fermi dyad is a function of pressure, temperature and mixture composition [43–49]. If spectrometers with poor spectral resolution are applied, the influence of pressure, temperature and mixture composition on the dyad's shape and peak positions is not always realisable. Figure 2.55 shows Raman spectra of the Fermi dyad of gaseous, liquid and supercritical CO_2. The exact pressure and temperature is mentioned in the legend of Fig. 2.55. The spectra were acquired with a spectrometer of 303 mm focal length equipped with a detector featuring 1600 pixels along the spectral axis, each pixels having an edge length of 16 μm. The slit width was set to 100 μm.

The second class comprises the example organic solvents acetone, dimethyl sulfoxide, ethyl acetate and EMIM $EtSO_4$ (ionic liquid). The pure substances are all liquid at ambient conditions. Class two solvents are separated from class three solvents, as they do not feature an OH vibration Raman signal I_{OH}.

The third class comprises the example solvents ethanol and water. They are separated from class two solvents, as the hydroxyl functional group—in contrast to the functional groups of the class two solvents—is known to potentially develop hydrogen bonds between the neighbouring molecules. Hydrogen bonds usually only develop in a dense phase, in which the mean distance between the hydroxyl groups of different molecules is not too far. In the gaseous phase of water or ethanol the mean intermolecular distance is

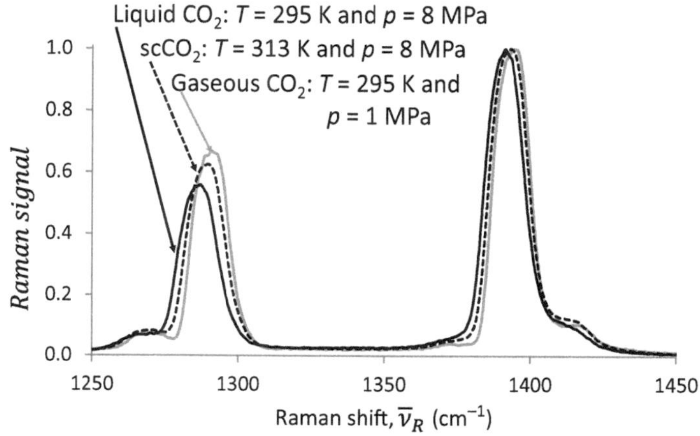

FIGURE 2.55 **Raman spectra of gaseous, liquid and supercritical CO_2.**

too far. Thus, usually hydrogen bonds are not formed in the gaseous phase. Summarising, hydrogen bonds develop in solids or liquids. They do not develop in gases. In supercritical fluids it depends on the density, whether hydrogen bonds are formed or not.

Figure 2.56 shows Raman spectra of the water stretch vibration in the liquid and in the gaseous phase. Both spectra provided in Fig. 2.56 show the water stretch vibration. Apparently the water stretch vibration of liquid water is a *band* and broad compared to the narrow *peak* of gaseous water. In liquid water hydrogen bonds strongly interact between the water molecules. As a consequence, the vibrational energy of each of the single water molecules is no more defined by the mass of the oxygen and the hydrogen nuclei and the binding force between them alone, but also by the development of the hydrogen bonds with neighbouring molecules. The hydrogen bonded water molecules mutually pull and push each other's vibration and with this cause a broad distribution of vibrational energies, meaning that different water molecules can vibrate with slightly different energies.

This distribution is reflected in the broad band of the stretch vibration Raman signal of liquid water. In contrast, the same stretch vibration of gaseous water is a narrow peak. In the gaseous phase the distance between the water molecules is as far as the water molecules cannot undergo hydrogen bonds. Consequently, the stretch vibration of each water molecule is *not* influenced by hydrogen bonds and thus all the water molecules in the gaseous phase feature the more or less same stretch vibration, which is mainly only defined by the masses of the nuclei and the binding force between them. At this point I want to underline that other functional groups also can potentially form hydrogen bonds. Ammonia would be another compound, which can form hydrogen bonds between the hydrogen and the nitrogen. If hydrogen bonds are pronounced, one can realise this from the significant broadening of a peak to a band.

The development of hydrogen bonds depends on the temperature, the density and on the composition of a fluid. The higher the temperature of the fluid, the higher is its thermal energy, or the more pronounced is the motion of the molecules within the fluid, which counteracts the development of hydrogen bonds. With increasing density the mean distance between the molecules decreases, which usually promotes the development of hydrogen bonds. Of course, the development

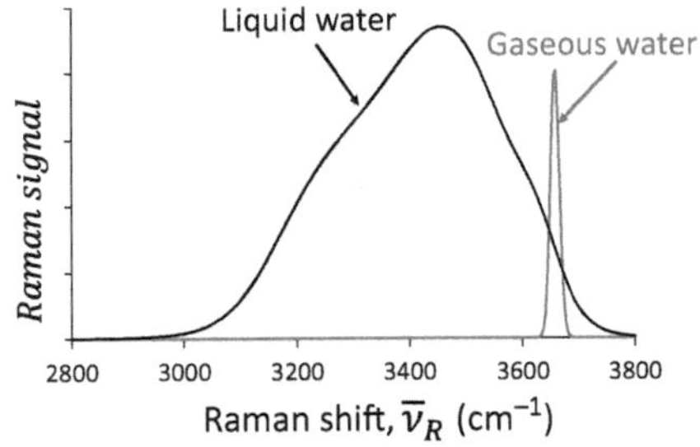

FIGURE 2.56 Raman spectra of the water stretch vibration of liquid water and of water vapour.

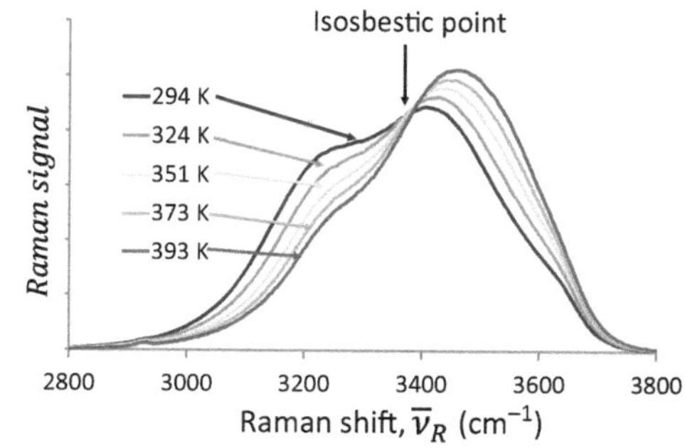

FIGURE 2.57 **Raman spectra of liquid water at 0.5 MPa for temperatures between 294 K and 393 K.**

of hydrogen bonds can be influenced if the composition of a fluid is varied, for example, if water is diluted with any compound that weakens the hydrogen bonds. Any change in the development of hydrogen bonds can be recognised in a change of the shape of the band of the water stretch vibration. Therefore from its shape, one first can extract information about the characteristics of the hydrogen bonding, from which in a second step one can deduce the temperature [50–56], the density [50,51,57,58], the cooperativity of the hydrogen-bonded molecules [59], or the composition of their environment [60]. As an example with respect to solely the temperature influence Fig. 2.57 shows Raman spectra of the stretch vibration of liquid water at 0.5 MPa and for five different temperatures. The isosbestic point is the joint point of intersection of all spectra, irrespectively of temperature. Therefore the Raman signal intensity at the isosbestic point does not change with temperature. With increasing temperature the Raman signal intensity left and right of the isosbestic point decreases and increases, respectively, while the overall Raman signal intensity (area under the spectra) remains constant.

The fourth class is represented by poly-ε-caprolactone, an amorphous biodegradable polyester with a melting temperature of 333 K and a glass transition temperature of 213 K. Thus, it is in a rubber-like state at ambient conditions, which is the main difference compared to the fluids contained in the other groups. With respect to the Raman spectrum of poly-ε-caprolactone a spectrometer was used which made accessible only the Raman spectrum in the fingerprint range.

2.4.2.3.3 Quantitative Nature of Raman Signals

In this section emphasis is laid on the application related Raman theory starting with the equations which directly provide information about the quantities of interest for its application in high-pressure process technology. Furthermore, a detailed description is provided of the parameters having influence on the magnitude of the Raman signals, which can best be achieved by using an optimal design of the applied Raman system. For interested readers a derivation of the equation describing the Raman scattering process can be found in different references [38,61,62].

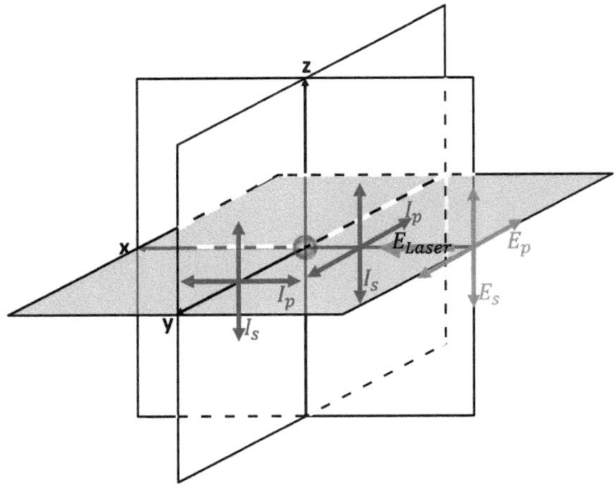

FIGURE 2.58 Sketch of the scattering geometry of a Raman experiment.

Figure 2.58 shows a sketch of the scattering geometry of the Raman experiments described further. The scatterer in the centre of the coordinate system *xyz* is excited with a laser beam of irradiance

$$E_{\text{Laser}} = \frac{c \cdot \varepsilon_0 \cdot E_0^2}{2}, \tag{2.100}$$

where c is the speed of light in vacuum in m/s, ε_0 is the permittivity in A s/V m and E_0 is the amplitude of the electric field in V/m. Thus, the irradiance E_{Laser} is given in W/m^2. The scattering plane is defined by the direction of the excitation with the irradiance E_{Laser} and the direction of observation of the point irradiance I scattered from the scatterer. In Fig. 2.58, the *xy*-plane is defined as scattering plane. The electric fields of the excitation irradiance E_{Laser} and the point irradiance I scattered from the scatterer are composed of a parallel ‘p’ and a perpendicular ‘s’ portion. This is expressed by the indices of the quantities $E_{\text{"index"}}$ and $I_{\text{"index"}}$, where the scattering plane acts as a reference. Consequently, if the electric field oscillates in the *xy*-plane, its polarisation is called parallel ‘p’. If it oscillates in the *zy*-plane or the *zx*-plane, it is called perpendicular ‘s’.

According to Long [38], the point irradiance I (Raman signal intensity) from molecules N_{ij} undergoing a Raman Stokes transition from the energy level i before the scattering process to j after the scattering process

$$I_{\frac{P}{S}} = \frac{\frac{P}{S}\mathrm{d}\Phi_{\frac{P}{S}}}{\mathrm{d}\Omega} = \frac{h}{8 \cdot \varepsilon_0^2 \cdot c} \cdot E_{\text{Laser}} \cdot L_{ij} \cdot N_{ij} \cdot (\mathrm{v}_i + 1) \cdot \frac{\left(\frac{1}{\lambda_{\text{L}}} - \bar{\nu}_{\text{R},ij}\right)^4}{\bar{\nu}_{\text{R},ij}} \cdot (\alpha'_{xyz,ij})^2, \tag{2.101}$$

in W(sr)$^{-1}$ is the radiant power Φ in W radiated by a point source (molecule) into a conical beam of a given direction in space and of solid angle dΩ in sr (steradian). For a Raman anti-Stokes transition, $(\mathrm{v_i}+1)$ and $((1/\lambda_\mathrm{L})-\bar{\nu}_{\mathrm{R},ij})^4$ in Eqn (2.101) have to be replaced by $(\mathrm{v_i})$ and $((1/\lambda_\mathrm{L})+\bar{\nu}_{\mathrm{R},ij})$. I is a function of Planck's constant h in J s, the permittivity ε_0, the speed of light in vacuum c, the excitation irradiance E_Laser, the local field L_{ij}, the number of scatterers (molecules) N_{ij} undergoing the respective Raman transition from i to j, the vibrational quantum number v_i of the energy level the molecules occupy before the Raman transition, the frequency of the excitation field (laser) $(1/\lambda_\mathrm{L})$ in $\mathrm{m^{-1}}$, the Raman shift of the Raman transition $\bar{\nu}_{\mathrm{R},ij}$ in $\mathrm{m^{-1}}$, Boltzmann's constant k in J/K, the temperature T in K and the derived polarisability $\alpha'_{xyz,ij}$ in $\mathrm{A\ m^2/V\ J^{0.5}}$. The unit of the derived polarisability $\alpha'_{xyz,ij}$ can also be given in $\mathrm{C\ m^2/s\ V\ J^{0.5}}$.

Of course only those molecules, which already before the transition from i to j occupy the initial energy level i, which is defined by the quantum numbers el, v and J are involved into the respective Raman transition. According to the Boltzmann Eqn (2.60) the number of molecules $N_{\mathrm{el,v},J}$ occupying a specific energy level with respect to the overall number of molecules N_total can be replaced by

$$N_{ij} = N_{\mathrm{el,v},J} = \frac{N_\mathrm{total}\cdot(2J+1)\cdot e^{\left(-\frac{T(\mathrm{el,v},J)\cdot h\cdot c}{k\cdot T}\right)}}{\sum_{\mathrm{el,v},J}\left[(2J+1)\cdot e^{\left(-\frac{T(\mathrm{el,v},J)\cdot h\cdot c}{k\cdot T}\right)}\right]} \quad (2.102)$$

which results in the signal intensity for *one single* Raman Stokes transition

$$I_{\frac{P}{S}} = \frac{h}{8\cdot\varepsilon_0^2\cdot c}\cdot E_\mathrm{Laser}\cdot L_{ij}\cdot\frac{N_\mathrm{total}\cdot(2J+1)\cdot e^{\left(-\frac{T(\mathrm{el,v},J)\cdot h\cdot c}{k\cdot T}\right)}}{\sum_{\mathrm{el,v},J}\left[(2J+1)\cdot e^{\left(-\frac{T(\mathrm{el,v},J)\cdot h\cdot c}{k\cdot T}\right)}\right]}\cdot(\mathrm{v}_i+1)$$
$$\cdot\frac{((1/\lambda_\mathrm{L})-\bar{\nu}_{\mathrm{R},ij})^4}{\bar{\nu}_{\mathrm{R},ij}}.(\alpha'_{xyz,ij})^2. \quad (2.103)$$

It should be mentioned here again that in Eqn (2.103) the expression (v_i+1) and $((1/\lambda_\mathrm{L})-\bar{\nu}_{\mathrm{R},ij})^4$ have to be replaced by (v_i) and $((1/\lambda_\mathrm{L})+\bar{\nu}_{\mathrm{R},ij})^4$ for Raman anti-Stokes transitions. It also has to be underlined that Eqn (2.103) provides only the Raman signal intensity for *one single* transition. This can be, for example, the specific Q-branch Raman Stokes transition from the energy level $\mathrm{el}_i=0$ and $\mathrm{v_i}=0$ and $J_i=3$ to the energy level $\mathrm{el}_j=0$ and $\mathrm{v_j}=1$ and $J_j=3$. Please be aware that in the detected Raman spectrum the Raman Stokes Q-branch signal is the summation of many single Q-branch transitions, as it is indicated for a rigid rotator and harmonic oscillator molecule in the sketch in Fig. 2.52. In Fig. 2.52 the Q-branch transitions between the vibrational states v_1'' and v_2'' or v_2'' and v_3'' and so on are not included, but also contribute additionally to the Q-branch signal intensity in the Raman spectrum at the same spectral position. This in general is also true for the Raman signals of the O- and S-branches, which are the summation of different transitions starting from different energy levels characterised by the same rotational J but different vibrational v quantum numbers. For the harmonic oscillator and the rigid rotator all S-branch transitions, irrespective of whether they started in the vibrational ground state or in an excited vibrational state will appear at the same spectral position in the Raman spectrum. The same is true for the O-branch transitions. From these considerations it can be derived as stated by Long [38] that the Raman signal intensity not of a single transition but, for example, of the entire Q-branch Raman Stokes signal is

$$I^{\text{Stokes Q}}_{\frac{P}{S}} = \frac{h \cdot E_{\text{Laser}} \cdot L_{ij}}{8 \cdot \varepsilon_0^2 \cdot c} \cdot \sum_{\text{v},J} \left[\frac{N_{\text{total}} \cdot (2J+1) \cdot e^{\left(-\frac{T(\text{el,v},J) \cdot h \cdot c}{k \cdot T}\right)}}{\sum_{\text{el,v},J} \left[(2J+1) \cdot e^{\left(-\frac{T(\text{el,v},J) \cdot h \cdot c}{k \cdot T}\right)} \right]} \cdot (\text{v}_i + 1) \right]$$
$$\cdot \frac{((1/\lambda_{\text{L}}) - \bar{\nu}_{\text{R},ij})^4}{\bar{\nu}_{\text{R},ij}} . (\alpha'_{xyz,ij})^2 \tag{2.104}$$
$$= \frac{h}{8 \cdot \varepsilon_0^2 \cdot c} \cdot E_{\text{Laser}} \cdot L_{ij} \cdot \frac{N_{\text{total}}}{1 - e^{\left(-\frac{\bar{\nu}_{\text{e}} \cdot h \cdot c}{k.T}\right)}} \cdot \frac{((1/\lambda_{\text{L}}) - \bar{\nu}_{\text{R},ij})^4}{\bar{\nu}_{\text{R},ij}} . (\alpha'_{xyz,ij})^2$$

and of the Q-branch Raman anti-Stokes signal is

$$I^{\text{Anti-Stokes Q}}_{\frac{P}{S}} = \frac{h \cdot E_{\text{Laser}} \cdot L_{ij}}{8 \cdot \varepsilon_0^2 \cdot c} \cdot \sum_{\text{v},J} \left[\frac{N_{\text{total}} \cdot (2J+1) \cdot e^{\left(-\frac{T(\text{el,v},J) \cdot h \cdot c}{k \cdot T}\right)}}{\sum_{\text{el,v},J} \left[(2J+1) \cdot e^{\left(-\frac{T(\text{el,v},J) \cdot h \cdot c}{k \cdot T}\right)} \right]} \cdot (\text{v}_i) \right]$$
$$\cdot \frac{((1/\lambda_{\text{L}}) + \bar{\nu}_{\text{R},ij})^4}{\bar{\nu}_{\text{R},ij}} . (\alpha'_{xyz,ij})^2 \tag{2.105}$$
$$= \frac{h}{8 \cdot \varepsilon_0^2 \cdot c} \cdot E_{\text{Laser}} \cdot L_{ij} \cdot \frac{N_{\text{total}}}{e^{\left(\frac{\bar{\nu}_{\text{e}} \cdot h \cdot c}{k.T}\right)} - 1} \cdot \frac{((1/\lambda_{\text{L}}) + \bar{\nu}_{\text{R},ij})^4}{\bar{\nu}_{\text{R},ij}} . (\alpha'_{xyz,ij})^2 .$$

In the following, mainly Eqn (2.101) but also Eqns (2.104) and (2.105) are grouped in order to show the impact of different quantities onto the Raman signal intensity (point irradiance) I. The first group

$$\frac{h}{8 \cdot \varepsilon_0^2 \cdot c} = \text{const} \tag{2.106}$$

is constant, and thus does not change during an experiment.

Impact of the excitation irradiance and the index of refraction: The second group in Eqn (2.101)

$$E_{\text{Laser}} \cdot L_{ij} \tag{2.107}$$

considers the amplitude of the excitation field. Thus, if the local field L_{ij}, which itself is a function of the index of refraction n_{refr}, is constant, the Raman signal intensity I scales linearly with the excitation irradiance E_{Laser}.

Usually, for applications at elevated pressures, the local field L_{ij} is not constant, as the index of refraction n_{refr} of the fluid under investigation changes due to pressure or composition alterations. Then, a quantification of *concentrations* in terms of 'number of molecules per volume' is not possible in general, as the influence of the local field L_{ij}, which usually is not known, has to be regarded. If the variation of the index of refraction n_{refr} over the wavelength range of the Raman signals is ignored, and if intermolecular interactions are not affected by the action, which induced the change in the index of refraction, the local field L_{ij} can be approximated [63,64]

$$L_{ij} = \frac{(n_{\text{refr}}^2 + 2)^4}{81}\ . \tag{2.108}$$

Eqn (2.108) predicts that a change of the index of refraction n_{refr} from 1 to 1.1 increases the local field L_{ij} from 1 to 1.31. Unfortunately the index of refraction is not known for many pressurised systems as a function of pressure, temperature and composition. This explains why *concentrations* in terms of 'number of molecules per volume' can be quantified from Raman signal intensities only for the few systems, for which the index of refraction and its behaviour as a function of pressure and temperature are exactly known. As this is usually not the case, the quantitative nature of Raman signal intensities—this will be shown later—is limited to the quantification of *compositions* in terms of molar or mass fractions or ratios.

Impact of the wavelength of the excitation radiation: The fourth group in Eqn (2.101)

$$\frac{((1/\lambda_{\text{L}}) - \bar{\nu}_{\text{R},ij})^4}{\bar{\nu}_{\text{R},ij}} \approx \frac{(1/\lambda_{\text{L}})^4}{\bar{\nu}_{\text{R},ij}} \tag{2.109}$$

shows the importance of the excitation laser wavelength. For excitation wavelengths ranging from the near infrared region towards the ultraviolet spectral region, $(1/\lambda_{\text{L}})$ is significantly larger than the Raman shift $\bar{\nu}_{\text{R},ij}$ and thus dominates the nominator in Eqn (2.109). The Raman signal intensity I increases with $(1/\lambda_{\text{L}})$ to the fourth power.

It should be mentioned already here that while the Raman signal intensity I increases with $(1/\lambda_{\text{L}})$ to the fourth power, care has to be taken, as the risk of resonant excitation instead of linear Raman scattering increases with decreasing excitation wavelength, too [65]. We saw in Sections 2.4.1.5 and 2.4.1.6 that the probability of absorption of UV/VIS radiation increases with decreasing wavelengths and that, thus, the probability of inducing fluorescence emission with the excitation radiation also increases. Then, the weak Raman signals would be interfered by rather strong and broadband fluorescence emissions rendering a quantification of the Raman signals complex and often impossible. Thus, the perfect excitation wavelength for linear Raman spectroscopy is always a compromise between Eqn (2.109) (going to smaller excitation wavelengths) and the desire to work in the linear Raman regime (going to larger excitation wavelengths in order to avoid resonance effects). Resonant Raman strategies on the other hand can also be exploited in order to either enhance the Raman signal intensity or separate fluorescence emissions spectrally from the Raman signals [66–68]. As in the latter case electronic states above the first excited one or highly excited vibrational states in the first excited electronic state have to be excited, resonant Raman spectroscopy requires UV excitation wavelengths. Resonant Raman strategies are not treated in this book.

Impact of temperature: The effect of temperature on the Raman signal intensity of one single transition is regarded in Eqn (2.101) in the expression N_{ij}. If Raman signal intensities or Raman signal peaks are considered, for which several transitions starting from different vibrational energy levels coincide, the influence of temperature is reflected in the summation of the intensities of the single transitions. This consideration resulted in Eqns (2.104) and (2.105) for the Q-branch transitions. Therefore the fifths group are

$$\frac{1}{1-\exp\left[\frac{-hc\bar{\nu}_{R,ij}}{kT}\right]}, \tag{2.110}$$

for the Stokes Q-branch and

$$\frac{1}{\exp\left[\frac{hc\bar{\nu}_{R,ij}}{kT}\right]-1}, \tag{2.111}$$

for the anti-Stokes Q-branch. This fifth group considers the influence of temperature T onto the Raman signal intensity I and is given as a function of temperature for different Raman shifts of 500, 1000, 2000 and 4000 cm^{-1} in Fig. 2.59 for the Stokes and the anti-Stokes Q-branch.

The larger the Raman shift of the scatterer $\bar{\nu}_{R,ij}$, the wider is the range in which the temperature does not influence Eqns (2.110) and (2.111), meaning the longer the curves in Fig. 2.59 remain horizontal with increasing temperature. As an example, for Raman shifts exceeding 2000 cm^{-1} the temperature influence on $\frac{1}{1-\exp\left[\frac{-hc\bar{\nu}_{R,ij}}{kT}\right]}$ is less than 5% for temperatures smaller

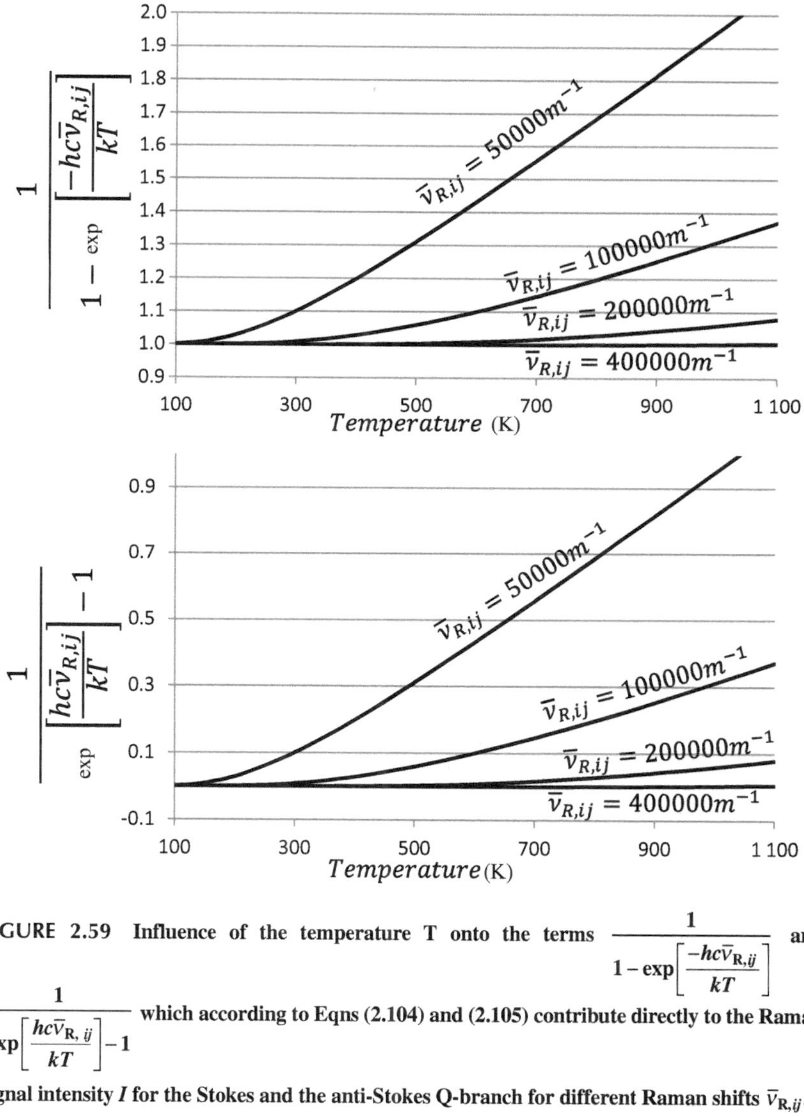

FIGURE 2.59 **Influence of the temperature T onto the terms $\frac{1}{1-\exp\left[\frac{-hc\bar{\nu}_{R,ij}}{kT}\right]}$ and $\frac{1}{\exp\left[\frac{hc\bar{\nu}_{R,ij}}{kT}\right]-1}$ which according to Eqns (2.104) and (2.105) contribute directly to the Raman signal intensity *I* for the Stokes and the anti-Stokes Q-branch for different Raman shifts $\bar{\nu}_{R,ij}$.**

than ~900 K. Taking into account one of the Raman shifts of carbon dioxide $\bar{\nu}_{R,CO_2} = 128{,}600\ m^{-1}$, the temperature influence on Eqn (2.110) is smaller than 5% for temperatures below ~600 K. For temperatures below 400 K the influence of temperature on Eqn (2.110) is even less than 1% for the aforementioned Raman shift of CO_2. The smaller the Raman shift $\bar{\nu}_{R,ij}$ is, the more pronounced is the influence of temperature on Eqns (2.110) and (2.111). Thus, it depends on the Raman shift of the Raman signal peak, whether or not the temperature influence on Eqns (2.110) and (2.111) has to be regarded.

Furthermore it has to be mentioned that the total number of molecules N_{total} also is a function of temperature, if, for example, due to a temperature variation of the sample, its density is varied. The influence of temperature on the density and thus on N_{total} is not regarded in Eqns (2.104) and (2.105) and consequently has to be regarded separately.

Impact of the derived polarisability: Finally, the derived polarisability $\alpha'_{xyz,ij}$ has to be considered. This quantity reflects the fact that, in general, the polarization of the Raman signal intensity (point irradiance) I is not the same one as that of the irradiance E_{Laser} producing it. In general, each polarisation component of I is a different linear combination of the polarisation components of E_{Laser}[5]. Table 2.4 lists—in the form of Eqn (2.112a–h)—the squared derived polarisability for different combinations of the polarization of the excitation field E_{Laser} and of the point irradiance I for detection of I along the y-axis and along the x-axis [38].

The quantities $(a')_{S,ij}$ and $(\gamma')_{S,ij}$ are the derived isotropic and anisotropic components of the polarisability of the scatterer undergoing the Raman transition from energy level i before the scattering process to j after the scattering process.

If the detection of the point irradiance I (Raman signal) does not follow polarisation resolved, the overall Raman signal intensity I

$$I = I_{\text{s}} + I_{\text{p}} \tag{2.113}$$

irradiated into a particular direction is the summation of both polarisation components I_{s} and I_{p}. In order to explain which configuration of polarisations would result in the highest Raman signal intensity I, it has to be considered that

$$\frac{45 \cdot (a')^2_{S,ij} + 4(\gamma')^2_{S,ij}}{45} > \frac{(\gamma')^2_{S,ij}}{15}! \tag{2.114}$$

In the following it is differentiated between detecting the Raman signals along the y-axis under 90° or along the x-axis under 0° with respect to Fig. 2.58.

2.4.2.3.4 Raman Signal Detection Configuration: Right Angle Versus Back Scattering

From the scattering geometry of a Raman experiment shown in Fig. 2.58 we can see that, if the Raman process is excited along the x-direction, the Raman signals I can be detected either in a right angle or in a back scattering configuration. In case of a right angle detection configuration the Raman signal can be either detected along the y- or the z-axis. The point irradiances I along the positive and negative z-axis as well along the negative y-axis (reverse to y-axis) are not sketched in Fig. 2.58 in order to keep the figure clearly arranged. In case of a back scattering detection

5. The polarisation of the laser is not always the polarisation of the excitation field. At high-pressure applications the excitation laser light usually has to pass through windows providing optical access to the high-pressure process. These windows can influence the polarisation of the laser beam. Thus we often do not know exactly the polarisation of the excitation field inside the high-pressure apparatuses, even if the polarisation of the laser beam, before it passes through the windows, is well known. This is discussed in detail in Section 2.4.2.3.4.

TABLE 2.4 List of the Squared Derived Polarisability $(\alpha'_{xyz,ij})^2$ for Different Combinations of Polarizations of the Excitation Irradiance E_{Laser} and the Point Irradiance I

Direction of detection	Polarisation of the signal (I)	Polarisation of the excitation (E_{Laser})	Squared derived polarisability $(\alpha'_{xyz,ij})^2$	Equations
y-axis (90°)	I_s	E_s	$=\frac{45\cdot(a')^2_{S,ij}+4(\gamma')^2_{S,ij}}{45}$	(2.112 a)
y-axis (90°)	I_s	E_p	$=\frac{(\gamma')^2_{S,ij}}{15}$	(2.112 b)
y-axis (90°)	I_p	E_s	$=\frac{(\gamma')^2_{S,ij}}{15}$	(2.112 c)
y-axis (90°)	I_p	E_p	$=\frac{(\gamma')^2_{S,ij}}{15}$	(2.112 d)
x-axis (0°)	I_s	E_s	$=\frac{45\cdot(a')^2_{S,ij}+4(\gamma')^2_{S,ij}}{45}$	(2.112 e)
x-axis (0°)	I_s	E_p	$=\frac{(\gamma')^2_{S,ij}}{15}$	(2.112 f)
x-axis (0°)	I_p	E_s	$=\frac{(\gamma')^2_{S,ij}}{15}$	(2.112 g)
x-axis (0°)	I_p	E_p	$=\frac{45\cdot(a')^2_{S,ij}+4(\gamma')^2_{S,ij}}{45}$	(2.112 h)

The indices are defined as those mentioned in the context of Figure 2.58.

configuration the Raman signal is detected from exactly that direction the Raman process also is excited from, so reverse to the x-axis (negative x-axis). Fig. 1.7 shows sketches of example experimental set-ups for signal detection in right angle or in back scattering configuration. Theoretically, the considerations we make here for the back scattering configuration can be transferred to a forward scattering configuration, meaning that the signals are detected along the x-axis. In practice forward scattering configurations are usually not realised with respect to Raman scattering, as then the weak Raman signals and the many orders of magnitude of more intense excitation radiation would take the same paths what would make the spectral purification of the Raman signals from laser radiation challenging.

According to Table 2.4 and Eqn (2.113) the intensity of the detected Raman signal is a function of the signal detection configuration (back scattering or right-angle scattering) and a function of the polarisations of the excitation laser radiation E_{Laser}. Therefore it will be derived here which combination of detection configuration and polarisations of excitation and detection will result in the highest signal levels and also is the most robust one with respect to high pressure applications.

Right angle signal detection: Table 2.4 and Eqn (2.113) say that the highest Raman signal intensity $I_{E_s}^{y\text{-axis}}$ along the *y*-axis can be realised for polarising the electric field of the excitation irradiance perpendicularly $E_{\text{Laser}} = E_s$.

$$I_{E_s}^{y\text{-axis}} = I_s + I_p \propto \frac{45 \cdot (a')_{S,ij}^2 + 4(\gamma')_{S,ij}^2}{45} + \frac{(\gamma')_{S,ij}^2}{15} \quad (2.115)$$

In this case I_s is the polarisation preserving component—preserving with respect to the polarisation of the excitation field E_{Laser}—and I_p the depolarised component with $I_s > I_p$. If the polarisation of the electric field of the excitation irradiance is turned to parallel $E_{\text{Laser}} = E_p$, the overall point irradiance $I_{E_p}^{y\text{-axis}}$ along the *y*-axis will decrease to

$$I_{E_p}^{y\text{-axis}} = I_s + I_p \propto \frac{(\gamma')_{S,ij}^2}{15} + \frac{(\gamma')_{S,ij}^2}{15}. \quad (2.116)$$

In this case I_p and I_s are both depolarised components with $I_s = I_p$.
As

$$I_{E_s}^{y\text{-axis}} > I_{E_p}^{y\text{-axis}}, \quad (2.117)$$

the polarisation of the electric field of the excitation irradiance influences the Raman signal intensity I detectable along the *y*-axis or respectively in right angle signal detection configuration. This statement is of utmost importance with respect to Raman measurements at elevated pressures, as the optical windows of high-pressure apparatuses can modify the polarisation of the excitation radiation unpredictably. Therefore a decrease or an increase of the Raman signal intensity I of one compound can be, for example, either due to an alteration of the concentration of this compound or due to a variation of the polarisation of the excitation field. This is different if the Raman signals $I^{x\text{-axis}}$ are detected in a back scattering configuration against the *x*-axis.

Back scattering signal detection: For detection in a back scattering configuration, the overall point irradiance $I^{x\text{-axis}}$ (Raman signal intensity) is not a function of the polarisation of the electric field of the excitation irradiance. If the electric field of the excitation irradiance is polarised perpendicular $E_{\text{Laser}} = E_s$, I_s is the polarisation-preserving component, while I_p is the depolarised component of the point irradiance $I_{E_s}^{x\text{-axis}}$.

$$I_{E_s}^{x\text{-axis}} = I_s + I_p \propto \frac{45 \cdot (a')_{S,ij}^2 + 4(\gamma')_{S,ij}^2}{45} + \frac{(\gamma')_{S,ij}^2}{15} \quad (2.118)$$

If the polarisation of the electric field of the excitation irradiance is turned to parallel $E_{Laser} = E_p$, I_s becomes the depolarised component, while I_p becomes the polarised component of the point irradiance $I_{E_p}^{x\text{-axis}}$.

$$I_{E_p}^{x\text{-axis}} = I_s + I_p \propto \frac{(\gamma')_{S,ij}^2}{15} + \frac{45 \cdot (a')_{S,ij}^2 + 4(\gamma')_{S,ij}^2}{45} \tag{2.119}$$

Summarising, for detection along the x-axis, there is always one polarised and one depolarised component of the point irradiance $I^{x\text{-axis}}$, making it

$$I_{E_s}^{x\text{-axis}} = I_{E_p}^{x\text{-axis}} \tag{2.120}$$

not a function of the polarisation of the electric field of the excitation irradiance. This has been different for the detection strategy along the y-axis (compare Eqns (2.117) and (2.120)).

I want to underline again that the influence of the polarisation of the electric field of the excitation irradiance onto the Raman signal intensity I becomes important for high-pressure applications. Windows installed into high-pressure vessels in order to guarantee optical access show the tendency of manipulating the polarisation of the electric field of the excitation irradiance.

From the considerations made here it becomes obvious that the back scattering configuration is the only configuration guaranteeing that the polarisation-manipulating-character of the windows does not influence the intensity of the detected Raman signal I. Furthermore, back-scattering set-ups are more robust with respect to misalignment of the excited and the detected volume due to changes of the index of refraction (compare Fig. 1.8), much easier to align and usually—due to a larger probe volume—provide higher Raman signal levels. Nevertheless, the spatial resolution of right angle detection set-ups is superior. If a Raman experiment does not need a special spatial resolution, I would therefore always prefer a Raman set-up in back scattering configuration. High spatial resolution also can be reached by setting up a confocal back scattering configuration.

2.4.2.3.5 How the Index of Refraction Impacts the Detected Portion of the Raman Signal

So far we assumed that the Raman signal intensity (point irradiance) I is detected through a pinhole positioned at either the y-axis or the x-axis. This configuration would result in the detection of very low Raman signal levels. In order to increase the detected Raman signal intensity, detection is usually accomplished through a large solid angle Ω. The solid angle Ω quantifies how much of the point irradiance I, which is given per steradian (sr), can be accumulated through a cone with its cone point at the scatterer and its opening direction along the direction of detection. Figure 2.60 shows the Raman scattering configuration for detection of the point irradiance I through a lens along the y-axis in right angle (90°) scattering direction and along the x-axis in back (0°) scattering direction.

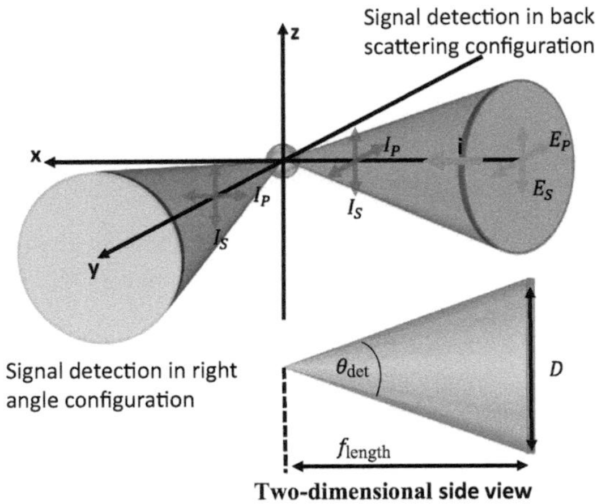

FIGURE 2.60 **Detection of the Raman signal through a lens along the *y*- and *x*-axis.**

The plane angle of detection

$$\theta_{det} = 2 \cdot \arctan\left(\frac{D}{2 f_{length}}\right) \tag{2.121}$$

is defined by the distance between the lens and the scatterer and the diameter D of the lens. Usually, the distance is set equal to the focal length f_{length} of the lens, as it can be seen from the experimental set-ups provided in Section 1.5.4. The solid angle

$$\Omega = 2\pi\left(1 - \cos\left(\frac{\theta_{det}}{2}\right)\right) \tag{2.122}$$

can be computed from the plane angle of detection θ_{det}. The larger the solid angle Ω, the stumper is the cone and the more Raman signal is collected by the lens. Thus the detected Raman signal intensity ${}^{\Omega}I$ in W (Watt), detected through a solid angle Ω is

$${}^{\Omega}I = (I_s \cdot k_s + I_p \cdot k_p) \cdot \Omega, \tag{2.123}$$

where k_s and k_p are experimental constants accounting for the detection efficiency of the experimental set-up for s- and p-polarised radiation.

Considering Raman applications at elevated pressures, the planar detection angle θ_{det} and with this the solid detection angle Ω cannot be computed based on Eqn (2.121) alone. One has to consider the influence of the index of refraction n_{refr} of the medium under investigation and the index of refraction of the windows providing optical access. Figure 2.61 shows the path of rays between a lens we use for signal detection and the origin of the Raman signals, which is the focal point of the lens (probe volume) for the scenario of detecting the Raman signals in air and for the scenario of detecting the Raman signals from the interior of a high pressure vessel (HPV).

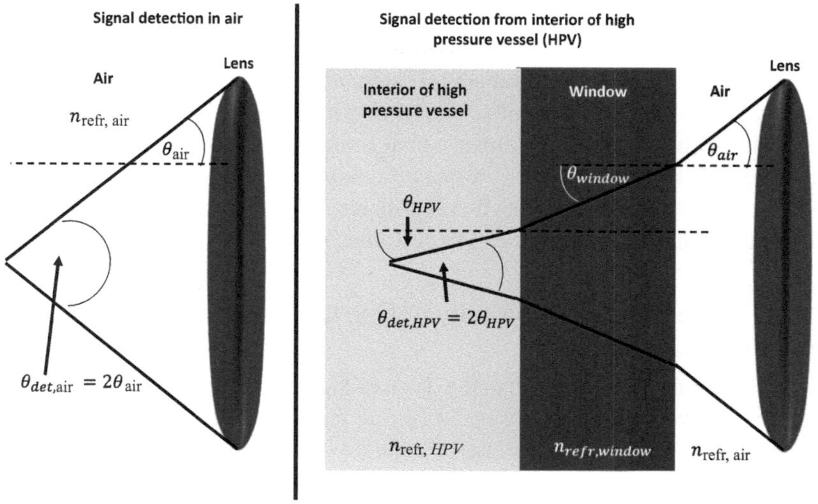

FIGURE 2.61 **Paths of rays focused by a lens into air (left) and through a glass window into the interior of a high pressure vessel filled with a fluid featuring an index of refraction larger than that one of glass (right) with $n_{\text{refr,HPV}} > n_{\text{refr,window}} > n_{\text{refr,air}}$.**

The index of refraction of the vessel interior $n_{\text{refr,HPV}}$ is supposed to be larger than that one of the glass window $n_{\text{refr,window}}$ and $n_{\text{refr,window}}$ is supposed to be larger than that one of air $n_{\text{refr,air}}$. Thus the plane detection angle $\theta_{\text{det,HPV}}$ inside the interior of the high pressure vessel is smaller than the plane detection angle $\theta_{\text{det,air}}$ is for a Raman experiment in ambient air. The refraction of the paths of the rays follows Snell's law

$$n_{\text{refr,air}} \sin(\theta_{\text{air}}) = n_{\text{refr,window}} \sin(\theta_{\text{window}}) = n_{\text{refr,HPV}} \sin(\theta_{\text{HPV}}). \quad (2.124)$$

Eqn (2.124) shows that the angle θ_{HPV}

$$\sin(\theta_{\text{HPV}}) = \frac{n_{\text{refr,air}}}{n_{\text{refr,HPV}}} \sin(\theta_{\text{air}}) \quad (2.125)$$

is a function of the refractive indices outside and inside the high pressure vessel, $n_{\text{refr,air}}$ and $n_{\text{refr,HPV}}$, respectively, and of the incident angle θ_{air}. θ_{HPV} is not a function of the index of refraction of the glass window material $n_{\text{refr,window}}$. Once the pressure, the temperature or the composition inside the high pressure vessel are changed, the only quantity changing in Eqn (2.125) is the index of refraction $n_{\text{refr,HPV}}$.

For a constant temperature of 313.15 K, Sun et al. [69] found an increase of the index of refraction of pure CO_2 from 1.0342 at 6 MPa to 1.1839 at 15 MPa. Assuming an experimental set-up as depicted in Fig. 2.61 with a spherical lens of diameter $D = 50\,\text{mm}$ and focal length $f_{\text{length}} = 100\,\text{mm}$, the pressure increase from 6 MPa to 15 MPa would then be accompanied with a decrease of the solid angle of detection from $\Omega = 0.1736$ sr $(= 0.0553\pi)$ to $\Omega = 0.1328$ sr $(= 0.0423\pi)$ corresponding to a decrease of 23%.

Summarising, the refractive index not only influences the intensity of the generated Raman signal I according to Eqns (2.101) and (2.108), but also affects the portion of the Raman signal intensity that can be collected $^{\Omega}I$. This again makes a quantification of *concentration* in terms of 'number of molecules per volume' from detected Raman signal intensities challenging and for most cases impossible. Nevertheless, a quantification of the mixture *composition* in terms of molar or mass fractions is still possible and is deduced when deriving the physical model for the quantification of Raman spectra in a subsequent section.

2.4.2.3.6 Raman Signal Intensity of One Compound in a Raman Spectrum of a Mixture

In this section we assume that the Raman signals are not interfered by any background, which potentially could be due to fluorescence, phosphorescence, Planck's radiation or any other interfering light source, such as computer screens, panel displays, LED lamps and so on. If the Raman spectrum is interfered, there are several experimental methods for interference rejection based on temporal or spectral interference discrimination or polarisation-based interference discrimination and mathematical strategies for interference correction in post-processing steps, which are summarised but not described in detail in Chapter 3. In the case of pure Raman spectra from one single compound, as they are shown for few example fluids in Fig. 2.54, the Raman signal intensity I_{compound}

$$I_{\text{compound}} = \int_{\text{left integration limit}}^{\text{right integration limit}} \text{Raman spectrum}_{\text{compound}} \tag{2.126}$$

of each compound, can be computed from the integral under the Raman spectrum between a left and a right integration limit. Of course the integration limits have to be chosen so that there are Raman transitions of this compound contained between the integration limits. But in general for the fluids shown in Fig. 2.54 the integration limits can be chosen between the very left and the very right margin of the spectra so that we can say that the Raman signal intensity is represented by the area under the detected Raman spectrum of this compound.

Figure 2.62 shows a (noisy) Raman spectrum of air. Fortunately in this case the Q-branch Stokes Raman signals[6] of nitrogen and oxygen do not spectrally overlap. Thus, according to Fig. 2.62 integration limits can be defined for nitrogen and oxygen separately, which guarantee that the integrals between these limits solely consider the Raman signals of either nitrogen I_{N_2} or oxygen I_{O_2}.

For mixtures of more complex molecules, the Raman signals of different compounds usually overlap spectrally as it is shown for the mixture acetone (ace) and carbon dioxide (CO_2) in Fig. 2.63 (top). In Fig. 2.63 (bottom) the Raman

6. The S- and O-branch transitions neighboring the Q-branch are both, too weak and too close to the Q-branch so that they are not visible.

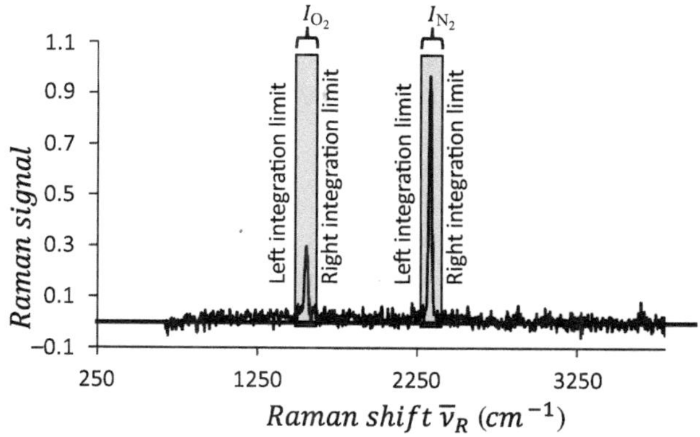

FIGURE 2.62 Raman signal intensities of oxygen I_{O_2} and nitrogen I_{N_2} as integrals of the Raman peaks of the oxygen and nitrogen between a left and a right integration limit. Oxygen and nitrogen peaks do spectrally not overlap.

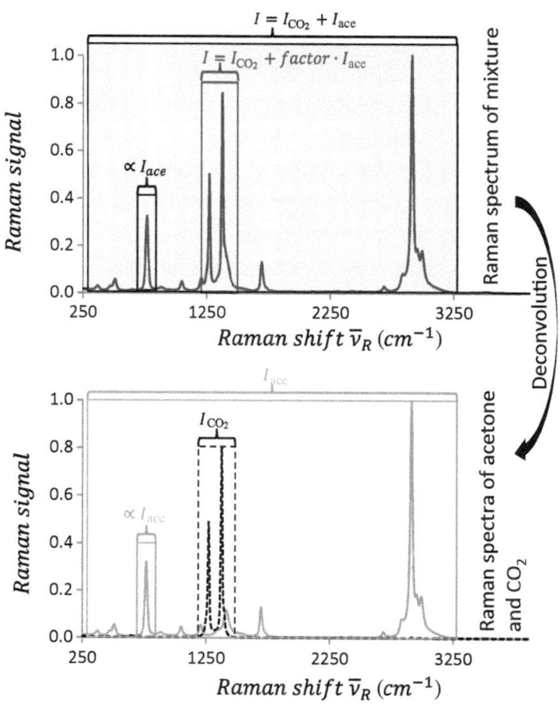

FIGURE 2.63 Raman signal intensities of carbon dioxide I_{CO_2} and acetone I_{ace} from the Raman spectrum of the mixture of carbon dioxide and acetone. In the Raman spectrum of the mixture the Raman signals of acetone and carbon dioxide overlap spectrally.

spectra of the pure compounds are given, from which the Raman spectrum of the mixture is composed of. It can be seen that in the Raman spectrum of the mixture there is not one single CO_2 Raman signal peak, which does not interfere with the Raman signal of acetone.

But there are several Raman signal peaks of acetone that do not interfere with the Raman signals of CO_2. This is shown as an example for one acetone peak whose signal intensity in Fig. 2.63 (top) is labelled $\propto I_{ace}$. Also other acetone peaks not interfering with CO_2 can be chosen in general for the representation of the Raman signal intensity of acetone I_{ace}; for CO_2 this here is more complex.

If the integration limits, which usually are applied for pure CO_2, are applied for the computation of the Raman signal of CO_2 from the Raman spectrum of the mixture, the obtained result $I = I_{CO_2} + \text{factor} \cdot I_{ace}$ would comprise the signal intensity of CO_2 plus the interference from acetone. Therefore a strategy has to be found which can isolate the Raman signal contributions of CO_2 from that one of acetone. On this purpose the Raman spectrum of the mixture can be deconvoluted into the Raman contributions of the compounds contained within the mixture. Different deconvolution methods are known which are described in the following. Then the Raman signal intensity I_{CO_2} of CO_2 can be computed from solely the isolated Raman spectrum of CO_2, as it is shown in Fig. 2.63 (bottom).

Deconvolution into rigid Raman spectra of pure compounds: With respect to Fig. 2.63 the Raman spectra of the pure compounds CO_2 and acetone are weighted with weighting factors till the least squares difference between the Raman spectrum of the mixture and the summation of the weighted pure compound Raman spectra is minimum.

This method requires the knowledge of the Raman spectra of the pure compounds, which should have ideally been acquired with the same experimental set-up the Raman spectra are recorded from the mixture afterwards. The advantage of this method is its rather low complexity. The disadvantage of this method is that the spectra of the pure compounds remain rigid during the deconvolution algorithm. In Figs 2.55 and 2.56 two examples were given where the Raman spectra change as a function of pressure or temperature. Additionally, also a variation of the mixture composition can induce peak shifts due to a change of intermolecular interactions. These effects are not regarded when the rigid Raman spectra of the pure compounds are considered for the deconvolution. Nevertheless, this deconvolution method is often sufficient, as the spectral changes are often small and can be neglected. Of course, the higher the spectral resolution of the Raman experiment, the more dramatically small spectral changes are realisable. Therefore sometimes 'poor' spectral resolution is beneficial. What has been described here for a binary mixture of course is also applicable for multi-component mixtures.

Deconvolution using indirect hard modelling: In the case of indirect hard modelling [70,71] and with respect to Fig. 2.63 the Raman spectra of CO_2 and of acetone are first replaced by the summation of different peaks. The CO_2 spectrum in Fig. 2.63 can be, for example, replaced by the summation of two peaks, each peak representing one of the two peaks of the CO_2 Raman signal (Fermi dyad). Each of these two peaks is characterised by a certain shape (Voigt profile, Lorentzian or Gaussian) and spectral position, including all the parameters required for the mathematical presentation of these peaks (see **Exercise 2.10**). These two peaks are thus assigned to CO_2. For the representation of the acetone Raman spectrum of course many more single peaks have to be defined, again each peak defined by a spectral position and a certain shape. These peaks are then assigned to acetone. Then each single peak of acetone and of CO_2 can be given the freedom of slightly changing its spectral position or shape during the deconvolution algorithm of the mixture spectrum. Peaks, which are known to remain at a constant spectral position with a constant shape, can be set rigid with respect to position and shape. Therefore in the case of indirect hard modelling spectral shifts of single peaks can be considered during the deconvolution. The advantage of this method is its flexibility and that not even the Raman spectra of the pure compounds have to be recorded with the same Raman experiment the mixtures are analysed afterwards. It is sufficient to know at which spectral positions (approximately) peaks have to be assigned either to CO_2 or to acetone. Their exact spectral position and their exact shape will be identified during the deconvolution algorithm. The disadvantage of this method is that many parameters (for each peak the spectral position, its shape and its intensity) have to be optimised during the deconvolution. Computation efforts can be reduced if the indirect hard modelling approach is only applied to spectral regions in which the Raman spectra of the different compounds overlap spectrally. An indirect hard modelling test software is available (http://www.s-pact.de). What has been described here for a binary mixture of course is also applicable for multi-component mixtures.

Deconvolution using hybrid approaches: Intermediate forms of the two deconvolution approaches described are possible and sometimes beneficial with respect to programming and computation time. As an example, during the deconvolution, the Raman spectrum of CO_2 can be assumed to be the rigid Raman spectrum of the pure CO_2 recorded at any condition while the indirect hard modelling approach is applied for acetone. Such a hybrid deconvolution strategy is described in detail for the mixture acetone, water and CO_2 in Section 2.5.6.

2.4.2.3.7 Quantification of the Composition of a Mixture

The concept of quantifying compositions is based on considering the ratio R of detected Raman signal intensities $I_{compound}$ radiated by molecules of different compounds (species). Therefore first the detected Raman signal intensities of different compounds have to be determined from the Raman spectrum of a mixture as it is described in the previous Section 2.4.2.3.6. According to Eqn (2.104), the ratio R of detected Q-branch Stokes Raman signal intensities I_a / I_b originating from two different compounds a and b undergoing Raman transitions from energy levels *i* or *k* before the Raman process to energy levels *j* or *l* after the Raman process, respectively, is

$$R=\frac{I_a}{I_b}=\frac{\left\{\frac{h\cdot E_{\text{Laser}}\cdot L_{a,ij}}{8\cdot\varepsilon_0^2\cdot c}\cdot\frac{N_a}{1-\exp\left(\frac{\bar{\nu}_{R,a,ij}\cdot h\cdot c}{k\cdot T}\right)}\cdot\frac{((1/\lambda_L)-\bar{\nu}_{R,a,ij})^4}{\bar{\nu}_{R,a,ij}}\cdot(\alpha'_{xyz,a,ij})^2\right\}\cdot k_a\cdot\Omega_a}{\left\{\frac{h\cdot E_{\text{Laser}}\cdot L_{b,kl}}{8\cdot\varepsilon_0^2\cdot c}\cdot\frac{N_b}{1-\exp\left(\frac{\bar{\nu}_{R,b,kl}\cdot h\cdot c}{k\cdot T}\right)}\cdot\frac{((1/\lambda_L)-\bar{\nu}_{R,b,kl})^4}{\bar{\nu}_{R,b,kl}}\cdot(\alpha'_{xyz,b,kl})^2\right\}\cdot k_b\cdot\Omega_b} \tag{2.127}$$

where k_a and k_b are experimental parameters considering the detection efficiency of the Raman signal originating from species a and b, respectively. From now on the indices *ij* and *kl* are skipped. Once the Raman signal intensities originating from species a and b are detected simultaneously with one and the same optical set-up in form of a Raman spectrum, the excitation irradiances E_{Laser}, the local fields L and the solid angles of detection Ω are identical for both species. Then—also cancelling out the constant groups—Eqn (2.127) simplifies to

$$R=\frac{I_a}{I_b}=\frac{\left\{\frac{N_a}{1-\exp\left(\frac{\bar{\nu}_{R,a,ij}\cdot h\cdot c}{k\cdot T}\right)}\cdot\frac{((1/\lambda_L)-\bar{\nu}_{R,a,ij})^4}{\bar{\nu}_{R,a,ij}}\cdot(\alpha'_{xyz,a,ij})^2\right\}\cdot k_a}{\left\{\frac{N_b}{1-\exp\left(\frac{\bar{\nu}_{R,b,kl}\cdot h\cdot c}{k\cdot T}\right)}\cdot\frac{((1/\lambda_L)-\bar{\nu}_{R,b,kl})^4}{\bar{\nu}_{R,b,kl}}\cdot(\alpha'_{xyz,b,kl})^2\right\}\cdot k_b}$$
$$=\frac{N_a\cdot K_a}{N_b\cdot K_b}=\frac{N_a}{N_b}\cdot K_{a/b} \tag{2.128}$$

with the capital letters K_a and K_b containing parameters comprising all influencing parameters except the number of species a and b molecules. The factor $K_{a/b}$ summarises the factors K_a and K_b. The function $R=\frac{I_a}{I_b}=\frac{N_a}{N_b}\cdot K_{a/b}$, which here has been derived for the ratio of two Q-branch Raman Stokes transition can also be derived in the identical form for the ratio of any transitions of species a and b. Then only the factors K_a and K_b would substitute different terms and would take other values. If $K_{a/b}$ can be considered a constant in the range of operational conditions (temperature, pressure, mixture composition, etc.) of the Raman experiment, the ratio R

$$R = \frac{I_a}{I_b} \propto \frac{N_a}{N_b} \tag{2.129}$$

is directly proportional to the ratio N_a / N_b of the numbers of species a and b molecules and can be used for the quantification of a mixture composition. In the following it will be demonstrated that the composition of not only binary but also multi-component mixtures can be quantified using this method. The proportionality factor $K_{a/b}$ can be determined by measuring the ratio $R = I_a / I_b$ of the Raman signal intensities of species a and species b in mixtures of known composition N_a / N_b.

The calibration measurements have to be carried out in binary mixtures of known composition. We will see later that calibration measurements carried out in the binary systems a/b and a/c qualify for the determination of the composition of the ternary system a/b/c. Of course, the accuracy and precision of the calibration later on determines the accuracy and precision of the evaluation of the composition of unknown mixtures. Therefore the calibration of the evaluation model is the most crucial step whenever the Raman techniques proposed here are to be applied. Here the generation of very defined mixtures of constituents a and b at elevated pressures is the most important issue. The composition of the mixture adjusted is defined by the molar fraction $x_{compound}$ of the compound a, b or c.

$$x_a = \frac{N_a}{N_a + N_b},\ x_b = \frac{N_b}{N_a + N_b},\ x_a + x_b = 1 \tag{2.130}$$

Combining Eqns (2.129) and (2.130), the molar fractions x_a and x_b can be given as a function of the ratio R of the Raman signal intensities of constituents a and b.

$$\begin{aligned} x_a &= \frac{N_a}{N_a + N_b} = \frac{1}{1+(N_b / N_a)} = \frac{1}{1+(I_b / I_a)\cdot K_{a/b}} = \frac{1}{1+(1/R)\cdot K_{a/b}} \\ x_b &= \frac{1}{(N_a / N_b)+1} = \frac{1}{(I_a / I_b)(1/K_{a/b})+1} = \frac{1}{R(1/K_{a/b})+1} \end{aligned} \tag{2.131}$$

Consequently a typical calibration diagram correlates the molar fraction of one constituent of the mixture with the ratio R of the Raman signal intensities I_a and I_b, like it is shown on the left side in Fig. 2.64 for acetone and CO_2 being the constituents a and b, respectively, and on the right side in Fig. 2.64 for acetone and water being—with respect to Eqn (2.131)—the constituents a and b, respectively.

In Fig. 2.64 the grey solid lines represent Eqn (2.131) with $K_{a/b}$ being 1.2411 and 2.5838 for K_{ace/CO_2} (left side) and $K_{ace/water}$ (right side), respectively. The circumstance that the measured data points in Fig. 2.64, which are represented by the circles and crosses, can be reflected by the physical model derived for the quantification of mixture compositions (Eqn (2.131), grey solid line) underlines

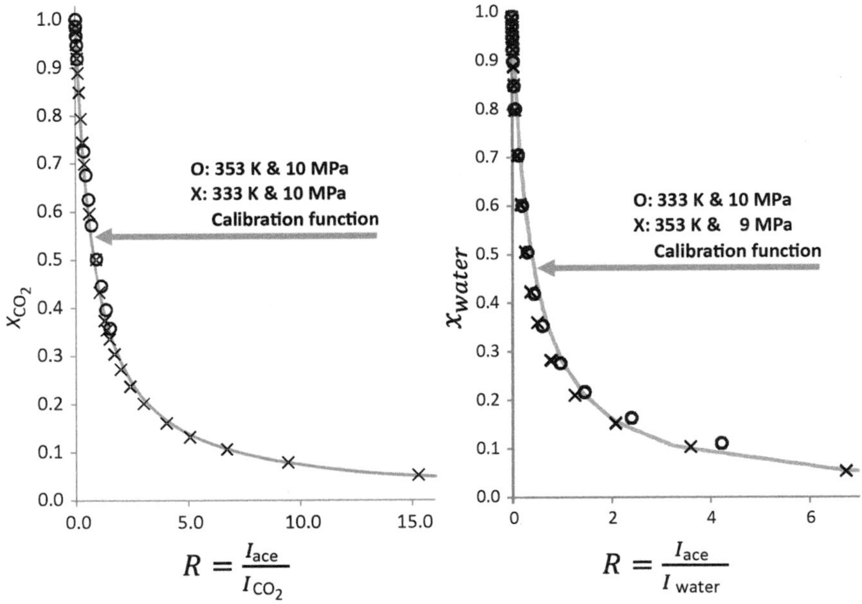

FIGURE 2.64 Calibration function of the binary systems composed of acetone and CO_2 (left side) and composed of acetone and water (right side).

the applicability and capability of the model. According to the fundamentals of the Raman theory (discussed in Section 2.4.2.3.3) the ratio R is expected not to be influenced by temperature variations in a temperature range below 400 K for CO_2. This expectation is confirmed by the fact that calibration measurements carried out at 333 K and 353 K can both be represented by Eqn (2.131) using only one calibration constant K_{ace/CO_2} or $K_{ace/water}$ for the two respective binary mixures.

It should be mentioned here again that the physical model developed and applied here is based on the assumption that the ratio R of Raman signal intensities of two different species is directly proportional to the ratio of the number of the species molecules according to Eqn (2.128). This assumption is valid only, if the ratio of the Raman scattering cross-sections of the species is a constant in the range of pressures, temperature and compositions. Especially for alcohol and water molecules, which tend to develop hydrogen bonds, a cooperative effect is known [59,72]. This means that the more the Raman scattering cross-section (for one molecule in average) increases, the more intense the hydrogen bonds are developed. As the model can be fitted to the experimental results according to Fig. 2.64 (right side) also for a binary mixture involving hydrogen bonds (with water), the 'cooperative effect' can be neglected and Eqn (2.131) may be applied also for the system containing water.

The composition of a ternary mixture is defined by at least two of the molar fractions x_a, x_b and x_c of the three constituents a, b and c the mixture is composed of. Eqns (2.132) to (2.134) provide the molar fractions of the three constituents as a function of the Raman signal intensities R. Again the proportionality of the Raman signal intensity I and its amount of substance in the mixture is considered by the calibration factors K.

$$\begin{aligned} x_a &= \frac{N_a}{N_a + N_b + N_c} = \frac{1}{1 + (N_b / N_a) + (N_c / N_a)} \\ &= \frac{1}{1 + (I_b / I_a) \cdot K_{a/b} + (I_c / I_a) \cdot K_{a/c}} \\ &= \frac{1}{1 + (1 / R_{a/b}) \cdot K_{a/b} + (1 / R_{a/c}) \cdot K_{a/c}} \end{aligned} \quad (2.132)$$

$$\begin{aligned} x_b &= \frac{1}{(N_a / N_b) + 1 + (N_c / N_b)} \\ &= \frac{1}{(I_a / I_b)(1 / K_{a/b}) + 1 + (I_c / I_b)(1 / K_{c/b})} \\ &= \frac{1}{(1 / K_{a/b}) R_{a/b} + 1 + (1 / K_{c/b}) R_{c/b}} \end{aligned} \quad (2.133)$$

$$\begin{aligned} x_c &= \frac{1}{(N_a / N_c) + (N_b / N_c) + 1} \\ &= \frac{1}{(I_a / I_c)(1 / K_{a/c}) + (I_b / I_c) K_{c/b} + 1} \\ &= \frac{1}{(1 / K_{a/c}) R_{a/c} + (1 / R_{c/b}) K_{c/b} + 1} \end{aligned} \quad (2.134)$$

$$x_a + x_b + x_c = 1 \quad (2.135)$$

The evaluation of the composition of a ternary mixture of unknown composition is achieved as follows:

1. The Raman signal intensities I_a, I_b and I_c of the three constituents have to be extracted from the Raman spectrum acquired from the ternary mixture of unknown composition according to one of the strategies introduced in Section 2.4.2.3.6.
2. From the Raman signal intensities I_a, I_b and I_c the ratios $R_{a/b}$, $R_{a/c}$ and $R_{c/b}$ can be computed.
3. Two of the three calibration constants $K_{a/b}$, $K_{a/c}$ and $K_{c/b}$ must be known. It is described for the binary mixtures acetone/CO_2 and acetone/water how to derive two of the calibration constants K_{ace/CO_2} and $K_{ace/water}$, which would be required for the measurement of the composition in a ternary system composed of acetone, water and CO_2.
4. The four Eqns (2.132) to (2.135) can be solved, as there are only four unknowns left, which are the three molar fractions x_a, x_b, and x_c and one not yet determined calibration constant.

Figure 2.65 shows in the core a Gibbs diagram for the ternary high-pressure phase diagram of the system acetone, water and CO_2. With respect to this ternary

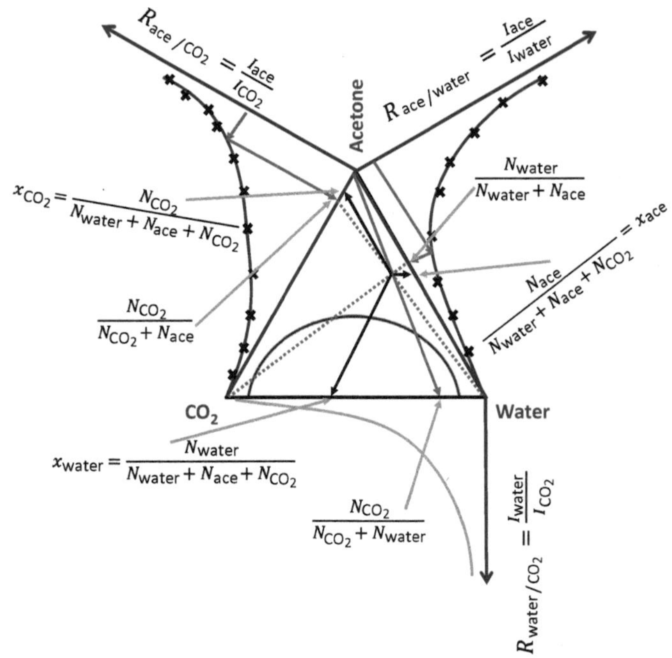

FIGURE 2.65 Graphical illustration how the composition of a ternary mixture can be determined after the calibration of two binary systems. The grey calibration line for water and CO_2 can be computed from the two other calibration lines.

system the binary systems water and acetone as well as CO_2 and acetone are completely miscible and are therefore suitable for the calibration. Example calibration results are shown in Fig. 2.64 for both systems. In Fig. 2.65 qualitative sketches of these calibration lines are tilted to the side-lines of the respective Gibbs diagram. After calibration of both binary systems the ratios $\frac{N_{CO_2}}{N_{CO_2}+N_{ace}}$ and $\frac{N_{water}}{N_{water}+N_{ace}}$ can be determined directly from the calibration lines. From the intersection of the iso-lines $\frac{N_{CO_2}}{N_{CO_2}+N_{ace}}=\text{const}$ and $\frac{N_{water}}{N_{water}+N_{ace}}=\text{const}$ (dashed in Fig. 2.65) the ratio $\frac{N_{water}}{N_{CO_2}+N_{water}}$ can be computed or determined graphically. Furthermore, from the intersection of the dashed iso-lines $\frac{N_{CO_2}}{N_{CO_2}+N_{ace}}=\text{const}$ and $\frac{N_{water}}{N_{water}+N_{ace}}=\text{const}$ also the molar fractions x_a, x_b, and x_c can be computed or determined graphically, as it is indicated by the black arrows.

Finally it has to be mentioned that by the division of both calibration functions

$$\frac{R_{\text{ace/CO}_2}}{R_{\text{ace/water}}} = \frac{I_{\text{ace}} / I_{\text{CO}_2}}{I_{\text{ace}} / I_{\text{water}}} = R_{\text{water/CO}_2} = \frac{(N_{\text{ace}} / N_{\text{CO}_2}) \cdot K_{\text{ace/CO}_2}}{(N_{\text{ace}} / N_{\text{Water}}) \cdot K_{\text{ace/water}}} = (N_{\text{water}} / N_{\text{CO}_2}) \cdot K_{\text{water/CO}_2} \tag{2.136}$$

a fictive calibration function can be determined for the system water and CO_2. This fictive calibration function cannot be measured directly, as water and CO_2 feature a wide miscibility gap. The fictive calibration function also covers the region of the miscibility gap as it has been computed from two subsystems, which are completely miscible. Due to the fact that it is a fictive calibration line, it is not based on data points and sketched grey in Fig. 2.65.

2.4.3 Brief Comparison of Absorption Techniques, Laser-Induced Fluorescence and Raman Scattering

In order to bring down Chapter 2 to a round figure, I want to compare absorption, emission and scattering techniques.

Spatial resolution: Absorption techniques typically work in transmission. Exceptions are attenuated total reflection techniques. Therefore one usually does not know where along the transmitted path the light has been attenuated. Thus, from absorption techniques one usually gets a line of sight information. Tomographic absorption techniques can reconstruct the transmitted object but can only be put into practice if the measurement object is accessible to light from many different directions what in high-pressure process technology is usually not the case. With attenuated total reflection absorption techniques spatially highly resolved information can be gathered, but from rather thin layers (see Chapter 6)

On the contrary, emission or scattering techniques can be realised as imaging, one-dimensional or zero-dimensional measurement techniques. Therefore they are typically applied when flow structures have to be resolved.

Temporal resolution: Meanwhile absorption, emission and scattering techniques can be realised with temporal resolutions shorter than few nanoseconds. In this context I want to mention the supercontinuum time of flight absorption technology (see Chapter 6).

Quantification of detected signals: Usually, from the absorption spectrum and from the Raman spectrum the composition of a mixture can be determined straightforward. This is mainly due to the fact that the events of absorption and of light scattering take place on a temporal scale, which is significantly shorter than the temporal scale of intermolecular interactions. This means that during the absorption event or during the Raman scattering event the molecules are not influenced by their environment. Consequently also the signal we detect from them is not influenced by their environment. This is very different for the laser-induced fluorescence mechanism. In this case, the excited molecule resides in the excited state before it fluoresces or depopulates other than by fluorescence. Consequently the signal we detect is very much influenced by parameters such as the chemical environment of the fluorescing molecules (quenching), which during a mixing experiment is not known but should be measured, and by temperature or pressure, due to collisions.

Rayleigh scattering is not suitable for the determination of the composition of mixtures, as the Rayleigh signal of different molecules appears at the same spectral position.

Temperature can be determined with absorption techniques, but not spatially resolved along the line of sight, and with Raman techniques, spatially resolved in one dimension or even two-dimensionally. In both cases the temperature is usually computed from the signal intensity of certain molecular transitions via the Boltzmann distribution. LIF thermometry strategies will be introduced in Chapter 5.

Pressure can in principle be determined with all techniques when the peak broadening is analysed. According to my knowledge pressure is usually determined with conventional techniques, as within one chamber the pressure can often be assumed to be 'constant' in space.

Signal intensity: Absorption cross-sections are typically high. Therefore samples often have to be attenuated in order to allow the incident light to pass through the sample to a detectable extent. The signal intensities detectable from laser-induced fluorescence experiments are also rather high. Raman signals are typically very low, due to the small cross-sections of the Raman scattering process. But the high molecular densities present in high-pressure applications compensate this drawback of Raman spectroscopy. Typical cross-sections per steradian in cm^{-2} $(sr)^{-1}$ are 10^{-27} for Rayleigh scattering, 10^{-31} for Raman scattering and 10^{-23} for the laser-induced fluorescence.

2.4.4 Exercises Corresponding to Section 2.4

Exercise 2.7: Composition Range Detectable With an Absorption Spectrometer

Let us assume that we have to measure the compositions of binary single-phase liquid mixtures with an absorption spectrometer. From the manual of the spectrometer we know that the reliable operation range of our device is guaranteed only for $0.046 \le A(\lambda) \le 1$. Furthermore we know that the absorption cross-sections of both compounds 1 and 2 within the mixture have the same magnitude $\varepsilon_{\text{compound 1}}(\lambda_1) = \varepsilon_{\text{compound 2}}(\lambda_2)$, but absorb at different wavelengths $\lambda_1 \ne \lambda_2$. Consequently, from the absorbance spectrum we can extract the absorbance for both compounds separately.

Task: Exercise 2.7

What is the range of mixture compositions that can be resolved with the available absorption spectrometer?

Answer: Exercise 2.7

Within one absorbance spectrum only absorbances between 0.046 and 1 can be evaluated reliably using the Beer–Lambert law

$$A(\lambda)=\varepsilon(\lambda)\cdot c_{\text{abs}}\cdot l. \quad (2.81)$$

This means that if the absorbance of compound 1 is $A_{\text{compound 1}}(\lambda_1)=1$, the absorbance of compound 2 must not be smaller than $A_{\text{compound 2}}(\lambda_2)=0.046$.
Therefore

$$\frac{A_{\text{compound 2}}(\lambda_2)=0.046}{A_{\text{compound 1}}(\lambda_1)=1}=\frac{c_{\text{compound 2}}\cdot\varepsilon_{\text{compound 2}}(\lambda_2)\cdot l}{c_{\text{compound 1}}\cdot\varepsilon_{\text{compound 1}}(\lambda_1)\cdot l}=\frac{c_{\text{compound 2}}}{c_{\text{compound 1}}}=0.046 \quad (2.137)$$

compound 2 must not be less concentrated than $c_{\text{compound 2}}=0.046\cdot c_{\text{compound 1}}$ which corresponds to a minimum compound 2 composition of

$$x_{\text{min,compound 2}}=\frac{c_{\text{compound 2}}}{c_{\text{compound 1}}+c_{\text{compound 2}}}=\frac{0.046}{1.046}=0.044. \quad (2.138)$$

Vice versa, if the absorbance of compound 2 is $A_{\text{compound 2}}(\lambda_2)=1$, the absorbance of compound 1 must not be smaller than $A_{\text{compound 1}}(\lambda_1)=0.046$ and thus

$$\frac{A_{\text{compound 2}}(\lambda_2)=1}{A_{\text{compound 1}}(\lambda_1)=0.046}=\frac{c_{\text{compound 2}}\cdot\varepsilon_{\text{compound 2}}(\lambda_2)\cdot l}{c_{\text{compound 1}}\cdot\varepsilon_{\text{compound 1}}(\lambda_1)\cdot l}=\frac{c_{\text{compound 2}}}{c_{\text{compound 1}}}=21.74 \quad (2.139)$$

compound 2 must not be higher concentrated than $c_{\text{compound 2}}=21.74\cdot c_{\text{compound 1}}$ which would correspond to a maximum compound 2 composition of

$$x_{\text{max,compound 2}}=\frac{c_{\text{compound 2}}}{c_{\text{compound 1}}+c_{\text{compound 2}}}=\frac{21.74}{22.74}=0.956. \quad (2.140)$$

Summarising, under the given circumstances the absorption spectrometer is suitable for reliably analysing the composition of binary mixtures in the range $0.044\leq x_{\text{compound 2}}\leq 0.956$.

Exercise 2.8: IR or Raman Active?

In Section 2.4 we saw that an IR-active molecule (meaning that it absorbs infrared light) has to feature a permanent dipole or a dipole which is induced during the vibration and that a Raman active molecule (meaning that a transition of a molecule can be seen in the Raman spectrum) has to feature a change of polarisability during the molecules motion (derived polarisability $\alpha'_{\text{vib}}>0$).

Task: Exercise 2.8

Which of the following molecular motions are IR-active and/or Raman active?

1. Symmetric stretch vibration of water
2. Asymmetric stretch vibration of water
3. Symmetric stretch vibration of methane
4. Asymmetric stretch vibration of methane
5. Symmetric stretch vibration of carbon dioxide
6. Asymmetric stretch vibration of carbon dioxide
7. Rotation of hydrogen
8. Rotation of methane

Answer: Exercise 2.8

1. Water is a permanent dipole → IR-active
   Polarisability changes during vibration → Raman active
2. Water is a permanent dipole → IR-active
   Polarisability changes during vibration → Raman active
3. Neither permanent nor induced dipole → *not* IR-active
   Polarisability changes during vibration → Raman active
4. Induced dipole → IR-active
   Polarisability changes during vibration → Raman active
5. Neither permanent nor induced dipole → *not* IR-active
   Polarisability changes during vibration → Raman active
6. Induced dipole → IR-active
   Polarisability does *not* change during vibration → *not* Raman active
7. Neither permanent nor induced dipole → *not* IR-active
   Polarisability changes during rotation as the 'one-dimensional' hydrogen changes its orientation with respect to the polarization of the external excitation field → Raman active
8. Neither permanent nor induced dipole → *not* IR-active
   Polarisability does *not* change during rotation as the 'spherical' molecule does effectively not change its orientation with respect to the polarization of the external excitation field → *not* Raman active

Exercise 2.9: Intensity Ratio Polarised and Depolarised Signal

Table 2.4 lists the terms of the squared derived polarisability $(\alpha'_{xyz,ij})^2$ for different combinations of the polarisations of the excitation field E_{Laser} and the detected Raman signal I. The quantities $(a')_{S,ij}$ and $(\gamma')_{S,ij}$ are the derived isotropic and anisotropic components of the polarisability of the scatterer and are assumed to be $\alpha'_{N_2} = 0.506 \cdot 10^{-16}\,\text{cm}^2$ and $\gamma'_{N_2} = 1.4 \cdot 10^{-16}\,\text{cm}^2$ for nitrogen.

Task: Exercise 2.9

What is the intensity ratio $I^s_{E_s} / I^p_{E_s}$ of the Raman signals polarised perpendicular $I^s_{E_s}$ and parallel $I^p_{E_s}$ if the Raman process is excited with perpendicularly polarised light E_s?

Answer: Exercise 2.9

$$\frac{I^s_{E_s}}{I^p_{E_s}} = \frac{\dfrac{45\cdot(a')^2_{N_2}+4(\gamma')^2_{N_2}}{45}}{\dfrac{(\gamma')^2_{N_2}}{15}} = \frac{\dfrac{45\cdot(0.506)^2+4(1.4)^2}{45}}{\dfrac{(1.4)^2}{15}} = 3.29$$

Exercise 2.10: Gaussian, Lorentzian and Voigt Profile Peak Shapes

It is mentioned in Section 2.4.2.3.6 that different mathematical peak shapes can be fitted to Raman signal peaks, which represent any acquired Raman spectrum. In this exercise the equations representing the Gaussian peak form

$$I_x^{\text{Gauss}} = I_0 \exp\left(-\frac{(x-x_0)^2}{2w^2}\right) \tag{2.141}$$

the Lorentzian peak form

$$I_x^{\text{Lorentz}} = I_0\left[\frac{w^2}{(x-x_0)^2+w^2}\right] \tag{2.142}$$

and the Voigt-Profile peak form

$$I_x^{\text{Voigt}} = \eta\cdot I_x^{\text{Gauss}} + (1-\eta)\cdot I_x^{\text{Lorentz}} \tag{2.143}$$

are discussed. They are all a function of x_0, which defines the centre of the respective peak or the spectral position at which the peak maximum appears, of w, which defines the width of the respective peak, of I_0, which defines the height of the peak and of x, which is the variable along the abscissa in Fig. 2.66. The Voigt profile is a mixture of the Gauss and the Lorentz peak, where $0 \le \eta \le 1$ represents the mixing factor in order to weight the contributions of both functions. Figure 2.66 shows an example Gaussian-, Lorentzian- and Voigt profile peak computed all with the parameters $x_0 = 100$, $w = 20$ and $I_0 = 1$. For the Voigt profile peak we used $\eta = 0.5$. One can see that the Lorentzian peak

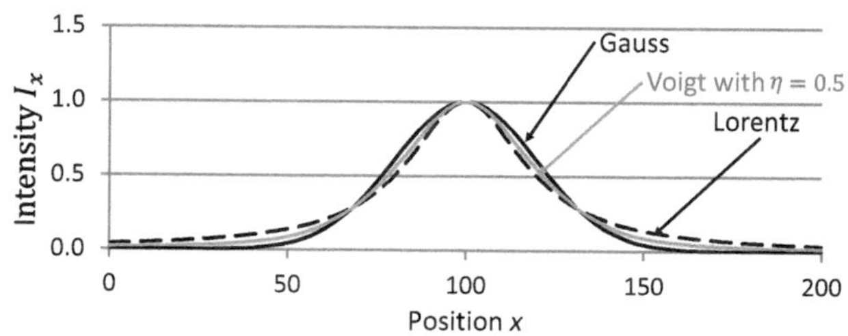

FIGURE 2.66 **Gaussian-, Lorentzian- and Voigt profile peaks.**

compared to the Gaussian peak is wider at the bottom and narrower at the top. Due to the mixing parameter of $\eta = 0.5$ the Voigt peak is exactly in between the Gaussian and the Lorentzian peak shapes.

Task: Exercise 2.10

Compute the area under each peak and find out which one, though we used the same parameters for the computation of them, features the largest area. On this purpose compute the peak functions with an increment of $\Delta x = 0.5$ between two data-points and then

$$I_{\text{Peak}} = \sum_{x=0}^{x=200} I_x \tag{2.144}$$

for each peak sum up all I_x between $x = 0$ and $x = 200$.

Answer: Exercise 2.10

$$I_{\text{Peak}}^{\text{Gauss}} = \sum_{x=0}^{x=200} I_x^{\text{Gauss}} = 100.27$$

$$I_{\text{Peak}}^{\text{Lorentz}} = \sum_{x=0}^{x=200} I_x^{\text{Lorentz}} = 109.91$$

$$I_{\text{Peak}}^{\text{Voigt}} = \sum_{x=0}^{x=200} I_x^{\text{Voigt}} = 105.09$$

Though we used the same parameters for the Gauss, the Lorentz and the Voigt peak the Lorentz peak covers the largest peak area (intensity).

Exercise 2.11: Raman Thermometry

Eqn (2.103) in Section 2.4.2.3.6 provides the Raman signal intensity of a certain transition. Based on this equation temperatures can be computed from the ratios of the Raman signal intensities of two different Raman transitions.

Task 1: Exercise 2.11

Given the pure rotational transitions S_1 and S_3 (compare Fig. 2.52) with their corresponding intensity ratio $I(S_1)/I(S_3) = 15.875$, calculate the temperature T in the probe volume. The excitation wavelength is 532 nm, the probe volume is filled with hydrogen ($B_{H_2} = 60.81\,\text{cm}^{-1}$, $\bar{\nu}_{e,H_2} = 4159.2\,\text{cm}^{-1}$) and only pure rotational transitions starting from the vibrational ground state $\text{v} = 0$ of the electronic ground state are considered. For the low rotational quantum numbers the centrifugal elongation can be neglected, meaning that the energy of the molecule can be computed according to the rigid rotator and harmonic oscillator models.

Answer: Exercise 2.11

$$I_{\frac{P}{S}} = \frac{h \cdot E_{\text{Laser}} \cdot L_{ij}}{8 \cdot \varepsilon_0^2 \cdot c} \cdot \frac{N_{\text{total}} \cdot (2J+1) \cdot e^{-\frac{T(\text{el,v},J) \cdot h \cdot c}{k \cdot T}}}{\sum_{\text{el,v},J} \left[(2J+1) \cdot e^{-\frac{T(\text{el,v},J) \cdot h \cdot c}{k \cdot T}} \right]} \cdot (\text{v}_i + 1) \cdot \frac{\left(\frac{1}{\lambda_{\text{L}}} - \bar{\nu}_{\text{R},ij} \right)^4}{\bar{\nu}_{\text{R},ij}} \cdot (\alpha'_{xyz,ij})^2$$

When creating the ratio of the intensities $I(S_1)/I(S_3)$, the first term of the Eqn (2.103) is independent of rotational level and thus disappears. N_{total} and the summation are also independent and cancel out as well. v_i is equal to zero since we only consider pure rotational transitions starting from $\text{v} = 0$. The polarisability term is also identical for both transitions. This leaves us with

$$\frac{I(S_1)}{I(S_3)} = \frac{(2 \cdot 1 + 1) \cdot e^{-\frac{(G_{\text{vib}}(\text{v}=0) + F_{\text{rot}}(J=1)) \cdot h \cdot c}{k \cdot T}}}{(2 \cdot 3 + 1) \cdot e^{-\frac{(G_{\text{vib}}(\text{v}=0) + F_{\text{rot}}(J=3)) \cdot h \cdot c}{k \cdot T}}} \cdot \frac{\bar{\nu}_{\text{R},S_3} \cdot ((1/\lambda_{\text{L}}) - \bar{\nu}_{\text{R},S_1})^4}{\bar{\nu}_{\text{R},S_1} \cdot ((1/\lambda_{\text{L}}) - \bar{\nu}_{\text{R},S_3})^4}$$

after forming the fraction of the intensities of the two rotational transitions. The Raman shift for the S_1 transition is $\bar{\nu}_{\text{R},S_1} = 10B$ and $\bar{\nu}_{\text{R},S_3} = 18B$ for the S_3 transition. We can further reduce the right side of the equation to

$$\frac{I(S_1)}{I(S_3)} = \frac{27}{35} e^{\frac{(F_{\text{rot}}(J=3) - F_{\text{rot}}(J=1)) \cdot h \cdot c}{k \cdot T}} \cdot \frac{((1/\lambda_{\text{L}}) - 10B)^4}{((1/\lambda_{\text{L}}) - 18B)^4}.$$

Inserting further values for λ_{L}, B and the intensities and then solving for the exponential function brings us to

$$\ln(18.464) = \frac{(F_{\text{rot}}(J=3) - F_{\text{rot}}(J=1)) \cdot h \cdot c}{k \cdot T}.$$

Since we are neglecting the centrifugal elongation due to the low rotational quantum number, the rotational energies are calculated as

$$F_{\text{rot}}(J) = BJ(J+1); F_{\text{rot}}(J=1) = 12{,}162\,\text{cm}^{-1}; F_{\text{rot}}(J=3) = 72{,}972\,\text{cm}^{-1}$$

We can now solve for T and insert the missing values, giving us the temperature as

$$T = \frac{(7{,}297{,}200\,\text{m}^{-1} - 1{,}216{,}200\,\text{m}^{-1}) \cdot h \cdot c}{k \cdot \ln(18.464)} = 300\,\text{K}$$

Task 2: Exercise 2.11

Given the Raman Stokes Q_{Stokes} and Raman anti-Stokes $Q_{\text{Anti-Stokes}}$ transitions with their corresponding intensity ratio $\frac{I(Q_{\text{Stokes}})}{I(Q_{\text{Anti-Stokes}})} = 3.95$, calculate the temperature in the probe volume from Eqns (2.104) and (2.105). The excitation wavelength is 532 nm and the probe volume is filled with nitrogen featuring a vibration constant $\bar{\nu}_{\text{e,N}_2} = 2330.7\,\text{cm}^{-1}$. Any higher vibrational contributions as well as anharmonicity are to be neglected.

Answer: Exercise 2.11

When creating the ratio $\frac{I(Q_{\text{Stokes}})}{I(Q_{\text{Anti-Stokes}})}$, some parts of the ratio of Eqns (2.104) and (2.105) disappear. N_{total} cancels out as well. The polarisability term is also identical for both transitions. This leaves us with

$$\frac{I^S}{I^{AS}} = \frac{e^{\frac{\bar{\nu}_{\text{e,N}_2}\cdot h\cdot c}{k\cdot T}}-1}{1-e^{-\frac{\bar{\nu}_{\text{e,N}_2}\cdot h\cdot c}{k\cdot T}}} \cdot \left(\frac{\nu_{i,S}+1}{\nu_{i,AS}}\right) \cdot \frac{((1/\lambda_{\text{L}})-\bar{\nu}_{\text{e,N}_2})^4}{((1/\lambda_{\text{L}})+\bar{\nu}_{\text{e,N}_2})^4} = e^{\frac{\bar{\nu}_{\text{e}}\cdot h\cdot c}{k\cdot T}} \cdot \left(\frac{\nu_{i,S}+1}{\nu_{i,AS}}\right) \cdot \frac{((1/\lambda_{\text{L}})-\bar{\nu}_{\text{e,N}_2})^4}{((1/\lambda_{\text{L}})+\bar{\nu}_{\text{e,N}_2})^4}.$$

We can solve for T and insert the missing values:

$$\ln\left(\frac{I^S}{I^{AS}} \cdot \frac{\nu_{i,AS}}{\nu_{i,S}+1} \cdot \left(\frac{(1/\lambda_{\text{L}})-\bar{\nu}_{\text{e,N}_2}}{(1/\lambda_{\text{L}})+\bar{\nu}_{\text{e,N}_2}}\right)^{-4}\right) = -\frac{\bar{\nu}_{\text{e}}\cdot h\cdot c}{k\cdot T}$$

$$T = -\frac{\bar{\nu}_{\text{e}}\cdot h\cdot c}{k\cdot \ln\left(\frac{I^S}{I^{AS}} \cdot \frac{1}{1+1} \cdot \left(\frac{(1/\lambda_{\text{L}})-\bar{\nu}_{\text{e,N}_2}}{(1/\lambda_{\text{L}})+\bar{\nu}_{\text{e,N}_2}}\right)^{-4}\right)} = 2000\,\text{K}$$

2.5 APPENDIX—CHAPTER 2

2.5.1 Translational Movement of Gas Molecules, the Mean Velocity of Gases and the Maxwell Velocity Distribution

2.5.1.1 Mean Thermal Energy of a Gas Molecule

Here we see that Eqn (2.26)

$$E_{\text{kin}} = \sum_{N_{\text{molec}}} \left(\frac{1}{2} m_{\text{molec}} \cdot s_{\text{molec}}^2\right) = N_{\text{molec}} \cdot \frac{f}{2} \cdot k \cdot T = E_{\text{thermal}} \qquad (2.26)$$

can also be derived by considering the mean kinetic energy of all the gas molecules and the pressure equation.

The kinetic energy E_{kin} of the gas,

$$E_{kin} = \sum_{N_{molec}} E_{kin,molec} = \sum_{N_{molec}} \left(\frac{1}{2} m_{molec} \cdot s_{molec}^2\right) = N_{molec} \cdot \overline{E_{kin,molec}} = N_{molec} \cdot \frac{1}{2} m_{molec} \cdot \overline{s_{molec}^2} \tag{2.145}$$

which is the summation of the kinetic energies $E_{kin,molec}$ of all the single gas molecules, can also be represented by considering the mean kinetic energy $\overline{E_{kin,molec}}$ of the molecules. As all the molecules in the gas volume have the same mass, Eqn (2.145) can be developed further to

$$E_{kin} = N_{molec} \cdot \frac{1}{2} m_{molec} \cdot \overline{s_{molec}^2} = \frac{1}{2} \cdot m_{gas} \cdot c_{gas}^2 \tag{2.146}$$

where m_{gas} is the mass of all the molecules contained in the volume and c_{gas} is the mean velocity of the gas molecules, with respect to the kinetic energy.

According to the gas theory, the pressure

$$P = \frac{2}{f} \cdot \frac{E_{kin}}{V} = \frac{2}{f} \cdot \frac{N_{molec}}{V} \overline{E_{kin,\,molec}} \tag{2.147}$$

in a volume V of a gas is a function of the degrees of freedom f of the translational movement of the gas molecules, the number density N_{molec} / V of the gas molecules and the mean kinetic energy $\overline{E_{kin,\,molec}}$ of a single molecule. Furthermore, according to the equation of state of an ideal gas

$$P = \frac{n_{subst} \cdot R \cdot T}{V}, \tag{2.148}$$

the pressure P is also a function of the temperature T and the volume V and the amount of substance n_{subst} contained in that gas volume. Bringing together Eqns (2.147) and (2.148) results in the correlation between the kinetic and the thermal energy

$$\overline{E_{kin,\,molec}} = \frac{f}{2} \cdot \frac{V}{N_{molec}} \cdot \frac{n_{subst} \cdot R \cdot T}{V} = \frac{f}{2} \cdot k \cdot T = \overline{E_{thermal,\,molec}} \tag{2.149}$$

with $n_{subst} / N_{molec} = 1 / N_{Avogadro}$ and $R / N_{Avogadro} = k$.

2.5.1.2 'Mean Velocity' of a Gas Molecule, With Respect to the Kinetic Energy

Inserting into (under the circumstances that $U_{chem} = U_{potential} = 0$)

$$U_{internal} = E_{kin} = E_{thermal} \tag{2.150}$$

the kinetic $E_{\text{kin}} = \frac{1}{2} \cdot m_{\text{gas}} \cdot c_{\text{gas}}^2$ and the thermal energy $E_{\text{thermal}} = \frac{(f=3)}{2} \cdot \frac{m_{\text{gas}}}{M_{\text{gas}}} \cdot R \cdot T$ of the gas, with M_{gas} being the molar mass of the gas, the 'mean velocity' of the gas molecules

$$c_{\text{gas}} = \sqrt{\frac{(f=3) \cdot R \cdot T}{M_{\text{gas}}}} \tag{2.151}$$

can be computed as a function of the molar mass of the gas molecules M_{gas} and the temperature T. I decided to put 'mean velocity' into quotation marks as strictly speaking c_{gas} is not the mean velocity of the gas molecules, but the velocity of a gas molecule whose kinetic energy $E_{\text{kin,molec}}$ corresponds to the mean kinetic energy of a single molecule $\overline{E_{\text{kin, molec}}}$.

2.5.1.3 Maxwell Velocity Distribution

It has been mentioned in Section 2.2.2 that the velocity distribution $F(s_{\text{molec}})$ of gas molecules at a certain temperature is described by the Maxwell distribution

$$F(s_{\text{molec}}) = 4\pi \left(\frac{M_{\text{gas}}}{2\pi \cdot R \cdot T} \right)^{\frac{3}{2}} s_{\text{molec}}^2 \cdot \exp\left(\frac{-M_{\text{gas}} \cdot s_{\text{molec}}^2}{2 \cdot R \cdot T} \right) \tag{2.152}$$

where M_{gas} is the molar mass of the gas species, T is the temperature, R is the ideal gas constant and s_{molec} is the translational velocity of a single gas molecule. Figure 2.67 illustrates the influence of the temperature and of the molar mass onto the velocity distribution of the gas molecules. With increasing temperature T and with decreasing molar mass M_{gas} the velocity distribution gets broader and the centre of the distribution moves to higher velocities s_{molec}. The

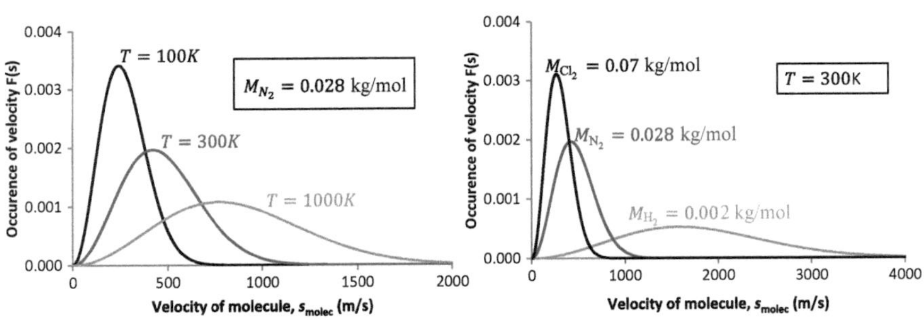

FIGURE 2.67 Velocity distribution of gas molecules computed according to the Maxwell distribution (Eqn (2.152)) for (left) nitrogen at three different temperatures and (right) at 300 K for three different gases. *(Source: Distributions computed by Jaypee Quino from the Erlangen Graduate School in Advanced Optical Technologies of the Friedrich-Alexander-Universität Erlangen-Nürnberg, Germany.)*

shift of the centre of the velocity distribution to higher velocities with increasing temperature or decreasing molar mass is also in agreement with Eqn (2.151), which describes the impact of the temperature and the molar mass onto the mean velocity.

2.5.2 Boltzmann's Distribution

Here it is shown how one can derive from the general form of the Boltzmann distribution (Eqn (2.60))

$$\frac{N_{\mathrm{el,v},J}}{N_{\mathrm{total}}}=\frac{(2J+1)\cdot e^{\left(-\frac{T(\mathrm{el,v},J)\cdot h\cdot c}{k\cdot T}\right)}}{\sum_{\mathrm{el,v},J}\left[(2J+1)\cdot e^{\left(-\frac{T(\mathrm{el,v},J)\cdot h\cdot c}{k\cdot T}\right)}\right]}$$
$$=\frac{(2J+1)\cdot e^{\left(-\frac{T(\mathrm{el})\cdot h\cdot c}{k\cdot T}\right)}\cdot e^{\left(-\frac{G_{\mathrm{vib}}(\mathrm{v})\cdot h\cdot c}{k\cdot T}\right)}\cdot e^{\left(-\frac{F_{\mathrm{rot}}(J)\cdot h\cdot c}{k\cdot T}\right)}}{\sum_{\mathrm{el,v},J}\left[(2J+1)\cdot e^{\left(-\frac{T(\mathrm{el})\cdot h\cdot c}{k\cdot T}\right)}\cdot e^{\left(-\frac{G_{\mathrm{vib}}(\mathrm{v})\cdot h\cdot c}{k\cdot T}\right)}\cdot e^{\left(-\frac{F_{\mathrm{rot}}(J)\cdot h\cdot c}{k\cdot T}\right)}\right]} \tag{2.60}$$

the simplified equations relevant for the population distribution over solely vibrational or rotational states. First we no more consider the ratio $N_{\mathrm{el,v},J}\,/\,N_{\mathrm{total}}$ but the ratio of the number of molecules occupying two (indices 0 and 1) distinct energy levels

$$\frac{N_{(\mathrm{el,v},J),1}}{N_{(\mathrm{el,v},J),0}}=\frac{N_{(\mathrm{el,v},J),1}\,/\,N_{\mathrm{total}}}{N_{(\mathrm{el,v},J),0}\,/\,N_{\mathrm{total}}}$$
$$=\frac{\left[(2J+1)\cdot e^{\left(-\frac{T(\mathrm{el})\cdot h\cdot c}{k\cdot T}\right)}\cdot e^{\left(-\frac{G_{\mathrm{vib}}(\mathrm{v})\cdot h\cdot c}{k\cdot T}\right)}\cdot e^{\left(-\frac{F_{\mathrm{rot}}(J)\cdot h\cdot c}{k\cdot T}\right)}\right]_1}{\left[(2J+1)\cdot e^{\left(-\frac{T(\mathrm{el})\cdot h\cdot c}{k\cdot T}\right)}\cdot e^{\left(-\frac{G_{\mathrm{vib}}(\mathrm{v})\cdot h\cdot c}{k\cdot T}\right)}\cdot e^{\left(-\frac{F_{\mathrm{rot}}(J)\cdot h\cdot c}{k\cdot T}\right)}\right]_0}. \tag{2.153}$$

If we consider that the two different energy levels are both in the electronic ground state $T(\mathrm{el})=T(X)=0$ we can simplify the equation to

$$\frac{N_{(\mathrm{el,v},J),1}}{N_{(\mathrm{el,v},J),0}}=\frac{N_{(X,\mathrm{v},J),1}}{N_{(X,\mathrm{v},J),0}}=\frac{\left[(2J+1)\cdot e^{\left(-\frac{G_{\mathrm{vib}}(\mathrm{v})\cdot h\cdot c}{k\cdot T}\right)}\cdot e^{\left(-\frac{F_{\mathrm{rot}}(J)\cdot h\cdot c}{k\cdot T}\right)}\right]_1}{\left[(2J+1)\cdot e^{\left(-\frac{G_{\mathrm{vib}}(\mathrm{v})\cdot h\cdot c}{k\cdot T}\right)}\cdot e^{\left(-\frac{F_{\mathrm{rot}}(J)\cdot h\cdot c}{k\cdot T}\right)}\right]_0}. \tag{2.154}$$

If we consider only the rotational energy levels featuring all the same vibrational quantum number and thus the same vibrational energy $\left[e^{\left(-\frac{G_{vib}(v)\cdot h\cdot c}{k\cdot T}\right)}\right]_1 = \left[e^{\left(-\frac{G_{vib}(v)\cdot h\cdot c}{k\cdot T}\right)}\right]_0$ we further can simplify the equation to

$$\frac{N_{(el,v,J),1}}{N_{(el,v,J),0}} = \frac{N_{(X,v_1=v_0,J),1}}{N_{(X,v_1=v_0,J),0}} = \frac{\left[(2J+1)\cdot e^{\left(-\frac{F_{rot}(J)\cdot h\cdot c}{k\cdot T}\right)}\right]_1}{\left[(2J+1)\cdot e^{\left(-\frac{F_{rot}(J)\cdot h\cdot c}{k\cdot T}\right)}\right]_0}. \tag{2.155}$$

If we now consider the ratio of molecules occupying any excited rotational $J = 1, 2, 3, \ldots$ and the rotational ground state $J = 0$, we reach the simplified Boltzmann equation

$$\frac{N_{(el,v,J),1}}{N_{(el,v,J),0}} = \frac{N_{(X,v_1=v_0,J),1}}{N_{(X,v_1=v_0,J),0}} = \frac{N_{X,v=const,J=i}}{N_{X,v=const,J=0}} = \frac{(2J+1)\cdot e^{\left(-\frac{F_{rot}(J=i)\cdot h\cdot c}{k\cdot T}\right)}}{(2(J=0)+1)\cdot e^{\left(-\frac{[F_{rot}(J=0)=0]\cdot h\cdot c}{k\cdot T}\right)}}$$
$$= (2J+1)\,e^{\left(-\frac{F_{rot}(J=i)\cdot h\cdot c}{k\cdot T}\right)}$$

already presented as Eqn (2.62) in Section 2.2.7.

If we want to derive the simplified Boltzmann equation describing the population distribution of molecules over energy levels featuring different vibrational quantum numbers we first have to sum up $N_{sum,el=const,v=const,J}$ all molecules occupying rotational energy levels within the same vibrational quantum number and relate it again to the overall number of molecules.

$$N_{sum,el=const,v=const,J} = \frac{\sum_{el=const,v=const,J=0}^{el=const,v=const,J=n} N_{el=const,v=const,J}}{N_{total}}$$
$$= \frac{\sum_{el=const,v=const,J=0}^{el=const,v=const,J=n}\left[(2J+1)\cdot e^{\left(\frac{T(el,v,J)\cdot h\cdot c}{k\cdot T}\right)}\right]}{\sum_{el,v,J}\left[(2J+1)\cdot e^{\left(\frac{T(el,v,J)\cdot h\cdot c}{k\cdot T}\right)}\right]} \tag{2.156}$$
$$= \frac{\sum_{el=const,v=const,J=0}^{el=const,v=const,J=n}\left[(2J+1)\cdot e^{\left(\frac{T(el)\cdot h\cdot c}{k\cdot T}\right)}\cdot e^{\left(\frac{G_{vib}(v)\cdot h\cdot c}{k\cdot T}\right)}\cdot e^{\left(\frac{F_{rot}(J)\cdot h\cdot c}{k\cdot T}\right)}\right]}{\sum_{el,v,J}\left[(2J+1)\cdot e^{\left(-\frac{T(el)\cdot h\cdot c}{k\cdot T}\right)}\cdot e^{\left(-\frac{G_{vib}(v)\cdot h\cdot c}{k\cdot T}\right)}\cdot e^{\left(-\frac{F_{rot}(J)\cdot h\cdot c}{k\cdot T}\right)}\right]}$$

Then we can form the ratio of the summation of all molecules $N_{\text{sum,el=const,v}=i,J}$ occupying energy levels within one vibrational quantum number $\text{v}=i=1,2,3,\ldots$ and the summation of all molecules $N_{\text{sum,el=const,v=0},J}$ occupying energy levels within the vibrational ground state $\text{v}=0$

$$\frac{N_{\text{sum,el=const,v}=i,J}}{N_{\text{sum,el=const,v=0},J}} = \frac{\sum_{\text{el=const,v}=i,J=0}^{\text{el=const,v}=i,J=n}\left[(2J+1)\cdot e^{\left(\frac{T(\text{el})\cdot h\cdot c}{k\cdot T}\right)}\cdot e^{\left(\frac{G_{\text{vib}}(\text{v}=i)\cdot h\cdot c}{k\cdot T}\right)}\cdot e^{\left(\frac{F_{\text{rot}}(J)\cdot h\cdot c}{k\cdot T}\right)}\right]}{\sum_{\text{el=const,v=0},J=0}^{\text{el=const,v=0},J=n}\left[(2J+1)\cdot e^{\left(\frac{T(\text{el})\cdot h\cdot c}{k\cdot T}\right)}\cdot e^{\left(\frac{G_{\text{vib}}(\text{v}=0)\cdot h\cdot c}{k\cdot T}\right)}\cdot e^{\left(\frac{F_{\text{rot}}(J)\cdot h\cdot c}{k\cdot T}\right)}\right]}. \tag{2.157}$$

For the electronic ground state $T(\text{el})=T(X)=0$ this equation simplifies to

$$\frac{N_{\text{sum},X,\text{v}=i,J}}{N_{\text{sum},X,\text{v}=0,J}} = \frac{\sum_{X,\text{v}=i,J=0}^{X,\text{v}=i,J=n}\left[(2J+1)\cdot e^{\left(\frac{G_{\text{vib}}(\text{v}=i)\cdot h\cdot c}{k\cdot T}\right)}\cdot e^{\left(\frac{F_{\text{rot}}(J)\cdot h\cdot c}{k\cdot T}\right)}\right]}{\sum_{X,\text{v}=0,J=0}^{X,\text{v}=0,J=n}\left[(2J+1)\cdot e^{\left(\frac{G_{\text{vib}}(\text{v}=0)\cdot h\cdot c}{k\cdot T}\right)}\cdot e^{\left(\frac{F_{\text{rot}}(J)\cdot h\cdot c}{k\cdot T}\right)}\right]}. \tag{2.158}$$

Furthermore, as we only sum up over the rotational energy levels, we can take the 'vibrational terms' out of the summation

$$\frac{N_{\text{sum},X,\text{v}=i,J}}{N_{\text{sum},X,\text{v}=0,J}} = \frac{e^{\left(-\frac{G_{\text{vib}}(\text{v}=i)\cdot h\cdot c}{k\cdot T}\right)}\cdot\sum_{X,\text{v}=i,J=0}^{X,\text{v}=i,J=n}\left[(2J+1)\cdot e^{\left(\frac{F_{\text{rot}}(J)\cdot h\cdot c}{k\cdot T}\right)}\right]}{e^{\left(-\frac{G_{\text{vib}}(\text{v}=0)\cdot h\cdot c}{k\cdot T}\right)}\cdot\sum_{X,\text{v}=0,J=0}^{X,\text{v}=0,J=n}\left[(2J+1)\cdot e^{\left(\frac{F_{\text{rot}}(J)\cdot h\cdot c}{k\cdot T}\right)}\right]} = \frac{e^{\left(-\frac{G_{\text{vib}}(\text{v}=i)\cdot h\cdot c}{k\cdot T}\right)}}{e^{\left(-\frac{G_{\text{vib}}(\text{v}=0)\cdot h\cdot c}{k\cdot T}\right)}} \tag{2.159}$$

and see that the summations over the rotational quantum numbers cancel out. The ratio of the two exponential functions can be rewritten as

$$\frac{N_{\text{sum},X,\text{v}=i,J}}{N_{\text{sum},X,\text{v}=0,J}} = \frac{N_{X,\text{v}=i}}{N_{X,\text{v}=0}} = e^{\left(-\frac{\left[G_{\text{vib}}(\text{v}=i)-G_{\text{vib}}(\text{v}=0)\right]\cdot h\cdot c}{k\cdot T}\right)},$$

which is Eqn (2.61) already introduced in Section 2.2.7.

2.5.3 Beer–Lambert Law

Figure 2.68 shows the exponential decay of the light intensity of a specific wavelength, which is absorbed on the way through a cuvette filled with the absorbing fluid.

Within the cuvette the intensity decay

$$\mathrm{d}I = -I(x)\cdot a(\lambda)\cdot \mathrm{d}x \tag{2.160}$$

over an infinitesimal small distance dx is proportional to the intensity $I(x)$ and the wavelength dependent attenuation coefficient $a(\lambda)$. An integration

$$\int_{I_{\text{incident},\lambda}}^{I_{\text{transmitted},\lambda}} \frac{\mathrm{d}I}{I(x)} = -\int_{x=0}^{x=l} a(\lambda)\cdot \mathrm{d}x. \tag{2.161}$$

results in the exponential equation

$$\frac{I_{\text{transmitted},\lambda}}{I_{\text{incident},\lambda}} = e^{-[a(\lambda)\cdot l]} \tag{2.162}$$

or in the logarithmic (normal) form

$$\ln\left(\frac{I_{\text{transmitted},\lambda}}{I_{\text{incident},\lambda}}\right) = -[a(\lambda)\cdot l]\,. \tag{2.163}$$

For the transformation into the decadal logarithmic form

$$\log\left(\frac{I_{\text{incident},\lambda}}{I_{\text{transmitted},\lambda}}\right) = \varepsilon(\lambda)\cdot c_{\text{abs}}\cdot l, \tag{2.164}$$

which is known as the Beer–Lambert law, the attenuation coefficient

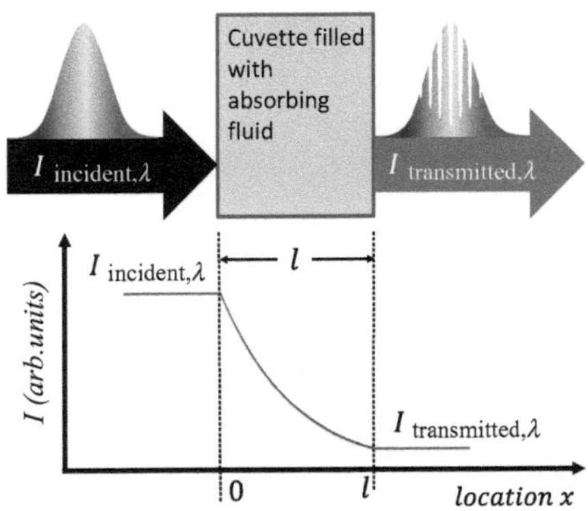

FIGURE 2.68 Exponential decay of the intensity of light of a specific wavelength passing through a medium, which absorbs light of this wavelength.

$$a(\lambda) = \varepsilon(\lambda) \cdot c_{abs} \cdot \left(\frac{\ln(10)}{\log_{10}(10)} = 2.3026 \right) \tag{2.165}$$

is substituted by the molar attenuation coefficient $\varepsilon(\lambda)$ and the molar concentration c_{abs} times a factor 2.3026, which is required for the transformation of the normal to the decadal logarithm.

2.5.4 Meaning of a Cross-Section

In spectroscopy the cross-section provides information about the probability of light–molecule interaction. We can find cross-sections not only for light–molecule interactions resulting in absorption of the incident photon but also in those resulting in scattering of the incident photon. But for now we will consider absorption only and describe the meaning of the cross-section with respect to the probability that a molecule absorbs an incident photon.

We consider that we launch the incident light within a cross-sectional area F through a cuvette, which is filled with the absorbing fluid. Of course only these molecules, which are present in that regions through which the incident light passes, can absorb the incident light. Therefore the more molecules $N_{pot,abs}$, which potentially could absorb, are present within the projection area F, the higher is the probability that the incident light is absorbed. Now we also have to consider the absorption cross-section $\varepsilon(\lambda)$ of the molecules, which defines the projection area around the molecule, through which a photon, which is absorbed, has to pass. If a photon passes the molecule outside this absorption area, it is too far away to be absorbed. Please be aware that the absorption cross-section and the spatial cross-section of the extension of a molecule are different. A nice illustration is to compare the absorption cross-section with the thickness of a baseball racket. The thicker the racket, which is the molecule, the higher is the probability that the baseball, which is the photon, will interact (collide) with the racket. Now, considering the cross-sectional area F of the incident light, the number of potentially absorbing molecules $N_{pot,abs}$ which are present within the projection area F and the absorption cross-section of each molecule $\varepsilon(\lambda)$, the probability that one incident photon will be absorbed

$$\text{Probability} = \frac{N_{pot,abs} \cdot \varepsilon(\lambda)}{F} = \frac{N_{Photon,abs}}{N_{Photon,incident}}. \tag{2.166}$$

can also be computed from the ratio of the numbers of absorbed $N_{Photon,abs}$ and incident photons $N_{Photon,incident}$.

2.5.5 Calibration of a Composition (Fuel/Air-Ratio) Indicating LIF Tracer

Before we can use an aromatic compound as a composition indicating LIF tracer we have to check whether or not the conditions $1 \ll k_{SV} \cdot c_Q$ and $k_{SV} = \text{const}$ hold (compare Eqns (2.88) and (2.90). On this account we calibrate the fluorescence signal intensity S_F as a function of the fuel/air-ratio c_F / c_Q. The index 'F' represents the potentially fluorescing molecules. As we know how many fluorescence tracer molecules we added to the non-fluorescing model fuel (e.g., isooctane) we can derive the fuel concentration from c_F. Let us assume that we use 1 wt% of tracer in isooctane. The index 'Q' represents the potentially quenching molecules. This is mainly oxygen, whose concentration in fresh air is known to be 21 vol%. Thus the ratio c_F / c_Q can be considered to indicate indirectly the fuel/air-ratio.

Two potential calibration routines, from which only one method is correct, are shown in Table 2.5 and Table 2.6. In both cases the amount of substance of LIF tracer (only 1 wt% in isooctane) is negligible compared to the amount of substances of fuel (isooctane), oxygen and air. In both tables the fuel/air-equivalence ratio is varied between infinite and zero by varying the composition of the mixture composed of fuel, oxygen and nitrogen, while the overall amount of substance is set constant at 60.5 mol (see right column). The fuel/air equivalence ratio

$$\phi = \frac{(c_F / c_Q)_{\text{actual}}}{(c_F / c_Q)_{\text{stoichiometric}}} \tag{2.167}$$

is defined as the ratio of the actual fuel/air-ratio and the fuel/air-ratio required for a stoichiometric combustion. For the stoichiometric combustion of 1-mol isooctane

$$C_8H_{18} + 12.5\,O_2 + 47.0\,N_2 \rightarrow 8\,CO_2 + 9\,H_2O + 47\,N_2$$

12.5-mol oxygen are required. As in air there are 3.762-mol nitrogen per 1-mol oxygen, 47-mol nitrogen accompany the combustion of 1-mol isooctane as inert gas.

TABLE 2.5 Wrong Calibration of a Composition LIF Tracer

Fuel/air-equivalence ratio	Model fuel isooctane (mol)	Oxygen (mol)	Nitrogen (mol)	Amount of substance total (mol)
∞	1	0	59.5	60.5
3	3	12.1	45.4	60.5
2	2	12.3	46.2	60.5
1.5	1.5	12.4	46.6	60.5
1	1	12.5	47.0	60.5
0.5	0.5	12.6	47.4	60.5
0.3	0.3	12.6	47.6	60.5
0	0	12.7	47.8	60.5

Mainly the fuel concentration is varied. The volume of the calibration chamber is 1 L.

TABLE 2.6 Correct Calibration of a Composition LIF Tracer

Fuel/air-equivalence ratio	Model fuel isooctane (mol)	Oxygen (mol)	Nitrogen (mol)	Amount of substance total (mol)
∞	1	0	59.5	60.5
3	1	4.17	55.33	60.5
2	1	6.25	53.25	60.5
1.5	1	8.3	51.2	60.5
1	1	12.5	47.0	60.5
0.5	1	25	35.5	60.5
0.3	1	41.67	17.83	60.5
0	0	12.5	47	60.5

Mainly the oxygen (quencher) concentration is varied. The volume of the calibration chamber is 1 L.

While in Table 2.5 the fuel/air-equivalence ratio is mainly varied by varying the amount of substance of fuel, and with this the concentration c_F inside the mixture, in Table 2.6 the fuel/air equivalence ratio is mainly varied by varying the amount of substance of oxygen and with this the concentration c_Q in the mixture.

Figure 2.69 shows sketches of calibration curves derived using the approaches described in the context of Table 2.5 and Table 2.6. The fluorescence

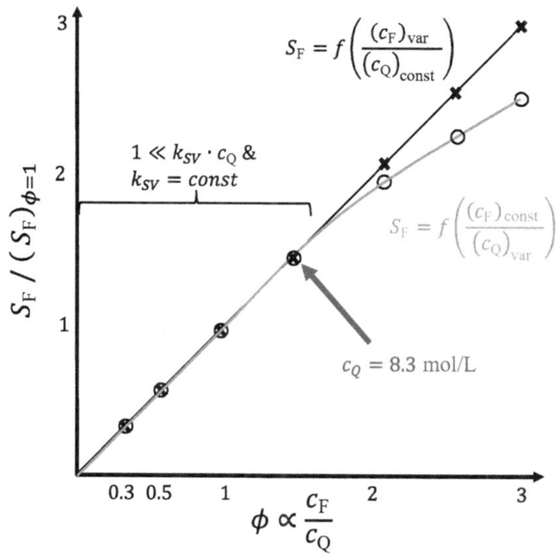

FIGURE 2.69 **Sketch of a calibration curve of a composition indicating tracer substance for two different calibration approaches.**

signal intensity of the ordinate is normalised to the fluorescence signal intensity corresponding to the fuel/air equivalence ratio of 1. This value was chosen for normalisation as for $\phi = 1$ the composition of the mixtures composed of fuel, oxygen and nitrogen are identical in both tables. While the black solid line represents the calibration curve acquired when mainly the concentration c_F was varied and c_Q was kept constant, the grey curve represents the calibration curve when mainly the concentration c_Q was varied and c_F was kept constant. According to Fig. 2.69, the composition LIF approach $S_F \propto c_F / c_Q$ seems to be applicable for fuel/air equivalence ratios smaller than approximately 1.5. This is only half of the truth. Recalling that the probability of quenching depends on the quencher concentration c_Q and not on the ratio c_F / c_Q makes clear that neither the fuel/air equivalence ratio ϕ nor the ratio c_F / c_Q but solely the quencher concentration c_Q defines where LIF signals can be evaluated according to the composition LIF strategy. Therefore one has to analyse the quencher concentrations for which $1 \ll k_{SV} \cdot c_Q$ and $k_{SV} = \text{const}$ holds.

According to Fig. 2.69 and Table 2.6 this is the case whenever the quencher (oxygen) concentration is equal or exceeds $c_Q \geq 8.3$ mol L^{-1}. (The volume of the calibration chamber is 1 L.) Only then the product of the Stern–Volmer *constant* and the concentration of quenchers is significantly larger than 1 (compare Eqn (2.90)). It should be underlined here that the quencher concentration c_Q in an internal combustion engine is of course a function of the compression. With increasing compression c_Q increases, as the volume decreases. Consequently, with increasing compression the fuel/air equivalence ratio regime in which the LIF signals can be evaluated according to the composition LIF strategy increases to larger ϕ-values, respectively larger ratios c_F / c_Q.

2.5.6 Hybrid Deconvolution of Mixture Spectra (Spectral Fit Approach)

Figure 2.70 shows the strategy how a 'best fit' spectrum, which finally best fits the measured Raman spectrum of the mixture, is composed of the Raman spectra of the compounds acetone, water and CO_2. The black dotted spectrum in the forefront of Fig. 2.70 shows the 'difference spectrum' between the 'measured spectrum' and the 'best fit' spectrum.

The integral of the squared 'difference spectrum' in the entire spectral range shown, is considered as the abort criterion of the Levenberg–Marquardt algorithm [74], which—solving a set of non-linear equations using the least-square method—fits the sum of the weighted Raman spectra of the pure compounds to the measured spectrum of the mixture.

In the further progress of this section, the origin of the Raman spectra of the pure compounds is explained, which during the fit procedure are weighted by multiplication with weighing factors.

Raman spectrum of the organic solvent: The Raman spectrum of pure acetone can be acquired experimentally from pure liquid acetone, in the best case at

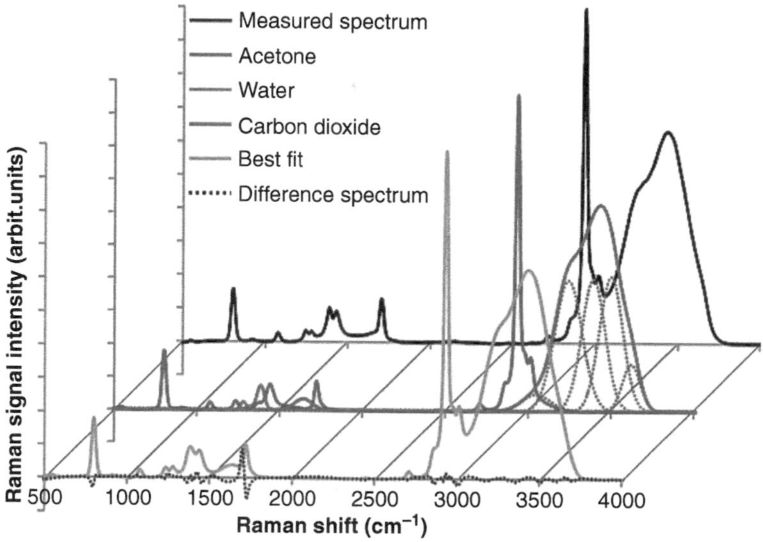

FIGURE 2.70 Illustration of the strategy for the isolation of the Raman signal contributions of spectrally overlapping compounds from the spectrum of the ternary mixture of acetone, water and CO_2. *(Source: [73].)*

the same operational conditions at which the acetone containing mixture should be analysed afterwards. In Fig. 2.70 one can see the maximum difference between the 'fit spectrum' and the 'measured spectrum' between 1500 cm^{-1} and 2000 cm^{-1} is very near the carbonyl Raman signal of acetone. When acetone gets in contact with CO_2 the CO_2 influences the energy of the carbonyl group and consequently the respective peak of acetone is spectrally shifted with respect to the pure acetone [75]. This CO_2–carbonyl interaction is responsible for the difference between the 'fitted spectrum' and the 'measured spectrum' between 1500 cm^{-1} and 2000 cm^{-1}. The intensity of the acetone Raman signal is the intensity of the entire Raman spectrum of acetone in the shown spectral range, meaning between wavenumbers of 500 cm^{-1} and 4000 cm^{-1}.

Raman spectrum of pure CO_2: The Raman spectrum of pure CO_2, which is used for the fit strategy, unfortunately cannot be taken from Raman measurements in pure CO_2. Once CO_2 is mixed with other compounds, the ratio of the intensities of the two main peaks the CO_2 Raman signal is composed of (they are called Fermi dyad, see Fig. 2.55) changes and thus becomes different to the ratio of these two dyads in a pure CO_2 phase. Additionally, the central wavelengths of the dyads can shift in mixtures and as function of the state of phase. Therefore the Raman spectrum of pure CO_2 is replaced by two Voigt profile peaks, which feature a variable central wavenumber and a variable intensity each. The intensity of the Raman signal of CO_2 is the summation of the intensities of the two Voigt profile peaks.

Raman spectrum of pure water: The Raman spectrum of pure water, which is used for the fit strategy, unfortunately can neither be taken from Raman

TABLE 2.7 Central Wavenumber Positions of the Seven Peaks the Spectrum of Liquid Water is Composed (cm^{-1})

Water (OH) bend vibration		Water (OH) stretching				
Two Gaussian peaks between 1500 cm^{-1} and 2000 cm^{-1}		Five Gaussian peaks between 2500 cm^{-1} and 4000 cm^{-1}				
1581	1642	3051	3233	3408	3500	3624

measurements in pure water. The shape of the Raman spectrum of water is influenced by the development of hydrogen bonds [57–59,76]. As hydrogen bonds change with a change in temperature (see Fig. 2.56), the degree of compression, the addition of other compounds, the Raman spectrum of water in a compressed mixture looks very much different from the Raman spectrum of pure water at a particular temperature. Fortunately it is known that the Raman spectrum of water between 1500 cm^{-1} and 2000 cm^{-1} (bending vibration of water) can be replaced by two Gaussian peaks and between 2500 cm^{-1} and 4000 cm^{-1} (stretch vibration of water) can be replaced by five Gaussian peaks. Additionally the central wavenumber positions of the Gaussian peaks of the two bending and the five stretching vibrations are known, which are listed in Table 2.7. In the centre plane of Fig. 2.70 only the five Gaussian peaks which together form the envelope of the stretch vibration Raman signals are given as dotted curves. The two small Gaussian peaks forming the envelope of the bending vibration Raman signal are not resolved, only its envelope is given. The intensity of the Raman signal of water is the summation of the intensities of the seven Gaussian single peaks.

REFERENCES

[1] Haken H, Wolf HC. Molecular physics and elements of quantum chemistry: introduction to experiments and theory. Berlin Heidelberg: Springer Science & Business Media; 2004.

[2] Hoffman D, Münch KU, Leipertz A. Two-dimensional temperature determination in sooting flames by filtered Rayleigh scattering. Optics Lett 1996;21:525–7.

[3] Most D, Dinkelacker F, Leipertz A. Direct determination of the turbulent flux by simultaneous application of filtered Rayleigh scattering thermometry and particle image velocimetry. Proc Combust Inst 2002;29:2669–77.

[4] Most D, Leipertz A. Simultaneous two-dimensional flow velocity and gas temperature measurements by use of a combined particle image velocimetry and filtered Rayleigh scattering technique. Appl Optics 2001;40:5379–87.

[5] Leipertz A, Obertacke R, Wintrich F. Industrial combustion control using UV emission tomography. Symp Int Combust 1996;26:2869–75.

[6] Higgins B, McQuay MQ, Lacas F, Rolon JC, Darabiha N, Candel S. Systematic measurements of OH chemiluminescence for fuel-lean, high-pressure, premixed, laminar flames. Fuel 2001;80:67–74.

[7] Snelling DR, Thomson KA, Smallwood GJ, lder ØĜL, Weckman EJ, Fraser RA. Spectrally resolved measurement of flame radiation to determine soot temperature and concentration. AIAA J 2002;40:1789–95.
[8] Amer E, Gren P, Sjödahl M. Stimulated laser induced fluorescence holography for imaging fluorescent species. Optics Commun 2013;311:124–8.
[9] Hell SW, Wichmann J. Breaking the diffraction resolution limit by stimulated emission: stimulated-emission-depletion fluorescence microscopy. Optics Lett 1994;19:780–2.
[10] Hindman J, Kugel R, Svirmickas A, Katz J. Stimulated fluorescence and fluorescence quenching in chlorophyll a and bacteriochlorophyll a. Chem Phys Lett 1978;53:197–200.
[11] Duarte FJ, Kelley P, Hillman LW, Liao PF. Dye laser principles. New York: Academic Press; 1990.
[12] Maiman TH. Ruby laser systems. Google Patents; 1967.
[13] Forster R, Frost M, Fulle D, Hamann H, Hippler H, Schlepegrell A, Troe J. High pressure range of the addition of HO to HO, NO, NO_2, and CO. I. Saturated laser induced fluorescence measurements at 298 K. J Chem Phys 1995;103:2949–58.
[14] Fulle D, Hamann H, Hippler H, Troe J. Temperature and pressure dependence of the addition reactions of HO to NO and to NO_2. IV. Saturated laser-induced fluorescence measurements up to 1400 bar. J Chem Phys 1998;108:5391–7.
[15] Magaldi M, Marrocco M. Negligibility of temporal laser wing effects in saturated LIF diagnostics. Appl Phys B 2003;76:699–705.
[16] Schäfer M, Ketterle W, Wolfrum J. Saturated 2D-LIF of OH and 2D determination of effective collisional lifetimes in atmospheric pressure flames. Appl Phys B 1991;52:341–6.
[17] Thurber MC, Grisch F, Hanson RK. Temperature imaging with single- and dual-wavelength acetone planar laser-induced fluorescence. Optics Lett 1997;22:251–3.
[18] Koban W, Koch JD, Hanson RK, Schulz C. Oxygen quenching of toluene fluorescence at elevated temperatures. Appl Phys B 2005;80:777–84.
[19] Nau WM, Scaiano JC. Oxygen quenching of excited aliphatic ketones and diketones. J Phys Chem 1996;100:11360–7.
[20] Ossler F, Alden M. Measurements of picosecond laser induced fluorescence from gas phase 3-pentanone and acetone: implications to combustion diagnostics. Appl Phys B 1998;64:493–502.
[21] Ware WR, Lee SK. Luminescent properties of hexafluoroacetone. II. Fluorescence quenching by oxygen, nitric oxide, and unsaturated hydrocarbons. J Chem Phys 1968;49:217–20.
[22] Faust S, Dreier T, Schulz C. Temperature and bath gas composition dependence of effective fluorescence lifetimes of toluene excited at 266 nm. Chem Phys 2011;383:6–11.
[23] Hickman CG, Gascooke JR, Lawrance WD. The S1-S0 (1B2-1A1) transition of jet-cooled toluene excitation and dispersed fluorescence spectra, fluorescence lifetimes, and intramolecular vibrational energy redistribution. J Chem Phys 1996;104:4887–901.
[24] Karolczak J, Komar D, Kubicki J, Wróżowa T, Dobek K, Ciesielska B, Maciejewski A. The measurements of picosecond fluorescence lifetimes with high accuracy and subpicosecond precision. Chem Phys Lett 2001;344:154–64.
[25] Ni T, Melton LA. Fluorescence lifetime imaging: An approach for fuel equivalence ratio imaging. Appl Spectrosc 1991;45:938–43.
[26] Lozano A, Yip B, Hanson RK. Acetone: a tracer for concentration measurements in gaseous flows by planar laser-induced fluorescence. Exp Fluids 1992;13:369–76.
[27] Hu H, Koochesfahani MM. A novel technique for quantitative temperature mapping in liquid by measuring the lifetime of laser induced phosphorescence. J Visual 2003;6:143–53.
[28] Liu JB, Pan Q, Liu CS, Shi JR. Principles of flow field diagnostics by laser induced biacetyl phosphorescence. Exp Fluids 1988;6:505–13.

[29] Okabe H. The relative intensities of fluorescence and phosphorescence in biacetyl vapor. J Am Chem Soc 1957;79:801–6.
[30] Pringsheim P. Fluorescence and phosphorescence. New York: Interscience Publishers; 1949.
[31] Stern O, Volmer M. Über die Abklingzeit der Fluoreszenz. Physikalische Zeitschrift 1919;20:183–8.
[32] Koban W, Schorr J, Schulz C. Oxygen-distribution imaging with a novel two-tracer laser-induced fluorescence technique. Appl Phys B 2002;74:111–4.
[33] Koban W, Koch JD, Hanson RK, Schulz C. Toluene LIF at elevated temperatures: implications for fuel–air ratio measurements. Appl Phys B 2005;80:147–50.
[34] Crosley DR. Laser probes for combustion chemistry. American Chemical Society; 1980.
[35] Gardiner WC. Gas-phase combustion chemistry. New York: Springer Science & Business Media; 2000.
[36] Tsang W, Hampson R. Chemical kinetic data base for combustion chemistry. Part I. Methane and related compounds. J Phys Chem Ref Data 1986;15:1087–279.
[37] Thurber MC, Grisch F, Kirby BJ, Votsmeier M, Hanson RK. Measurements and modeling of acetone laser-induced fluorescence with implications for temperature-imaging diagnostics. Appl Optics 1998;37:4963–78.
[38] Long DA. Raman spectroscopy. New York: McGraw-Hill; 1977.
[39] Garnier A, Chanin M. Description of a Doppler Rayleigh lidar for measuring winds in the middle atmosphere. Appl Phys B 1992;55:35–40.
[40] Miles R, Lempert W. Two-dimensional measurement of density, velocity, and temperature in turbulent high-speed air flows by UV Rayleigh scattering. Appl Phys B 1990;51:1–7.
[41] Herzberg G. Spectra of diatomic molecules. Malabar, Florida: Krieger Publishing; 1989.
[42] Atkins PW, Friedman RS. Molecular quantum mechanics. Oxford: Oxford University Press; 1997.
[43] Wang X, Chou IM, Hu W, Burruss RC, Sun Q, Song Y. Raman spectroscopic measurements of CO_2 density: experimental calibration with high-pressure optical cell (HPOC) and fused silica capillary capsule (FSCC) with application to fluid inclusion observations. Geochim Cosmochim Acta 2011;75:4080–93.
[44] Rodriguez-Meizoso I, Lazor P, Turner C. *In situ* Raman spectroscopy for the evaluation of solubility in supercritical carbon dioxide mixtures. J Supercrit Fluids 2012;65:87–92.
[45] Lavorel B, Millot G, Saint-Loup R, Berger H, Bonamy L, Bonamy J, Robert D. Study of collisional effects on band shapes of the ν1/2ν2 Fermi dyad in CO_2 gas with stimulated Raman spectroscopy. I. Rotational and vibrational relaxation in the 2ν2 band. J Chem Phys 1990;93:2176–84.
[46] Kachi Y, Tsukahara T, Kayaki Y, Ikariya T, Sato J, Ikeda Y. Raman spectral shifts of CO_2 as measure of CO_2-philicity of solutes in supercritical carbon dioxide. J Supercrit Fluids 2007;40:20–6.
[47] Cabaço MI, Danten Y, Tassaing T, Longelin S, Besnard M. Raman spectroscopy of CO_2–acetone and CO_2–ethanol complexes. Chem Phys Lett 2005;413:258–62.
[48] Ikeda-Fukazawa T, Kita D, Nagashima K. Raman spectroscopic study of CO_2 sorption process in poly methyl methacrylate. J Polym Sci Part B 2008;46:831–42.
[49] Pastore Carbone MG, Di Maio E, Musto P, Braeuer A, Mensitieri G. On the unexpected non-monotonic profile of specific volume observed in PCL/CO_2 solutions. Polymer 2015;56:252–5.
[50] Becucci M, Cavalieri S, Eramo R, Fini L, Materazzi M. Accuracy of remote sensing of water temperature by Raman spectroscopy. Appl Optics 1999;38:928–31.
[51] Müller T, Grünefeld G, Beushausen V. High-precision measurement of the temperature of methanol and ethanol droplets using spontaneous Raman scattering. Appl Phys B 2000;70:155–8.

[52] Gorbaty YE, Bondarenko GV, Venardou E, Garcia-Verdugo E, Sokolova M, Ke J, Poliakoff M. High-pressure high-temperature Raman spectroscopy of liquid and supercritical fluids. Appl Spectrosc 2003;57:1300–3.
[53] Hankel RF, Günther A, Wirth K-E, Leipertz A, Braeuer A. Liquid phase temperature determination in dense water sprays using linear Raman scattering. Optics Express 2014;22:7962–71.
[54] Walrafen GE, Hokmabadi MS, Yang WH. Raman investigation of the temperature dependence of the bending v2 and combination v2+vl bands from liquid water. J Phys Chem 1988;92:2433–8.
[55] Buback M, Schulz KR. Raman scattering of pure ammonia at high pressures and temperatures. J Phys Chem 1976;80:2478–82.
[56] Knauer OS, Lang MC, Braeuer A, Leipertz A. Simultaneous determination of the composition and temperature gradients in the vicinity of boiling bubbles in liquid binary mixtures using one-dimensional Raman measurements. J Raman Spectrosc 2011;42:195–200.
[57] Auer B, Kumar R, Schmidt JR, Skinner JL. Hydrogen bonding and Raman, IR, and 2D-IR spectroscopy of dilute HOD in liquid D_2O. Proc Natl Acad Sci 2007;104:14215–20.
[58] Carey DM, Korenowski GM. Measurement of the Raman spectrum of liquid water. J Chem Phys 1998;108:2669–75.
[59] Barlow SJ, Bondarenko GV, Gorbaty YE, Yamaguchi T, Poliakoff M. An IR study of hydrogen bonding in liquid and supercritical alcohols. J Phys Chem A 2002;106:10452–60.
[60] Hankel RF, Rojas PE, Cano-Sarabia M, Sala S, Veciana J, Braeuer A, Ventosa N. Surfactant-free CO_2-based microemulsion-like systems. Chem Commun 2014;50:8215–8.
[61] Seeger T. Moderne Aspekte der linearen und nichtlinearen Raman-Streuung zur Bestimmung thermodynamischer Zustandsgrößen in der Gasphase. Erlangen: ESYTEC; 2006.
[62] Leipertz A. Temperaturbestimmung in Gasen mittels linearer und nichtlinearer Raman - Prozesse; 1984.
[63] Eckhardt G, Wagner WG. On the calculation of absolute Raman scattering cross sections from Raman scattering coefficients. J Molec Spectrosc 1966;19:407–11.
[64] Ahmad SR, Foster VG. The local-field effect in Raman scattering. J Phys D 1997;30:2677.
[65] Egermann J, Seeger T, Leipertz A. Application of 266-nm and 355-nm Nd:YAG laser radiation for the investigation of fuel-rich sooting hydrocarbon flames by Raman scattering. Appl Optics 2004;43:5564–74.
[66] Petry R, Schmitt M, Popp J. Raman spectroscopy—a prospective tool in the life sciences. ChemPhysChem 2003;4:14–30.
[67] Efremov EV, Ariese F, Gooijer C. Achievements in resonance Raman spectroscopy: review of a technique with a distinct analytical chemistry potential. Anal Chim Acta 2008;606:119–34.
[68] Schmitt M, Popp J. Raman spectroscopy at the beginning of the twenty-first century. J Raman Spectrosc 2006;37:20–8.
[69] Sun Y, Shekunov BY, York P. Refractive index of supercritical CO_2-ethanol solvents. Chem Eng Comm 2003;190:1–14.
[70] Alsmeyer F, Koß H-J, Marquardt W. Indirect spectral hard modeling for the analysis of reactive and interacting mixtures. Appl Spectrosc 2004;58:975–85.
[71] Kriesten E, Mayer D, Alsmeyer F, Minnich CB, Greiner L, Marquardt W. Identification of unknown pure component spectra by indirect hard modeling. Chemometr Intel Lab Syst 2008;93:108–19.
[72] Sokolova M, Barlow SJ, Bondarenko GV, Gorbaty YE, Poliakoff M. Comparison between IR absorption and Raman scattering spectra of liquid and supercritical 1-butanol. J Phys Chem A 2006;110:3882–5.
[73] Schuster JJ, Will S, Leipertz A, Braeuer A. Deconvolution of Raman spectra for the quantification of ternary high pressure phase equilibria composed of carbon dioxide, water and organic solvent. J Raman Spectrosc 2014;45:246–52.

[74] Marquardt DW. An algorithm for least-squares estimation of nonlinear parameters. J Soc Indus Appl Math 1963;11:431–41.
[75] Sala S, Danten Y, Ventosa N, Tassaing T, Besnard M, Veciana J. Solute-solvent interactions governing preferential solvation phenomena of acetaminophen in CO_2-expanded organic solutions: a spectroscopic and theoretical study. J Supercrit Fluids 2006;38:295–305.
[76] Dougherty RC. Temperature and pressure dependence of hydrogen bond strength: a perturbation molecular orbital approach. J Chem Phys 1998;109:7372–8.

Chapter 3

Raman Spectroscopy From an Engineering Point of View

I hope that Raman inexperienced researchers feel motivated to build up their own Raman measurement device once they have read this chapter. Therefore, I describe in the first section which type of Raman set-ups are applicable in general and what measurement purposes they are suitable for. In the following section, I provide information regarding the identification of the optimum components for the respective measurement purpose and Raman set-up. Here the Sections 3.2.1–3.2.4 can be considered as a guidance for those who really intend to build up their own Raman sensor. The last section more or less mentions some unexpected complications I had to face with respect to Raman sensor development in the past. Of course, having built up a Raman sensor, one also wants to carry out Raman measurements. Then very often one has to face undesired interferences in the Raman spectra, which make an evaluation of the Raman signals challenging. The origin of these interferences can be manifold. In many research papers there is not much reported about how the raw spectra were processed in order to get the very nice ones which are shown in the publication. For sure not all very nicely looking Raman spectra shown in publications have been detected as such. Sometimes the raw spectra are processed, though it is neither explicitly mentioned that they have been processed nor how they have been processed. Therefore the second to last section of this chapter covers methods suitable for the rejection of broadband interferences. The last section before the appendix reports some case studies where Raman measurement techniques brought valuable insights into high-pressure process technology. I also tried to cover here some fields of application which might not be so obvious to researchers coming from the field of supercritical fluids. In the Appendix that follows this chapter, I provide first information about the working principle of quality-switched pulsed lasers and which special characteristics result from them and why these characteristics are so beneficial if fast processes have to be resolved temporally. Then I provide a comprehensive description of the family of CCD detectors, the 'pros' and 'cons' of different CCD architectures and under which circumstance which CCD architecture is suitable for which measurement purpose. The pulsed laser and the CCD detectors are part of the appendix of this chapter, as this is the first chapter where

Supercritical Fluid Science and Technology, Vol 7. http://dx.doi.org/10.1016/B978-0-444-63422-1.00003-1

the respective information is relevant. But of course the information provided in this appendix is also relevant for the subsequent chapters and will therefore act as reference beyond this chapter.

3.1 THREE BASIC RAMAN SENSOR DESIGNS

Each Raman sensor introduced in this section is composed of

- an excitation light source, which usually is a laser,
- a dispersive unit for the spectral separation of the Raman signals,
- a detection unit, which usually is a charge-coupled device (CCD) and
- light forming or manipulating components, such as lenses, mirrors and prisms.

All these components are independent of each other and can be obtained in a big variety of specifications from different manufacturers. Nowadays – thanks to USB interfaces – the combination of different stand-alone components to one sensor system is straightforward and finally all required components can be controlled simultaneously from only one computer. The big variety of components available on the market enables the experimenter to adopt and fit the Raman sensor system to the needs of the high-pressure experiment.

In an ideal case, the Raman sensor is applied to measure *in situ* in the real process. Then the respective apparatus has to be equipped with windows for the provision of optical access to the interior of the process and consequently the measurements can be performed under real technical process circumstances. If the process itself was adapted to a rigid and already preconfigured 'black-box' Raman sensor system, for example by the installation of a bypass from the real apparatus to the Raman sensor, the obtained results would not directly reflect the mechanisms taking place inside the real process (compare Section 1.3).

In the following three, basic example experimental set-ups are introduced and described in detail. They were chosen in order to cover an as broad as possible range of measurement purposes. This includes

- one of the most straight-forward set-ups for measuring the Raman signal irradiated from a point source in back scattering direction described in Section 3.1.1,
- one very flexible set-up for measuring the Raman signal irradiated from many point sources aligned along a line simultaneously and spatially resolved described in Section 3.1.2 and
- one set-up suitable for performing two-dimensional (2-D) measurements with a measurement duration short enough to temporally 'freeze turbulence' in Section 3.1.3.

Although the 0-D Raman set-up is suitable for measuring for example the temperature or the solubility of a component or the composition of a mixture at one point, the one-dimensional (1-D) Raman set-up is suitable for measuring composition- or temperature profiles and how they behave in time. From the temporal evolution of temperature and composition profiles especially, heat and mass transport phenomena can be analysed. The 2-D Raman set-up is suitable for analysing rapid and turbulent mixing processes especially.

3.1.1 Signal Detection From One Point (0-D Raman Spectroscopy)

In Fig. 1.7 already two examples of 0-D optical measurement techniques are given, one for signal detection in back scattering configuration and one for signal detection in right-angle configuration. Whenever there is no need to measure with a high spatial resolution, the application of a back scattering sensor configuration in a cage system, as it is shown in Fig. 3.3, is recommended because of three main advantages. These advantages are based on the considerations we already made in Section 2.4.2.3.

No Misalignment due to a Variation of the Index of Refraction (see Fig. 1.8):

It can be seen in Figs 3.3 and 3.4 that the excitation laser or the excitation fibre, the optical components and the glass fibre detection bundle are all mounted together in one connected cage system. This assures that the focal spot of the excitation laser and the probe volume which is imaged onto the plane round end of the fibre bundle always remain congruent after an initial alignment, even if the cage system is moved.

The following example is reflected by the sketch we already discussed in the context of Fig. 1.8. With the back scattering Raman sensor one can first measure the composition of the light gas phase in a view cell, then move the cage system downwards and measure the composition of the denser liquid phase, without any effort in re-aligning the set-up. Even the change of the index of refraction between the liquid- and the gas-phase system does not cause any trouble. If the index of refraction inside the measurement cell changes, both, the location of the excited volume and the probed volume experience approximately the same spatial shift. This means that an overlap between the excited volume and the probed volume is always assured. Assuming the right-angle signal detection configuration, the probed volume and the excited volume are no more congruent after the change of the index of refraction. Therefore, without re-aligning the right angle signal detection configuration, no Raman signals or significantly reduced Raman signals intensities will be detected (compare Fig. 1.8).

The Detectable Signal Intensities are High:

Figure 1.8 indicates by the dashed lines the depth of focus of the detection system. For the case of the back scattering–detection configuration Raman signals can be detected not only from the centre of the excitation beam waist but also from locations in front and behind the beam waist. This is not possible for the case of the right angle–detection configuration, as the depth of focus of the detection system only partially overlaps with the waist of the excitation beam. Therefore significantly (up to one order of magnitude and dependent on the focal length of the used lenses) more intense Raman signals are detectable with the back-scattering configuration. If the focal length of that lens is changed, through which the excitation beam is focused and at the same time the Raman signals are collimated, the solid angle of detection as well as the depth of focus is manipulated. At first glance, one would expect higher detectable signal levels using a lens with a shorter focal length. Because then the solid angle of signal detection would be increased (Section 2.4.2.3.5 and Eqn (2.121)). Nevertheless, if the focal length of the lens is decreased, the depth of focus is also decreased and therefore the Raman signals detectable then originate from a smaller volume. The decrease of the depth of focus compensates the increase of the solid angle of signal detection and thus the detected signal intensity remains unchanged.

It should be mentioned here that if the focal length of the focusing lens in a back-scattering configuration is chosen too long, the very long depth of focus can overlap spatially with the window material of the high-pressure vessel. Then signals originating from the windows could interfere with the Raman spectrum of the fluid inside the measurement chamber.

Depolarisation Effects Do Not Influence the Raman Signal Intensity (compare Section 2.4.2.3.4):

According to the descriptions made in Section 2.4.2.3.4 for signal detection in back-scattering direction (along the reverse x-axis), the summation of s- and p-polarised Raman signals is constant, irrespective of the polarisation of the excitation source (Eqn (2.120)). This also means that the birefringence of the windows does not influence the intensity of the detected Raman signals. The birefringence of the windows describes their tendency to manipulate the polarisation of an electric field (laser light) while it passes through the window. The birefringence of the windows can change due to a change of stress, which can be due to a change of pressure or temperature. Even if the birefringent windows manipulate the polarisation of the excitation radiation in an unknown way, the detected Raman signal intensities will remain the same (compare Eqns (2.118) and (2.119)). In the case of the right angle signal detection set-up, a change of the birefringence of the windows will impact the intensity of the detected Raman signals (compare Eqns (2.115) and (2.116)).

In order to demonstrate this experimentally, four normalised spectra shown in Fig. 3.1 were recorded at four different operational conditions by pressurising and/or heating a mixture of acetone and CO_2 with a constant CO_2 molar fraction of 0.82.

According to the theory derived in Section 2.4.2.3.7, the normalised Raman spectra of the mixture of constant composition but at different conditions should be identical, if the factor K in Eqn (2.128) in fact is a constant in the range of varied operational conditions. The four spectra in Fig. 3.1 can be assumed to be identical. Thus we can conclude that in accordance to theory the change of the birefringence of the windows of the vessel (due to pressure and temperature

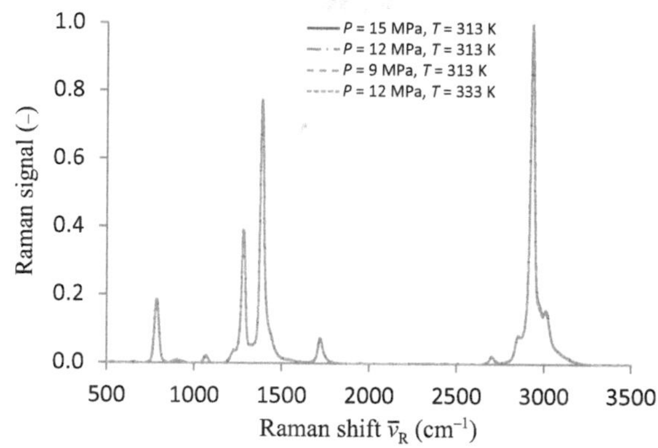

FIGURE 3.1 **Raman spectra of a mixture of acetone and CO_2 with a CO_2 molar fraction of 0.82 at four different operational conditions (three different pressures and two different temperatures).**

variation) might indeed have changed the polarisation of the excitation laser radiation, but did not influence the detected Raman spectra. This also indicates that 'back scattered' calibration results which came from measurements carried out at a particular temperature and pressure may be transferred for the evaluation of the composition of mixtures also from 'back scattered' Raman spectra at operational conditions other than the calibration conditions.

3.1.1.1 The Very Basic Configuration

In the following, a 0-D Raman sensor working in a back–scattering signal detection configuration is described, as it is shown in Fig. 3.2. Many types of lab microscopes are designed according to the same configuration. This type of sensor design is especially suitable for measuring compositions of mixtures

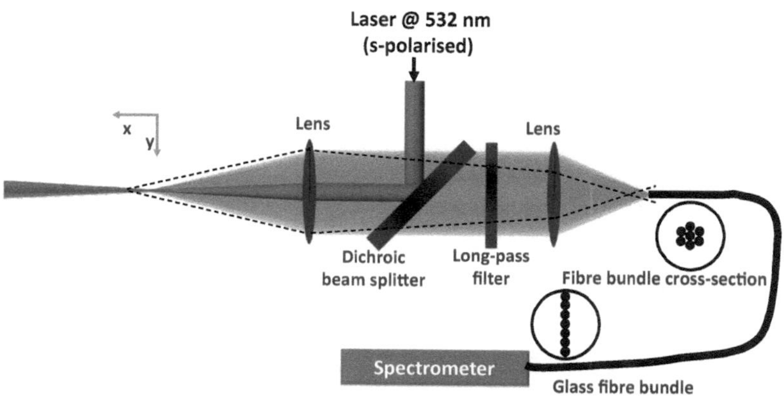

FIGURE 3.2 **Sketch of a zero-dimensional (0-D) Raman sensor with signal detection in back-scattering direction.** *(Source: Ref. [1].)*

in the liquid, the supercritical or the gaseous state of phase, when the systems are in thermodynamic equilibrium. Under this circumstance, there is no need to realise neither a certain temporal- nor a spatial resolution. The only prerequisite is the detection of Raman spectra with a high signal-to-noise ratio.[1] Consequently a 'low-budget' excitation laser can be utilised. In the following, this Raman sensor configuration is described based on a frequency-doubled continuous-wave (cw) Nd:YAG excitation laser with an emission wavelength of 532 nm and an output power of 200 mW. In general other lasers can be used instead, but then also other filters and dichroic mirrors have to be taken which are not mentioned here.

The excitation beam is reflected via a dichroic beam splitter, which features a high reflection for wavelengths shorter than 540 nm and high transmission for wavelengths longer than 540 nm, through the focusing lens towards the laser focal point. The excited sampling volume (laser beam focus) is imaged onto the entrance interface of the fibre bundle (where the fibres are assembled like a rosette) according to the optical path of the red-shifted Raman scattered signals. Instead of the glass fibre bundle, a single glass fibre can also be utilised. The dashed lines indicate the path of Raman signals originating from behind the focal spot. Only a reduced portion of them is launched into the fibre bundle, as the major portion misses the cross-section of the rosette assembly of the fibres. Summarising, the Raman signals are not sampled from the focal spot alone but from a certain volume along the waist the laser beam forms around the focal spot. The length of this volume is said to be the depth of focus.

The dichroic beam splitter in combination with the long-pass filter assures that elastically scattered laser light is efficiently rejected. Elastic light scattering interference originates from Rayleigh, Mie or optical scattering (compare sections 2.3.4 and 2.4.2.2) in the sampling volume or due to reflections from glass surfaces or the vessel walls. Without their rejection, the intensity of the elastic light scattering interferences would exceed the intensity of the Raman signals by many orders of magnitude and would make a detection of Raman spectra impossible. Having purified the Raman signals from the elastic light-scattering interference, they are guided through the fibre bundle towards the spectrometer. Here USB compatible spectrometers, which are equipped with an integrated CCD chip already, can be integrated straightforward into each set-up.

The 0-D Raman sensor sketched in Fig. 3.2 is shown as a photograph in Fig. 3.3.

The beam diameter of the excitation laser is expanded using a telescope for two reasons:

- The power of the laser radiation reaching the probe volume can be tuned by opening or closing the variable aperture, which is installed behind the telescope in Fig. 3.3, only if the beam has been expanded before. If not, the aperture can act as a binary 'on–off' switch, only.

---

1. The meaning of the signal-to-noise ratio and how it is influenced is described in the appendix of this chapter.

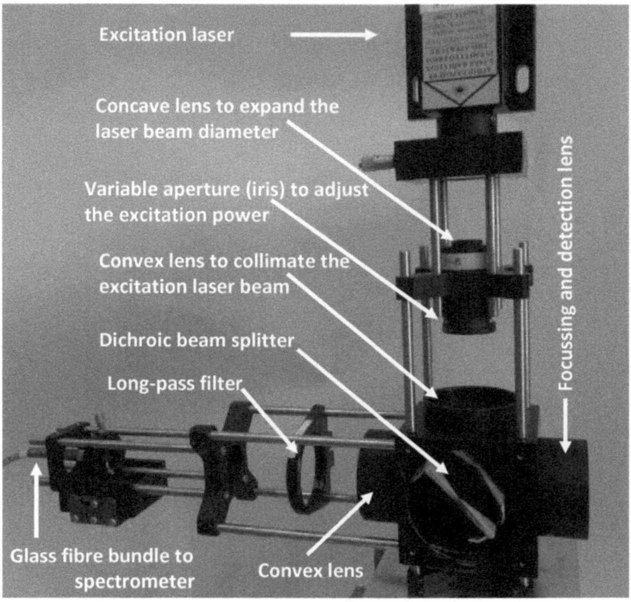

FIGURE 3.3 **Photograph of the 0-D Raman sensor with integrated excitation laser source.**

- The focal point (focal beam waist) of the excitation laser beam can be visually identified by the experimenter the better, the more the laser beam has been expanded before it passes through the final focusing lens. This also simplifies aligning the cage-system's housing of the 0-D Raman sensor in order to exactly image the excited volume (excitation laser focal waist) onto the 'detection fibre' bundle.

3.1.1.2 The High Power Raman Sensor With Continuous Excitation

The laser devices which can provide excitation powers above 200 mW are typically too big and too heavy to integrate them directly into the cage system of the Raman sensor. Then the laser beam has first to be launched into a separate 'excitation fibre' guiding the excitation radiation from the laser to the cage system. Consequently, as it is shown in Fig. 3.4, the 'excitation fibre' is mounted into the cage system instead of the laser.

Fields of application are in general whenever the signal-to-noise ratio of the Raman sensor design described in Section 3.1.1.1 is not sufficient and in particular for

- the detection of molecules which are rare in the mixture,
- the detection of molecules if the density of the fluid under investigation is low, such as in low-pressure gases at high temperatures, or
- the detection of molecules if the experiment needs acquisition times shorter than those realisable with a lower laser power.

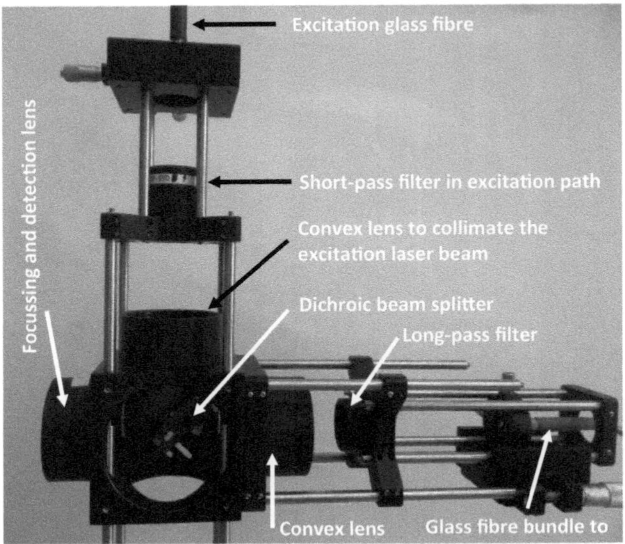

FIGURE 3.4 **Photograph of the 0-D Raman sensor with external excitation laser source.**

The laser light coming out of the 'excitation fibre' diverges according to the fibre's numerical aperture. Therefore, one lens is required for the collimation of the diverging laser radiation.

One has to consider that the radiation coming out of the 'excitation fibre' – is spectrally broadened compared to the radiation (laser emission) initially launched into the 'excitation fibre'. This is due to light–matter interactions between the laser radiation and the glass fibre material taking place all the path length the radiation travels through the glass fibre. Consequently a short-pass filter has to be integrated into the excitation part of the cage system which blocks all wavelengths red-shifted with respect to the laser wavelength. Otherwise, the red-shifted excitation radiation, would be detected as elastically scattered signals from the sampling volume and would also interfere spectrally with the red-shifted Raman signals. This interference would make a reliable evaluation of Raman signals impossible. A more comprehensive description of this circumstance is provided in Section 3.2.5.

3.1.1.3 The Pulsed Version With ns Temporal Resolution

Once experiments with microsecond temporal resolution have to be carried out, the continuous-wave laser has to be replaced by a typically quality-switched pulsed excitation laser.[2] This might be the case when turbulent processes have to be frozen temporally [2]. A pulsed laser provides high excitation energies in

2. The differences between continuous-wave and Q-switched pulsed lasers are described in the appendix of this chapter.

FIGURE 3.5 Photograph of a laser-induced breakdown (ionisation of air) at the focal spot of a pulsed laser. In addition to the breakdown spark the converging and diverging laser beam left and right of the focal beam waist is nicely visible. One also can see bright spots, which are due to elastic light scattering from dust particles. *(Source: Photo made by Daniel Bassing from the Erlangen Graduate School in Advanced Optical Technologies of the Friedrich-Alexander-Universität Erlangen-Nürnberg, Germany.)*

a short light pulse. Here, pulsed Nd:YAG lasers are first choice, as they feature a high robustness, high single-shot energies from few Millijoules up to several Joules and short full-width-half-maximum pulse durations of approximately 10 ns. 1 J contained in one single 10 ns lasting laser pulse corresponds to pulse powers of 100 MW. The treatment of such high pulse powers is challenging with respect to two aspects:

- If focused to one small spot, the high irradiance ($W\,m^{-2}$) causes ionisation (breakdown) of the matter in the focal spot. The breakdown that follows is shown as a photo in Fig. 3.5. It is a bright spark which completely obscures the desired Raman spectrum and makes a quantification of the Raman signals impossible. The breakdown spark can be as bright as saturating the detectors used for the acquisition of the Raman signals and might even destroy the CCD chips. If image intensifiers[3] based on the external photoelectric effect, such as an intensified camera, are used in the detection path, even more care has to be taken, as the risk of intensifier damage in the case of a breakdown spark is even much higher than the risk of CCD damage.
- Furthermore, the extremely high peak power of a laser pulse can exceed the damage threshold of the optical windows and destroy them which for measurement applications in high-pressure systems has to be prevented. Figure 3.6 shows a crack in the centre of a window induced by high-power laser pulses. If the laser focus is located close to the optical windows, the cross-section of the pulsed laser beam is rather small when it passes through the window. Then the damage threshold of the window material can be exceeded already at peak powers far (orders of magnitude) below 100 MW [3].

Therefore, in order to avoid both, ionisation of the matter in the focal spot as well as exceeding the laser-induced damage threshold of the window material, a

3. The working principle of image intensifiers is described in the appendix of this chapter.

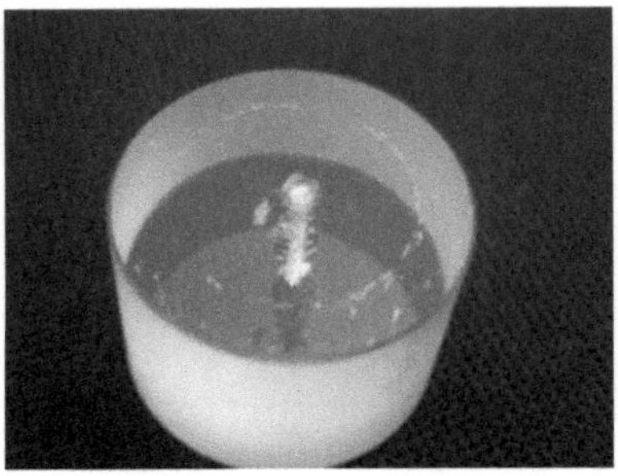

FIGURE 3.6 Photograph of a window broken by laser pulses having exceeded the damage threshold of the window material.

pulse stretcher is essential to be used in combination with common Q-switched high power pulsed Nd:YAG lasers. The pulse stretcher stretches the laser pulse in time, whereas its pulse energy – in the ideal case – remains constant and thus reduces the pulse peak power [4]. In reality, due to the reflectivity of the mirrors of less than one, the use of a pulse stretcher is always accompanied with energy losses. The working principle of a pulse stretcher is described in Section 3.2.2.

Pulsed high-power lasers are too big and too heavy to be installed directly into a cage system as one is shown in Fig. 3.3. They also cannot be integrated indirectly by using an 'excitation fibre' as it is shown in Fig. 3.4. This is not possible, as the intensity of the laser pulses (ns-regime) inside the small cross-section of the light guiding core of the fibre usually exceeds the damage threshold of the glass fibres. Hollow-core fibres are not a real alternative as they suffer from high bending losses (they are not flexible) and/or high in-coupling losses. This means that for the pulsed Raman sensor the excitation laser cannot be directly connected with the detection system in a flexible and small cage system. A direct connection is possible, if all the components are fixed on one table (or small optical bread board), which would result in a rather large sensor system significantly reduced in its flexibility.

If there is no demand of performing measurements with very short exposure time (high temporal resolution), continuous-wave lasers should be used especially for safety reasons.

3.1.2 Detection of Line Profiles

For the description of this 1-D Raman sensor shown in Fig. 3.7. I also chose a continuous-wave Nd:YAG excitation laser with an emission wavelength of 532 nm. Again, one also can use a different type of an excitation laser, which has to be suitable for the selective excitation of the Raman process (see Section 3.2.1). Of course in this case, as the wavelengths of the Raman signals according to Eqn (2.99) are a function of the excitation wavelength, also the wavelength-specific components (dichroic mirrors and filters) mentioned in the following have to be modified accordingly.

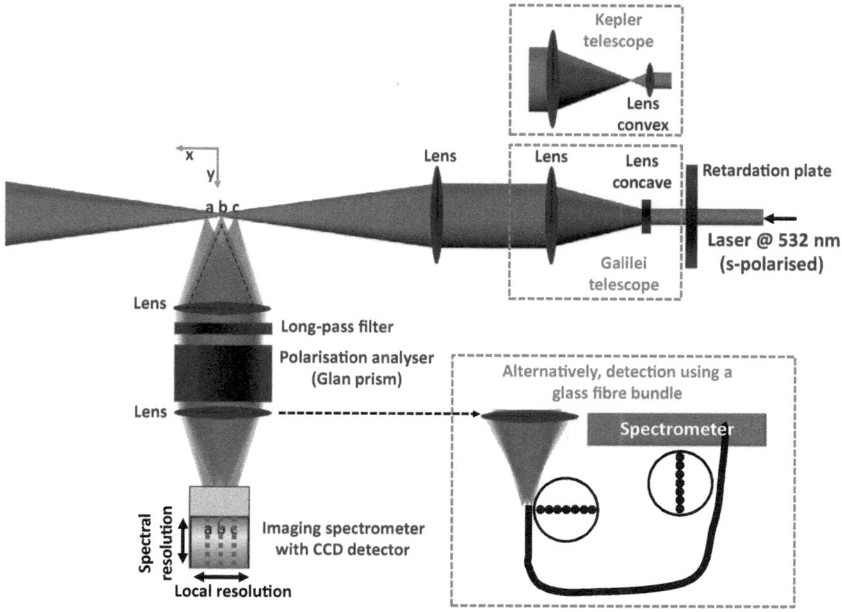

FIGURE 3.7 Sketch of a 1-D Raman sensor with signal detection in right angle direction. If a glass fibre bundle is used for signal detection in combination with a spatially not resolving spectrometer, the 1-D Raman sensor becomes 0-*D*.

So, a frequency-doubled continuous-wave (cw) Nd:YAG laser with an emission wavelength of 532 nm and an output power of 5 W (available with much higher powers) can be used as excitation source. For many experiments, only a small portion of the available laser power is sufficient for the generation of Raman signal levels with signal-to-noise ratios exceeding 100. The polarisation of the laser beam is adjusted using a retardation plate, either a half-wave plate in order to adjust the plane of polarisation between 's' and 'p' or a quarter-wave plate in order to convert linear polarised light into circular polarised light (see Eqn (2.6) in Section 2.1.2). Then a telescope is used for the expansion of the cross-section of the excitation laser beam. The laser beam is expanded in order to increase the numerical aperture of the focussed excitation beam. The higher this numerical aperture is (the stumper the excitation beam cone is), the smaller is the cross-section of the focal waist.

Two types of telescopes are sketched in Fig. 3.7. Both are suitable for the expansion of the cross-section of continuous-wave laser beams. Once the cross-section of a pulsed laser beam should be expanded, the Galilei type telescope is the only option. If a Kepler type telescope was used instead, the laser intensity in the focal spot between the two lenses would exceed the ionisation threshold of air. Within the generated plasma the excitation radiation is absorbed which significantly reduces the excitation power of the laser pulse left for the excitation of the probe volume. Irrespectively of this, it should be mentioned here again that a pulse stretcher has to be applied, when using high-energy short-pulsed lasers. A pulse stretcher set-up is not contained in Fig. 3.7 but sketched in Section 3.2.2.

Along the positions 'a', 'b' and 'c' in Fig. 3.7, the beam waist may be assumed to be constant in diameter (constant cross-section), resulting in a cylinder-like excited volume. The depth of focus of that part of the set-up detecting the Raman signals in y-direction does not influence the local resolution of the Raman sensor. The Raman signals are generated only in the volume excited by the excitation laser. Thus the local resolution of the 1-D Raman sensor in y- and z-direction is determined by the excitation part of the set-up alone and is the cross-section of the laser focal beam waist. It should be mentioned that high numerical apertures of the focused beam on the one hand side decrease the cross-section of the beam waist, but on the other hand side shorten the length in x-direction along which the beam waist may be assumed to be shaped cylinder-like with a constant cross-section.

Only the local resolution in x-direction is determined by the detection part of the sensor. Two lenses image the beam waist shown in Fig. 3.7 onto the entrance slit of an imaging spectrometer. The installation of a long-pass filter assures again the efficient rejection of elastically scattered light, while the red-shifted Raman signals pass to the spectrometer. The Raman signals are dispersed spectrally on their way through the spectrometer, but the information about their local origin is preserved. Consequently, one dimension of the 2-D CCD chip of the camera mounted onto the imaging spectrometer is required to record the spectral information, whereas the other dimension is required to record the spatial information. Considering again the example already given in Fig. 1.10, the CCD chip of 35×35 pixels, can record spectra from maximum 35 locations along the waist of the laser beam. There are 5 pixels binned along the spatial axis in Fig. 1.10. Therefore only 7 instead of 35 increments can be resolved spatially, which are indicated by the seven cylindrical slices, which all together compose the measurement volume. The pixels along the spectral axis should be binned adequately in order to the requirements of the evaluation procedure of the spectra.

Figure 3.7 also shows the alternative of launching the Raman signals into the spectrometer using a fibre bundle. In this case it is recommended to use a fibre bundle configured 'linear to linear'. Then, the waist of the excitation laser beam can be imaged onto the fibre bundle interface with the fibres aligned parallel to the x-axis. Consequently, each fibre guides Raman signals originating spatially resolved from one location to the spectrometer. If an imaging spectrometer with a 2-D CCD chip is used then, and if the alignment of the fibre bundle coupled to the entrance slit is perpendicular to the spectral axis of the spectrometer, Raman signals can be recorded spatially resolved exactly how it has been described previously for the 1-D Raman sensor not employing a fibre bundle. If the fibres of the bundle-end which are coupled to the spectrometer are not aligned but configured like a rosette or if the spectrometer is equipped with a 1-D array CCD detector only, the 1-D Raman sensor becomes 0-D.

The retardation plate integrated in the sketch shown in Fig. 3.7 is required for the compensation of the depolarisation effects of the windows of the high-pressure chamber due to their birefringence. This, as shown in Section 3.1, is only relevant for signal detection using the right-angle configuration. This is not relevant for the signal detection using the 0-D back-scattering Raman sensor. According to my experimental experience, the best compensation can be achieved when the polarisation of the excitation laser is set to be linear at 45°. Then the influence of the birefringence of the windows of the high-pressure vessel on the Raman spectrum is minimised. Unfortunately I cannot give a clear physical explanation for this.

3.1.3 Raman Imaging (Ramanography)

The authors who first reported the detection of spatially resolved Raman signals from one plane during one single measurement instant called this technique Ramanography [5]. This is why next to the more frequently used term 'Raman imaging' the term 'Ramanography' still exists. A 2-D Raman sensor is sketched in Fig. 3.8. It can take snapshots (temporally resolved photos) of the mixing

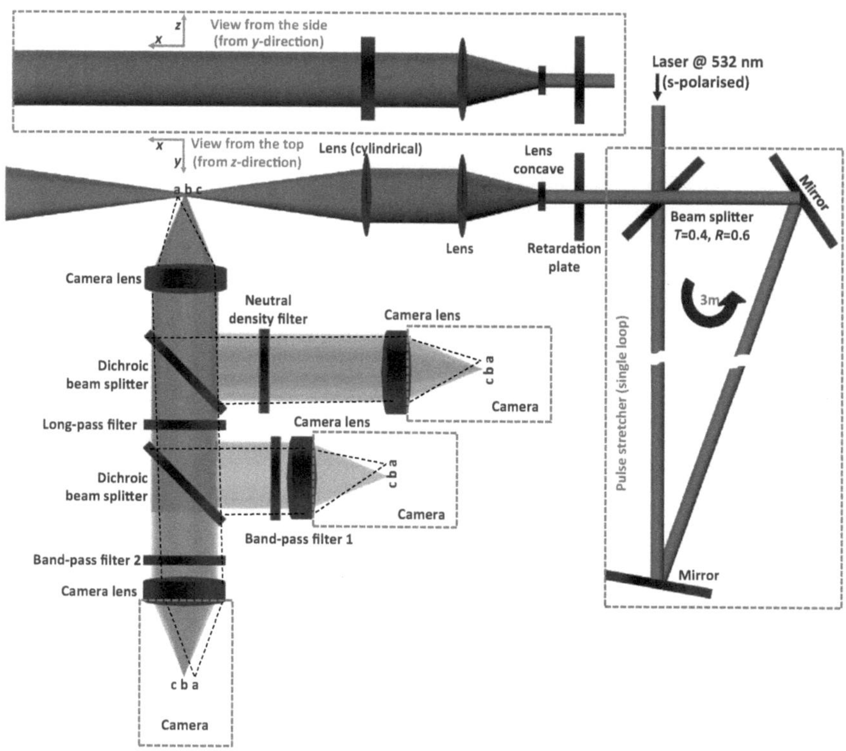

FIGURE 3.8 Sketch of a 2-D Raman sensor with signal detection in right angle direction.

process when for example a solution is injected into an anti-solvent or when fuel and oxidiser are mixed in a rocket engine. Related mixing processes are usually turbulent. Thus, the acquisition of one photo has to be short enough to temporally 'freeze' the turbulent motion. If the acquisition time is too long, the acquired images will look blurred. Of course Raman imaging can also be applied if stationary flow processes have to be analysed. But in this case, instead of the Raman imaging set-up one also can use a 0-D Raman experiment and scan the probe volume through the flow and then put together the 2-D field from the many point measurements.

In general there are again two options for the realisation of short acquisition times. The first option is using a Q-switched pulsed laser instead of a continuous-wave excitation source. Pulsed lasers provide high excitation energies at rather short pulse durations (typically $\sim$10 ns). Raman signals are generated quasi instantaneous with the excitation pulse. Thus, even if the acquisition time of the camera is long compared to the laser pulse duration, Raman signals can be detected only while the laser pulse passes the sampling volume. Then the effective image acquisition time, or the temporal resolution, is solely determined by the excitation pulse duration. The second option is using a camera which can realise short acquisition times. This strategy only leads to a sufficient signal-to-noise ratio if also enough of the Raman signal is produced during this short acquisition time.

Therefore using a camera with short acquisition time and at the same time exciting the Raman process with a weak laser would be useless. Then not enough Raman signal would be acquired during the short acquisition of one image. Therefore the camera should be combined and synchronised with Q-switched pulsed laser which can generate a sufficiently intense Raman signal during the excitation pulse. As it has been mentioned previously several times, it is strongly recommended to use a pulse stretcher in order to decrease the single-shot power of the laser pulses by increasing their pulse duration [6,7]. The working principle of a single-loop pulse stretcher is provided in Section 3.2.2.

A Galilei telescope expands the cross-section of the laser beam because of three reasons.

1. The height (in z-direction) of the laser light sheet formed by the subsequent cylindrical lens is determined by the cross-section of the expanded beam.
2. The thickness (in y-direction) of the laser light sheet formed by the subsequent cylindrical lens is influenced by the cross-section of the beam before it is focused.
3. The risk of window damage is decreased, as the cross-section of the beam passing the windows is increased.

The detection part of the 2-D Raman sensor is composed of several cameras, each equipped with a camera lens, one additional camera lens positioned directly in front of the excitation light sheet, and several filters and beam splitters in order to spectrally separate the desired signals from each other and to purify them from undesired signals. A camera lens instead of a common achromatic lens is used for the collection of the Raman signals originating from the excitation light sheet. This measure assures optimum corrections for chromatic aberrations and coma. The camera lens is positioned in its shortest focal distance from the light sheet. Consequently, the rays originating from one point out of the excitation light sheet are collimated. In Fig. 3.8, the paths of collimated rays originating from point 'b' are given in colour, whereas the paths of collimated rays originating from point 'a' are given by dashed lines. With respect to the example set-up we used for the analysis of the particle generation during the supercritical anti-solvent process (see Section 1.7.2) [8] a first dichroic beam splitter with high reflectivity for wavelengths shorter than 540nm and high transmission for wavelengths longer than 540nm, reflects the elastically scattered green laser light towards the first camera. A neutral density filter attenuates the strong elastically scattered signals by more than 6 orders of magnitude and brings them down to a signal level in the order of the Raman signals. A camera lens mounted onto the camera then focusses the collimated green signal paths onto the CCD chip. Green photons are scattered elastically from the excitation light sheet, if the incident excitation photons are scattered from interfaces (particles or droplets in a bulk fluid phase). Thus, the first camera records an image of the distribution of phase boundaries and from now on is referred to as 'elastic light scattering camera'. Of course this camera can also be employed for the detection of the signals corresponding to any other wavelength.

The second dichroic beam splitter features a high reflectivity for wavelengths shorter than 600nm and a high transmission for wavelengths longer than 600nm. Therefore in this case the Raman signals of the anti-solvent CO_2 which arise at the wavelengths of 571.1nm and 574.4nm (colloquially the two strong Raman signals of CO_2 are also called Fermi dyad) are reflected towards the second camera, whereas the strong Raman signal of the organic solvent (C-H vibration) propagates to the third camera. From now on they are referred to as 'anti-solvent camera' and 'solvent camera'. Both cameras are equipped with camera lenses in order to focus the collimated signal paths onto their CCD chips. Furthermore, each of the cameras are equipped with one long-pass filter suppressing the excitation wavelength and one band pass filter assuring the detection of only the desired Raman signals. This is for the anti-solvent camera a band pass filter with a central wavelength of 574.4nm and a (FWHM) band width of 1nm at a maximum transmission of ~73%. Consequently, the CO_2 Raman signal arising at 571.1nm is not detected. For the solvent camera the band pass filter has a central wavelength of 630nm and a (FWHM) bandwidth of 10nm at a transmission of ~68%. Summarising, the second and the third camera record the distribution of anti-solvent and solvent molecules in the excitation light sheet, respectively.

Meanwhile there are band pass filters with even higher transmission available, though they also feature a narrow bandwidth. The selection of the appropriate band-pass filter is again a trade-off between the maximum transmission at the wavelength the desired signals arise and the bandwidth, which should be narrow enough to allow only the desired signals to pass to the detector. Furthermore, it is recommended to use band-pass filters which block the spectral region outside their transmission range as efficiently as possible. The band-pass filters used in the introduced Raman sensor feature an optical density of 4 outside their transmission range. This assures that only the desired Raman signals find their way to the detector. What has been accomplished by the spectrometers in the set-ups for the 0-D and the 1-D Raman sensors is here realised by the dichroic beam splitters and the band pass filters. It enables imaging the distribution of phase boundaries (elastic light scattering camera), the distribution of the anti-solvent molecules (anti-solvent camera) and the distribution of the solvent molecules (solvent camera) temporally resolved. As it will be demonstrated in the Section 3.4.8 this information is sufficient in order to extract valuable and quantitative information about the processes governing particle formation in the supercritical anti-solvent processes.

3.2 ENGINEERING OF A RAMAN SENSOR

3.2.1 What Is the Suitable Excitation Wavelength for Linear Raman Spectroscopy

Uniformly in Section 3.1 'green' excitation lasers with an emission wavelength of 532 nm were considered for the descriptions of the experimental set-ups. Other excitation wavelengths can be used instead. It should be mentioned here that the identification of a suitable excitation wavelength is always a trade-off between high signal intensity and the interference of undesired fluorescence emissions. Eqn (2.109) shows that the Raman signal intensity can be increased significantly, if the excitation wavelength is shifted to shorter wavelengths. Nevertheless, shifting the excitation wavelength towards the ultraviolet spectral region increases the risk of resonantly exciting some electronic transitions (see Section 2.4.1.5). Then the excitation laser radiation might be absorbed by the molecules and the thus excited molecules might fluoresce. As the absorption cross-section and the fluorescence quantum yield exceed the Raman scattering cross-section by many orders of magnitude, even a trace amount of molecules fluorescing can completely overload the desired Raman spectrum and will make a quantitative interpretation of the Raman signals impossible. Considering this, an excitation wavelength in the green region of the spectrum (intermediate between the 'ultraviolet' and the 'infrared') is very often a good choice for four reasons (not for biological or medical applications!):

- Green radiation is visible, which makes an alignment of the Raman sensor straightforward and also simplifies the assignment of the probe volume to a location inside the measurement object.
- Robust and high power continuous-wave (cw) lasers are available with emission wavelengths of 532 nm or 514.5 nm, such as the Nd:YAG laser (second

harmonic) or the argon ion laser, respectively, with powers of up to 20 W (often a 200 mW continuous-wave laser – as it is integrated in Fig. 3.3 – is sufficient to detect Raman spectra with signal-to-noise ratios exceeding 100 in 100 ms).

- Once the Raman process is excited in the green spectral region, the Stokes-shifted Raman signals are generated ranging from the green (for small Raman shifts) to the red (for large Raman shifts) spectral region. Between 'green' and 'red', commercially available charge-coupled device (CCD) detectors feature their highest quantum yield, which is prerequisite for detecting the signals efficiently. Although nowadays in the visible region back-illuminated CCDs feature quantum efficiencies exceeding 90%, front-illuminated CCDs reach quantum efficiencies of less than 60%.
- Spectral filters and dichroic mirrors which are essential for the separation of the Raman signals from the elastically scattered laser light (compare Fig. 3.2) show a much better performance (transmission for the Raman wavelengths and suppression of excitation wavelength) in the spectral region from 'green' to 'red', than for wavelengths shorter than 'green'.

Summarising, the visibility of green radiation, the availability of (high power) cw lasers, the available high efficiency detectors and the high performance of spectral filters in the spectral region from 'green' to 'red' very often (exceptions are described in the subsequent paragraph) make the identification of a 'green' excitation laser reasonable.

Sometimes 'green' is not the proper excitation wavelength. Usually, the more complex the electronic systems of the molecules in the system under investigation are, the longer excitation wavelengths are sufficient to cause the undesired resonant excitation (see Section 2.4.1.5), which might be followed by interfering emissions. Therefore, compounds interesting for biomedical studies, such as carotenoids, proteins, yeast, blood, anatomic tissue and many more absorb the green excitation light and as a consequence fluoresce. The same is true for inorganic but complex molecules such as some ionic liquids or dye molecules for example [9]. Figure 3.9 shows raw (not processed) Raman spectra of the ionic liquid 'EMIM $EtSO_4$' excited with wavelengths of either 532 nm or 785 nm. Additionally a photograph of the green laser beam passing through the cuvette filled with 'EMIM $EtSO_4$' is shown. The strong fluorescence, which makes the laser beam look yellow when it passes the cuvette, can be seen with the naked eye. Once the green excitation laser is switched off, 'EMIM $EtSO_4$' is a transparent ionic liquid with a slight yellowish (hardly recognisable) undertone. The yellowish undertone is due to any impurities which are the residuals of the synthesis process of ionic liquids. Absolutely clean 'EMIM $EtSO_4$' most probably would not fluoresce.

In order to suppress the interfering fluorescence, the excitation wavelength must be shifted towards or even into the infrared (IR) spectral region. Figure 3.9 shows that the fluorescence interference can be discriminated against largely by using an excitation wavelength in the near infrared region, such as 785 nm. This

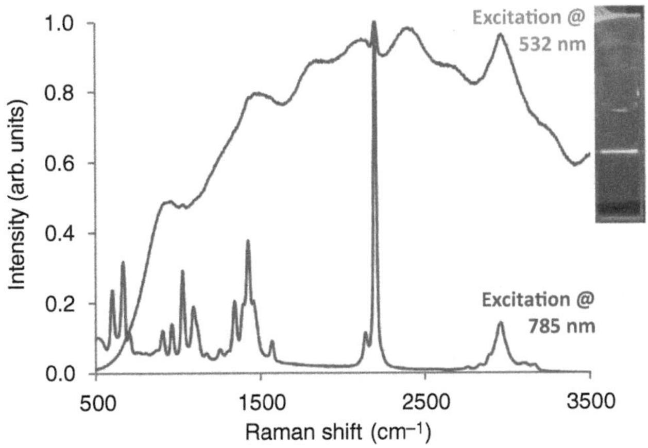

FIGURE 3.9 Raman spectra (not processed) of the ionic liquid 'EMIM $EtSO_4$' using as excitation wavelengths either 532 nm or 785 nm.

explains why studies in biological systems usually use excitation wavelength in the red or even in the (N)IR spectral region [10]. For the red spectral region the Helium–Neon (HeNe) laser with one of its emission wavelengths at 633 nm (other emission wavelength are possible) and different diode lasers with tuneable emission wavelengths are known [11]. The most frequently used diode laser wavelengths for studies in biological systems are between 780 and 800 nm, which again is a trade-off between different factors:

- Wavelengths shorter than 780 nm are still absorbed by many aqueous biological systems due to electronic resonances and cause undesired fluorescence.
- Wavelengths longer than 800 nm might be absorbed by rotational–vibrational resonances of water (aqueous solution) or by the molecules dissolved in water. This is why the region around 785 nm is called 'optical window' of biological systems and aqueous solutions.

Some biological substances even absorb wavelengths between 780 nm and 800 nm and longer. This is why excitation wavelengths in the (N)IR are used sometimes, too. In the (N)IR region – next to others – the Nd:YAG or Nd:YLF lasers operated at their fundamental wavelength of 1064 nm are available. It should be mentioned that in the wavelength region longer than 900 nm the common silica-based CCD material (Si) shows weak quantum efficiencies. Therefore detectors of Indium–Gallium–Arsenide (InGaAs) have to be used, instead. The use of a CCD detector can be avoided at all, if the Raman spectrum is not detected using a diffraction- or refraction-based spectrometer, but applying the Fourier-transform (FT) Raman technique [12,13], which is not further explained here. With respect to absorption spectroscopy FT spectrometers are frequently applied in the (N)IR spectral range and therefore described in

Section 6.1.2 of this book. It should be mentioned that the FT Raman technique is not first choice for *in situ* measurements in high-pressure process technology, as FT Raman sensors lack in flexibility to adopt the sensor to the system under investigation and suffer from a rather long signal integration time. Whenever neither flexibility nor short integration times are required, the advantage of exciting no undesired fluorescence background with FT-Raman spectroscopy in the IR spectral range makes – in my opinion – this technique superior to the here reported and described Raman sensors using diffraction-based spectrometers instead of interferometer-based ones.

3.2.2 Stretching the Laser Pulse in Order to Prevent Window Damage

Figure 3.10 shows a sketch of a single-loop pulse stretcher together with the temporal shape of the incoming and the outgoing laser pulse. A pulse stretcher has to be used when high power Q-switched pulsed lasers are operated and when high excitation energies are required in short excitation times in order to carry out temporally resolved measurements (compare Sections 3.1.1.3 and 3.1.3).

In Fig. 3.10, the incoming laser pulse is assumed to have a full width at half maximum (FWHM) pulse duration of 10 ns and a single-pulse energy of 1 J. Once the incident laser pulse hits the beam splitter of the pulse stretcher, 40% of the incident energy are reflected, while 60% enter the loop of the pulse stretcher. Accordingly, the first pulse of the outgoing pulse train carries 40% of the energy of the incident laser pulse. The 60% portion of the pulse energy, which entered the loop, reaches the dichroic beam splitter the next time from the back side. Then 60% of the pulse, which have been inside the loop now leave the pulse stretcher. Therefore the second pulse of the outgoing pulse train carries 36% (60% × 60%) of the energy of the incident pulse. Still inside the loop is 24% of the energy of the incident pulse. The next time the pulse reaches the dichroic beam splitter, 60% of the still remaining energy (24%) are transmitted and leave the pulse stretcher. Therefore the energy of the third pulse of the outgoing pulse train is 14% (24% × 60%). As a consequence, the peak power (peak maximum) of the stretched laser pulse is only 40% of the peak power of the incident laser pulse, while the energy of the stretched and the incident pulse are nearly the same (in reality there are losses of approximately 15% due to a not perfect alignment and due to the reflection losses at each mirror).

The reduction of the peak intensity is due to the temporal stretching of the laser pulse. This enables the experimenter to launch high pulse energy at moderate pulse intensity and in still rather short pulse duration into the probe volume. Consequently, exceeding the laser-induced breakdown of the fluid under investigation or exceeding the intensity threshold of the optical windows can be prevented. High excitation energies are required, as the scattering cross-section of the Raman process is weak.

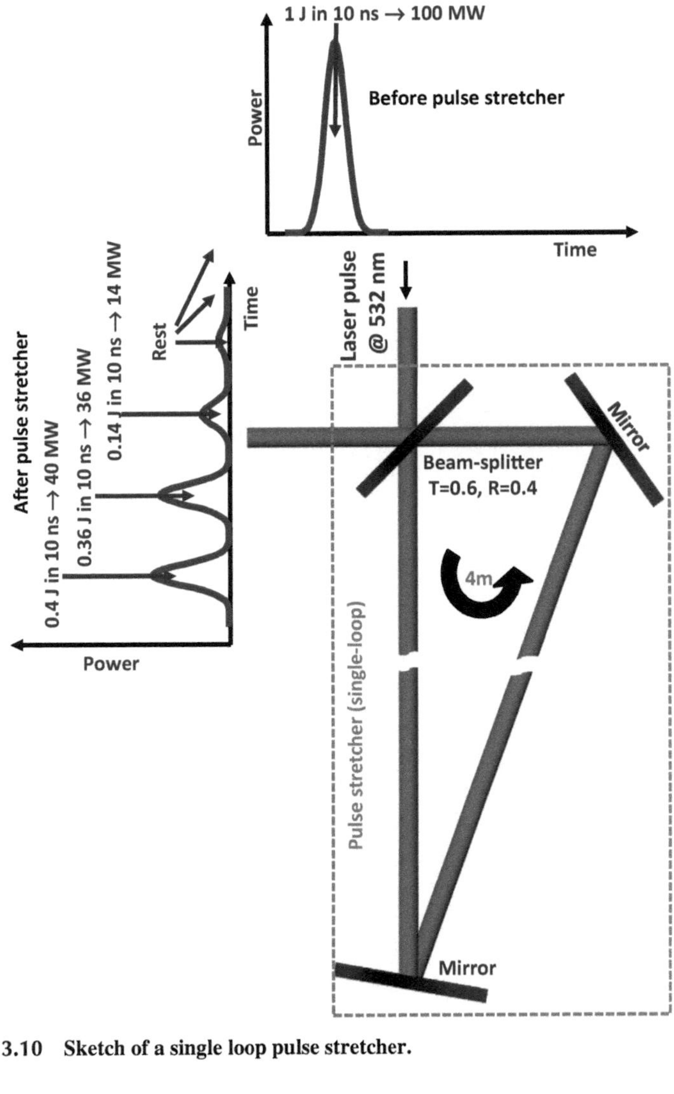

FIGURE 3.10 **Sketch of a single loop pulse stretcher.**

The length of the loop of the pulse stretcher is supposed to correspond to the length of the incident laser pulse. Light travels a distance of approximately 3 m in air in 10 ns. Therefore the loop in Fig. 3.10 is minimum 3 m in length. This assures that the second pulse of the outgoing pulse train does not overlap within its FWHM with the first pulse of the outgoing pulse train.

Multiple-loop pulse stretchers can be applied in order to further stretch the incident laser pulse. Reference [7] provides the fundamentals required for the development of multiple loop pulse stretchers. Designs of pulse stretchers other than the one sketched in Fig. 3.10 are described elsewhere [4,6,7].

3.2.3 What Is the Suitable Detector

The ideal detector[4] for spectrally resolved Raman signals excited with a 'green' laser is a back-illuminated CCD chip. Back-illuminated CCDs have quantum efficiencies (QEs) of up to approximately 95% in the visible part of the spectrum. The QE quantifies the probability of the conversion of one photon incident to the CCD into one electron according to the internal photoelectric effect. The generated electrons are accumulated and stored within the photo sensitive layer of the semiconductor material of the CCD as long as the CCD is exposed to light. Applying a voltage to the electrodes, which are positioned below the thin photo sensitive layer (in the front-illuminated configuration the electrodes are positioned above the photo sensitive layer), the electrons can be moved within this layer to the read-out-register, where the charges (number of electrons) stored in the flying range of one electrode are converted into voltage. In contrast to the back-illuminated CCD configuration, the front-illuminated CCDs reach a QE of approximately 55% in the visible. This is due to the fact that the incident photons have to pass the electrode structure, which is now assembled on top of the photo sensitive layer, before they can be converted into electrons and stored in the sensitive layer. Passing the electrodes is accompanied with some losses. Whenever the signal levels are sufficiently high, the front-illuminated CCD chips are a cheaper alternative to back-illuminated CCD chips. To give an example, though in Fig. 3.8 cameras equipped with front-illuminated CCD chips have been used, sufficiently intense Raman signals were acquired even with the 2-D set-up.

Furthermore it should be mentioned that it is advantageous to use CCD chips with frame-transfer (FTO) or interline-transfer (ITO) option. FTO and ITO ensure that the acquired image can be shifted rather fast into a secured section (frame-transfer area) of the CCD chip, which is not exposed to light. Consequently, during the long lasting read-out process, the image is not smeared by photons still incident to the active area of the CCD chip. Chips without frame-transfer option always suffer from smearing of the images, as the incident photons continuously produce charges in the CCD chip while the image is shifted line by line into the read-out register. If no frame-transfer option is available, the following are required for the prevention of smearing effects:

- A mechanical shutter in front of the detector (or spectrometer),
- The discrimination of any photon sources (absolute darkness) during the read-out process or
- A very fast read-out process

Very fast read-out can be realised, if only few pixels have to be read out. For example, the 2-D detector in Fig. 3.2, which is incorporated into the spectrometer,

4. The appendix of this chapter provides information about the working principle of CCD detectors.

can be considered to be 1-D. By binning[5] all pixels of either one column or one line together to one 'long' superpixel, the 2-D resolution becomes 1-D, as the resolution along the binned axis (now superpixel) is lost. The Raman spectrum is then dispersed along the axis, which has not been binned. Summarising, the 2-D CCD chip is converted into a 1-D CCD array with its spectral resolution along the remaining dimension. This reduces the number of pixels which have to be read out from 1024×1024 for the case without binning to 1024×1 for the case of full vertical binning and consequently reduces the time required to read out the spectrum by a factor of approximately 1024^{-1}. It is still recommended to switch off the light during the measurements in order to reduce the background interferences.

For the experiments shown in Fig. 3.7, a back-illuminated CCD detector without FTO has been mounted onto the imaging spectrometer. The chip is composed of 400 pixels along the horizontal axis, which is the spatial dimension, and 1600 pixels along the vertical axis, which is the spectral dimension. Then at a time 20 pixels along the spatial dimension were binned, which reduces the spatial dimension from 400 pixels to 20 superpixels and at the same time reduces the time required to read out the 20 spectra (instead of 400 spectra without binning) by a factor of 20^{-1}. The spectral axis is still composed of 1600 pixels. The time required to read the 20 spectra can be reduced further, if the resolution along the spectral axis is reduced. This can be made by binning at a time 2 pixels to one superpixel. Then the spectral axis is composed of 800 instead of 1600 pixels. This measure reduces the time required to read out the 20 spectra by a factor of 2^{-1}. Again it is recommended to switch off the light in the lab during the measurements to reduce the background light. While scientific cameras are rather flexible regarding different possibilities of binning, the CCD chips, which are already incorporated into ready-to-use spectrometers, do usually not offer the possibility of manipulating the binning options set by the manufacturer. In return, they are cheaper.

3.2.4 Guiding the Signals to the Detector

The cross-section of Raman scattering is weak [14]. Therefore care should be taken in order to detect as many of the Raman scattered photons as possible. Losses can be minimised at diverse components within the detection path of a Raman sensor. Here the measures suggested are with respect to the 0-D Raman sensor but can be transferred to the 1- and 2-D sensor systems.

Usually the spectrometer with integrated CCD chip or array is the most expensive component of the 0-D Raman sensor. Thus, in most cases the detection path is optimised towards an already existing spectrometer. This is why the description of the optimisation chain of the detection path of the Raman sensor starts with the spectrometer, then treats the choice of a suitable 'detection fibre' and finally discusses the specifications of the lenses, mirrors and filters mounted into the Raman sensor.

5. 'Binning' is described in the appendix of this section.

3.2.4.1 The Spectrometer and the Detection Glass Fibre

Regarding the dispersive element, which is called grating, diffraction spectrometers can be equipped with the following:

- A ruled reflection grating,
- A ruled transmission grating,
- A holographic reflection grating or
- A holographic transmission grating

In Fig. 2.27, the path of light is shown in a spectrometer equipped with a reflection grating. While ruled gratings feature a higher efficiency, holographic gratings feature low stray light resulting in sharper diffraction orders. Transmission gratings are usually used in very compact spectrometers.

In order to achieve the best performance of the spectrometer, light has to be launched into the spectrometer with the numerical aperture $\mathrm{NA}_{\mathrm{Spectrometer}}$ of the spectrometer

$$\mathrm{NA}_{\mathrm{Spectrometer}} = n_{\mathrm{refr}}\, sin\beta_A = n_{\mathrm{ref}\,r}\, sin\left(\arctan\frac{D}{2f_{\mathrm{length}}}\right) \approx n_{\mathrm{refr}}\frac{D}{2f_{\mathrm{length}}} = (n_{\mathrm{refr}}=1)\frac{D}{2f_{\mathrm{length}}} \quad (3.1)$$

According to the upper part of Fig. 3.11, the numerical aperture of the spectrometer in air (refractive index $n_{\mathrm{refr}} \approx 1$) is defined by the diameter D and the focal length f_{length} of the entrance pupil (first concave mirror or convex lens in the spectrometer) of the spectrometer.

The numerical aperture of the glass fibre defines the divergence of the light coming out of the fibre. The lower part of Fig. 3.11 shows one scenario for which the numerical aperture of the fibre is smaller and another scenario for which the numerical aperture of the fibre is larger than that one of the spectrometer. If the numerical aperture of the fibre is smaller than the one of the spectrometer, the grating of the spectrometer will not be illuminated completely. Consequently the performance of the spectrometer goes down, as the contrast between diffraction maxima and minima is worsened. If the numerical aperture of the fibre is larger than the one of the spectrometer, a portion of the signal cannot be collimated by the entrance pupil of the spectrometer and thus is lost. In the worst case this lost portion of the signal is scattered inside the spectrometer and reaches the detector as stray light, which would heavily obscure the detected spectra and make a reliable evaluation of the spectra impossible.

Summarising, the numerical aperture of the fibre and of the spectrometer should be identical. If this combination is not available, then it is recommended to choose the numerical aperture of the fibre smaller than that one of the spectrometer in order to avoid signal losses and stray light interferences.

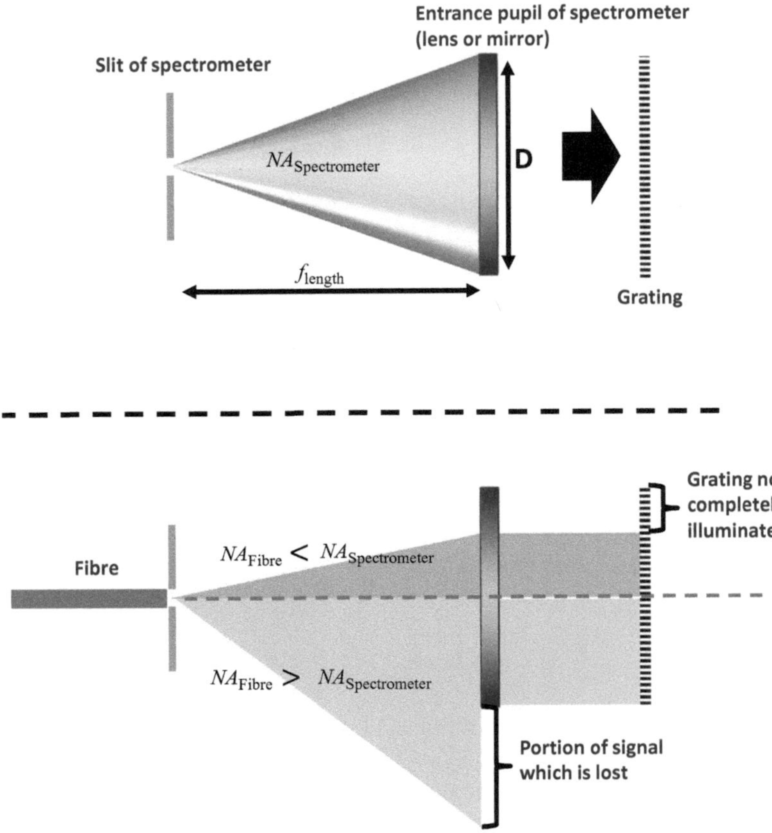

FIGURE 3.11 Sketch of the numerical aperture $NA_{Spectrometer}$ of the spectrometer, which is defined by the diameter D of the first entrance pupil and its focal length (upper part) and sketch of the NA_{Fibre} of the fibre connected to the spectrometer (lower part).

Figure 3.12 compares the use of a single fibre and a fibre bundle as signal light guide. Taking the diameter of 380 μm as a reference it is obvious that the fibre bundle in contrast to the single glass fibre suffers from a lot of void space in between the light guiding cores. The area of the light guiding core of the single fibre is in the given example approximately two times the area of all fibre cores of the bundle. Therefore approximately half of the signal is lost in the void spaces when the light is coupled into the fibre bundle. But at the other end of the fibres, where the fibre couples the light into the spectrometer, things are different. Assuming that the slit of the spectrometer is set narrower than the diameter of the single fibre's core, then some portion of the light coming out of the fibre is blocked at the blades of the spectrometer's entrance slit, which are given in Fig. 3.12 as grey blades.

Assuming a slit width of 100 μm, approximately 50% of the light coming out of the single fibre are blocked from the blades. In contrast, all the light

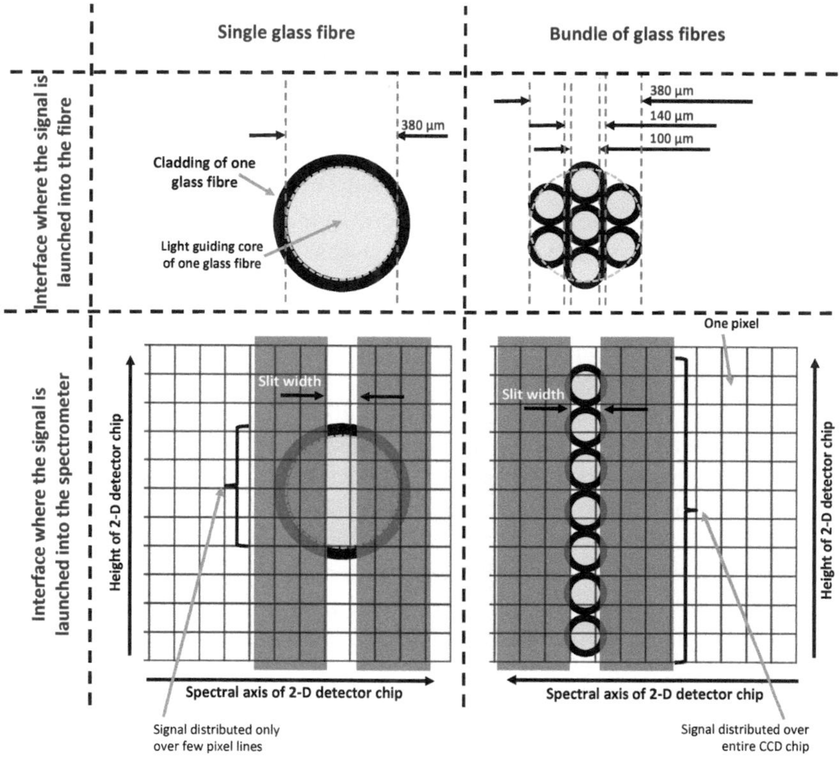

FIGURE 3.12 **Sketch of a single glass fibre composed of a light guiding core und a cladding (left), and a bundle of glass fibres (right).**

coming out of the small fibres of the glass fibre bundle can pass through the slit. Summarising, regarding the void spaces of the fibre bundle at the entrance side of the light guide and the losses of the single fibre at the exit side of the light guide, or at the slit of the spectrometer, both, the single thick fibre and the fibre bundle launch approximately the same amount of light into the spectrometer. But one also has to consider how the light launched into the spectrometer will finally be detected on the CCD chip. Let us assume that the detector of the spectrometer is a 2-D CCD chip, which in Fig. 3.12 is indicated as the grid. Only few pixel lines will be occupied for the resolution of the spectrum of the signal coupled into the spectrometer through the single fibre. On the contrary, the fibre bundle distributes the in-coupled signal on more pixel lines. Considering that each pixel can only detect a limited number of photons, a lot more signal can be detected when the signals are launched into the spectrometer via the linear end of the glass fibre. Summarising again, it depends on the slit width and the architecture of the detector, whether one better uses a single fibre or a fibre bundle.

3.2.4.2 Specifications of the Two Detection Lenses in the 0-D Set-Up

Regarding the lens focusing the Raman signal into the 'detection fibre', the numerical aperture NA_{Lens} and its diameter D are the two key parameters to be considered. As the Raman signals can cover a broad range of wavelengths, it is recommended to use an achromatic lens here.

In the previous section, we derived according to which criteria the numerical aperture of the 'detection fibre' should be chosen. Once the numerical aperture of the 'detection fibre' is determined, the numerical aperture of the lens focusing the Raman signals onto the 'detection fibre' has to be chosen accordingly or smaller (maximum equal to the numerical aperture of the fibre). If the numerical aperture of the lens exceeds that one of the 'detection fibre' the condition for 'total reflection' inside the fibre is no more fulfilled for a certain portion of the light which results in signal losses. Summarising, according to Eqn (3.1) the ratio of diameter D and focal length f_{Lens} of the respective lens is determined by the numerical aperture of the 'detection fibre'.

It can be seen from Fig. 3.2 that the diameter of the lens focusing the Raman signals onto the 'detection fibre' and the lens focusing the excitation radiation and at the same time collecting the Raman signals from the probe volume should be the same. Therefore the diameter D of the respective lens cannot be chosen arbitrarily. Giving one example: If the numerical aperture of the 'detection fibre' is 0.125, the numerical aperture of the lens should maximum be 0.125. The diameter D of the lens is set 2' (~50 mm). Therefore – according to Eqn (3.1) – the focal length of the lens should be at least 200 mm.

The lens focusing the excitation laser beam in Fig. 3.2 is also the lens detecting the scattered signals in backward direction. Therefore its diameter D and its focal length f_{length} together determine the solid angle of detection according to Eqns (2.121) and (2.122). The focal length of this lens cannot be chosen arbitrarily small. It must at least be long enough to position the focal spot (probe volume) inside the measurement object, while the Raman sensor is positioned outside the measurement object. The diameter of this lens also cannot be chosen arbitrarily large, as then the bore diameter of the glass windows of the high-pressure apparatuses had to be increased. It should be the diameter of the lens and not the diameter of the glass windows which effectively determines the solid angle of detection. As the Raman signals cover a wide spectral range, it is recommended to use an achromatic lens here. It should be underlined that in combination with a pulsed laser, achromatic lenses must not be used in the excitation path of the Raman sensor. Achromatic lenses are composed of different lenses made of different materials which are typically glued together. The power of pulsed laser pulses can easily damage the glue and consequently blur the achromatic lens (not-glued achromatic lenses are an alternative).

As already reported earlier, at first glance it might be preferably to use lenses with short focal lengths in order to detect the signal through a large solid angle.

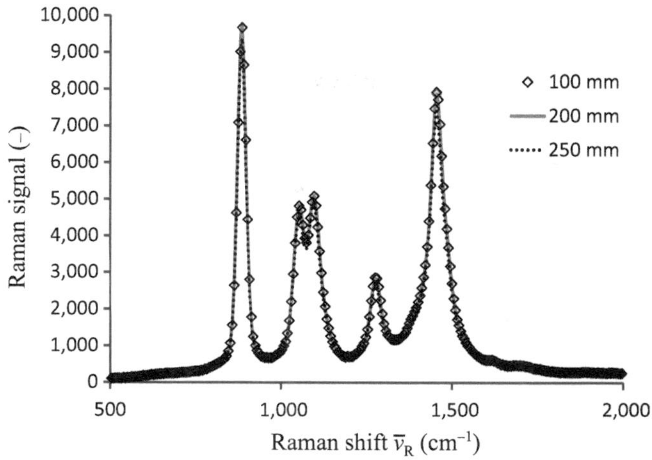

FIGURE 3.13 Raman spectra of liquid ethanol acquired with the 0-D Raman sensor introduced in Figure 3.2 changing the focal length of the lens focusing the excitation laser beam to the probe volume from 100 mm to 250 mm. The focal length of the lens focusing the signals onto the 'detection fibre' has been 200 mm.

Figure 3.13 shows Raman spectra of liquid ethanol acquired with the 0-D Raman sensor introduced in Fig. 3.2 changing the focal length of the lens focusing the excitation laser beam to the probe volume from 100 mm to 250 mm. The focal length of the lens focusing the signals onto the 'detection fibre' has been 200 mm. Irrespective of the focal length of the first lens (the one focusing the excitation laser beam), the detected signal intensity is the same. This means that the reduction of the solid angle of detection with increasing focal length has to be compensated by other effects. By increasing the focal length of the first lens, not only the solid angle of detection is decreased but also three other dependencies are changed.

1. Increasing the focal length of the first lens (the one focusing the excitation laser beam) increases the cross-section of the waist of the excitation laser beam focus in an approximation proportionally. This effect must not influence the detected signal intensity, as the increased cross-section of the probe volume (focal beam waist) is compensated by a respectively lower excitation photon density (per area). Both effects compensate each other.
2. Increasing the focal length of the first lens reduces the magnification of the lens system. The lens system comprises the first lens and the lens used for focusing the signals onto the 'detection fibre'. The magnification is indirectly proportional to the focal length of the first lens. Considering the aforementioned effect (the cross-section of the focal beam waist increases directly proportional with the focal length), the image of the focal beam waist at the entrance interface of the 'detection fibre' has the same diameter, irrespectively of the focal length of the first lens. Therefore the efficiency of

coupling the signals into the 'detection-fibre' is not influenced by the focal length of the first lens and thus, the effect of magnification also should not impart on the intensity of the detected Raman spectra.

3. Therefore the last effect must be responsible for the compensation of the impact of the reduction of the solid angle of detection with increasing focal length. This is the depth of focus, which increases with increasing focal length. Both effects, the increasing depth of focus and the decreasing solid angle of detection compensate the impact of the focal length of the first lens onto the intensity of the detected spectra.

3.2.5 Helpful Suggestions One Might Not Consider: Unexpected and Undesired Interferences

Undesired interferences can have diverse origins and are recognisable as broadband underground or as unknown peaks in the Raman spectra. In general interferences can have their origin inside the system under investigation or outside the system of investigation.

- Interferences having their origin outside the system of investigation can be due to light sources in the laboratory, such as computer screens, lamps or even illuminated instrument panels, and due to the excitation source itself.
- Interferences from inside the system under investigation might be due to emissions of compounds in the system. As emissions such as fluorescence or phosphorescence are several orders of magnitude more intense than the Raman scattering process, already traces of emitting compounds can cause immense underground signals which make a reliable evaluation of the spectra impossible or at least challenging.

3.2.5.1 'Excitation Fibre' as Interference Source

In the context of Fig. 3.4, it is mentioned that glass fibres which are used in order to guide the excitation radiation from the laser head to the Raman sensor system manipulate the spectrum of the excitation radiation. This is due to the interactions between the excitation radiation and the material the glass fibres are made of. Therefore an excitation beam, which is quasi-monochromatic before it is launched into the 'excitation-fibre', suffers from a broadband underground when it leaves the excitation fibre according to Fig. 3.14. Consequently the probe volume does not only experience quasi-monochromatic excitation radiation, but also the spectrally broadband underground. In case of a green excitation radiation the Raman scattering process is sketched in a term scheme of a diatomic molecule in Fig. 3.15. Simultaneously the broadband underground is scattered according to the Rayleigh process. Figure 3.15 shows that the spectrum (energy of scattered photons) of the Raman scattered photons (with respect to the initial mono-chromatic excitation radiation) interferes with the spectrum of the Rayleigh scattered photons (with respect to the broadband underground). Thus it is

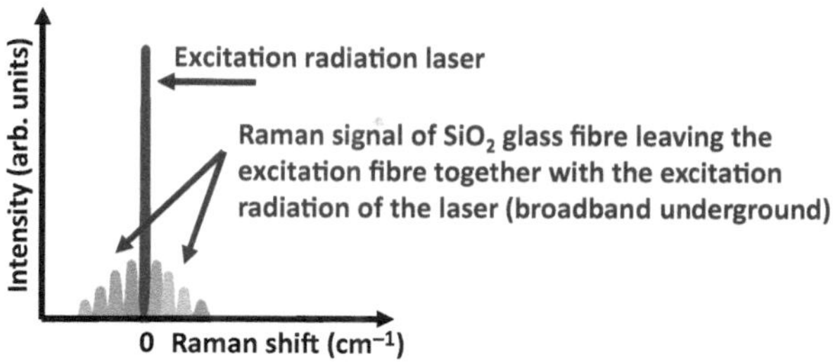

FIGURE 3.14 Sketch of the spectrum of the light coming out of a glass fibre once a quasi-monochromatic laser is launched into the glass fibre. The broadband background originates mainly from the Raman process taking place inside the glass fibre due to interactions between light and glass.

essential to mount a short-pass filter, which suppresses the wavelengths longer than the initial excitation wavelength behind the exit of the 'excitation fibre' as it is shown in Fig. 3.4, already before the excitation radiation reaches the measurement volume.

3.2.5.2 Diode Lasers as Interference Source

Figure 3.16 sketches that the radiation emitted by laser diodes is spectrally not monochromatic. It is composed of the amplified 'laser wavelength' which is considered to be quasi-monochromatic, and a small portion of the broadband emission spectrum of the diode itself. The emission spectrum of the diode can be as broad as 100 nm. The intensity ratio between the 'laser-wavelength' and the broadband underground is quantified according to laser diode manufacturers

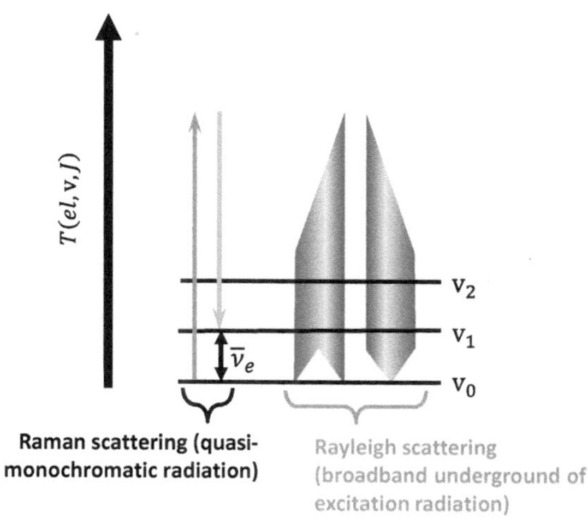

FIGURE 3.15 Sketch of the interference between the Raman process excited by the green excitation laser beam and the Rayleigh process excited by the broadband underground of the excitation radiation.

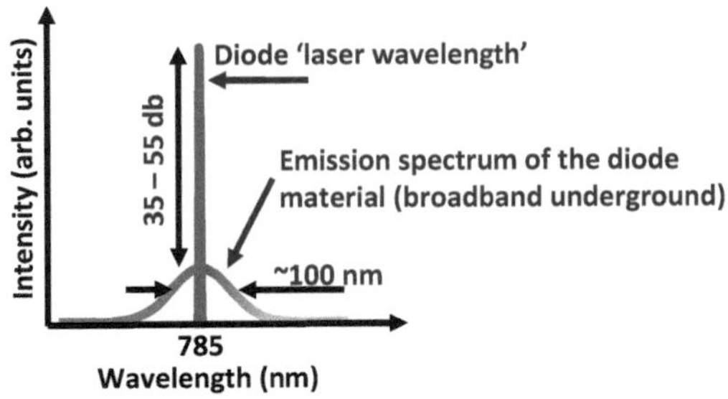

FIGURE 3.16 Sketch of the spectrum of light coming out of a diode laser having a 'laser wavelength' of 785 nm.

between 35 db and 55 db, which corresponds to ratios between 3162 and 316,227. As Raman scattering is several orders of magnitude weaker than Rayleigh scattering, the Raman signals excited by the diode laser wavelength interfere with the Rayleigh signals excited by the broadband emission of the diode material. Figure 3.17 shows a sketch illustrating the interference between the Raman and the Rayleigh spectrum in a term scheme of a diatomic molecule.

Thus it is essential to mount a short-pass filter behind the laser, which suppresses the wavelengths longer than the diode laser wavelength. Figure 3.18 compares spectra acquired from liquid ethanol using a 0-D Raman sensor with a laser diode excitation source operated at 785 nm, once with a short-pass filter (blocks wavelengths longer than 800 nm) and once without a short-pass filter installed in the excitation path of the Raman sensor. When no short-pass filter is installed, the acquired spectrum suffers from strong interferences in the wavenumber region shorter than 1500 cm^{-1}. These interferences originate from the spectrally broad emissions of the diode, which are elastically scattered in the probe volume and then detected. Unfortunately the broad emissions of the diode spectrally overlap with the Raman signals of ethanol.

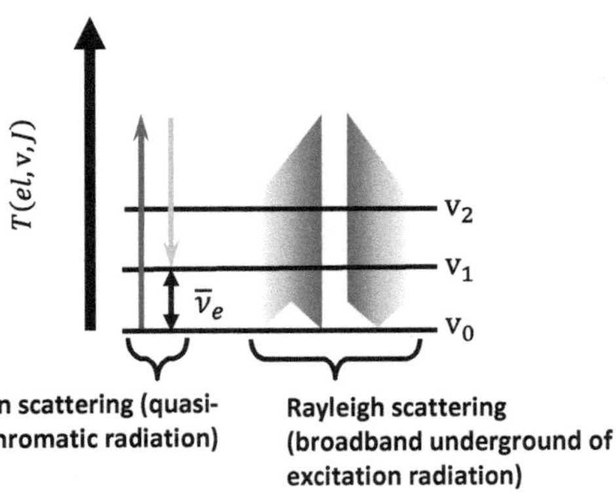

FIGURE 3.17 Sketch of the interference between the Raman process excited by the 785 nm excitation laser diode and the Rayleigh process excited by the broadband underground of the diode material emission.

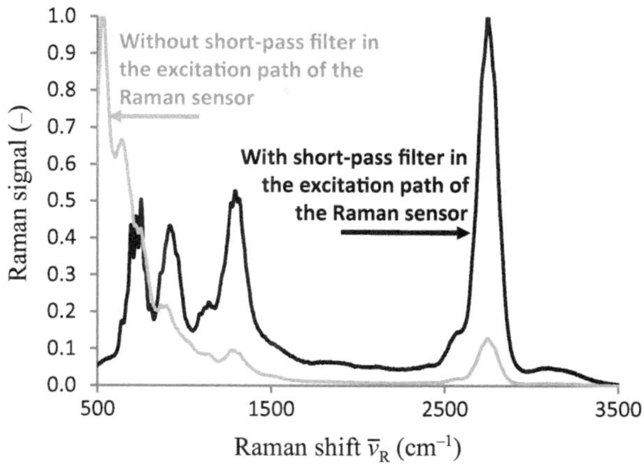

FIGURE 3.18 Spectra of liquid ethanol acquired with a 0-D Raman sensor with a diode laser excitation source having an emission wavelength of 785 nm with a short-pass filter installed (black spectrum) and without a short-pass filter installed (grey spectrum) in the excitation path.

Therefore the usage of a short-pass filter, which blocks wavelengths longer than the 'laser wavelength' of the laser diode in the excitation path assures that the broad emissions of the diode are not guided to the probe volume and therefore cannot be scattered elastically there. As a result, the spectrum acquired with the short-pass filter installed, does not suffer from the undesired interference.

3.2.5.3 Traces of Impurities as Interference Source

The major drawback of Raman scattering is that its cross-section is orders of magnitude smaller than the probability of absorption–emission interactions such as 'laser-induced fluorescence' and 'laser-induced phosphorescence'. This implies that whenever the excitation laser source excites emission processes inside the system under investigation, the Raman signal will suffer from a strong emission background. Of course the wavelength at which the Raman process is excited has to be chosen in order to not excite emission processes. Nevertheless, there might be traces of impurities in the system under investigation, which the experimenter does not know, and by chance these impurities might be excited to emit photons. As the emissions are also red-shifted with respect to the excitation wavelength, they spectrally overlap with the desired Raman signal and therefore cannot be discriminated against experimentally, once they are in the system. In order to underline how sensitive Raman spectroscopy is with respect to interfering emissions from impurities, few examples are provided.

- In high-pressure process technology seals are used which are made of polymer. Sometimes the polymer seals carry softeners or colorant. Softeners or colorant feature rather complex electron systems which can easily be excited to fluoresce. Thus, if these softeners or colorants are extracted from the seal into the fluid under investigation, this trace amount of impurity obscures the desired Raman spectra and might make a reliable evaluation of them impossible. Therefore all the armatures, pipes, apparatuses and containers in contact with the fluids under investigation are recommended to be free of polymer seals carrying softeners or colorants (compressed CO_2 is an excellent extraction agent). Metal-based seals or softener-free Teflon seals are recommended to be used.

- Also paper towels the experimenter uses in order to clean beaker glasses can cause undesired interferences. If the paper towels are coloured, a small portion of the colorant will always stick to the surface of the glass after the drying process. Once the beaker, which for the naked eye looks perfectly clean is filled with the fluid to be analysed, the colorant molecules dissolve and cause a broadband fluorescence background which makes a reliable evaluation of the Raman spectra complex. Thus, coloured paper towels should not be used in a 'Raman laboratory'.

3.2.5.4 Window of the High-Pressure Chamber as Interference Source

Again, the fact that the Raman scattering cross-section is orders of magnitude smaller than the probability of other light–matter interactions explains, why even impurities captured inside the window material can cause undesired broadband underground signals making a reliable evaluation of Raman spectra challenging.

Figure 3.19 shows a sketch of a combination of a 0-D Raman sensor with signal detection in back-scattering direction and a second 0-D Raman sensor with signal detection in right angle direction. This sketch explains why especially the back-scattering configuration is sensitive to impurities captured inside the window material. The excitation laser is focused through one window of a high-pressure apparatus into the probe volume of the Raman sensor, which in Fig. 3.19 is positioned in the centre of the respective measurement chamber. The excitation laser beam then leaves the measurement chamber through a second window.

If the depth of focus of the detection path spatially overlaps with the windows, undesired photons originating from the window material will be collected next to the desired Raman signals coming from the fluid in between the windows. Therefore, if the impurities contained inside the glass material are excited to fluoresce or phosphoresce, the respective emissions will be detected, interfere with the Raman spectrum and cannot easily be eliminated experimentally due to the spectral overlap of the Raman signals and the fluorescence and phosphorescence emissions.

Reducing the focal length of the lens, which focusses the excitation laser into the measurement chamber also reduces the depth of focus. Nevertheless I made the experience that if there are fluorescing or phosphorescing impurities contained in the window material, the measure of reducing the depth of focus, only reduces the interfering emissions, but does not eliminate them. Thus,

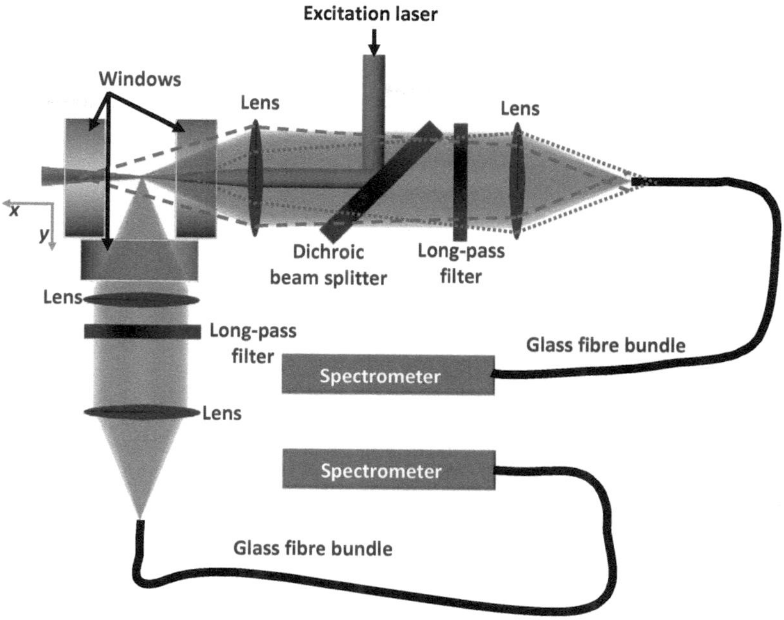

FIGURE 3.19 Sketch combining two 0-D Raman sensors, one with signal detection in back-scattering direction (along the negative *x*-direction) and one with signal detection in right-angle direction (in *y*-direction). The dashed and dotted lines indicate the depth of view of the detection path and are only given for the back-scattering configuration.

especially for the configuration in back-scattering direction, window material (including its impurities) has to be used which cannot be excited to fluorescence or phosphorescence. According to my experience unfortunately especially sapphire, which is frequently used in optically accessible high-pressure apparatuses, suffers from impurities which can easily be excited to emissions with a green excitation source. Ideally sapphire without impurities (at least without those which can be excited to emit) is used, but which is not easy to get. Therefore, borosilicate glass or quartz glass windows (not as pressure resistant as sapphire and quartz glass can shiver!) are an alternative when operating with green excitation lasers. On the other hand I observed strong interferences coming from borosilicate glass when operating with a 785 nm diode laser as excitation source. Then the sapphire windows, exactly those which caused strong interference for the green excitation, did not show any interference.

Figure 3.20 shows spectra acquired from liquid ethanol inside a high-pressure variable volume cell (HPVVC) with the 0-D Raman sensor sketched in Fig. 3.2 with an excitation wavelength of 532 nm. Once the HPVVC has been equipped with sapphire windows (grey spectrum) and once with borosilicate glass windows (black spectrum). The grey spectrum shows a broadband underground coming from the emissions of the impurities contained inside the sapphire windows.

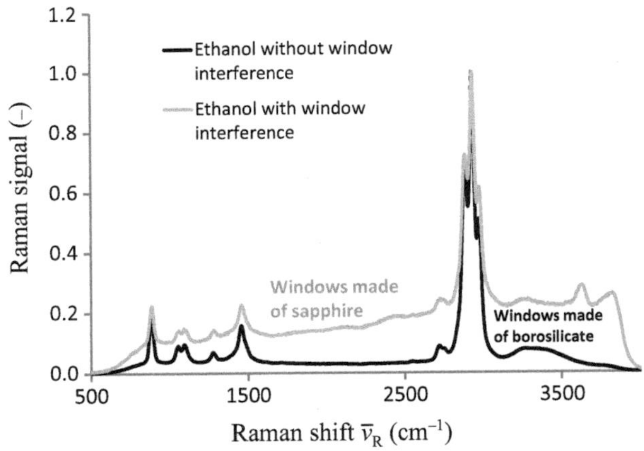

FIGURE 3.20 Spectra acquired from liquid ethanol inside a high-pressure variable volume cell (HPVVC) with the 0-D Raman sensor sketched in Figure 3.2 with an excitation wavelength of 532 nm. Once the HPVVC was equipped with sapphire windows (grey spectrum), and once with borosilicate glass windows (black spectrum). The grey spectrum is interfered with the emissions of the impurities contained inside the sapphire windows.

These impurities absorb the excitation laser light, and subsequently emit photons, which are also red-shifted with respect to the excitation wavelength. When the sapphire windows have been replaced by borosilicate windows, the broadband underground is significantly reduced, which enables a reliable evaluation of the detected Raman spectra. Figure 3.19 also shows that the depth of focus of the Raman sensor detecting in right angle direction (depth of focus is not indicated in Fig. 3.19) does not spatially overlap with the locations where the excitation laser beam passes through the window. Therefore the emissions originating from the two windows through which the excitation laser is guided, are not launched directly into the detection glass fibre bundle and consequently do not interfere as much with the desired Raman spectra. Still, even in the case of the right-angle configuration, the emissions originating from the windows can find their way as stray light into the detection fibre and cause minor interferences.

3.2.5.5 Birefringence of Windows and Its Impact on the Raman Signal Intensity Ratio R

Figure 3.21 shows that the windows of the high-pressure chamber may become birefringent, once they are clamped into the chamber and/or the chamber is pressurised. Most probably this is due to the pressure-induced stress and due to the stress caused by the clamping of the windows. Therefore the initial polarisation of the excitation laser is manipulated in an unpredictable way when the laser passes through the windows. In order to demonstrate this effect, the excitation radiation in Fig. 3.21 outside the chamber, so before it passes the windows, has been adjusted to s-polarisation. Then a polarisation filter has been

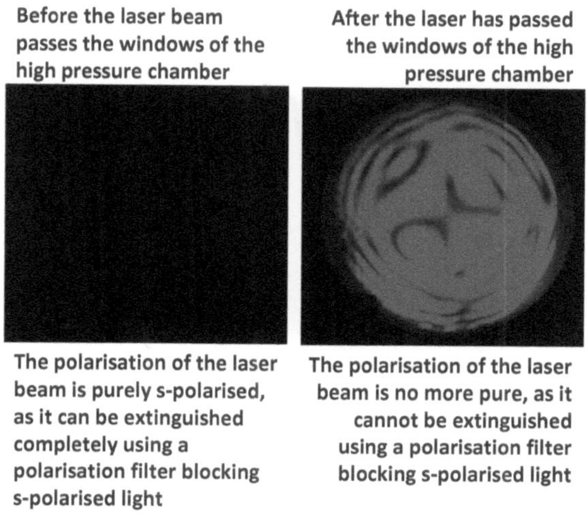

FIGURE 3.21 Images of the laser beam cross-section behind the high-pressure chamber, once the excitation laser beam of the 1-D Raman sensor given in Figure 3.7 has passed two windows (aligned line in sight) of the high-pressure vessel. *(Source: Photo made by Oliver Knauer from the Lehrstuhl für Technische Thermodynamik of the Friedrich-Alexander-Universität Erlangen-Nürnberg, Germany.)*

positioned into the excitation beam path in front of the window of the chamber. The polarisation filter was set to block s-polarisation completely and to let pass p-polarised light only. With the polarisation filter positioned in front of the high-pressure chamber it is possible to completely extinguish the s-polarised excitation laser beam. This is indicated by the black image given in Fig. 3.21. Once the polarisation filter – still set to block s-polarised light and to transmit p-polarised light – is positioned behind the high-pressure chamber (so when the laser has passed the two windows), one can see the intensity profile of the expanding laser beam cross-section (right side in Fig. 3.21).

Laser light can pass the polarisation filter only, if the laser beam, which is purely s-polarised before it passes the window of the chamber, is no more purely s-polarised when it leaves the chamber through the second window (opponent to the first window). In other words, the windows of the chamber must have manipulated the polarisation of the excitation laser beam. Unfortunately – according to Section 2.4.2.3.4 – the overall Raman signal intensity is affected by the polarisation of the excitation radiation for signal detection in right angle direction (not for signal detection following the back-scattering configuration). Furthermore, the signal intensity of different Raman signal bands (lines) are affected differently by the change of the excitation polarisation, as it is shown in Fig. 3.22 for the C-H stretch vibration Raman signal of acetone. This means that for a constant mixture composition one can measure different Raman signal intensity ratios R (Eqn (2.129)) just because of a change of the polarisation of the excitation ratio.

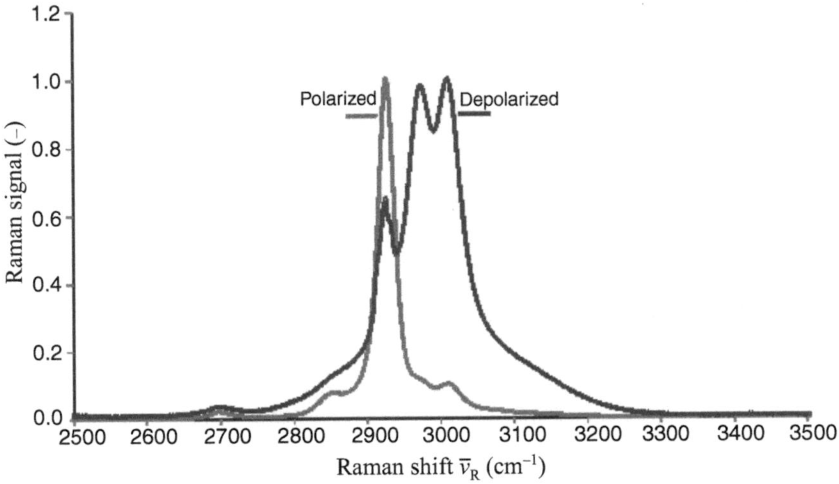

FIGURE 3.22 Polarised and depolarised Raman signal of the CH stretch vibration of acetone detected in right-angle configuration. Polarised, excitation s-polarised and detection s-polarised; depolarised, excitation s-polarised and detection p-polarised.

As this depolarisation effect is unpredictable, its impact onto R cannot be corrected for by post processing. Thus, experimental measures have to be installed, which minimise the windows' depolarisation effect onto R. Based on my experimental experience (trial and error in this case) I recommended setting the polarisation of the excitation laser to be linear at 45°. This corresponds to equal portions of perpendicular (50%) and parallel (50%) polarisation. Then the depolarisation of the initially perpendicular portion towards the parallel polarisation is supposed to be partially compensated by the depolarisation of the initially parallel portion towards the perpendicular polarisation. Both effects alone would result in a decrease or an increase of the Raman signal strength of particular Raman bands (lines), respectively. The physical model which is used for the extraction of the composition of the mixture from the Raman signal intensity ratio R (see Section 2.4.2.3.7) does not account for depolarisation effects. Therefore the fact that the best agreement between the model and the calibration experiments has been achieved for the initial linear polarisation at 45° clearly indicates that depolarisation effects can be neglected in this case. For an initial polarisation of purely parallel, purely perpendicular and purely circular no acceptable agreement between the model and the experiment could be found.

3.3 PURIFICATION OF RAMAN SIGNALS FROM UNDESIRED INTERFERENCES

Sometimes, though a lot of care is taken and though experimental efforts are made in order to not excite any interfering emissions, the Raman spectrum still can suffer from them. In this section I will provide a brief review on possibilities of how an interfering broadband underground can be eliminated from Raman spectra mainly in post-processing steps. In general, the reported strategies can be applied for any kind of broadband underground, including fluorescence or phosphorescence emissions, as long as the background is broad compared to the Raman signal peaks. What is meant by this will become clear further. Of course, the risk of also altering the Raman spectrum is inherent to any post-processing step for background elimination. Therefore the better alternative to post-processing is always to get around the interfering background physically, meaning experimentally already before the final spectra are recorded. Experimental methods for physical background elimination are addressed only shortly further. If there is interfering background, usually the experimental methods alone can only help reduce the background but cannot eliminate it completely, often a combination of experimental methods for background reduction and post-processing methods for the elimination of the remaining background is recommended. In Ref. [15] my group provides a detailed description of the combination of experimental and post-processing methods for the elimination of background, unfortunately not with respect to high pressure. Here only a short summary is provided. Nevertheless I am convinced that the approaches reported in [15] can also be applied in high-pressure process technology and science. This is also the reason why we recently started employing the shifted-excitation difference Raman spectroscopy (SERDS) technique in order to quantitatively monitor the extraction process of natural products using compressed CO_2 as extraction agent. Natural products are known to be colourful and consequently their fluorescence has to be eliminated from their Raman spectrum.

3.3.1 Background Elimination via Post-Processing

As mentioned previously, the interfering background that masks the Raman spectrum has to be broadband. Then, algorithm-based baseline correction methods which include frequency filtering using fast Fourier-transformation (FFT) [16], wavelet transformation [17,18], polynomial fitting [19–22] and baseline correction using asymmetric least squares (AsLS) [23–25] are proposed. The frequency filtering method involves frequency content separation in order to extract the Raman spectra from the fluorescence background [24]. This method can cause artefacts to the Raman spectra due to the similarity of the frequencies of the Raman signals and the noise. Wavelet transformation is a powerful tool in signal processing [26]. Broadband background rejection based on wavelet transformation highly depends on the decomposition method used and the shape of the background [23].

Polynomial fitting is one of the most frequently used methods for fluorescence rejection [27]. Usually, a polynomial of fourth- to sixth-order is fitted to the fluorescence background. Then, this polynomial function is subtracted from the raw spectrum to receive the purified Raman spectrum. However, due to the variation of the fluorescence background from sample to sample, polynomial

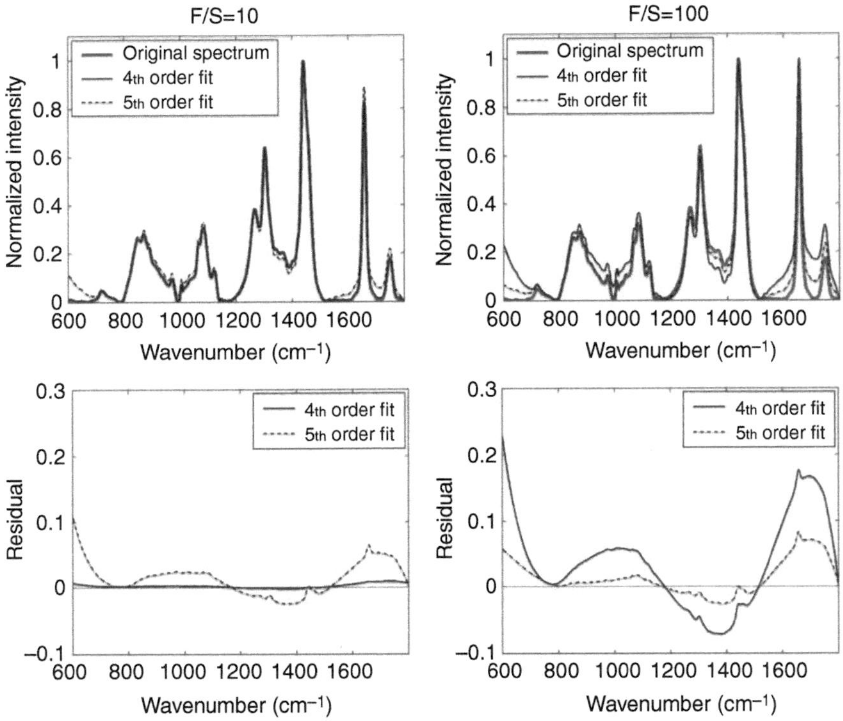

FIGURE 3.23 The top row shows the subtracted spectra compared with the original spectrum, whereas the bottom row shows the residuals. The left and right columns are for F/S = 10 and F/S = 100, respectively. *(Source: Ref. [22].)*

fitting depends on the spectral range and the chosen polynomial order [27]. Cao et al. [22] showed the effect of fluorescence background intensity variation on the best polynomial fit. They added simulated Gaussian background to pure Raman spectra in order to generate synthetic fluorescence/Raman signal ratios (F/S) of 10 and 100. As it can be seen in Fig. 3.23 from the purified spectra after the subtraction of the polynomial fit, the subtraction of the fourth-order polynomial fit performs better at F/S = 10 whereas the fifth-order polynomial fit after subtraction shows the better result at F/S = 100. Thus, they developed a method which automatically chooses a polynomial order to avoid user intervention which is called 'adaptive minmax'. It involves multiple fits with different polynomial orders and chooses the one which gives the minimum error.

In order to improve the performance of single polynomial fitting and in order to reduce its limitations, Lieber and Mahadevan-Jansen [28], proposed an iterative polynomial fitting (ModPoly) method. The initial raw spectrum (grey) is fitted with a polynomial function (dotted black) as shown in Fig. 3.24. Then the raw spectrum and the polynomial-fit spectrum are compared along the spectrum. If the intensity of the raw spectrum exceeds the intensity of the

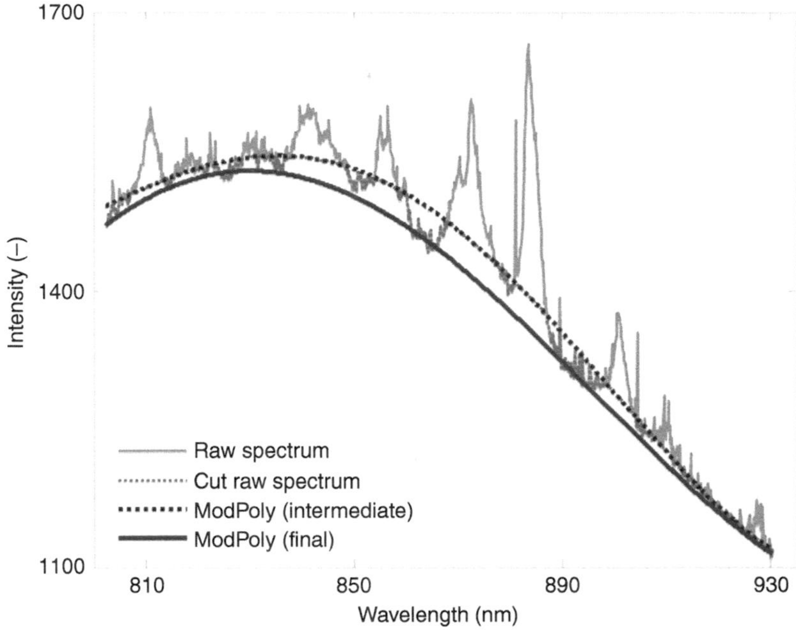

FIGURE 3.24 An intermediate fitting using the ModPoly method, demonstrating the 'cut raw spectrum' (dotted line (dotted red line in the web version)) to be used for the next round of polynomial fitting and the final polynomial fit function (solid line (solid blue line in the web version)). *(Source: Graph prepared by Medhanie Gebrekidan from the Erlangen Graduate School in Advanced Optical Technologies of the Friedrich-Alexander-Universität Erlangen-Nürnberg, Germany.)*

polynomial fit, the raw spectrum value at this wavenumber position is replaced by the polynomial fit value. If not, the value of the raw spectrum at this wavenumber position is kept. In other words, all peaks exceeding the polynomial fit are cut and brought to the polynomial fit (red). In a next step a new polynomial is fitted to the 'cut raw spectrum' and again the intensities of new polynomial fit and the 'cut raw spectrum' are compared. And again at each wavenumber position where the intensity of the 'cut raw spectrum' exceeds the intensity of the new polynomial fit, the raw spectrum value at this wavenumber position is replaced by the polynomial fit value. This process is repeated until the final polynomial fit function shown as blue line in Fig. 3.24 is obtained, which strikes the raw spectrum only at some local minima.

Eiler and Boelens [25] also presented another technique of fluorescence rejection based on asymmetric least squares curve fitting which uses Whittaker smoother [29]. The fit function is computed by penalising the fitting error and its smoothness. Thus, the fit function is as close as possible to the background as well as it is as smooth as possible. Asymmetric weights are utilised in order to penalise the fitting error. Lower weights are assigned to regions where Raman peaks exists and higher weight are assigned to peak-free regions. These

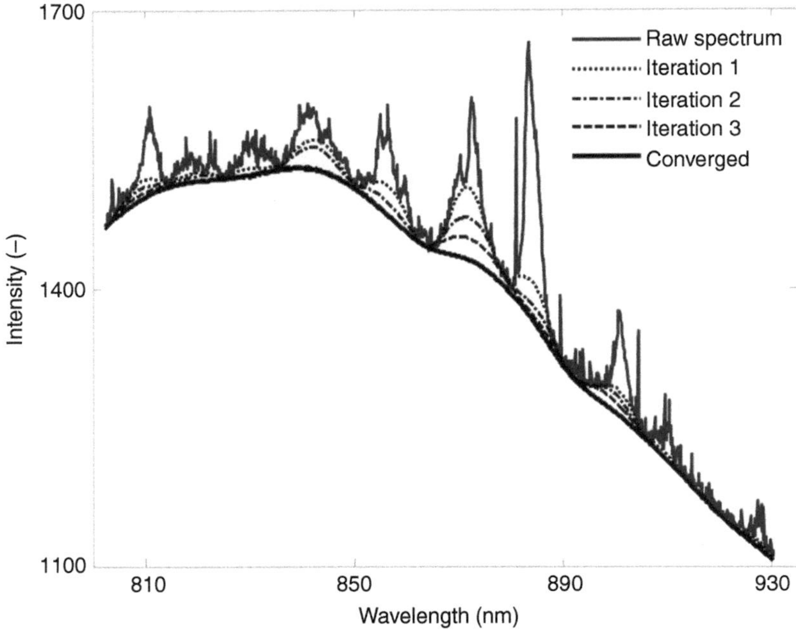

FIGURE 3.25 Raw spectrum of nerve tissue of pig (grey line (blue line in the web version)), the first three iterations' estimated baseline by asymmetric least squares (dotted and broken lines) and the final converged baseline (solid black line) estimated by asymmetric least squares. *(Source: Graph prepared by Medhanie Gebrekidan from the Erlangen Graduate School in Advanced Optical Technologies of the Friedrich-Alexander-Universität Erlangen-Nürnberg, Germany.)*

weights are updated iteratively to bring the fit function closer to the background as shown in Fig. 3.25. However, the weighting parameters involved must be tuned. Thus human judgement will generally be needed especially for samples with strong fluorescence. He et al. [23] further improved this method by including an additional term which penalises the first derivative of the fit function and reported a 16 times improvement of the root mean square error (RMSE).

3.3.2 Experimental Background Elimination Using SERDS

Next to the SERDS technique there are other experimental techniques known for background suppression or at least for Raman signal enhancement with respect to the background, such as surface enhanced Raman spectroscopy (SERS without 'D'), resonant Raman spectroscopy or time-gating methods [30]. Here especially the SERDS technique is described, as this one still is completely non-invasive, provides straightforward quantifiable Raman spectra and thus is most suitable for quantitative *in situ* measurements in real high-pressure process technology.

Eqn (2.99) states that the Raman spectrum can be shifted with respect to the wavelength (not with respect to the Raman shift wavenumber), if the wavelength of the excitation laser is changed. On the contrary, Fig. 2.44 shows that due to the rapid relaxation mechanism. The fluorescence emission starts from the vibrational ground state in the first excited state, irrespectively of the energy level occupied directly after absorption. Consequently – this is what is known as Kasha's rule [31] – the fluorescence emission ideally is supposed to be not affected by a slight change of the excitation wavelength. Exactly on these facts the SERDS technique is based on, namely that the Raman spectrum shifts with the excitation wavelength shift but that the fluorescence spectrum does not. The fluorescence emission can be eliminated by subtraction of two spectra from each other acquired with slightly different excitation wavelengths. This gave rise to the development of the SERDS [32–34]. Shreve et al. [34] analysed Raman scattering from $CHCl_3$ in the presence of added fluorescent laser dye and described the simplicity and effectiveness of SERDS. They were able to observe Raman peaks which are three orders of magnitude weaker than the fluorescence background. Volodin et al. [35] detected methanol in red wine using SERDS and reported its accuracy as compared to baseline fitting methods. Appiah et al. [36] demonstrated the effectiveness of SERDS as compared to polynomial methods by extracting the ethanol spectrum from dark Jamaican Rum. As it can be seen from Fig. 3.26 the processed spectrum (red line) is not a pure Raman spectrum of ethanol which can be caused by incomplete fluorescence rejection. However, a single polynomial fit can in principle eliminate the still remaining fluorescence background via post-processing.

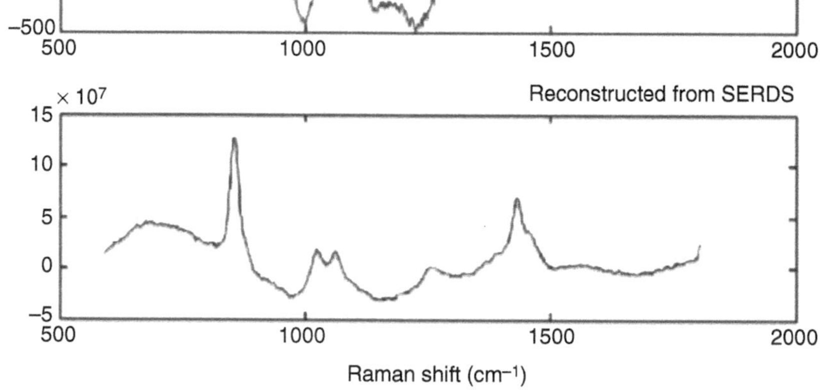

FIGURE 3.26 Raman spectra of dark rum: reconstructed based on SERDS data (dashed line (dashed red line in the web version)), and the conventional Raman spectrum obtained with polynomial baseline fitting (dotted line (dotted blue line in the web version)). *(Source: Ref. [36].)*

Martins et al. [37] and Sowoidnich et al. [38] also applied the SERDS technique in order to analyse biological tissues and reported the capability of SERDS to eliminate undesired fluorescence background in a systematic and reproducible way. Moreover, Noack et al. [32] applied SERDS in combination with support vector regression for online analysis of algae production and showed its potential as a non-intrusive tool for online monitoring of biotechnological processes.

3.4 CASE STUDIES

When searching databanks or the internet for 'Raman' and 'high pressure' most results refer to Raman measurements in solids at high pressures and are strongly related to material sciences. As this book '*In situ* Spectroscopic Techniques at High Pressure' will appear in the frame of the book series '*Supercritical Fluid Science and Technology*' I decided to report on few case studies relevant especially for process technology. Goncharov [39] provided one of the latest – according to my knowledge – reviews on Raman measurements at elevated pressure, putting the focus of his review on scientific disciplines including physics and chemistry of materials under extremes. Previous reviews laid their focus on mineralogy and environmental applications [40–42]. Readers who are not only interested in *in situ* Raman measurement relevant for process technology I want to refer to these reviews.

The sequence of the following case studies is organised according to the complexity of the claims the measurement task raises on the Raman experiment. Therefore I will first start describing 0-D measurements for the study of vapour liquid equilibria (VLE) or investigations into gas hydrates. These case studies have in common that no claims are raised with respect to a temporal resolution. This will change when in the second type of case studies the Raman measurement techniques have to temporally resolve the progress of reactions, or of heat and mass transport processes. This is relevant for the fuel/air mixture generation in internal combustion engines, extraction, impregnation and dissolution processes as well as all types of reaction. The respective type of case studies comprises 0- as well as 1-D Raman experiments. Finally, temporally and spatially resolved as well as 2-D Raman measurements (Raman imaging) are described, which have been made in order to analyse mixing processes between fluids at elevated pressure. This is relevant, for example for the fuel/oxygen mixture generation in rocket engines as well as for the particle generation according to the supercritical anti-solvent (SAS) process.

3.4.1 Vapour Liquid Equilibria

One major impediment for modelling and dimensioning new processes is the often not adequate availability of thermodynamic properties, especially at elevated pressure. These include the phase equilibrium compositions of the system at the respective process conditions [43]. Generally, VLE data are determined via synthetic or analytical methods [44]. Primarily in synthetic methods the phase transition is detected while the exact composition in the system is known. Here, a precise optical detection of this transition point is required. In contrast, the analytical methods are mostly sampling methods. It has to be underlined that a sampling step always leads to a perturbation of the equilibrium and furthermore to a change in the overall composition of the system [45]. Thus, sampling is quoted as the main error source [46] during the quantitative analysis of a sampled volume. Additionally, these methods are very time consuming due to the circumstance that after each sampling event the equilibrium cell has to be evacuated, cleaned and filled with the new composition to ensure reproducibility. A non-invasive measuring technique, such as Raman spectroscopy, avoids the mentioned drawbacks inherent to sampling methods.

In Section 2.4.2.3 the 0-D Raman set-up with signal detection in back-scattering direction is described as a quantitative measurement device. Furthermore it is derived that the ratio of the Raman signals originating from different compounds within a mixture reflects the ratio of the number of substances of the respective compounds in the mixture. Based on this dependency the composition of a mixture can be computed from its Raman spectrum after a thorough calibration.

According to my knowledge Stratmann and Schweiger [47] were the first to exploit Raman spectroscopy for the determination of high-pressure VLE. They built up the Raman experiment with signal detection in right angle direction shown in Fig. 3.27 and spilt the excitation laser beam in order to being able to excite the dense and the light phase inside the high-pressure view cell (HPVC) without moving neither the Raman experiment nor the HPVC. This also presumed that the signal paths had to be combined in order to being able to measure the composition of one phase after the other with the same spectrometer.

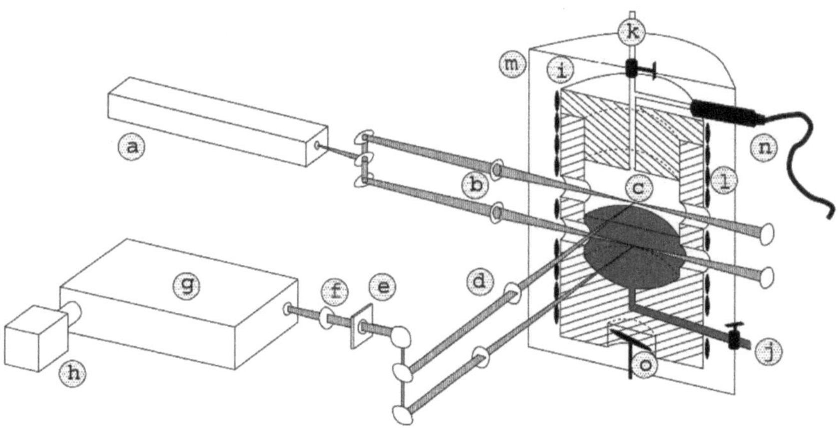

FIGURE 3.27 Experimental set-up of the first Raman experiment for the determination of VLE. *(Source: Ref. [47].)*

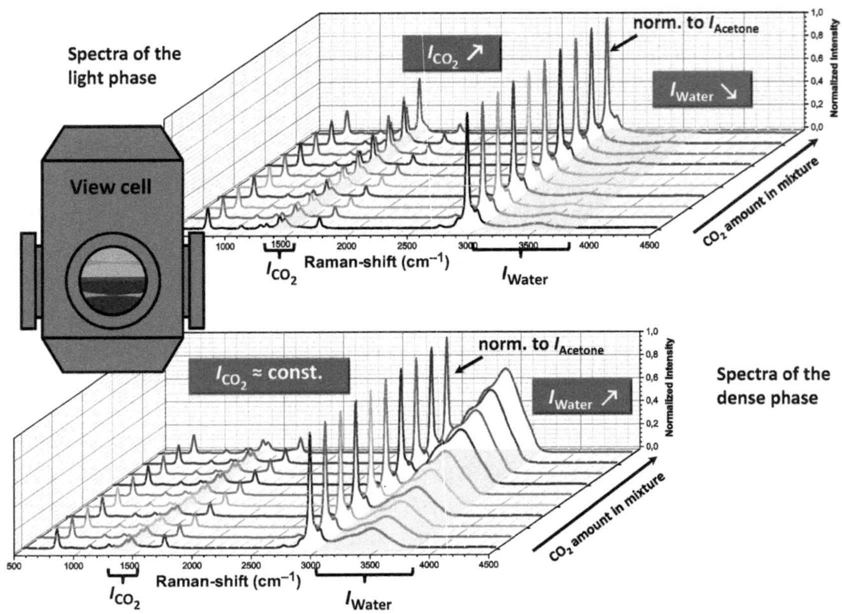

FIGURE 3.28 Series of Raman spectra detected from the light and from the dense phase of a mixture of acetone, water and CO_2 during the addition of CO_2 at 313 K and 9 MPa. *(Source: Graphic prepared by Julian Schuster from the Lehrstuhl für Technische Thermodynamik of the Friedrich-Alexander-Universität Erlangen-Nürnberg, Germany.)*

Their results have been in excellent agreement with the results published in literature by others on the same system at identical operational conditions.

Since the work of Stratmann and Schweiger, the progress in electronics lead to a miniaturisation of the electronic equipment installed in such an experiment. Therefore today significantly smaller and cheaper lasers, spectrometers and detectors are available.

They are small enough to arrange them all together in one coherent cage system as it is shown in Figs 3.3 and 3.4. Then the entire cage system can be moved up and down in order to measure subsequently the composition of the light and the dense phase inside the HPVC. In several contributions [48–51] we demonstrated the good agreement between the VLE composition data we obtained via Raman spectroscopy and others obtained using conventional methods. Figure 3.28 shows series of Raman spectra detected from the dense (bottom) and from the light phase (top) of the mixture of acetone, water and CO_2 during the isothermal and isobaric (variable volume cell) addition of CO_2 at 9 MPa and 313 K. The CO_2 content inside the HPVC is enriched from the foreground to the background. All Raman spectra are normalised to the dominant Raman signal peak of acetone $I_{Acetone}$ (CH stretch vibration). The spectral regions of the appearance of the Raman signals representing the CO_2 and the water are highlighted.

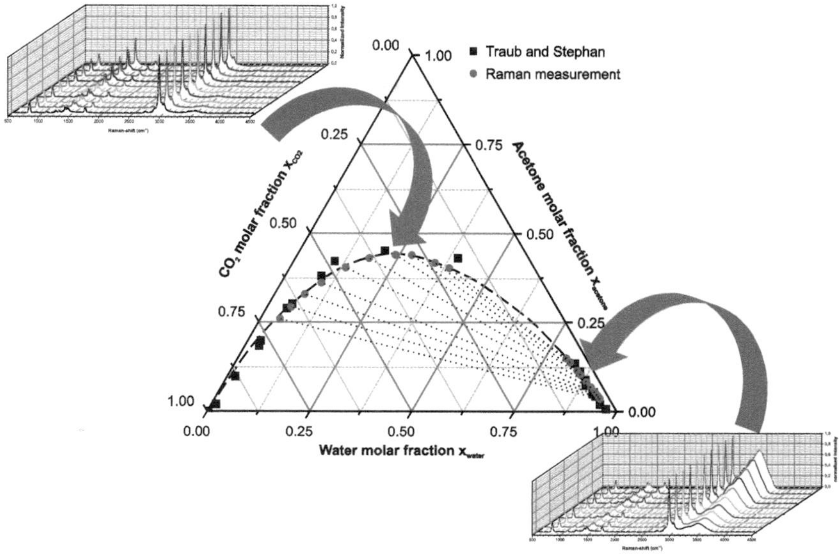

FIGURE 3.29 Gibbs diagram of the ternary mixture of acetone, water and CO_2 at 313 K and 9 MPa. Raman measurements (discs) are compared with data resulting from Traub and Stephan [52] who used conventional measurement techniques (squares). *(Source: Graphic prepared by Julian Schuster from the Lehrstuhl für Technische Thermodynamik of the Friedrich-Alexander-Universität Erlangen-Nürnberg, Germany.)*

Obviously in the dense phase the ratio of the Raman signal intensities of CO_2 I_{CO_2} and acetone $I_{Acetone}$ are not affected by the addition of CO_2 to the two-phase mixture. But one can see clearly that the Raman signal intensity of water I_{water} grows with respect to the one of acetone $I_{Acetone}$. Simultaneously in the light phase the Raman signal intensities of CO_2 I_{CO_2} and of water I_{Water} grow and shrink, respectively, with respect to the one of acetone $I_{Acetone}$. Summarising, the isobaric and isothermal addition of CO_2 to the ternary mixture caused a depletion of acetone and CO_2 in the dense phase, while at the same time and as a consequence the acetone is transferred more and more from the dense to the light phase. The Gibbs diagram which resulted from the evaluation of the Raman spectra is shown in Fig. 3.29, where also the Raman measurement results are compared with data provided by Traub and Stephan [52].

At this point I want to highlight one decisive advantage of Raman measurements compared to conventional methods. Applying the non-invasive Raman method does never require the removal of a probe volume from the HPVC for subsequent analysis, which would disturb the thermodynamic equilibrium. Therefore the Raman measurements can at identical conditions first be made in the dense and then in the light phase, or vice versa. The connecting line between the compositions of the dense and the light phase measured at identical conditions directly provides the tie line.

The more compounds are contained within a mixture, the more complex the quantitative evaluation of the Raman spectrum of the respective mixture gets (see Section 2.4.2.3.7). Therefore it is rather not the Raman experiment itself, but the processing of the Raman spectra which amounts to a successful determination of high-pressure VLEs. Different possibilities of processing the Raman spectra are summarised in Section 2.4.2.3.6. The exact method used for the processing of the spectra shown in Figs 3.28 and 3.29 is described in detail in the appendix of Chapter 2 in Section 2.5.6.

3.4.2 Gas Hydrates

Also with respect to investigations into gas hydrates there is no need for a temporally resolving measurement technique. Once the clathrates are formed they are rather stable and – if the operational conditions are kept constant – are available for the experiment for hours.

Gas hydrates are solid networks of cavities, whose frameworks are formed of hydrogen bonded water molecules [53]. In typical hydrates most cavities house one hydrate former – the so-called guest molecule – for example small hydrocarbons [54], noble gases [55], fluorinated compounds or small ether molecules [56]. With respect to the energy and environmental sectors gas hydrates are considered as boon, but also as bane. They are considered boon for several reasons.

- Huge amounts of methane hydrates are stored in the sea sediments and permafrost deposits. They represent a potential source of natural gas [57,58].
- The concentration of guest molecules captured inside the gas hydrates is high. Thus gas hydrates can act as storage medium during transportation of 'gases', for example of hydrocarbons or hydrogen.
- Due to the same reason they may be utilised for the 'ultimate' disposal of environmentally problematic gases such as carbon dioxide (CO_2) [59,60].

But they are also considered as bane, as the undesired formation of gas hydrates in oil and gas pipelines has been responsible for pipeline blockage many times. The annual costs for chemicals added as hydrate formation 'inhibitors' are in the range of 100s of millions of US$, not comprising the convey stop and drill time loss due to hydrate pipe blockage [61,62]. Summarising, the comprehensive understanding of the hydrate formation mechanisms and the knowledge of the properties of hydrates is essential for the environmental and energy sectors.

Most probably due to the complex nature of water and due to the complex interplay of water and the hydrate-forming molecules, the formation mechanisms of hydrates (kinetics, meta-stability, formation of different kinds of cavities) as well as the mechanisms promoting or inhibiting the formation of hydrates are not fully understood [63]. Spectroscopic approaches have provided some insights into the hydrate formation mechanisms and into the generation of different kinds of cavities on a microscopic level [64–67].

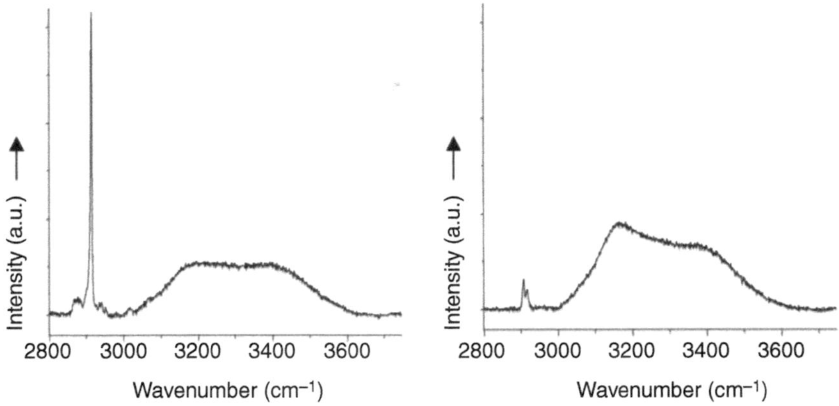

FIGURE 3.30 OH band of CH_4, C_2H_6 or C_3H_8 structure II hydrate (left) and OH band of CO_2 CH_4 structure I hydrate (right). *(Source: Ref. [68].)*

For example Sloan et al. [64] analysed the spectral position of the CH stretch vibration of the hydrogen former methane. Depending on the environment of the methane molecules (methane can be in the pure gaseous phase, dissolved in liquid water or captured in hydrate cages) the CH stretch vibration features different energies and thus different central peak positions in the Raman spectrum. From the peak intensities conclusions can be drawn about the occurrences of different hydrate cages within the hydrate from which the structure of the formed hydrate is derived. Sloan et al. [64] also exploited this technique in order to follow the evolution of hydrate formation with and without hydrate-inhibiting molecules being present. Due to the slow generation of gas-hydrates the required temporal resolution is rather low. With respect to the experiments reported by Sloan et al. [64] one spectrum was recorded in 25 s.

Schicks et al. [68] analysed the differences in the Raman spectra of differently structured hydrates and ice. Their comparison was not only based on the analysis of the methane Raman signal peaks but also on the Raman band of water, as it is shown in Fig. 3.30.

Lu et al. [69] analysed the ratio of the Raman signal intensity ratios of methane and water, as it is shown in Fig. 3.31, and from this derived the methane concentration in water in the presence of sI (s-one) hydrate. Very nicely it can be seen that the shape of the water Raman signal band between 3000 cm^{-1} and 3800 cm^{-1} is a function of especially temperature. This dependency has already been described in detail in the context of Fig. 2.57.

The mentioned studies have two things in common. Firstly, the gas hydrates themselves have been analysed and not the liquid water-rich phase. Secondly, the focus of the spectral investigations into hydrates has been placed on the hydrogen former (mostly methane), but not on the water molecules, which form the cavities. Few other studies are known that put more emphasise onto the role

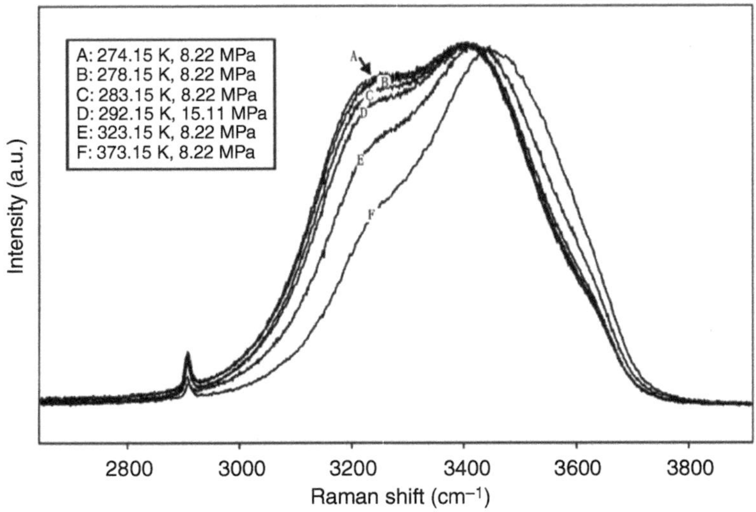

FIGURE 3.31 Raman spectra of dissolved CH_4 (near 2910 cm^{-1}) and water at various temperatures and pressures. *(Source: Ref. [69].)*

of the water molecules – not in the liquid water-rich-phase – but in the solid gas hydrate phase [68–70].

My current research attempts are towards the analysis of the working principle of gas hydrate inhibiting or promoting substances [71]. On this account we especially focus on the shape of the Raman signal of the water stretch vibration and from this can deduce information how inhibiting or promoting substances influence the development of hydrogen bonds before the onset of the hydrate formation.

3.4.3 Reactions

As mentioned in Section 1.1, supercritical and compressed fluids can act as advantageous media in reactions, mainly due to the tunability of their unique properties. Due to the restricted access to high-pressure processes for any type of measurement equipment not much was known about molecular-level dynamics of chemical reactions in supercritical media, when in the year 1993 Masten et al. [72] exploited Raman spectroscopy in order to monitor the decay of hydrazine and the production of ammonia in supercritical water. Though it took them 60 s to acquire one Raman spectrum, they were still able to resolve the reaction temporally by applying a trick. They had the reaction medium continuously flowing through their miniature high-pressure view cell. By changing the flow rate, the residence time of the reaction medium in the pipes before it passes the location of Raman measurements in the view cell could be varied, which finally enabled a temporal resolution much better than the time required for the acquisition of high-quality spectrum. After 1 year in 1994 Myrick et al. [73] used a fibre probe for the

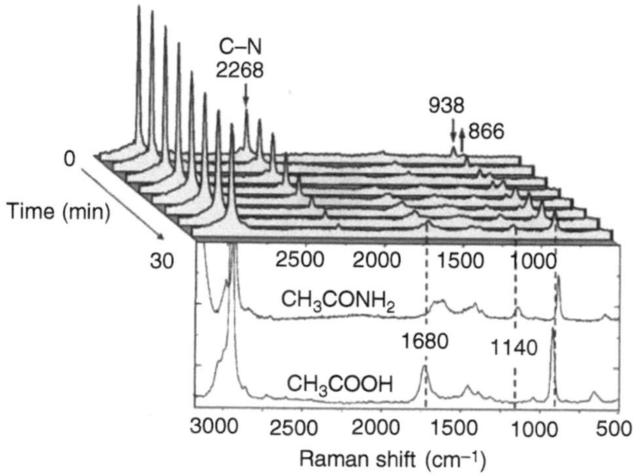

FIGURE 3.32 Raman spectra of acetonitrile in water at 300°C and 30 MPa, showing clearly the disappearance of the peak at 2268 cm^{-1}, due to the CN vibration. Downward arrows indicate a decreasing concentration of a particular species with time, whereas upward arrows indicate an increasing concentration. The spectra were collected during the course of 30 min. The spectra are compared with standard spectra (front panel) of the intermediate acetamide, and the final product, acetic acid, in water under ambient conditions. *(Source: Ref. [77].)*

provision of optical access to the high-pressure reactor. Within the fibre probe one fibre was utilised for the excitation of the Raman process whereas the other one was utilised for the collection of the Raman signals. Raman spectroscopic measurements for the oxidation in supercritical water [74,75] and for hydrothermal reactions [76] followed. Venardou et al. [77] used a 0-D Raman experiment with signal detection in back-scattering configuration in order to monitor the hydrolysis of acetonitrile in near-critical water. Figure 3.32 shows the Raman spectra Venardou et al. recorded during the 30 min course during the hydrolysis experiment.

The decreasing peak intensity at 2268 cm^{-1} indicates the depletion of the reacting CH_3CN. Other peaks rise due to the formation of CH_3COOH.

Picard et al. [78] applied Raman spectroscopy as reaction monitoring tool to high-pressure bio process technology. They quantitatively analysed the alcoholic fermentation by Saccharomyces cerevisiae under high pressure.

3.4.4 Mass Transfer During Sorption or Extraction Processes

If mass- or heat transfer processes are stationary, a series of 0-D measurements (point measurements) acquired from several locations can reveal the concentration or temperature profile through the transfer layer. From this the heat- or mass-transport properties can be derived. Davies et al. [70] for example followed this approach using confocal Raman microscopy in order to quantify the mass transport of water and methane through a thin hydrate film formed at the interface between the water-rich and the gas-rich phase.

Also for non-stationary heat- and mass transport processes the respective properties can be derived based on point measurements. This is possible if the non-stationary heat- or mass-transport processes are slow compared to the temporal resolution of the measurement technique. Then a series of successively acquired point measurements can reveal temporal temperature or concentration profiles from one location from which again the heat- and mass transport properties can be derived.

Nevertheless I here want to demonstrate the capability of the 1-D Raman measurement technique with signal detection in right angle direction, as I described them in Section 3.1.2. With this technique the two dimensions of a CCD detector chip are used to simultaneously capture spectroscopic information along a line. So one dimension – from now on referred to as spatial dimension – of the detector is reserved for resolving the spatial information along a 1-D measurement line. This can be either the horizontal or the vertical one. The other dimension, which is then normal to the spatial dimension and which from now on is referred to as spectral dimension resolves the spectral information originating from each spatially resolved point along the measurement line. Therefore it is suitable for the acquisition of many Raman spectra, each Raman spectrum originating from a different location along the measurement line, within one single measurement event. In other words, one measurement does not provide the temperature or the composition from one point only, but profiles of temperature or composition. Thus, one can follow the evolution of temperature or composition profiles, if several measurements are made successively with a respective temporal resolution. This circumstance makes this technique very attractive for the analysis of heat and mass transport mechanisms. Its potential for measurements at ambient pressure has been demonstrated several times [79–82].

In collaboration with researchers from the University of Naples Federico II, we analysed the sorption of CO_2 into molten polymer poly-ε-caprolactone (PCL) when pressurised with CO_2. An example Raman spectrum of pure molten PCL is shown in Fig. 2.54. For this purpose a pendant drop of molten PCL was positioned in an optically accessible high-pressure cell. Figure 3.33 shows photographs of the same drop pendant at 353 K once in vacuum and once in a 7.1 MPa CO_2 atmosphere.

At 7.1 MPa the drop is bigger, as it swelled due to the sorption of CO_2 into the polymer. Figure 3.33 also shows example Raman spectra of molten polymer saturated with CO_2 at different pressures. The CO_2 Raman signal increases with pressure, as with pressure also the solubility of CO_2 in PCL increases. In order to analyse the sorption kinetics we applied a 1-D Raman experiment. The excitation laser beam was positioned to pass through the pendant drop as shown in Fig. 3.33 (top right). Figure 3.34a shows how the detected Raman signals are distributed over the CCD chip of the camera. The horizontal axis of the CCD chip reflects the spatial dimension. The measurement positions are provided with respect to the ratio r/R which is specified in Fig. 3.33 (top left). The vertical pixel columns reflect the Raman spectra belonging to the different

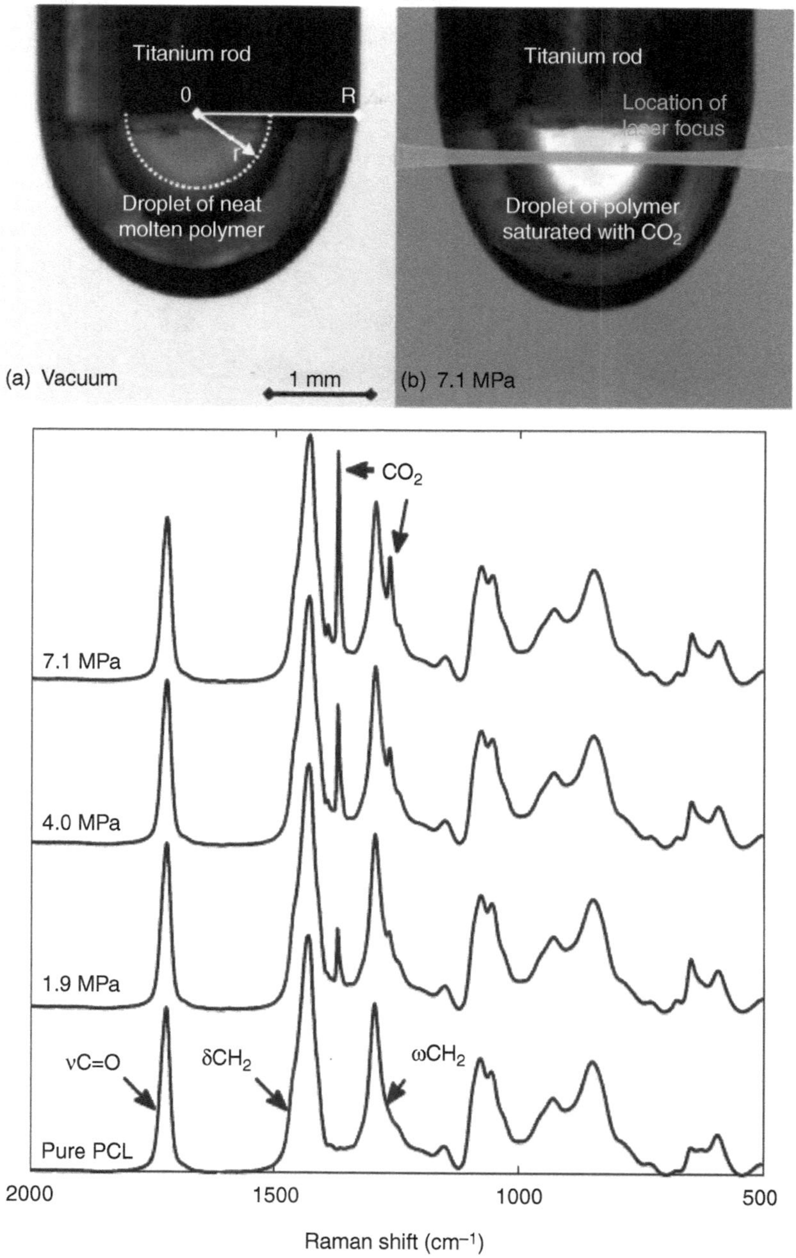

FIGURE 3.33 (top) Hemispherical poly-ε-caprolactone (PCL) drop at 80°C with (a) definition of radial position in vacuum and (b) location of measurement volume in compressed CO_2 at 7.1 MPa. *(Source: Ref. [83].)*

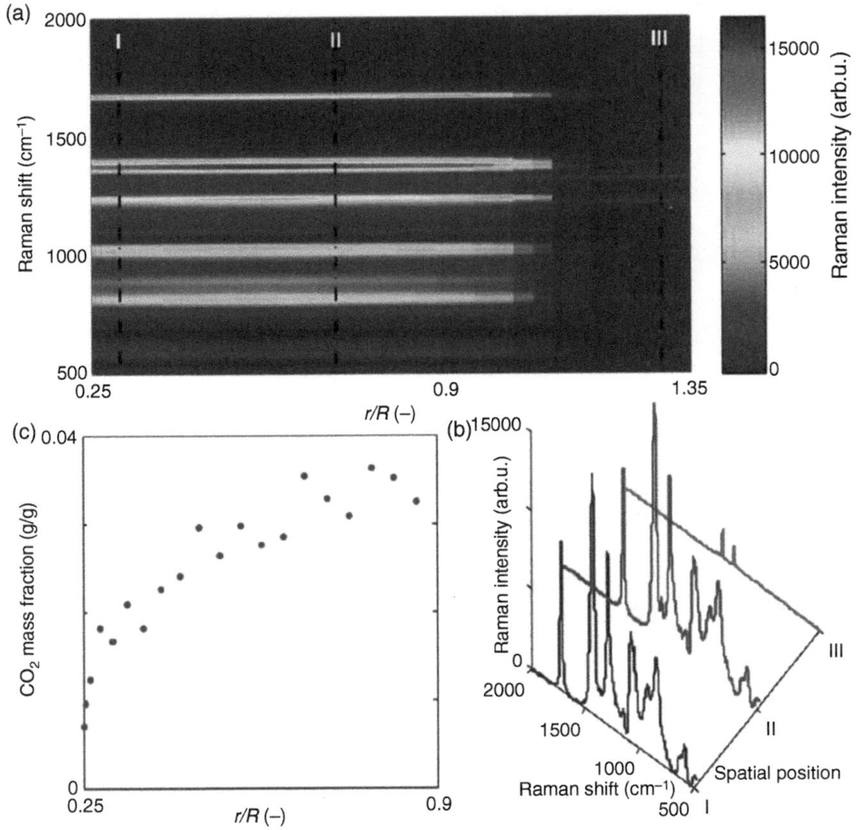

FIGURE 3.34 (a) Image acquired at $t = 180$ s. (b) Raman spectra at selected radial positions. (c) Mass fraction profile computed from the Raman spectra. *(Source: Ref. [83].)*

measurement locations *r/R*. For the measurement locations labelled with 'I', 'II' and 'III' extracted example Raman spectra are shown in Fig. 3.34b. Measurement location 'III' is outside the PCL drop. Therefore, its spectrum only shows the characteristic Fermi dyad of CO_2. At the measurement location 'I' and 'II' the Raman spectra are dominated by PCL. At location 'I' the Raman spectrum visually does not show any contributions of CO_2, whereas at measurement location 'II' a small CO_2 Raman signal contribution can be noticed between the two dominant PCL peaks. Of course, CO_2 can be detected the earlier inside the PCL drop, the closer the measurement location is located to the surface of PCL. Therefore it is already detectable at location 'II' but not yet at location 'III'. Figure 3.34c shows the CO_2 mass fraction profile, which corresponds to the Raman signals shown in Fig. 3.34a. Thus with only one measurement an entire composition profile can be provided. In a next step we fitted a diffusion model to the temporal evolution of the composition profiles and from this approximated the diffusion coefficient.

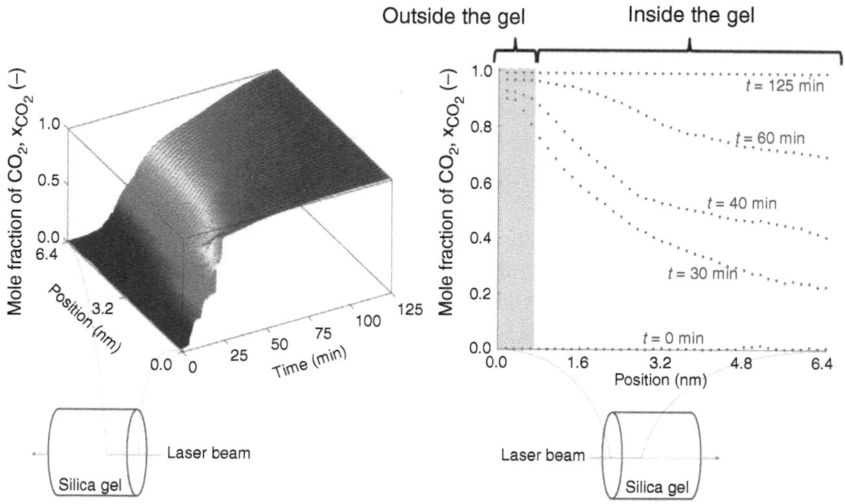

FIGURE 3.35 Measurement of composition profiles inside a silica-based aerogel during the supercritical drying process. *(Source: Graphic prepared by Jaypee Quino from the Erlangen Graduate School in Advanced Optical Technologies of the Friedrich-Alexander-Universität Erlangen-Nürnberg, Germany.)*

A very similar approach we applied successfully in order to quantify the diffusivity for the system dimethylsulphoxide and CO_2 [84].

Our latest activities in the field of mass transport analysis at elevated pressures refer to the supercritical drying of silica-based aerogels. Supercritical CO_2 is taken as extraction agent. The aerogel, whose pores before the drying process are filled with the organic solvent ethanol, is put into an optically accessible high-pressure chamber. The chamber is closed and then pressurised with CO_2 above the mixture critical point of the mixture of ethanol and CO_2. The laser beam is launched as indicated in Fig. 3.35 through the centre of the gel cylinder.

Raman signals are recorded from a certain length within the gel and also outside. The left part of Fig. 3.35 shows the temporal evolution of the composition profiles during the drying process. On the right side of Fig. 3.35 some examples composition profiles are shown, which are characteristic for measurement times before the drying process started, during the drying process and when the drying process can be considered completed. Again one can fit mass transport models to the measurement results in order to approximate the binary diffusion coefficient.

3.4.5 Fuel-Air Mixture Generation Analysis in Internal Combustion Engines Using 1-D Raman Spectroscopy

The combustion of fuel in an internal combustion process for sure is the most frequently applied and widest spread high-pressure process. In an internal combustion engine, irrespectively of whether it is operated as Diesel or Otto process, the mixture generation of fuel and air always precedes the combustion process. It is well known that the quality of the mixture largely determines the combustion. Badly mixed loads might not be ignitable (misfire) or produce a huge amount of toxic exhaust gas or soot. Therefore the control and the improvement of the mixture generation process in internal combustion processes is one of the major issues in internal combustion engine research. On this account Miles [85] and later Schütte et al. [86,87] carried out 1-D composition measurements in internal combustion engines during the fuel/air mixing process. This technique has been extended by Egermann et al. [88] for the direct gasoline injection and later by Taschek et al. [89] for the Diesel combustion to a polarisation-resolving Raman experiment as it is shown in Fig. 3.36. I will discuss this experimental Raman set-up in more detail further as the combination of the equipment used here is perfectly optimised to the measurement task. In order to make the internal combustion engine optically accessible a glass ring is installed between the cylinder head and the cylinder. The piston is elongated accordingly. Furthermore the piston features a transparent piston top window and a nut is cut through the piston cylinder wall. So one can look vertically through the piston top window and horizontally through the region where the piston material was cut away. Exactly there a mirror is mounted. This mirror is not combined with the piston, but stands still, while the piston moves up and down when the optical engine runs.

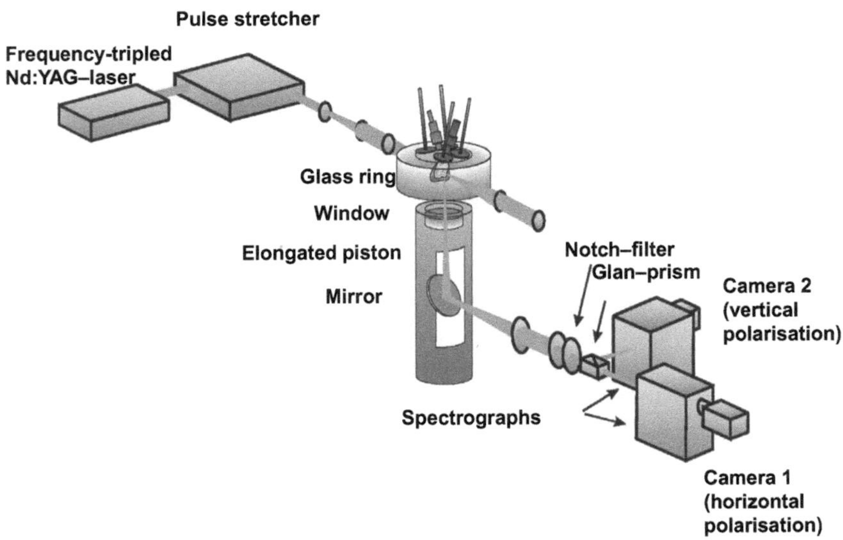

FIGURE 3.36 Schematic of the polarisation-resolving 1-D Raman experiment for the analysis of the mixture generation between the injected fuel and the air. *(Source: Figure provided by the Lehrstuhl für Technische Thermodynamik of the Friedrich-Alexander-Universität Erlangen-Nürnberg, Germany; Figure prepared by Ingo Schmitz and Wolfgang Ipp [88].)*

Summarising, optical access is provided from the side through the glass ring and via the mirror in the slotted piston from the bottom. Figure 3.36 shows that the Raman process is excited from the side through the glass ring, while the Raman signals are detected through the piston window. For the efficient excitation of the Raman process a frequency triplet pulsed Nd:YAG laser emitting at 355 nm is used. We saw in Section 3.2.1 that an excitation wavelength in the UV spectral region brings the advantage of an efficient excitation of the Raman process but that at the same time the risk of also exciting interfering fluorescence emissions rises. The diesel and gasoline fuel one can get from the fuel stations would terribly fluoresce if excited with 355 nm. Therefore the experimenters for their Raman investigations replaced the standard gasoline or diesel fuel with non-fluorescing model fuels. In order to prevent the ionisation (breakdown) of the fuel/air mixture in the focal spot of the laser pulse inside the internal combustion engine, the laser pulse is stretched temporally using a triple-loop pulse stretcher. This decreases the peak power of the laser pulse, while the energy contained in one laser pulse remains ideally unchanged. The concept of pulse stretching is described in Section 3.2.2. Pulse stretching also brings the advantage that the risk of laser-induced glass damage of the rather expensive glass rings is excluded. So the experimenters were able to locate the laser beam waist close to the glass ring material. This also enabled the analysis of the influence of the combustion engine's walls onto the mixture generation. Behind the pulse stretcher two lenses form a Galileo-type telescope and expand the laser beam diameter. The third and last lens in the excitation path focusses the laser beam through the measurement volume, which is the beam waist of the excitation pulse. The generated Raman signals are collected vertically through the piston window and via the still standing mirror by the first lens in the detection path. Elastically scattered stray light is suppressed by a following notch filter, before the detected signal is focussed via a last lens through a polarisation analyser onto the entrance slits of two imaging spectrometers. Horizontally polarised light simply passes through the polarisation filter to one spectrometer, whereas vertically polarised light is reflected to the other one. In Section 2.4.2.3.4 I described that in the case of right angle signal detection the majority of the generated Raman signal features the polarisation of the excitation radiation. A smaller portion is depolarised, meaning that its polarisation is turned by 90° with respect to the polarisation of the excitation radiation. So the spectrometer assigned to the polarisation preserving light detects more Raman signal than the spectrometer assigned to the depolarised light. But why does it make sense to split the Raman signal according to its polarisation? This makes sense as soon as considering that the excitation wavelength of 355 nm even in non-fluorescing model fuels excites some fluorescence emissions. They can originate from lubricants or from compounds that were dissolved from the pipes and hoses into the model fuel. In Section 2.4.1.6 I mentioned that the fluorescence is a completely depolarised process, meaning that irrespectively of the polarisation of the excitation source the intensities of the polarised and the depolarised

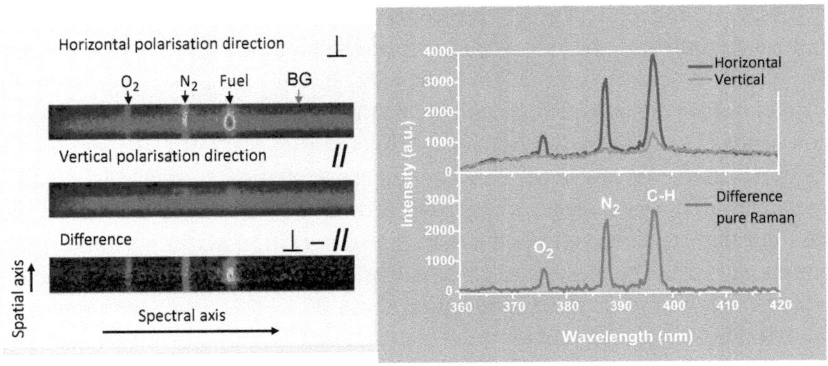

FIGURE 3.37 Image of vertical (V) and horizontal (H) polarisation direction and resulting background corrected image. *(Source: Figure provided by the Lehrstuhl für Technische Thermodynamik of the Friedrich-Alexander-Universität Erlangen-Nürnberg, Germany [89].)*

fluorescence signal are the same. So the polarisation analyser splits the interfering fluorescence emissions into equal portions so that both spectrometers, the one assigned to the strong polarisation-preserving Raman signal compound and the one assigned to the weak depolarised Raman signal compound, detect the same amount of fluorescence signal. Figure 3.37 therefore shows that the broadband interfering fluorescence emissions can be eliminated experimentally from the Raman spectrum, if the signals assigned to the depolarised spectrometer are subtracted from those assigned to the polarisation-preserving spectrometer. Of course not the spectrometers but more precisely the cameras the spectrometers are equipped with detect the Raman signals. The experimenters decided to choose ICCD detectors for the signal acquisition for three reasons.

Firstly, as already mentioned, the mixture generation is followed by the luminous combustion and in the case of the Otto process already before the bright spark is fired. Consequently detectors have to be chosen which can realise short exposure times and which can be closed completely before other undesired luminous processes would cause interference with the desired detected signals. Secondly, detectors have to be used which feature an acceptable quantum efficiency in the transition region between the VIS and the UV spectral regions. This is where the corresponding Raman signals show up when excited at 355 nm. Thirdly, due to the small scattering cross-sections of the Raman process a detector type is beneficial which can amplify the detected signal before read-out. At the time when these experiments were carried out only the ICCD camera fulfilled all these requirements. For those readers who want to know more about detectors and for what measurement purposes they are ideally suitable for I provide some information about CCD detectors in Section 3.5.2.

Due to the smart arrangement and selection of equipment in the Raman experiment the experimenters – as an example – were able to measure fuel/air-ratio profiles in a vaporising diesel spray stream in the vicinity of the walls of the glass windows at different crank angles (CA), as it is shown in Fig. 3.38.

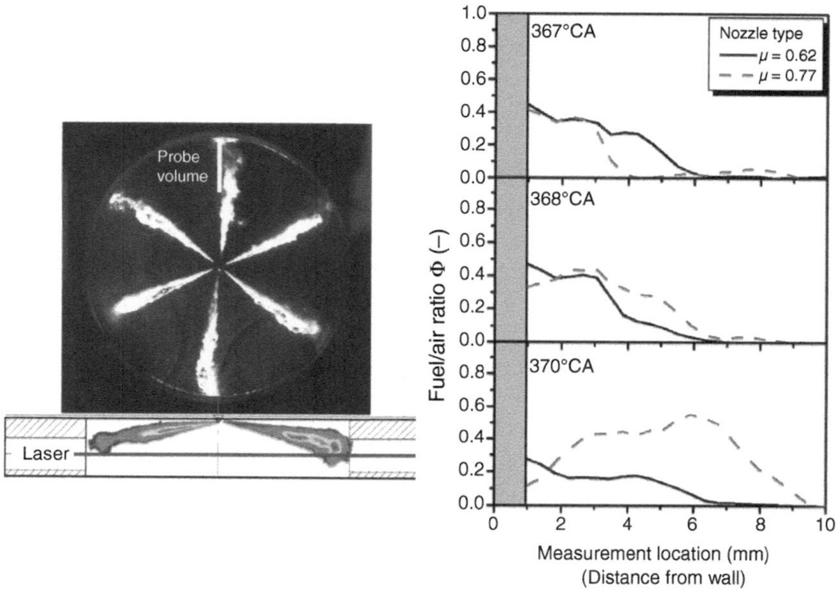

FIGURE 3.38 (Left) Position of the measurement volume inside the combustion bowl of the DI diesel engine. (Right) Influence of the nozzle orifice geometry on the average fuelair ratio in the measurement volume. *(Source: Ref. [89].)*

3.4.6 Fuel/Oxygen Mixture Generation Analysis in Rocket Engines Using Raman Imaging

The development of the 1-D Raman experiments for combustion research was motivated by the desire to understand the mixing process between fuel and air. Exactly the same desire motivated the development of Raman imaging techniques. The alignment of a Raman imaging experiments is rather simple compared to a 1-D Raman experiment. The Raman signals do not have to be aligned through the narrow entrance slit of the spectrometers, as simply no spectrometers are used in Raman imaging experiments. The spectral separation of different Raman signals is achieved installing either dichroic beam splitters or band-pass filters (or the combination of both) into the detection path of the Raman experiment. The camera detectors simply take images of the excitation light sheet. A typical Raman imaging experimental set-up is described in detail in Section 3.1.3. Next to Raman imaging experiments relevant for ambient pressure combustion [90–92], Meier et al. [93] in the year 1996 already carried out temporally resolved 2-D Raman measurements in a modified internal combustion engine. With respect to the title of this book series, especially those experiments have to be mentioned, which were carried out in order to better understand the mixing between a fuel and an oxidiser at conditions close or exceeding the critical point of the mixture. The respective investigations were especially relevant for the development of rocket engines [94–98]. In rocket combustion chambers the fuel and the oxygen are injected at pressures exceeding the critical pressure of the pure components and also exceeding that one of the mixture. But they are injected at temperatures below the critical mixing temperature of the binary mixture. During the mixing and combustion process the temperature rises resulting in a transcritical mixing process. Under these circumstances neither the mixing models considering the mixture generation between a liquid and a gas (liquid atomisation models) nor the ones considering the mixture generation between two gases (gas–gas mixing) are suitable for the description of the real mixing processes. Therefore new models which can cope with transcritical or supercritical mixing processes had to be developed. These models are not only unavailable still but also would be desperately required for the modelling of the diesel mixture generation. Also in a diesel combustion chamber most probably transcritical mixing takes place which cannot be adequately modelled using liquid atomisation or gas–gas mixing models. Transcritical mixing also takes place in the supercritical anti-solvent (SAS) process for particle generation (see Section 1.7.2). Raman imaging experiments for the analysis of the mechanisms taking place in the SAS process are reported further.

3.4.7 Heat and Mass Transfer Analysis in Hydrogen Jets Injected Into Compressed Nitrogen Using Raman Imaging

As last but one case study I want to report on the simultaneous temperature and composition field measurements during hydrogen/nitrogen mixing using Raman imaging. I developed this technique mainly in order to provide to others a tool suitable for studying the mixture generation process in hydrogen powered internal combustion engines or rocket combustion chambers.

For this purpose a Raman imaging experiment was set up around an optically accessible injection chamber operating at 1 MPa, as it is shown in Fig. 3.39. Hydrogen can be preheated and fed through the injector into the chamber. Nitrogen is continuously flushed through side ports, which are not indicated in Fig. 3.39.

A high-power frequency-doubled Nd:YAG laser (see Section 3.5.1) emitting at 532 nm with a single-pulse energy of 650 mJ and full-width half-maximum pulse width of approximately 10 ns is used as excitation source. A green excitation wavelength assures – at least with respect to the system composed of hydrogen and nitrogen – that no interfering fluorescence is excited. A single-loop pulse stretcher (see Section 3.2.2) is installed in order to decrease the risk of laser-induced ionisation of the gases inside the chamber and in order to avoid laser-induced window damage. Furthermore a Galilei-type telescope expands the laser beam profile. The last lens in the excitation path is a cylindrical lens that forms the light sheet through the hydrogen jet. One can see that the Raman signals are detected with three cameras. One camera detects the Q-branch vibrational Raman–Stokes signal (see Fig. 2.52) of nitrogen. Due to the Raman

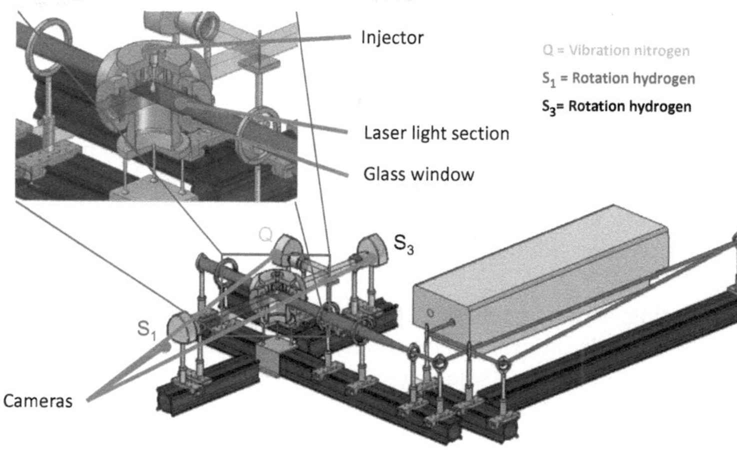

FIGURE 3.39 Sketch of the experimental set-up utilised for the simultaneous acquisition of temperature and composition profiles during hydrogen/nitrogen mixing.

shift of nitrogen (see Table 2.3) it appears at a wavelength of 607 nm when excited with 532 nm. Therefore this camera is equipped with a band-pass filter for 607 nm and a long-pass filter which suppresses the elastically-scattered stray light. In Section 2.2.7 and in the context of Fig. 2.59 it is described that the population of the vibrational ground state and with this also the Q-branch Raman–Stokes signal in a wide range of temperature can be considered to be not a function of temperature. Thus the respective camera detects a spatial distribution of the nitrogen molecules across the light sheet, irrespectively of the temperature. The detected nitrogen Raman signal is thus solely dependent on the concentration of the nitrogen molecules, but not on the temperature. In contrast to the rather temperature insensitive vibrational population distribution, the rotational distribution is known to be rather temperature sensitive (see Section 2.2.7). Thus rotational Raman spectroscopy can be a nice tool for temperature determination. The ratio of two rotational Raman signals of hydrogen has to be evaluated for the temperature determination in order to make the temperature measurement technique independent of shot-to-shot intensity fluctuations of the laser and independent of the still unknown concentration field of hydrogen. Referring to Fig. 2.59 we detected the S_1 and S_3 pure rotational Raman–Stokes signals, as due to the Boltzmann distribution firstly both levels are sufficiently populated in the analysed temperature range and secondly, the ratio of both populations provides a rather temperature sensitive response. The Raman shifts relevant for the S_1 and S_3 transitions are 587 cm^{-1} and 1033 cm^{-1} and can be computed for a real molecule according to Section 2.2.3. According to their Raman shifts these Raman signals appear at 549 nm and 563 nm, respectively, if excited at 532 nm. Consequently the two cameras detecting the hydrogen rotational Raman signal have to be equipped with the suitable band-pass filter and each one additionally with a long-pass filter for the suppression of elastically scattered stray light. Exercise 2.11 shows in detail how temperature can be computed from the ratio of rotational Raman signals. Once the temperature is known, one can combine the Raman signal intensity of either the S_1 or the S_3 signal with the derived temperature information in order to compute from both the number of hydrogen molecules. Once the number of hydrogen molecules is known, it can be put in correlation with the number of nitrogen molecules. The mixture composition can be computed from both concentrations. The number of nitrogen molecules we already computed from the Q-branch Raman signal of nitrogen.

Summarising, two cameras detecting pure rotational signals are required for measuring the 'hydrogen' temperature. When the temperature is known, a temperature independent and solely concentration dependent 'hydrogen' signal can be computed from them. The third camera detecting the vibrational Raman signal of nitrogen is required for measuring also a temperature independent and solely concentration dependent 'nitrogen' signal. The ratio of the two temperature independent but concentration dependent signals provides the mixture composition. Figure 3.40 shows resulting single-shot and averaged composition and temperature fields.

For this experiment we did not use ICCD detectors, but EMCCD detectors. The 'pros' and 'cons' of the different detector types are described in Section 3.5.2. Here it was not required to realise a very short exposure of the imaging section of the detector.

The detector chips are only then illuminated when one laser pulse passes through the measurement chamber and consequently the Raman signals pass through the band-pass filters to the detector chips. Before and after the laser pulse the chips are quasi not exposed to light. Therefore the advantages of using an EMCCD with a better quantum efficiency compared to the ICCD prevail. Of course the cameras and the laser are synchronised temporally. The EMCCD detectors' chips featured a frame-transfer architecture. Therefore the acquired image can be secured rather fast in the frame-transfer section of the chips what enables the comfort of measuring with the room-light switched on in the laboratory.

Finally I want to make clear that the temperature imaging concept described previously only works for hydrogen. Due to the large rotational constant of hydrogen the Raman shifts of the pure rotational transitions are rather large. This brings the advantage that the rotational Raman lines of hydrogen firstly are spectrally rather far away from the Rayleigh signal and secondly also the spectral distance between the different rotational lines is large. Only because of this we were able to spectrally separate the Raman lines from each other and from the dominant Rayleigh signal in an imaging set-up using long-pass filters, band-pass filters and dichroic mirrors. Of course in a non-imaging set-up also the rotational Raman lines of other molecules can be spectrally resolved. Then spectrometers have to be employed and 2-D imaging is no more possible.

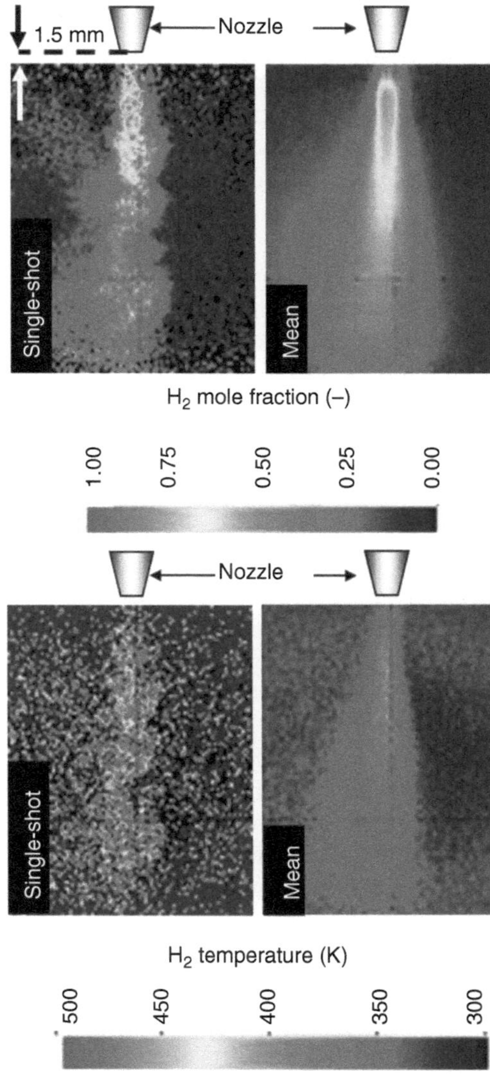

FIGURE 3.40 Mean and single-shot hydrogen molar fraction and hydrogen temperature images of a hot hydrogen jet that is injected into room temperature nitrogen at an absolute pressure of 1.1 MPa. *(Source: Ref. [99].)*

3.4.8 SAS Raman Imaging

The supercritical anti-solvent (SAS) technology, which is described in detail in Section 1.7.2, is very much suited for the execution of Raman imaging experiments. There is no luminosity coming from the measurement object itself. Therefore there is no requirement to use expensive ICCD cameras. The concentration of the different species is high due to the high pressures (8–20 MPa).

Therefore, rather strong Raman signals can be expected, which can be detected using a front-illuminated camera. Front-illuminated cameras are cheaper than back-illuminated cameras. Furthermore the laser pulse energy required for the excitation of fairly enough Raman signal levels is moderate.

A detailed description of the Raman imaging experiment we used for the analysis of the SAS-process is provided in Section 3.1.3 already. Therefore in the following I will briefly describe which information can be deduced from the acquired images.

Figure 3.41 shows the raw mean images of 50 single shot images detected simultaneously with three cameras. The images in the left column were acquired during the injection of the pure solvent dimethylsulphoxide (DMSO) (without the solute yttriumacetate (YAc)), whereas the images in the right column were acquired during the injection of the solution (1 wt% YAc in DMSO). For both experiments the operational conditions were 16 MPa and 313 K, for which DMSO and CO_2 are completely miscible [100]. The DMSO Raman images, which are similar in the left and the right column, illustrate the distribution of the injected DMSO. The raw CO_2 Raman images have to be corrected for a DMSO crosstalk, which arises from DMSO Raman shifts close to the ones of the CO_2 Raman signals (compare Figs 2.54 and 2.63). The crosstalk was quantified during the calibration experiments, when Raman measurements were carried out in defined homogeneous mixtures of DMSO and CO_2 at 16 MPa and 313 K with the identical experimental set-up. The elastic-light-scattering images indicate the presence of phase boundaries, and as a consequence, are affected by the presence of YAc. For the pure solvent injection, phase boundaries can be identified only within a short distance downstream the nozzle exit. They originate from the interface between the injected liquid DMSO and the bulk supercritical CO_2 [101], which under miscible conditions diminishes rapidly downstream the nozzle exit. Because of the same reason, phase boundaries of the same intensity can be identified downstream the nozzle exit in the right column for the injection of the solution. Further downstream, the elastic-light-scattering signals do not completely disappear. This is, as the YAc starts to precipitate and generates a solid/fluid phase boundary, which is detected as elastic light scattering signals.

From the crosstalk corrected CO_2 Raman images the CO_2 partial density ρ_{CO_2} or the CO_2 concentration can be calculated. For this the CO_2 partial density outside the operational sphere of the DMSO jet is assumed to be equal to the density of pure CO_2 at 16 MPa and 313 K. The CO_2 molar fraction x_{CO_2} was calculated from the ratio of the raw CO_2 Raman images and the sum of the raw CO_2 Raman image and the raw DMSO image, according to the calibration procedure. Example single-shot images are shown in Fig. 3.42.

Concluding, one can say that the measurement technique developed and introduced qualifies for the highly resolved, quantitative and 2-D measurement of the anti-solvent partial density ρ_{CO_2} and of the molar fraction x_{CO_2} which both govern particle generation. Simultaneously, elastic light scattering imaging

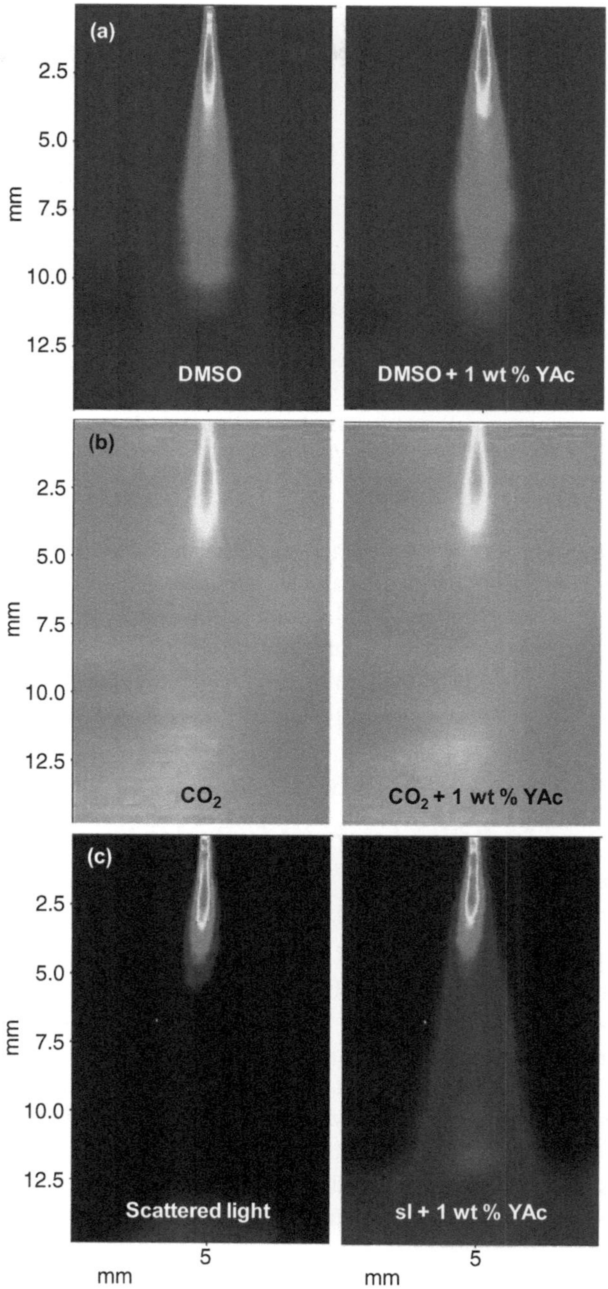

FIGURE 3.41 Raw mean images of the detected Raman signals of DMSO (a) and CO_2 (b) and of the elastically scattered light signals (c) at a chamber pressure of 16 MPa (CO_2 flow rate 53 g min^{-1}) and an injection flow rate of 2.2 g min^{-1} at 313 K. Left column, injection of pure DMSO into CO_2; right column, injection of a YAc/DMSO (1 wt %) solution into CO_2. *(Source: Ref. [102].)*

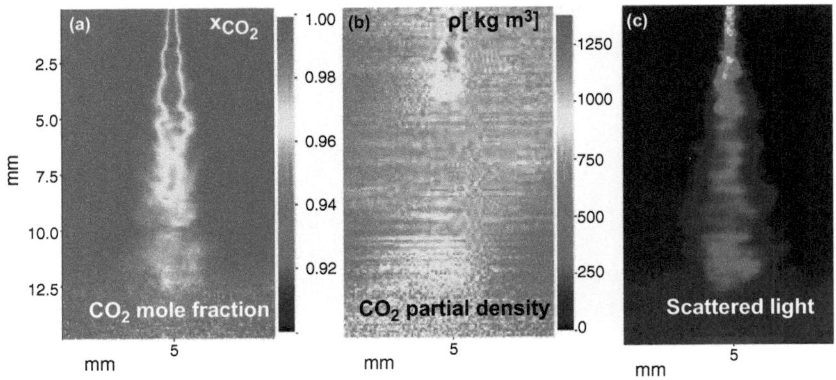

FIGURE 3.42 Simultaneously acquired single-shot images of the CO_2 molar fraction x_{CO_2} (a), of the CO_2 partial density distribution (b) and of the phase boundary distribution indicating particle formation (c). *(Source: Ref. [102].)*

yields information about the presence of phase boundaries. Thus, all parameters influencing the SAS process can be correlated spatially resolved and on a single-shot basis with high temporal resolution.

If the saturation concentration of YAc in DMSO/CO_2 mixtures at 16 MPa and 313 K is additionally known, even the supersaturation [103], which is the driving force for nucleation and particle growth, can be visualised.

3.5 APPENDIX – CHAPTER 3

3.5.1 Pulsed Excitation Lasers (Q-Switching Using a Pockels Cell)

In Sections 3.1.1, 3.1.2 and 3.1.3 pulsed lasers are integrated as excitation source into the experimental set-ups. As it has already been mentioned in the corresponding sections, this might be useful if very short measurement times (shorter than the shortest acquisition time of the detector) have to be realised in order to either suppress interfering background such as Planck radiation or in order to temporally freeze turbulent flows. It should become clear here especially that a pulsed laser is very different from switching a continuous-wave laser alternating on and off. The knowledge of the working principle of a continuous-wave laser [104] is assumed and therefore not described here.

A pulsed laser is designed in order to accumulate and store the pump energy in the gain medium of the laser during the non-lasing periods and then to release the stored energy rapidly in a short laser pulse of high power. There are active and passive methods known for operating a laser in a pulsed mode [104]. Here, the working principle of a pulsed Q-switched Nd:YAG laser (neodymium-doped yttrium aluminium garnet; Nd:$Y_3Al_5O_{12}$) equipped with an electrically switchable quarter-wave ($\lambda/4$) Pockels cell is described, as this

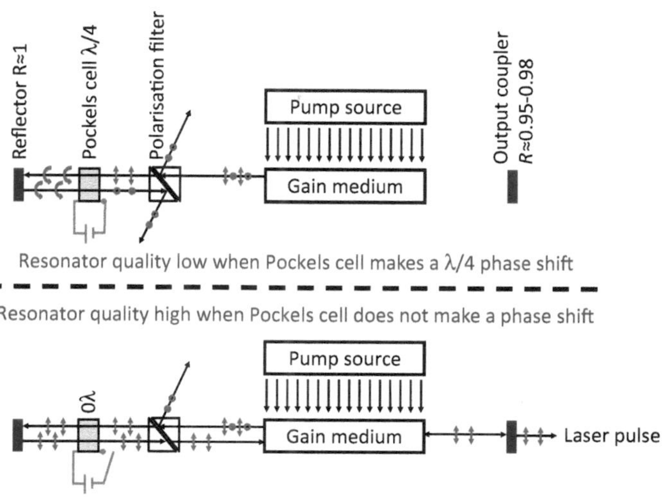

FIGURE 3.43 Configuration of a pulsed-laser equipped with a quarter-wave Pockels cell as Q-switch. Top, Pockels cell acts as quarter-wave plate and thus the resonator is blocked or said to feature a low quality factor; bottom, Pockels cell does not change the polarisation of the light and thus the resonator features a high quality. The Pockels cell is switched between quarter-wave phase shift and no phase shift operation within nanoseconds electrically.

laser – to my knowledge – is one of the most frequently used pulsed lasers in laser-diagnostics for chemical engineering purposes. From the following description also the basic idea behind Q-switching and the pulsed operation of lasers will become clear.

Figure 3.43 shows a sketch of the configuration of a Q-switched laser, once when the laser can lase (bottom) and once when the laser resonator is blocked. The pump source, which with respect to a Nd:YAG laser usually is a flash lamp or a diode laser provides the pump energy. The pump energy is absorbed by the gain medium. Due to absorption the particles (molecules) the gain medium is composed of, are excited to the upper laser level. As already described in Section 2.4.1.6.1, the excited particles fluoresce in order to depopulate back into the electronic ground state. The fluorescence is a spontaneous emission process that is non-directional. Thus the excited particles of the gain medium spontaneously emit in all directions of the room. But in order to achieve stimulated emission, light rays have to be reflected between the two resonator mirrors. Only then they can provide feedback to the gain medium. Therefore, the black arrows in the top part of Fig. 3.43 only represent SPONTANEOUS emission directed towards the left reflector, which potentially can be reflected back and forth in the resonator several times, without leaving the resonator. Spontaneously emitted light directed directly towards the right reflector would also be reflected and therefore can provide feedback to the gain medium one time. But when leaving the gain medium afterwards to the left side, the black arrow also represents these emissions.

Due to the combination of a polarisation filter and an electrically switchable quarter-wave plate, which is called Pockels cell, the resonator is blocked for any polarisation of the spontaneous emission. The polarisation filter first filters out the s-polarised light. The remaining p-polarised light is converted into circularly polarised light when it passes the quarter-wave plate the first time. The circularly polarised light is reflected at the reflector and passes the quarter-wave plate a second time. Consequently, the light effectively experienced a phase shift of half a wave (two times a quarter of a wave), which corresponds to a rotation of the initially p-polarised light to s-polarisation (compare Section 2.1.2). Again the s-polarised light is filtered out. The SPONTANEOUSLY emitted light cannot provide feedback in order to stimulate STIMULATED emission. In other words, the resonator is blocked or is said to feature a low quality factor (Q-factor). Summarising, as long as the Pockels cell is electrically switched to act as a quarter-wave plate the laser does not work, as it is sketched in the top part of Fig. 3.43. In the bottom part of Fig. 3.43 the Pockels cell is electrically 'switched off', meaning that it does not influence the polarisation of the p-polarised light passing it. Consequently the p-polarised light after reflection at the reflector passes the Pockels cell another time and still as p-polarised light can pass the polarisation filter and provide feedback to the gain medium. Then, the SPONTANOEUSLY emitted light can stimulate directional STIMULATED emissions in the gain medium, and the stimulated emissions form a standing wave between the resonator mirrors. The resonator is then said to feature a high quality factor (Q-factor). As the output coupler mirror's reflectivity is smaller than one, a portion of the light reflected inside the resonator is coupled out, which is the laser radiation, or the laser emission.

Pockels cells can be switched within few nanoseconds to act as zero-wave and quarter-wave plates. This gives rise to the Q-switch operation of pulsed lasers, which is illustrated as an example in Fig. 3.44. In Fig. 3.44 the temporal course of the energy provided by the pump source (here it is a flash lamp) and the following population of the upper laser level (excited state) of the particles in the gain medium is sketched. After some delay the population of the upper laser

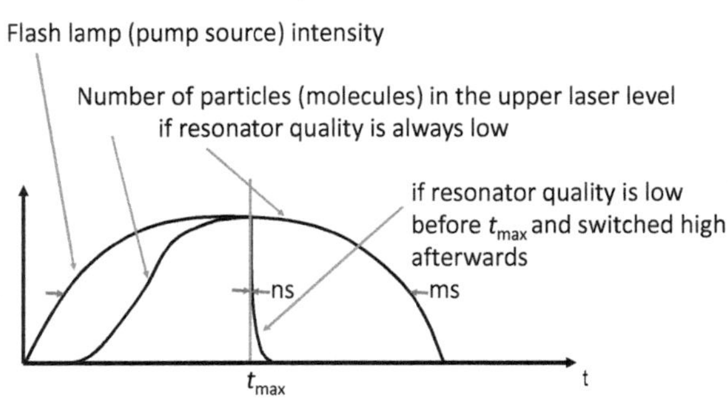

FIGURE 3.44 Process of populating and depopulating the upper laser level during one flash lamp pump pulse.

TABLE 3.1 Comparison of the Pulse Energies and the Pulse Durations of Q-Switched and Not-Q-Switched Pulsed Lasers and the Flash Lamp

Pulse energy of flash lamp in energy/time	100 J/ms	100 kW
Pulse energy/time of laser not Q-switched (Q-factor always high)	1 J/ms	1 kW
Pulse energy/time of laser Q-switched	1 J/10 ns	100 MW

lever starts to grow, as the flash lamp provides energy that can be absorbed by the particles in the gain medium. As long as more particles absorb than depopulate (mainly due to spontaneous emission as long as the Q-factor of the resonator is low), the population of the upper laser level grows. At the time t_{max} the maximum number of particles in the upper laser level is reached. If the Q-factor of the resonator is kept low, the population of the upper laser level shrinks afterwards as more excited particles depopulate (still mainly due to spontaneous emission) than not-excited particles absorb and get excited. This is mainly due to the intensity decrease of flash lamp's pump pulse towards the end of the pulse. In the case of a permanent low Q-factor the laser will not emit a laser pulse.

In the pulsed operation the Q-factor of the resonator is switched from low to high ideally just at the instant when the highest number of particles occupies the upper laser level, or in other words when the largest amount of energy is stored in the gain medium. Then all the energy stored in the gain medium is released in a short laser pulse, of approximately 10 ns (typical for Nd:YAG lasers). Thanks to the availability of ultrafast switchable Pockels cells the Q-factor of the resonator can in fact be switched in few nanoseconds form low to high. Next to the Pockels cell concept, there are other concepts of switching rapidly the Q-factor of resonators in order to form giant power laser pulses [104], which are not treated here.

Table 3.1 compares the energy and the pulse duration of a Q-switched and a not-Q-switched laser with the pulse characteristics of the pump source. Obviously a lot of pump energy is not converted into laser emission. The Q-switched operation clearly results in the highest pulse powers just due to the fact of a short pulse duration compared to the not-Q-switched laser. The pulse energy is the same for both operation modes.

3.5.2 Detectors: CCD, EMCCD and ICCD

Choosing the ideal detector is in my opinion the most important step when preparing an experiment. Here I recommend to carefully consider the 'pros' and 'cons' of different detector types and if possible to consult an experienced person. The quantum efficiencies of different detector types can easily be different by a factor of two. A factor of two in the detected signal can easily decide on the success of the experiments. One might argue that a detector featuring low quantum efficiency can be compensated by an excitation source featuring more excitation energy. Yes, this is true for some cases. But be aware of that for some experiments there is an upper limit for the excitation energy, determined for example by the acceptable laser-induced heating of the material, the laser damage threshold of the optical windows, the threshold for break down or simply by the fact that no excitation source with higher excitation energy is available.

Consequently, I hold the opinion that at least a short description of the most frequently employed variants of the charged coupled device (CCD) detectors must not be missing in this book.

Photomultipliers, photodiodes and avalanche diodes are not covered here. Also complementary metal-oxide semiconductor (CMOS) detectors are not covered here, as still today they cannot keep up with the performance of CCD detectors with respect to sensitivity, dynamic range and noise. Therefore CCD detectors are still first choice in science and machine vision. Nevertheless I want to mention that CMOS detectors for years have captured the consumer market (standard digital cameras and cameras in mobile phones) due to the higher frame rate and significantly lower power consumption (longer battery lifetime) compared to CCDs.

3.5.2.1 Pixel Design: Front and Back Illumination

Figure 3.45 shows the pixel design of a front-illuminated (left) and a back-illuminated CCD chip (right). In both cases the chip is mainly composed of bulk Silicon (Si), an insulation layer and the electrode structure. When a photon hits the Si material it can be converted according to the internal photoelectric effect into one electron e^{-1}. This conversion efficiency is called quantum efficiency (QE). A QE of 50% simply means that 5 out of 10 incident photons are converted each into one electron. The electrons are stored in the Si-bulk material. We will see later that the stored electrons have to be moved within the Si-bulk material in order to read out how many electrons have been initially stored in which region of the CCD chip. The electrode structure is required for the shift of the stored electrons. As long as no potential is applied to the electrodes, the electrons are not moved. If the electrodes are brought to a positive potential, the negatively charged electrons move towards the electrodes, until they reach the insulation layer, which is installed as a barrier to prevent the electrons from moving into the electrode structure. If electrons are located too far away from the electrode structure (out of the sensitive region), they will not experience the positive potential of the electrodes and thus cannot be influenced using the electrodes. Therefore only those electrons which are initially stored within the sensitive layer can later be detected as electrons.

In order to shift the electrons, different electrodes can be brought to different potentials. The negatively charged electrons tend to move towards the more positive potential. Thus electrons can be shifted within the sensitive region (mainly at the surface) from the region below one electrode to the region below

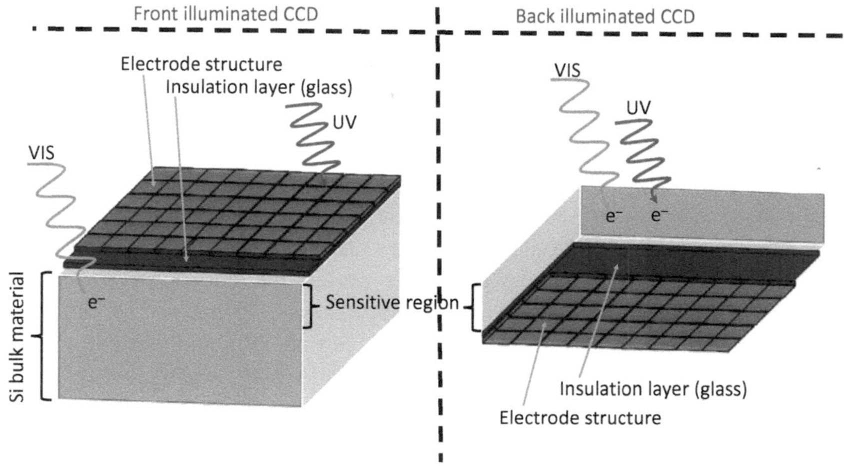

FIGURE 3.45 Design of a front-illuminated and a back-illuminated charge-coupled device (CCD) chip.

the neighbouring electrode, if the neighbouring electrode is brought to a more positive potential.

In the case of the front-illuminated pixel design the incident photons have to pass the electrode structure and the insulation layer before they can be converted into electrons. Unfortunately the electrode structure (and the micro-lens arrays which might be installed above the electrode structure but which are not integrated in Fig. 3.45) absorbs UV radiation. Therefore standard front-illuminated CCDs feature a small QE in the UV spectral region. Also not 100% of the visible spectrum reaches the sensitive region of the Si-bulk material resulting in QEs in the VIS spectral range of typically below 60%. In order to improve especially the QE in the UV spectral range special pixel designs such as the open electrode design or the addition of a phosphorescence layer have been developed, which are not treated further here.

In the case of the back-illuminated pixel design the incident photons do not have to pass neither the electrode structure nor the insulation layer. This results in QEs in the UV spectral range between ~30% and ~40%. In the VIS spectral range QEs exceeding 90% can be realised. Therefore with respect to performance the back-illuminated design is superior to the front-illuminated one. Unfortunately the front-illuminated pixel design cannot simply be turned in order to get a back-illuminated pixel design. If the front-illuminated pixel design was simply turned, the incident photons now entering the Si bulk material from the back side would be converted into electrons in regions which are too far away from the electrodes. Consequently these electrons would not be stored in the sensitive region of the Si bulk material and therefore cannot be read out. This is why the Si layer in the back-illuminated pixel design has to be much thinner than in the front-illuminated design. The generation and the treatment of such thin Si layers for the back-illuminated pixel design is costly compared to the thicker Si layers suitable for the front-illuminated pixel design. Consequently detectors equipped with back-illuminated chips are more expensive than those equipped with front-illuminated chips.

3.5.2.2 Read-Out of a Full Frame CCD Chip

Figure 3.46 shows the top view of a full frame CCD chip, which is composed of an imaging section, a read-out register and a charge-to-voltage conversion unit. The different fields in the imaging section represent the interaction range of the different electrodes. In Fig. 3.46 the imaging section is composed of 36 pixels, as there are six times six electrodes arranged horizontally and vertically. Each electrode defines one pixel. Real CCD chips feature many more pixels.

The size of one pixel defines how many electrons can be stored within one pixel. We will see later that it is beneficial if more electrons can be stored in the pixels of the read-out register than in the pixels of the imaging section. Therefore the pixels in the read-out register are sketched bigger. The maximum number of electrons that can be stored in one pixel is said to be the full well capacity. For pixels with an area of approximately 26 μm times 26 μm (edge length) the full well capacity can exceed 500,000 e^{-1}. Smaller pixels with an edge length of 10 μm can feature a full well capacity of 25,000 e^{-1}. We will also see later the full well capacity in combination with the read-out noise defines the dynamic range of the detector.

Figure 3.47 demonstrates the interplay of the imaging section, the read-out register and the charge-to-voltage conversion unit. The numbers contained within the pixels do not define the pixel number, but simply show how many electrons are stored in the respective pixels. So at the pixel left in the top line 1 electron is stored, whereas in the pixel in the centre of the CCD chip 5 electrons are stored. Assuming an ideal detector with a QE of 100% means that as an example 8 photons must have been incident on the pixel area of the pixel in the middle in the bottom line. Therefore 8 electrons are stored within the pixel after

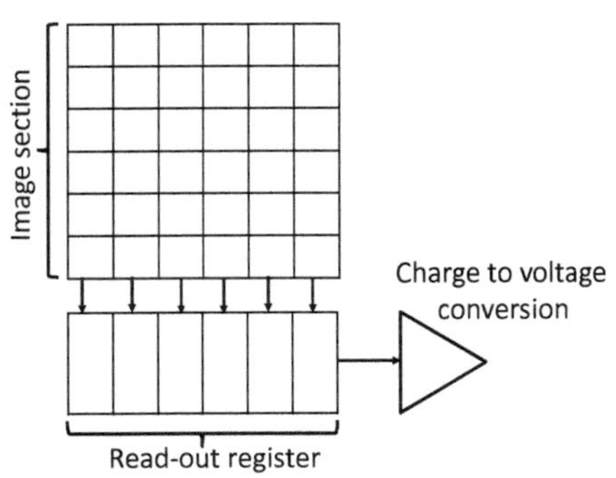

FIGURE 3.46 Sketch of the architecture of a full frame CCD chip. The electronic circuits are not integrated.

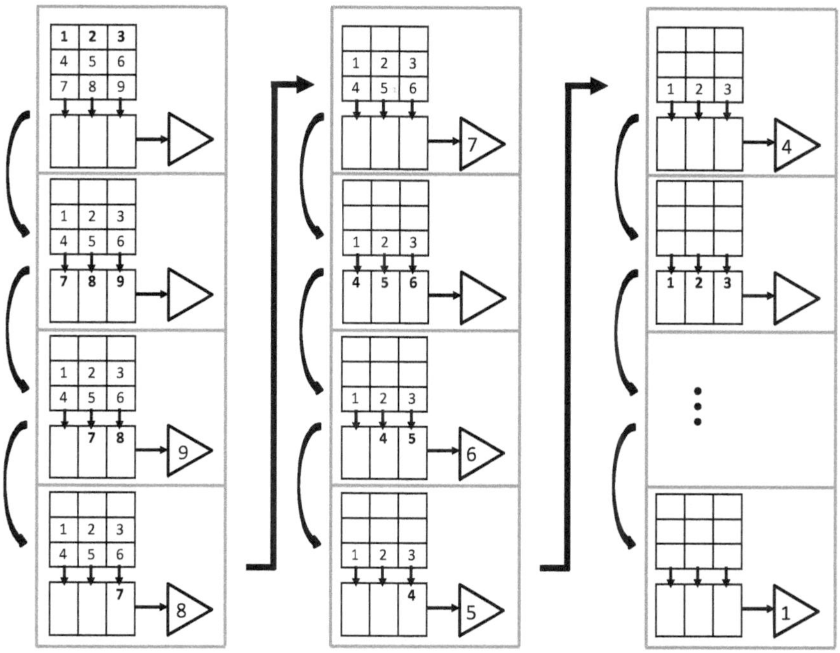

FIGURE 3.47 **Read-out sequence of a CCD chip whose imaging section is composed of nine pixels.**

the illumination of the CCD chip. From now on the illumination of the CCD chip is stopped, meaning that the number of stored signal-electrons (we only consider electrons which have been converted from a signal photon) remains constant. If different potentials are applied to the electrodes including the read-out register, the electrons stored in the pixels can be shifted vertically one line towards the bottom. Consequently the bottom line of the imaging section of the CCD is shifted vertically into the read-out register.

In the read-out register the electrons are shifted step by step horizontally into the charge-to-voltage conversion unit. Here it is read out how many electrons have been stored within the respective pixel. Having read out the read-out register completely, the next line from the imaging section can be shifted vertically into the read-out register. Then again the electrons stored in the pixels are shifted step by step horizontally and read out in the charge-to-voltage conversion unit. This procedure is repeated until the electrons stored in the last pixel are read out. With respect to the chip shown in Fig. 3.47 the complete read-out of the chip comprises three vertical line shifts into the read-out register and nine horizontal pixel shifts into the charge-to-voltage conversion unit. This reveals the main disadvantage of the CCD technology which is the circumstance that the read-out of a CCD chip is rather time consuming (numbers will follow later). Previously – before I described the read-out mechanism – I explicitly

FIGURE 3.48 Smeared image of a section of the flyer of the institution the author is affiliated with (SAOT). Smearing occurs as during the read-out the CCD chip is still exposed to light coming from the object (the flyer). *(Source: Acquisition made by Sebastian Luther from the Erlangen Graduate School in Advanced Optical Technologies of the Friedrich-Alexander-Universität Erlangen-Nürnberg, Germany.)*

mentioned that we assumed that the CCD chip is not illuminated any more during the long lasting read-out mechanism. If the CCD chip was exposed to light during the read-out mechanism, the still impinging photons would still be converted into electrons, while the image is already shifted stepwise from the imaging section into the read-out register. This effect is known as smearing and is illustrated in Fig. 3.48.

For the acquisition of the image shown in Fig. 3.48 a camera equipped with a CCD chip composed of 1600 times 400 pixels is used (only an excerpt is shown). The chip was read out with a frequency of 1 MHz. Therefore it takes approximately 640 ms to read out the 640,000 pixels, when neglecting the time required to shift the pixel-lines vertically from the imaging section into the read-out register. During this time the object is still imaged onto the CCD chip, meaning that there is still light impinging on the CCD chip and converted into electrons. Smearing can be eliminated by either installing a mechanical shutter preventing the light to propagate to the CCD chip during read-out or by more advanced CCD chip designs, which are shown in the following. Mechanical shutters usually reduce the numerical aperture of the detector and thus reduce the amount of light that can be detected on the CCD chip, which of course is undesired.

3.5.2.3 Architectures Other Than the Full Frame CCD Chip

Figure 3.49 shows the sketch of the architecture of a frame-transfer CCD chip.

Here a frame-transfer section is installed next to the imaging section. The frame-transfer section is covered so that no light can be converted into electrons there. But the electrons captured in the imaging section can be shifted into the frame-transfer section before the pixels are read out. Therefore, before the electrons stored in the 36 pixels of the imaging sections are read out, they are shifted during six vertical shifts into the frame-transfer section. Then the acquired image is secured from new photons during the time-consuming read-out. Of

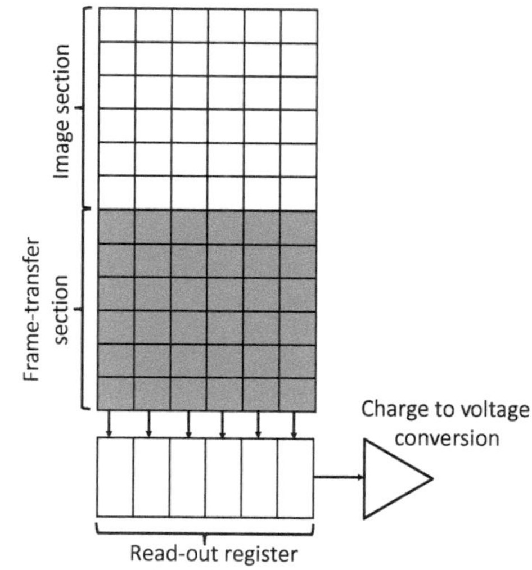

FIGURE 3.49 **Sketch of the architecture of a frame-transfer CCD chip.** The electronic circuits are not integrated.

course the frame-transfer chip design can reduce smearing significantly compared to the full frame CCD chip. Nevertheless the time required to shift the image completely into the secured frame-transfer section might still be too long to completely eliminate smearing.

The interline-transfer CCD chip architecture (Fig. 3.50) can further reduce the smearing. Here each second pixel column is covered (grey). Incident light can be converted into electrons only if it impinges on the not-covered pixels. Often micro lens arrays are used on top of the pixels directing the incident light from the covered pixels to the non-covered pixels. Once the image is captured in the non-covered pixels of the imaging section, it can be completely secured in the covered pixels by only one single horizontal shift. Therefore this chip

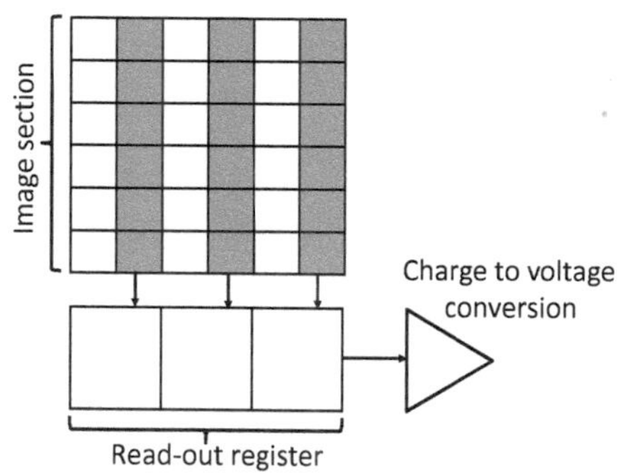

FIGURE 3.50 **Sketch of the architecture of an interline-transfer CCD chip. The electronic circuits are not integrated.**

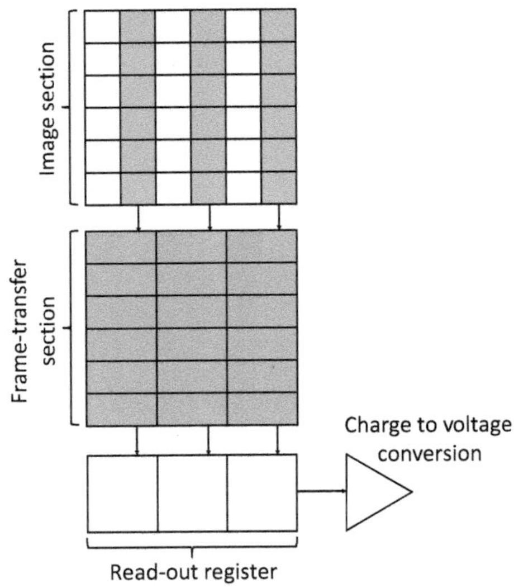

FIGURE 3.51 Sketch of the architecture of an interline frame-transfer CCD chip. The electronic circuits are not integrated.

architecture is the one suffering least from smearing. Once the image is shifted into the secured pixel columns the time-consuming read-out mechanism can be initialised. While the image is read out a second image can be captured already in the not-covered pixel columns. Therefore cameras equipped with interline-transfer CCD chips are in general suitable to acquire two snapshots directly one after the other. Thus, if the required electronics enables this option the respective cameras are also called double shutter cameras. But one also has to keep in mind that the second image cannot be shifted into the secured area as long as the read-out of the first image has not yet been completed. This results in exposure times of the first image of down to 0.5 μs but in exposure times of the subsequent image of milliseconds.

Figure 3.51 shows the combination of a frame-transfer and an interline-transfer CCD chip. With such a chip architecture two images can be detected shortly one after the other, similar to the interline-transfer architecture. But the interline frame-transfer architecture makes possible additionally a short exposure of the second image.

A first image is captured within the not-covered pixels in the imaging section. This image can be shifted during only one shift-step into the covered pixels columns and is secured from smearing. At the same time the not-covered pixel columns can detect the second image. This secondly acquired image can also be shifted into the secured pixel columns as soon as the firstly acquired image has been shifted from the secured pixel columns completely into the frame-transfer section. Summarising, the minimum exposure time of the first image is as long as the one for the interline-transfer chip, but the minimum exposure time of the second image is significantly reduced compared to the pure interline-transfer chip. In general the acquisition of a third image is possible as soon as the second image has been shifted into the secured pixel columns. But then again the exposure time of the third image is as long as it takes to read out the first image and to then shift the second image into the frame-transfer section.

3.5.2.4 Signal-to-Noise Ratio (SNR)

With respect to CCD detectors the signal is the number of signal electrons. Signal electrons are those electrons which – according to the internal photoelectric effect – result from the conversion of one incident photon into one signal electron within the sensitive layer of the pixel. The number of signal electrons

$$N_{\text{Signal}} = N_{\text{Photons}} \cdot \text{QE} \tag{3.2}$$

is the product of the number of incident photons N_{Photons} and the quantum efficiency of the detector QE. On the contrary the total noise

$$\sigma_{\text{total}} = \sqrt{\sigma^2_{\text{read-out}} + \sigma^2_{\text{thermal}} + \sigma^2_{\text{signal}}} \tag{3.3}$$

is the square root of the summation of the quadrats of the read-out noise $\sigma_{\text{read-out}}$, the thermal noise σ_{thermal} (also called dark noise) and the signal noise σ_{signal} (also called shot noise).

The **read-out noise** $\sigma_{\text{read-out}}$ emerges in the charge-to-voltage conversion unit when the number of electrons stored in the pixels are converted into a voltage. A certain number of signal electrons N_{signal} stored within the pixel

$$N_{\text{read}} = N_{\text{signal}} \pm \sigma_{\text{read-out}} \tag{3.4}$$

is not always read out correctly. More as well as less electrons can be read out N_{read}. If many pixels contain the same number of electrons, the standard deviation of the read out electrons from the mean value would determine the read-out noise. The faster the read-out is set in the camera settings, the higher the read-out noise will be. We also tend to make more reading mistakes when reading a text fast and less reading mistakes when reading a text slowly. Therefore, if no fast read-out is required, it is always recommended to choose a slow read-out rate in order to minimise the read-out noise. The read-out noise of scientific CCD detectors varies from few electrons at low read-out rates (~50 kHz) to several 10s of electrons for fast read-out rates (3 MHz). In fact whenever the number of read out electrons is not larger than the read-out noise, we cannot be sure whether any signal electrons have been stored inside the respective pixel before read-out. The number of read out electrons could then completely be due to the read-out noise. Therefore, only if the number of read-out electrons exceeds the read-out noise we can be sure that electrons have been stored inside the pixel before the read-out. This also means that if we choose fast read-out rates in the camera settings and with this increase the read-out noise, the lower detection limit rises. We will see later that in order to go round this problem the electron multiplying CCD detector has been developed.

The **thermal noise** σ_{thermal} emerges from the fact that due to thermal effects (not due to the photoelectric effect) there are permanently electrons set free in the silicon layer of the CCD chips. As the origin of these electrons is totally thermal, they are also referred to as dark current. The dark current is a function of the temperature of the CCD chip. At room temperature and depending on the chip material and pixel size for example between 100 and 1000 thermal electrons are set free per second and pixel. On the contrary less than 0.0002 electrons per second and pixel are produced due to dark current if the CCD chip is cooled down to 183 K (−90°C). Of course also the dark current is subject to some noise, meaning that not in each pixel on the CCD chip exactly the same number of dark current electrons are set free. The noise of the dark current

$$\sigma_{\text{thermal}} = \sqrt{N_{\text{thermal}}} \tag{3.5}$$

is the square root of the number of electrons set free in the chip due to thermal effects. If the CCD chip is cooled the thermal noise can usually be neglected compared to the read-out noise.

The **signal noise** σ_{signal} emerges from the statistical variation of the arrival rate of the photons on the CCD chip and is

$$\sigma_{\text{signal}} = \sqrt{N_{\text{signal}}} \tag{3.6}$$

the square root of the signal electrons. The signal noise cannot be influenced externally such as the read-out noise via the read-out rate or the thermal noise via the chip's temperature. The more signal electrons are generated, the larger the signal noise grows. But as the signal noise (square root of the signal) grows slower than the signal itself, the signal-to-noise ratio of course also increases with increasing number of signal electrons.

Summarising, the signal-to-noise ratio (SNR) of the CCD detector is

$$\text{SNR} = \frac{N_{\text{signal}}}{\sigma_{\text{total}}} = \frac{N_{\text{signal}}}{\sqrt{\sigma^2_{\text{read-out}} + \sigma^2_{\text{thermal}} + \sigma^2_{\text{signal}}}}. \tag{3.7}$$

For an ideal detector, or when the read-out noise $\sigma_{\text{read-out}}$ and the thermal noise σ_{thermal} can be neglected with respect to the signal noise, the SNR simplifies to

$$\text{SNR}_{\text{ideal}} = \frac{N_{\text{signal}}}{\sqrt{\sigma^2_{\text{signal}}}} = \frac{N_{\text{signal}}}{\sqrt{N_{\text{signal}}}} = \sqrt{N_{\text{signal}}}. \tag{3.8}$$

3.5.2.5 Dynamic Range of a CCD Detector

The dynamic range *DR* is the ratio of the largest and the smallest signal detectable simultaneously on one CCD chip or – with respect to signal electrons – the ratio of the highest and the smallest number of signal electrons detectable simultaneously on one CCD chip. The highest number of signal electrons that can be stored within one pixel is limited by the full well capacity of the pixel. The smallest number of signal electrons detectable as signal is limited by the read-out noise $\sigma_{\text{read-out}}$ (if thermal noise can be neglected). Therefore the dynamic range of a CCD detector in decibel

$$\mathrm{DR}_{\mathrm{db}} = 20 \cdot log \underbrace{\left(\frac{\text{full well capacity}}{\sigma_{\text{read-out}}} \right)}_{\mathrm{DR}} \tag{3.9}$$

is defined by the ratio of the *full well capacity* and the read-out noise $\sigma_{\text{read-out}}$. The factor 20 and the log are required to convert the DR into decibel $\mathrm{DR}_{\mathrm{db}}$. If for example a pixel features a full well capacity of 200,000 electrons while the read-out noise is 10 electrons per pixel, then the dynamic range of the respective CCD chip is DR=20,000. With respect to this example the ratio of the maximum and the minimum signal detected on two different pixels must not exceed 20,000, if both signals are supposed to be detected simultaneously on one CCD chip. If the ratio of the maximum and the minimum signal exceeds 20,000, then either the high signal will saturate the full well capacity of the pixels when the low signal just exceeds the read-out noise, or the low signal will disappear in the read-out noise when the high signal is kept just below the full well capacity.

I want to make this more clear giving an example which might be relevant when fine sprays have to be imaged. Let us assume that the pixel size of the CCD chip is 20 μm × 20 μm, that the read-out noise again is 10 electrons per pixel and that the full well capacity of one pixel again is 200,000. Therefore the dynamic range of this CCD chip again is 20,000. We are going to image the spray on the CCD chip with a magnification of 0.5, meaning that a drop with a real diameter of 1 μm in the spray would be imaged with a diameter of 0.5 μm on the CCD chip. We assume to know that the smallest and the biggest drops in the spray feature a diameter of 0.5 μm and 30 μm. Due to the magnification of 0.5 they will be imaged with diameters of respectively 0.25 μm and 0.15 μm. Furthermore we assume in order to make this example more clearly that per pixel only one drop is detected, irrespectively of whether the drop is small or big. In order to image the spray we choose arbitrarily the laser-induced fluorescence as imaging technique. As the laser-induced fluorescence signal originate from the volume of the drop, the signal resulting from the 30 μm drops will be 216,000 ($=30^3/0.5^3$) times more than the signal resulting from the 0.5 μm drops. Unfortunately, this intensity ratio by more than a factor of ten exceeds the dynamic range of the CCD chip. This means that if the signal integration time of the CCD chip is chosen long in order to lift the signals originating from the small drops above the read-out noise, the strong signal originating from the big drops will overflow the full well capacity of the pixels. On the contrary, if the signal integration time of the CCD detector is chosen short in order to keep the signals originating from the big drops just below the full well capacity of the pixel, the weak signals originating from the small drops will disappear in the read-out noise. Within the dynamic range of the detector the ratio of diameters of large and small droplets that can simultaneously be detected is approximately 27 (=$20{,}000^{1/3}$).

3.5.2.6 Detectors for Signal Amplification Before Read-Out (EMCCD and ICCD)

In the following we will assume that due to cooling of the CCD chip we can neglect thermal noise at all. Then the remaining noise sources are due to read-out and due to the signal itself.

I already mentioned previously that if the number of read out electrons does not exceed the read-out noise

$$N_{\text{read}} \leq \sigma_{\text{read-out}}, \tag{3.10}$$

one cannot definitely state whether the read out number of electrons is really due to signal electrons N_{signal} or solely due to the noise generated during read-out. It also has been aforementioned that the read-out noise increases with increasing read-out rate. Therefore experimenters run into problems whenever they have to acquire little signals at high acquisition rates. The small signal levels would need long signal integration times in order to accumulate enough signal electrons in the pixels before read-out. But long signal integration times are in conflict with high data acquisition rates (frame rates). On the contrary high acquisition rates demand high read-out rates and are in conflict with long signal integration times. Therefore the CCD chip architectures introduced previously (full frame, frame transfer, interline transfer and interline frame transfer) inherently are not suitable for the acquisition of low level signals at high frame rates. Especially on this purpose the electron multiplying charge-coupled device (EMCCD)-chip has been developed, whose architecture is shown in Fig. 3.52.

The number of electrons captured within the pixels can be multiplied by an amplification factor when they are shifted into the multiplication register. Due to this multiplication the number of electrons captured inside the pixels can be lifted above the read-out noise level. I want to make this clear giving one example. If the number of signal electrons in one pixel is five at a read-out noise of 20 electrons (high read-out noise due to the fast read-out), the signal electrons will be eaten up by the noise level, whenever the number of read-out electrons does not exceed 20 ($N_{\text{read}} \leq \sigma_{\text{read-out}}$). If the five signal electrons are multiplied within the multiplication register with an amplification factor of nine, 45 instead of five electrons are captured within the respective pixel before read-out. Though, the read-out noise is still 20 electrons, the number of read out electrons will always exceed the read-out noise level ($N_{\text{read}} = N_{\text{signal}} \pm \sigma_{\text{read-out}} = 45 \pm 20 > \sigma_{\text{read-out}}$) and thus the experimenter can be sure that there have been signal electrons captured within the pixel before the multiplication and before the read-out. If one signal electron is supposed to be lifted by multiplication above the read-out

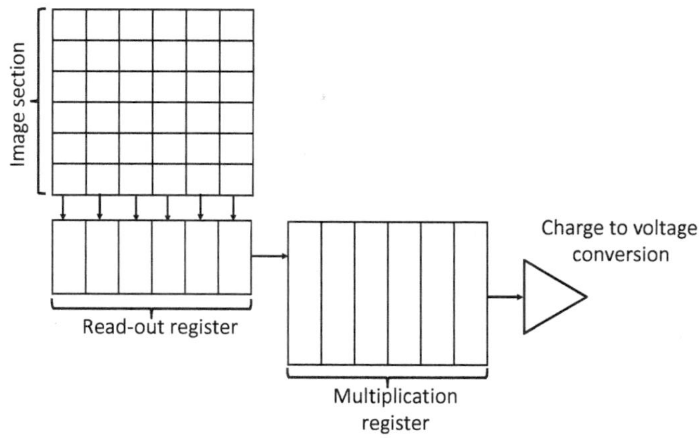

FIGURE 3.52 Sketch of the architecture of an electron multiplying charge-coupled device-chip. The electronic circuits are not integrated.

noise level an amplification factor of minimum 41 is required ($N_{read}=N_{signal}\pm\sigma_{read\text{-}out}=41\pm20>\sigma_{read\text{-}out}$).

Summarising, the electron multiplication before the read-out provides the possibility to lift low signal levels, which originally are below the read-out noise level, above the read-out noise level and thus to identify them surely as signal. This is especially of great benefit when due to high desired frame rates fast read-out rates result in high read-out noise levels.

Finally it has to be mentioned that the amplification factor cannot be chosen freely for each pixel. One fixed amplification factor has to be set for the entire read-out process. Consequently the number of signal electrons is multiplied not only in pixels containing few signal electrons but also in those containing many signal electrons. Thus, the full well capacity of the pixels in the multiplication register has to exceed the one in the read-out register, which in Fig. 3.52 is represented by the bigger pixel size.

Another detector type which amplifies the signal before the read-out process is the intensified charge-coupled device (ICCD) detector, which is sketched in Fig. 3.53. I want to make clear below that this type of detector is especially useful when very short exposure times down to 0.5 ns have to be realised. Within an ICCD detector the external and the internal photoelectric effect are combined. The incident photon illustrated by the wave in Fig. 3.53 hits the photocathode. There, due to the external photoelectric effect, an electron is set free, which in contrast to the internal photoelectric effect leaves the photocathode material. If a voltage is applied between the photocathode, the microchannel plate (MCP) and the phosphor screen, the set free electron moves towards the MCP. Within the microchannels the number of electrons is multiplied. Similar to the EM-gain of the EMCCD detector the amplification factor can be adjusted by the experimenter between 1 and typically 1000. Consequently and according

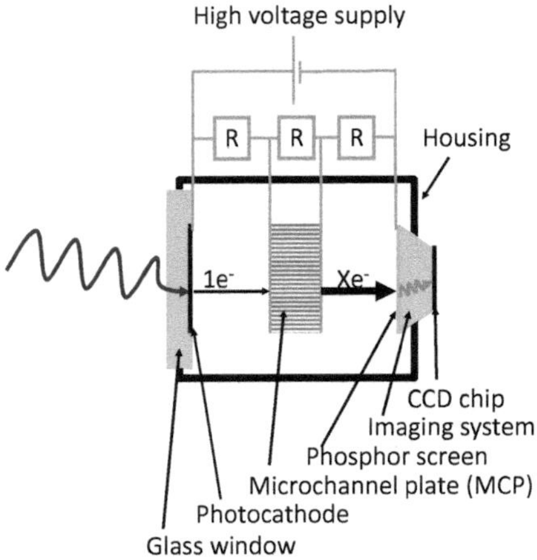

FIGURE 3.53 Sketch of the set-up and the working principle of an intensified charge-coupled device (ICCD) detector.

to the set amplification factor X times more electrons leave the MCP towards the phosphor screen than were reaching the MCP from the photocathode. At the phosphor screen the electrons are converted back into photons. I want to make clear that the wavelength of the photon hitting the photocathode and the wavelengths of the photons converted from the electrons at the phosphor screen are not identical. The wavelength information is completely lost.

The phosphor screen emits the phosphorescence into all directions of the coordinate system. Only those photons emitted into the direction of the imaging system, which is installed between the phosphor screen and the CCD chip, are guided to the CCD chip and can be detected there due to the internal electric effect. The imaging system can be either a bundle of many glass fibres or based on lenses. It has to be mentioned that if no voltage is applied between the photocathode, the MCP and the phosphor screen, the electrons set free at the photocathode will not move from left to right. In other words, the ICCD detector only works if a voltage is applied. We already saw in the section on the Q-switched pulsed laser that voltages can be switched rapidly. Therefore the switched-off ICCD detector can be switched on and off again in less than 2 ns. This duration is referred to as minimum gating time. It then does not matter how long it takes to read-out the CCD chip, as signal is forwarded to the CCD chip only during the gating time of the intensifier.

In my opinion the name intensified CCD is rather misleading with respect to the field of application of the ICCD cameras. The attribute 'Intensified' inheres that this detector type should be used when low signal levels are to be intensified. This is simply not the case. The quantum efficiency (QE) of the photocathode material – and depending on the photocathode material – nowadays is below 50% and for many cases even less than 20%. Therefore, it is much more reasonable to use a back-illuminated CCD chip with QEs above 90% for the detection of marginal signal levels. With the back-illuminated CCD chip many more incident photons are converted into signal electrons than at the photocathode of the ICCD. Herewith the risk that the number of signal electrons is below the detection limit of the entire detector (let it be defined by the read-out noise of the CCD chip) is higher in the case of the ICCD. If the number of signal electrons is below the read-out noise, the signal levels can in both cases be lifted above the read-out level, once via the electron multiplication in the MCP (ICCD camera) and once (EMCCD camera) via electron multiplication in the multiplication register. But in any case, due to the higher QE of the back-illuminated CCD chip compared to the photocathode the back-illuminated EMCCD camera will also be superior to the ICCD. There is only one circumstance when – due to my opinion – the ICCD camera performs better than CCD- or EMCCD cameras. This is the case when short exposure times have to be realised, which cannot be realised with the CCD technology alone. Referring to the interline-transfer chip described in the context of Fig. 3.50, I already mentioned that with this chip architecture exposure times down to 0.5 μs are achievable. I want to provide one example further where exposure times of 0.5 μs are not enough in order to discriminate against interfering illuminations. Assume for example that Raman measurements should be made during the combustion of the fuel in an optical accessible diesel engine. In this case the flame itself is very luminous. Of course one would use a pulsed laser for the excitation of the Raman process for two reasons. Firstly the experimenter wants to freeze the turbulence during the short moment of Raman signal generation. Secondly – and this is the main reason – the experimenter has to assure that during this short period of Raman signal generation the generated Raman signal significantly exceeds the flame luminosity of the combustion. Only then the detected Raman signals are not interfered by the flame luminosity. Of course this concept of discriminating against the flame luminosity works out only if the detector can be gated (temporally synchronised) to the excitation pulse. Thus the detector is supposed to be closed before the laser pulse excites the Raman signals. When the laser pulse propagates through the diesel combustion chamber and generates the Raman signals there, the detector is supposed to be open and to detect the Raman signals, but only as long as the Raman signals are generated. This is approximately equal to the pulse duration of the excitation pulse (typical pulse duration of Q-switched Nd:YAG lasers are approximately 10 ns or less than 100 ns if pulse stretched). Directly after the laser excitation pulse the detector is supposed to be closed again in order to prevent the detection of further flame luminosity. Therefore, the best measurement results, meaning minimum interference of the desired Raman signals with the undesired flame luminosity, can be achieved if an ICCD camera is temporally synchronised to the excitation laser pulse.

Finally it should be mentioned that whenever signal electrons are multiplied either in the MCP or in the multiplication register an additional source of noise comes up, which is called amplification noise. It simply means that a certain number of signal electrons does not always result in the same number of electrons after the multiplication with a constant amplification factor. This noise contribution has not been considered in Section 3.5.2.4.

3.5.2.7 Binning of Several Pixels to Superpixels

Sometimes it is reasonable to bin single pixels together to form superpixels, as it is shown in Fig. 3.54, where in each single pixel four signal electrons are stored. Binning worsens the resolution of the CCD chip, but sometimes this is not only disadvantageous.

In many applications the spatial resolution ideally achievable with respect to the size of the pixels of the CCD chip is better than the spatial resolution achievable in practice (compare Section 1.5.3), or the experiment simply does not require a very fine spatial or spectral resolution. Exactly then binning is the possibility of improving the SNR and reducing the read-out time.

After a 2×2 binning (two horizontally times two vertically) the formed superpixels in Fig. 3.54 contain the 16 signal electrons which before binning where contained in the four binned single pixels. With respect to Eqn (3.8) the SNR of an ideally assumed detector was two in the case of four signal electrons per pixel but is four in the case of 16 signal electrons per superpixel. Consequently 2×2 binning improves the SNR by a factor of two. Analogical 3×3 binning would improve the SNR by a factor of three and so on. Simultaneously, binning reduces the number of pixels that have to be read out. In the case of binning 2×2 single pixels to superpixels, the number of pixels that have to be read out is reduced by a factor of four. Consequently also the time required for read-out is approximately reduced by a factor of four. Figure 3.55 shows that binning is effectively realised in the read-out register and not in the imaging section of the CCD chip. In order to realise the binning of two pixels along the vertical axis the signal electrons captured in the second-to-last pixel line are simply added to the signal electrons captured in the last pixel line. The horizontal binning is realised by adding the signal electrons in two neighbouring pixels of the read-out register before read-out.

In the context of Fig. 3.46 I already mentioned that the full well capacity of the pixels in the read-out register has to exceed the full well capacity of the pixels in the imaging section. This is due to the fact that in the case of binning many signals electrons can accumulate in one pixel of the read-out register.

FIGURE 3.54 Illustration of 2×2 binning the pixels on the CCD chip to superpixels.

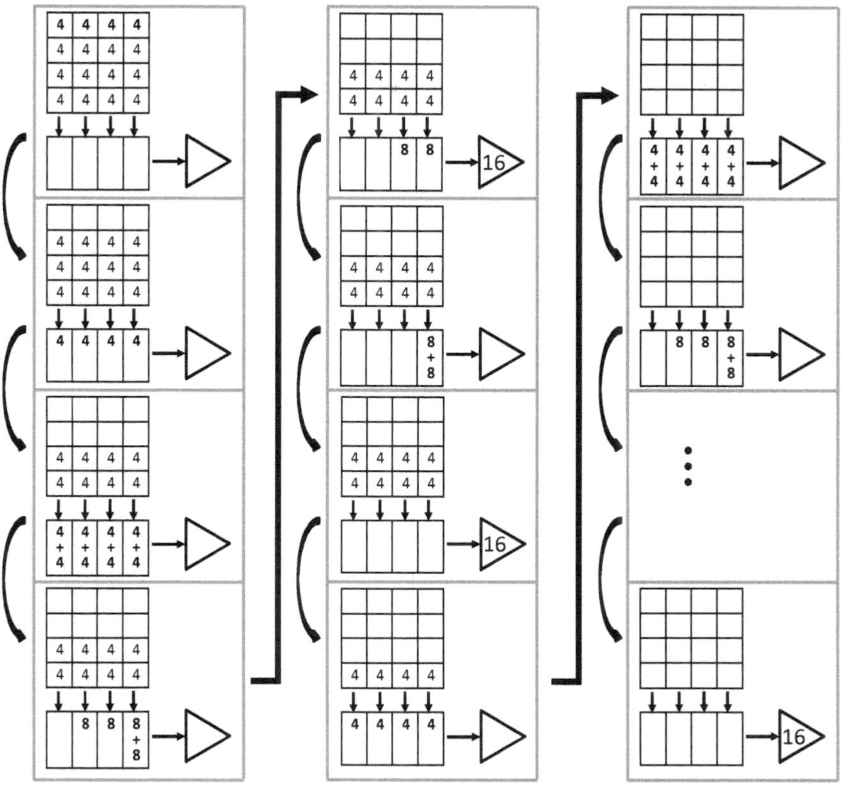

FIGURE 3.55 **Read-out sequence of a CCD chip when 2 × 2 binning is realised on the chip.**

I also want to mention that 'full vertical binning' simply refers to adding up all signal electrons captured within one pixel column in the imaging section in the corresponding pixel in the read-out register.

Finally I want to make clear that binning next to the SNR and next to the read-out time also improves the dynamic range of the detector. The read-out noise, which in Eqn (3.9) is in the denominator, is not affected by binning. But in case of binning, the effective full well capacity which in Eqn (3.9) is in the nominator, might be increased. The full well capacity of the pixels in the read-out register usually exceeds the one of the pixels in the imaging section. One has to check then carefully whether the maximum number of signal electrons that can be captured in the pixel of the read-out register really is limited by its full well capacity. Just imagine the full well capacity of the pixels in the read-out register being 1,000,000 electrons and that one of the pixels in the imaging section being 200,000 electrons. Then a 2 × 2 binning can maximum result in 800,000 electrons in one read-out pixel. Thus in this case the maximum number of signal electrons capable was not the full well capacity of the read-out pixel (1,000,000), but the summation of the full well capacities of the pixels in the imaging section (800,000).

REFERENCES

[1] Luther SK, Schuster JJ, Leipertz A, Braeuer A. Non-invasive quantification of phase equilibria of ternary mixtures composed of carbon dioxide, organic solvent and water. J Supercrit Fluids 2013;84:146–154.

[2] Braeuer A, Dowy S, Leipertz A, Schatz R, Schluecker E. Injection of ethanol into supercritical CO_2: determination of mole fraction and phase state using linear Raman scattering. Opt Express 2007;15:8377–82.

[3] Batteh JH. Damage thresholds in laser-irradiated glass. J Appl Phys 1982;53:7537–44.

[4] Braeuer A, Beyrau F, Weikl MC, Seeger T, Kiefer J, Leipertz A, Holzwarth A, Soika A. Investigation of the combustion process in an auxiliary heating system using dual-pump CARS. J Raman Spectrosc 2006;37:633–40.

[5] Hartley DL. In: Lapp M, Penney CM, editors. Laser Raman gas diagnostics. New York: Plenum; 1974. p. 1151–7.

[6] Khare R, Shukla PK, Mishra GK, Mukherjee C, Talwar S, Dubey VK, Saxena P, Mittal JK. A novel confocal optical pulse stretcher for laser pulses. Opt Commun 2009;282:3850–3.

[7] Kojima J, Nguyen Q-V. Laser pulse-stretching with multiple optical ring cavities. Appl Optics 2002;41:6360–70.

[8] Braeuer A, Dowy S, Torino E, Rossmann M, Luther SK, Schluecker E, Leipertz A, Reverchon E. Analysis of the supercritical antisolvent mechanisms governing particles precipitation and morphology by in situ laser scattering techniques. Chem Eng J 2011;173:258–66.

[9] Noack K, Leipertz A, Kiefer J. Molecular interactions and macroscopic effects in binary mixtures of an imidazolium ionic liquid with water, methanol, and ethanol. J Mol Struct 2012;1018:45–53.

[10] Petry R, Schmitt M, Popp J. Raman spectroscopy – a prospective tool in the life sciences. ChemPhysChem 2003;4:14–30.

[11] Aydin Ö, Kahraman M, Kiliç Eu, Çulha M. Surface-enhanced Raman scattering of rat tissues. Appl Spectrosc 2009;63:662–8.

[12] Schrader B, Dippel B, Fendel S, Keller S, Löchte T, Riedl M, Schulte R, Tatsch E. NIR FT Raman spectroscopy – a new tool in medical diagnostics. J Mol Struct 1997;408–409:23–31.

[13] Schrader B, Schulz H, Andreev GN, Klump HH, Sawatzki J. Non-destructive NIR-FT-Raman spectroscopy of plant and animal tissues, of food and works of art. Talanta 2000;53:35–45.

[14] Long DA. Raman spectroscopy. New York: McGraw-Hill; 1977.

[15] Gebrekidan MT, Knipfer C, Stelzle F, Popp J, Will S, Braeuer A. A shifted-excitation Raman difference spectroscopy (SERDS) evaluation strategy for the efficient isolation of Raman spectra from extreme fluorescence interference. J Raman Spectrosc 2015. Available from: http://onlinelibrary.wiley.com/doi/10.1002/jrs.4775/abstract

[16] Schulze G, Jirasek A, Yu MM, Lim A, Turner RF, Blades MW. Investigation of selected baseline removal techniques as candidates for automated implementation. Appl Spectrosc 2005;59:545–74.

[17] Ramos PM, Ruisánchez I. Noise and background removal in Raman spectra of ancient pigments using wavelet transform. J Raman Spectrosc 2005;36:848–56.

[18] Barclay V, Bonner R, Hamilton I. Application of wavelet transforms to experimental spectra: smoothing, denoising, and data set compression. Anal Chem 1997;69:78–90.

[19] Huang Z, McWilliams A, Lui H, McLean DI, Lam S, Zeng H. Near-infrared Raman spectroscopy for optical diagnosis of lung cancer. Int J Cancer 2003;107:1047–52.

[20] Gao M, Lewis G, Turner GM, Soubret A, Ntziachristos V. Effects of background fluorescence in fluorescence molecular tomography. Appl Optics 2005;44:5468–74.

[21] Brennan JF, Wang Y, Dasari RR, Feld MS. Near-infrared Raman spectrometer systems for human tissue studies. Appl Spectrosc 1997;51:201–8.

[22] Cao A, Pandya AK, Serhatkulu GK, Weber RE, Dai H, Thakur JS, Naik VM, Naik R, Auner GW, Rabah R. A robust method for automated background subtraction of tissue fluorescence. J Raman Spectrosc 2007;38:1199–205.

[23] He S, Zhang W, Liu L, Huang Y, He J, Xie W, Wu P, Du C. Baseline correction for Raman spectra using an improved asymmetric least squares method. Anal Methods 2014;6:4402–7.

[24] Peng J, Peng S, Jiang A, Wei J, Li C, Tan J. Asymmetric least squares for multiple spectra baseline correction. Analytica chimica acta 2010;683:63–8.

[25] Eilers PH, Boelens HF. Baseline correction with asymmetric least squares smoothing, Leiden University Medical Centre Report, 2005.

[26] Zhang ZM, Chen S, Liang YZ, Liu ZX, Zhang QM, Ding LX, Ye F, Zhou H. An intelligent background-correction algorithm for highly fluorescent samples in Raman spectroscopy. J Raman Spectrosc 2010;41:659–69.

[27] Zhao J, Lui H, McLean DI, Zeng H. Automated autofluorescence background subtraction algorithm for biomedical Raman spectroscopy. Appl Spectrosc 2007;61:1225–32.

[28] Lieber CA, Mahadevan-Jansen A. Automated method for subtraction of fluorescence from biological Raman spectra. Appl Spectrosc 2003;57:1363–7.

[29] Eilers PH. A perfect smoother. Anal Chem 2003;75:3631–6.

[30] De Luca AC, Dholakia K, Mazilu M. Modulated Raman spectroscopy for enhanced cancer diagnosis at the cellular level. Sensors 2015;15:13680–704.

[31] Kasha M. Characterization of electronic transitions in complex molecules. Discuss Faraday Soc 1950;9:14–9.

[32] Noack K, Eskofier B, Kiefer J, Dilk C, Bilow G, Schirmer M, Buchholz R, Leipertz A. Combined shifted-excitation Raman difference spectroscopy and support vector regression for monitoring the algal production of complex polysaccharides. Analyst 2013;138:5639–46.

[33] Kiefer J, Noack K, Leipertz A. Background supression for Raman analysis of pharmaceutically active compounds in fluorecing media. Am Pharma Rev 2013;16:32–5.

[34] Shreve AP, Cherepy NJ, Mathies RA. Effective rejection of fluorescence interference in Raman spectroscopy using a shifted excitation difference technique. Appl Spectrosc 1992;46:707–11.

[35] Volodin B, Dolgy S, Ban VS, Gracin D, Juraic´ K, Gracin L. Application of the shifted excitation Raman difference spectroscopy (SERDS) to the analysis of trace amounts of methanol in red wines. Proceeding of SPIE 8939; 2014, 89390Y-89390Y-89310.

[36] Appiah B, Dolgy S, Ban V, Melnik E, Volodin B. A novel method for practical implementation of the shifted excitation Raman difference spectroscopy (SERDS), Special Issues, 2012;27:1–9.

[37] Martins MAdS, Ribeiro DG, Pereira dos Santos EA, Martin AA, Fontes A, Martinho HdS. Shifted-excitation Raman difference spectroscopy for in vitro and in vivo biological samples analysis. Biomed Opt Express 2010;1:617–26.

[38] Sowoidnich K, Kronfeldt H-D. Shifted excitation Raman difference spectroscopy at multiple wavelengths for *in-situ* meat species differentiation. Appl Phys B 2012;108:975–82.

[39] Goncharov AF. Raman Spectroscopy at high pressures. Int J Spectrosc 2012;2012.

[40] Gillet P, Hemley RJ, McMillan PF. Vibrational properties at high pressures and temperatures. Rev Mineral Geochem 1998;37:525–90.

[41] McMillan PF, Dubessy J, Hemley R. Chapter 7 – Applications in earth, planetary and environmental sciences. In: Corset GT, editor. Raman microscopy. London: Academic Press; 1996. p. 289–365.

[42] McMillan PF. Raman spectroscopy in mineralogy and geochemistry. Ann Rev Earth Planet Sci 1989;17:255–79.

[43] Brunner G. Applications of supercritical fluids. Ann Rev Chem Biomol Eng 2010;1:321–42.

[44] Fonseca JMS, Dohrn R, Peper S. High-pressure fluid-phase equilibria: experimental methods and systems investigated (2005–2008). Fluid Phase Equilibria 2011;300:1–69.

[45] Peper S, Dohrn R. Sampling from fluid mixtures under high pressure: review, case study and evaluation. J Supercrit Fluids 2012;66:2–15.

[46] Schwedt G. Analytische chemie: grundlagen, methoden und praxis. Weinheim: Wiley-VCH; 2008.

[47] Stratmann A, Schweiger G. Fluid phase equilibria of ethanol and carbon dioxide mixtures with concentration measurements by Raman spectroscopy. Appl Spectrosc 2002;56:783–8.

[48] Adami R, Schuster J, Liparoti S, Reverchon E, Leipertz A, Braeuer A. A Raman spectroscopic method for the determination of high pressure vapour liquid equilibria. Fluid Phase Equilibria 2013;360:265–73.

[49] Luther SK, Schuster JJ, Leipertz A, Braeuer A. Non-invasive quantification of phase equilibria of ternary mixtures composed of carbon dioxide, organic solvent and water. J Supercrit Fluids 2013;84:146–54.

[50] Luther SK, Will S, Braeuer A. Phase-specific Raman spectroscopy for fast segmented microfluidic flows. Lab Chip 2014;14:2910–3.

[51] Schuster JJ, Will S, Leipertz A, Braeuer A. Deconvolution of Raman spectra for the quantification of ternary high pressure phase equilibria composed of carbon dioxide, water and organic solvent. J Raman Spectrosc 2014;45:246–52.

[52] Traub P, Stephan K. High-pressure phase equilibria of the system CO_2 – water – acetone measured with a new apparatus. Chem Eng Sci 1990;45:751–8.

[53] Sloan Jr ED, Koh C. Clathrate hydrates of natural gases. New York: CRC Press; 2007.

[54] Tsimpanogiannis IN, Economou IG, Stubos AK. Methane solubility in aqueous solutions under two-phase (H–Lw) hydrate equilibrium conditions. Fluid Phase Equilibria 2014;371:106–20.

[55] Papadimitriou NI, Tsimpanogiannis IN, Economou IG, Stubos AK. Influence of combining rules on the cavity occupancy of clathrate hydrates by Monte Carlo simulations. Mol Phys 2014;112:2258–74.

[56] Ruffine L, Trusler JPM. Phase behaviour of mixed-gas hydrate systems containing carbon dioxide. J Chem Thermodyn 2010;42:605–11.

[57] Sum AK, Koh CA, Sloan ED. Clathrate hydrates: from laboratory science to engineering practice. Ind Eng Chem Res 2009;48:7457–65.

[58] Sloan ED. A changing hydrate paradigm – from apprehension to avoidance to risk management. Fluid Phase Equilibria 2005;228:67–74.

[59] Kerr RA. Gas hydrate resource: smaller but sooner. Science 2004;303:946–7.

[60] Nogami T, Oya N, Ishida H, Matsumoto H. Development of natural gas ocean transportation chain by means of natural gas hydrate (NGH). In: 6th International Conference on Gas Hydrates (ICGH 2008), Vancouver, British Columbia, Canada; 2008.

[61] Sloan ED. Fundamental principles and applications of natural gas hydrates. Nature 2003;426:353–63.

[62] Joshi SV, Grasso GA, Lafond PG, Rao I, Webb E, Zerpa LE, Sloan ED, Koh CA, Sum AK. Experimental flowloop investigations of gas hydrate formation in high water cut systems. Chem Eng Sci 2013;97:198–209.

[63] Sum AK, Koh CA, Sloan ED. Developing a comprehensive understanding and model of hydrate in multiphase flow: from laboratory measurements to field applications. Energ Fuel 2012;26:4046–52.

[64] Sloan E, Subramanian S, Matthews P, Lederhos J, Khokhar A. Quantifying hydrate formation and kinetic inhibition. Ind Eng Chem Res 1998;37:3124–32.

[65] Brewer PG, Ussler W, Peltzer E, Walz P, Kirkwood W, Hester K, Xin Z, Accurate *in situ* observation of deep-sea sediment dissolved methane profiles in hydrate bearing provinces. In: 7th International Conference on Gas Hydrates (ICGH 2011), Edinburgh, Scotland, United Kingdom; 2011.

[66] Schicks JM, Naumann R, Erzinger J, Hester KC, Koh CA, Sloan ED. Phase transitions in mixed gas hydrates: experimental observations versus calculated data. J Phys Chem B 2006;110:11468–74.

[67] Qin J, Kuhs WF. Quantitative analysis of gas hydrates using Raman spectroscopy. AIChE J 2013;59:2155–67.

[68] Schicks JM, Erzinger J, Ziemann MA. Raman spectra of gas hydrates – differences and analogies to ice Ih and (gas saturated) water. Spectrochim Acta A Mol Biomol Spectrosc 2005;61:2399–403.

[69] Lu W, Chou IM, Burruss RC. Determination of methane concentrations in water in equilibrium with sI methane hydrate in the absence of a vapor phase by *in situ* Raman spectroscopy. Geochimica et Cosmochimica Acta 2008;72:412–22.

[70] Davies SR, Sloan ED, Sum AK, Koh CA. In Situ Studies of the Mass Transfer Mechanism across a Methane Hydrate Film Using High-Resolution Confocal Raman Spectroscopy. J Phys Chem C 2009;114:1173–80.

[71] Braeuer A, Hankel RF, Mehnert MK, Schuster JJ, Will S. A Raman technique applicable for the analysis of the working principle of promoters and inhibitors of gas hydrate formation. J Raman Spectrosc 2015. Available from: http://onlinelibrary.wiley.com/doi/10.1002/jrs.4744/abstract

[72] Masten DA, Foy BR, Harradine DM, Dyer RB. In situ Raman spectroscopy of reactions in supercritical water. J Phys Chem 1993;97:8557–9.

[73] Myrick ML, Kolis J, Parsons E, Chike K, Lovelace M, Scrivens W, Holliday R, Williams M. In situ fiber-optic Raman spectroscopy of organic chemistry in a supercritical water reactor. J Raman Spectrosc 1994;25:59–65.

[74] Hunter TB, Rice SF, Hanush RG. Raman spectroscopic measurement of oxidation in supercritical water. 2. Conversion of isopropyl alcohol to acetone. Ind Eng Chem Res 1996;35:3984–90.

[75] Rice SF, Hunter TB, Rydén ÅC, Hanush RG. Raman spectroscopic measurement of oxidation in supercritical water. 1. Conversion of methanol to formaldehyde. Ind Eng Chem Res 1996;35:2161–71.

[76] Schoppelrei JW, Kieke ML, Brill TB. Spectroscopy of hydrothermal reactions. 2. Reactions and kinetic parameters of $[NH_3OH]NO_3$ and equilibria of $(NH_4)_2CO_3$ determined with a flow cell and FT Raman spectroscopy. J Phys Chem 1996;100:7463–70.

[77] Venardou E, Garcia-Verdugo E, Barlow SJ, Gorbaty YE, Poliakoff M. On-line monitoring of the hydrolysis of acetonitrile in near-critical water using Raman spectroscopy. Vib Spectrosc 2004;35:103–9.

[78] Picard A, Daniel I, Montagnac G, Oger P. In situ monitoring by quantitative Raman spectroscopy of alcoholic fermentation by Saccharomyces cerevisiae under high pressure. Extremophiles 2007;11:445–52.

[79] Bardow A, Marquardt W, Goke V, Koss HJ, Lucas K. Model-based measurement of diffusion using Raman spectroscopy. AIChE J 2003;49:323–34.

[80] Kriesten E, Voda MA, Bardow A, Goke V, Casanova F, Blumich B, Koss HJ, Marquardt W. Direct determination of the concentration dependence of diffusivities using combined model-based Raman and NMR experiments. Fluid Phase Equilibria 2009;277:96–106.

[81] Blesinger C, Beumers P, Buttler F, Pauls C, Bardow A. Temperature-dependent diffusion coefficients from 1D Raman spectroscopy. J Solution Chem 2014;43:144–57.
[82] Bardow A, Goke V, Koss HJ, Lucas K, Marquardt W. Concentration-dependent diffusion coefficients from a single experiment using model-based Raman spectroscopy. Fluid Phase Equilibria 2005;228:357–66.
[83] Knauer OS, Pastore Carbone MG, Braeuer A, Di Maio E, Leipertz A. Investigation of CO2 sorption in molten polymers at high pressures using Raman line imaging. Polymer 2013;54:812–8.
[84] Braeuer A, Knauer OS, Quino J, Leipertz A. Quantification of the mass transport in a two phase binary system at elevated pressures applying Raman spectroscopy: pendant liquid solvent drop in a supercritical carbon dioxide environment. Int J Heat Mass Transfer 2013;62:729–40.
[85] Miles PC. Raman line imaging for spatially and temporally resolved mole fraction measurements in internal combustion engines. Appl Optics 1999;38:1714–32.
[86] Schütte M, Finke H, Grünefeld G, Krüger S, Andresen P, Stiebels B, Block B, Meyer H, Hentschel W. Spatially resolved air-fuel ratio and residual gas measurements by spontaneous Raman scattering in a firing direct injection gasoline engine. In: SAE, SAE; 2000.
[87] Schütte M, Grünefeld G, Andresen P, Hentschel W, Homburg A, Nassif-Pugsley D. Fuel/air-ratio measurements in direct injection gasoline sprays using 1D Raman scattering. In: SAE, SAE; 2000.
[88] Egermann J, Koebcke W, Ipp W, Leipertz A. Investigation of the mixture formation inside a GDI engine by means of linear Raman spectroscopy. In: International Symposium on Combustion, Combustion Institute; 2000. pp. 1145–1152.
[89] Taschek M, Egermann J, Schwarz S, Leipertz A. Quantitative analysis of the near-wall mixture formation process in a passenger car direct-injection Diesel engine using linear Raman spectroscopy. Appl Optics 2005;44:6606–15.
[90] Long MB, Levin PS, Fourguette DC. Simultaneous two-dimensional mapping of species concentration and temperature in turbulent flames. Optics Lett 1985;10:267–9.
[91] Masri AR, Dibble RW, Barlow RS. Raman–Rayleigh scattering measurements in reacting and non-reacting dilute two-phase flows. J Raman Spectrosc 1993;24:83–9.
[92] Marran DF, Frank JH, Long MB, Stårner SH, Bilger RW. Intracavity technique for improved Raman/Rayleigh imaging in flames. Optics Lett 1995;20:791–3.
[93] Meier F, Wiltafsky G, Köhler J, Stolz W. Quantitative time resolved 2-D fuel–air ratio measurements in a hydrogen direct injection SI engine using spontaneous Raman scattering. In: SAE, SAE; 1996.
[94] Anderson TJ. Oxygen concentration measurements in a high pressure environment using Raman imaging. In: American Institute of Aeronautics and Astronautics; 1995.
[95] Woodward RD, Talley DG. Raman imaging of transcritical cryogenic propellants. In: American Institute of Aeronautics and Astronautics; 1996.
[96] Mayer W, Schick AHA, Vielle B, Chauveau C, Gökalp I, Talley DG, Woodward RD. Atomization and breakup of cryogenic propellants under high-pressure subcritical and supercritical conditions. J Propul Power 1998;14:835–42.
[97] Decker M, Schik A, Meier UE, Stricker W. Quantitative Raman imaging investigations of mixing phenomena in high-pressure cryogenic jets. Appl Optics 1998;37:5620–7.
[98] Mayer W, Telaar J, Branam R, Schneider G, Hussong J. Raman measurements of cryogenic injection at supercritical pressure. Heat Mass Transfer 2003;39:709–19.
[99] Braeuer A, Leipertz A. Two-dimensional Raman mole-fraction and temperature measurements for hydrogen-nitrogen mixture analysis. Appl Optics 2009;48:B57–64.

[100] Andreatta AE, Florusse LJ, Bottini SB, Peters CJ. Phase equilibria of dimethyl sulfoxide (DMSO) + carbon dioxide, and DMSO + carbon dioxide + water mixtures. J Supercrit Fluids 2007;42:60–8.
[101] Reverchon E, Torino E, Dowy S, Braeuer A, Leipertz A. Interactions of phase equilibria, jet fluid dynamics and mass transfer during supercritical antisolvent micronization. Chem Eng J 2010;156:446–58.
[102] Braeuer A, Dowy S, Leipertz A. Simultaneous Raman and elastic light scattering imaging for particle formation investigation. Optics Lett 2010;35:2553–5.
[103] Dowy S, Torino E, Luther SK, Rossmann M, Braeuer A. Imaging the supersaturation in high-pressure systems for particle generation. Chem Eng J 2011;168:896–902.
[104] Eichler J, Eichler HJ. Laser. Berlin: Springer; 2003.

Chapter 4

Shadowgraph and Schlieren Techniques

The shadowgraph technique as well as the schlieren technique both rely on the refraction of light rays when they pass a medium which features gradients of the index of refraction. In this context Sections 2.3.1 and 2.3.2 provide the fundamentals for both techniques and already described how light rays are refracted when they pass an interface between two fluids of different index of refraction. It is also described briefly that light rays are deflected from their original path when they pass locations featuring a gradient of the index of refraction pointing to a direction other than the original direction of propagation of the light rays. In the following section first a more detailed description about the refraction of light rays at a schliere is provided. In Section 2.3.2 one also can find in Figure 2.24 a first shadowgram, which I acquired using a pocket lamp as light source and detecting the shadowgram with the camera of my mobile phone. To this effect I point out in the following subsection the simplicity of a shadowgraph experimental set-up. Section 4.1.3 will make clear that schlieren techniques are more complex than shadowgraph techniques, they require much more efforts with respect to the alignment of the optical path and many more optical components. Though I hope to have made clear the differences between shadowgraph and schlieren techniques in their respective subsections already, I underline them again by comparing both techniques in a separate section. I avoid providing information about how the composition, the density or the temperature of a fluid can be computed from its index of refraction. One can find these descriptions in textbooks [1–3] on shadowgraph or schlieren techniques, but not for high pressures. At high pressure the correlations between composition, temperature and density are more complex and mostly unknown. Therefore the use of shadowgraph and schlieren techniques here is described solely for qualitative analysis. Quantitative analysis would depend on a detailed knowledge of the aforementioned correlations.

Then Section 4.2 describes the working principle of a very special shadowgraph technique, which is called ballistic spray imaging. The experimental set-up required for the realisation of such an experiment is complex and comprehensive, but – according to my knowledge – there is no other optical

Supercritical Fluid Science and Technology, Vol 7. http://dx.doi.org/10.1016/B978-0-444-63422-1.00004-3

measurement technique which can provide similar insights into dense sprays close to the nozzle exit. Non-optical alternatives are based on the use of pulsed X-ray beams [4] which can be generated by a synchrotron. Of course, compared to synchrotrons the experimental efforts for ballistic spray imaging are ridiculously small. Ballistic spray imaging is furthermore based on the clever combination of several optical effects and because of this it deserves to get introduced in a separate section instead of in an example case study.

In Section 4.3 few case studies show examples of high-pressure applications of the shadowgraph and schlieren techniques. The shadowgraph case studies are chosen in order to make clear that nearly all shadowgraph studies reported in literature do not exactly follow the principle of shadowgraph techniques described before in Section 4.1.2. Usually the reported shadowgraph techniques are hybrid version between the schlieren and the real shadowgraph technique. In literature it is absolutely accepted to call these hybrid forms 'shadowgraph techniques'. But in the context of this book in order to highlight that these hybrid forms do not exactly follow the principles of real shadowgraph techniques, I decided to refer to them as 'pseudo-shadowgraph techniques'.

4.1 HOW SHADOWGRAPH AND SCHLIEREN TECHNIQUES WORK

This section provides the fundamentals of refraction of light rays from schlieren and the working principles of the shadowgraph and the schlieren technique, and with this provides the essentials for the understanding of the case studies reported in Section 4.3. It also emphasises the differences of both measurement techniques.

4.1.1 Light Refraction by a Two-Dimensional Schliere

Figure 4.1 shows the interaction between a light ray (electromagnetic wave) and a schliere. A schliere can be considered as gradient of the index of refraction (its

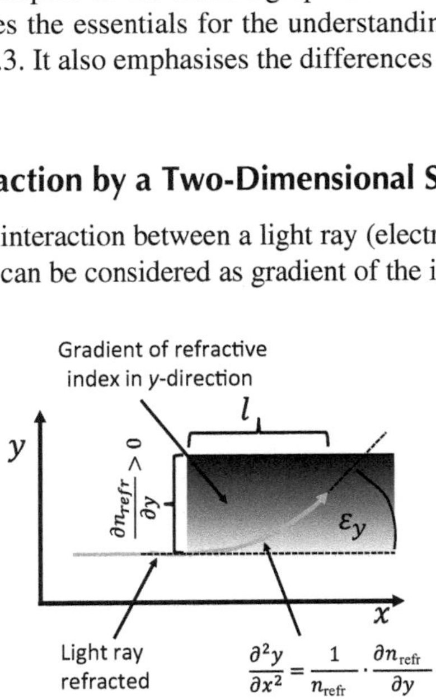

FIGURE 4.1 Geometric theory of light refraction of a two-dimensional schliere with a gradient of the index of refraction in *y*-direction.

real part) in a transparent medium. The gradients can cover one, two or three dimensions. In Fig. 4.1 a two-dimensional schliere with a positive gradient of the index of refraction $\partial n_{\mathrm{refr}}/\partial y > 0$ in y-direction is sketched. Before the light ray interacts with the schliere it moves straight and parallel into the x-direction. When it travels within the schliere it is bent towards the regions of higher index of refraction (here the y-direction) with a curvature [3,5]

$$\frac{\partial^2 y}{\partial x^2} = \frac{1}{n} \cdot \frac{\partial n_{\mathrm{refr}}}{\partial y} \tag{4.1}$$

and after travelling a distance of l it is bend by the angular ray deflection ε_y.

In the books of Feldmann and Mayinger [2], of Settles [3] and also in works of others [5–7] the angular-ray deflection is derived for small angles of deflection. At elevated pressure intense schlieren can cause a rather large angular deflection. Therefore the 'classical' derivation assuming small angles of deflection is not covered in this book. Furthermore the dependencies between the index of refraction and the density of a fluid or the composition of a mixture, which are well known for ambient pressure conditions, are not comprehensively available at elevated pressure. This makes the computation of densities, temperature or compositions from angular ray defractions, which are effectively measured with schlieren experiments, challenging and would require thorough pre-investigations into the influence of density, temperature and mixture composition onto the index of refraction. Therefore for the contents treated in this chapter we can settle with knowing that gradients of the index of refraction deflect light and that the steeper the gradient of the index of refraction is, the more the light is deflected.

4.1.2 Working Principle of Shadowgraph Techniques

Figure 4.2 shows a light ray diagram of a direct shadowgraph experiment. 'Direct' in this context means that there are no lenses installed behind the pinhole light source. Only one condenser lens is used in order to collect and guide a large portion of the light emitted by the original light source through the pinhole. Since the pinhole is small it acts like a point light source. This assures that different rays passing the pinhole do not intersect each other afterwards, as long as there is no schliere present. Consequently the screen in the upper part of Fig. 4.2 is 'homogeneously' illuminated. If there is a schliere between the pinhole and the screen, the light rays passing a gradient of the index of refraction are bent by the angular ray deflection ε_y. The index of refraction of the two-dimensional schliere in Fig. 4.2 is represented by the purple dashed line, meaning that the highest index of refraction is at the top of the schliere, then decreases linearly $\partial n_{\mathrm{refr}}/\partial y = \mathrm{const} < 0$ and takes a constant value $\partial n_{\mathrm{refr}}/\partial y = 0$ over the major region of the schliere. The light rays passing the negative gradient of the index of refraction are bent up towards the higher index of refraction.

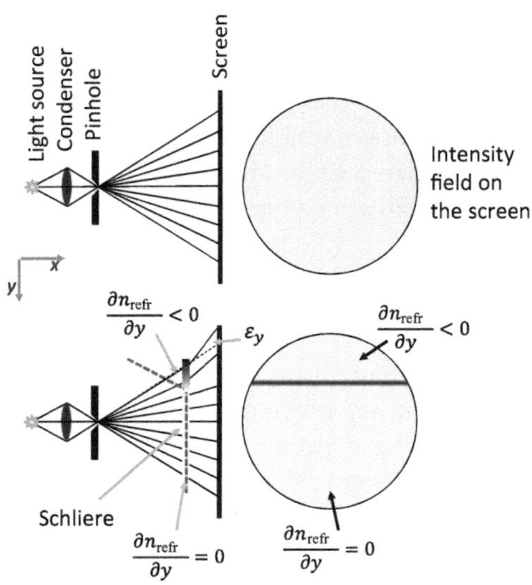

FIGURE 4.2 Sketch of the light paths in a diverging direct shadowgraph, without schliere at the top and with schliere at the bottom.

Consequently they also hit the screen further up, which results in a dark region on the screen indicating the transition from the not refracted light rays to the refracted ones.

It should be underlined here that the areas above and below the dark region feature the same level of brightness they would feature without schliere, though the corresponding rays once passed (upper part) and once did not pass regions of a gradient of the index of refraction (bottom part). Therefore we can conclude that the shadowgraph technique does not provide information about the gradient of the index of refraction $\partial n_{refr}/\partial y$, but only about where the gradients change spatially, so about the second derivative of the index of refraction $\partial^2 n_{refr}/\partial y^2$. Furthermore the position and the width of the dark region on the screen is a function of the distance between the schliere and the screen. In a shadowgraph the schliere is not imaged onto the screen! Only if the screen is positioned directly behind the schliere, the shadowgraph represents exactly the schliere according to $\partial^2 n_{refr}/\partial y^2$. The farther the screen is away from the schliere, the more intense different rays will overlap that are refracted from different regions of the schliere into different directions. This effect is not sketched in Fig. 4.2.

4.1.3 Working Principle of Schlieren Techniques

The general working principle of the schlieren experiment is described first using the example of a lens based schlieren system with a pinhole as light source. This brings the advantage that the paths of the different light rays in the sketches can be presented clearer. We will then see that using a slit instead of a pinhole results in essential advantageous features. Finally a schlieren experiment which is based on mirrors instead of lenses is briefly described. Figure 4.3 illustrates in several sketches the working principle of a schlieren experiment.

All the sketches have in common that a condenser lens collects and guides a large portion of the light emitted by the light source through the pinhole, so that the pinhole can be considered to act like a point light source. The first lens (lens 1) is positioned in the distance of its focal length f_{length} from the pinhole and consequently collimates the light rays. The second lens (lens 2) in Fig. 4.3 is equal to the first lens and focuses the collimated light rays in the distance of its focal length f_{length}.

Behind the focus the light rays diverge again and illuminate the screen. I made the sketch in the upper row of Fig. 4.3 in order to make clear two things. Firstly, all the light passing the pinhole arrives at the screen (losses at the lenses are neglected). Secondly, the two lenses – lens 1 and lens 2 – image the pinhole

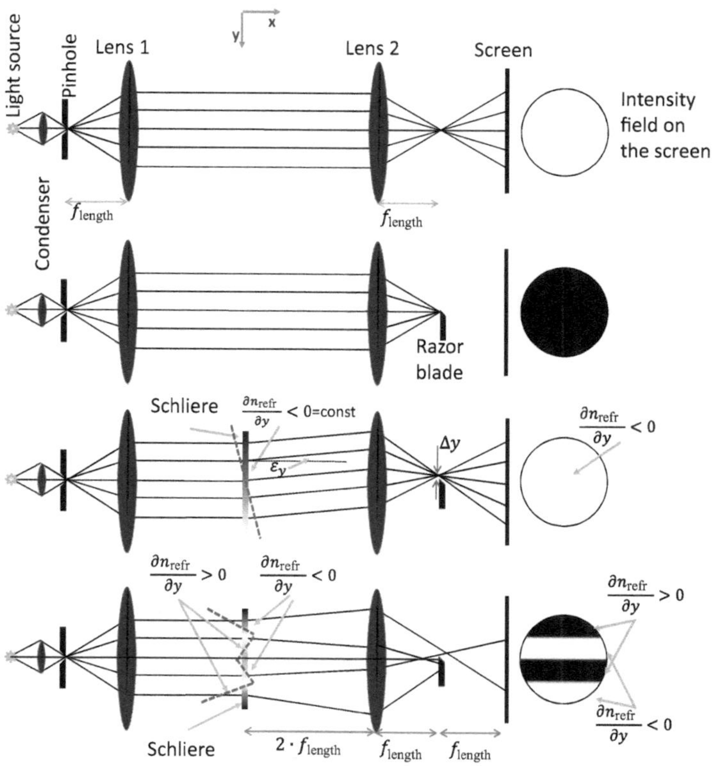

FIGURE 4.3 Sketch of the light paths in a schlieren experiment using a point light source (pinhole). (Top row) Light passing the pinhole hits the screen. (Second row) Razor blade blocks the light at the plane of the image of the pinhole and thus no light hits the screen. (Third row) A schliere with constant gradient of index of refraction tilts the light paths and thus shifts the image of the pinhole up by Δy above the razor blade so that light can hit the screen. (Fourth row) Schliere of different gradients of the index of refraction tilts light passing different schlieren regions differently onto or above the razor blade so that only light passing certain regions of the schliere hits the screen.

plane in a distance of f_{length} behind the second lens. I made the sketch in the second row of Fig. 4.3 in order to show that all light rays can be blocked if a knife-edge or a razor blade is inserted into that plane where the pinhole plane is imaged. The blade is inserted from the bottom so that the image of the pinhole is just blocked completely at the edge of the razor blade. Consequently the screen is no more illuminated. The third row in Fig. 4.3 shows a schliere in the schlieren region between lens 1 and lens 2. The purple dashed line represents the index of refraction. In this case the gradient of the index of refraction is constantly negative. All the light rays passing this schliere are therefore bent up towards regions of a higher index of refraction and as the gradient of the index of refraction is constant everywhere in the schliere, all the light rays behind the schliere feature the same angular ray deflection ε_y. Due to the angular ray deflection ε_y upwards, the image of the pinhole – as it is shown in Fig. 4.3 – is also lifted up by

$$\Delta y \propto \varepsilon_y \cdot f_{\text{length}}. \tag{4.2}$$

Consequently all the light passing the pinhole now also can pass the razor blade and illuminates the screen. As the razor blade was aligned to just cut the image of the pinhole for $\varepsilon_y = 0$ any $\varepsilon_y < 0$ (negative because of the y-directions points from the top to the bottom) lifts the image of the pinhole above the razor blade. The magnitude of ε_y is not relevant. If the gradient of the index of refraction was positive, refraction would bent the light rays down and thus also the image of the pinhole would be shifted down and would still be blocked by the razor blade. Also in this case the magnitude of bending is not relevant. Any rays not bent or bent to the bottom will not pass the razor blade. Summarising, I made the sketch in the third row of Fig. 4.3 in order to show that the sign (positive or negative) of the gradient of the index of refraction determines whether the image of the pinhole is shifted upwards above the razor blade or downwards onto the razor blade.

The sketch in the fourth row of Fig. 4.3 shows a schliere in the schlieren region with varying gradients of the index of refraction. Two regions of the schliere feature a positive and two other regions a negative gradient of the index of refraction. The positive and the negative gradients feature different magnitudes. From the previous sketch we recall that it solely depends onto the sign of the gradient of the index of refraction, whether or not light passing the schliere can also pass the razor blade. Thus it is logic that only these regions in the schliere, which bend the light rays upwards appear bright on the screen. The regions not bending at all or bending the light rays downwards appear dark on the screen. In this context it has to be clarified that the lens 2 images the schliere onto the screen with a magnification of minus one, if – as it is sketched in Fig. 4.3 – the distance between lens 2 and the schliere and the screen is two times the focal length of the lens each time. A magnification of -1 means that the object – in this case the schliere – is imaged with its original size (in two-dimensions) but upside down. As the pattern of the gradient of the index of refraction from top to bottom is negative, then positive, then negative and then positive again, the corresponding schlieren-image shows the pattern now from the bottom to the top like bright, then dark, then bright and then dark again. I prepared the sketch in the fourth row of Fig. 4.3 in order to demonstrate that a schlieren experiment using a point light source (pinhole) can act only as a binary measurement instrument. It can only measure the sign of the gradient of the index of refraction, but not its magnitude. We will see from the following descriptions that a schlieren experiment also can measure the magnitude of the gradient of the index of refraction, if the pinhole is replaced by a slit. On this account the following Figure 4.4 is discussed.

Figure 4.4 shows a lens-based schlieren experiment using a vertical slit instead of a pinhole. The slit has a certain extension – 1.5 mm for schlieren experiments in gases is a typical value – in the y-direction. The extension in z-direction is not relevant for the following discussions. The condenser lens now images the light source onto the backside of the slit-plane. The image of the light source is supposed to illuminate the slit homogeneously and to fill it out completely. Due to the spatial extension of the slit in y-direction light rays representative for the upper and the lower edge of the slit are indicated as dashed and solid lines, respectively.

Again the first lens (lens 1) is positioned in the distance of its focal length f_{length} from the slit and collimates the light rays. The two lenses – lens 1 and lens 2 – are identical and image the slit with a magnification of minus one onto the plane positioned in a distance of the focal length f_{length}. A third optional lens, which in Fig. 4.4 is equal to the other lenses, is positioned in the distance of its focal length from the image of the slit and collimates the light rays again. The sketch in the first row in Fig. 4.4 shows that all the light passing the slit illuminates the screen if the razor blade is positioned outside the image-region of the slit.

In the sketch shown in the second row in Fig. 4.4 the razor blade is moved into the image region of the slit. Due to the vertical extension of the image of the slit the razor edge can now be aligned to only cut a certain portion of the slit. In Fig. 4.4 approximately 50% of the image of the slit are blocked so that only the other 50% of the light rays can proceed to the screen. Consequently the screen is illuminated with less light. Therefore it is filled grey in Fig. 4.4. This level of illumination from now on is referred to as background illumination level. I want to emphasise here that the screen is still illuminated homogeneously. Summarising, one can adjust the background illumination level of the screen by moving the razor blade to a greater or lesser extent into the image region of the slit. Moving the razor blade to a greater or lesser extent into the image of the slit darkens or brightens the screen homogeneously in a perfectly aligned schlieren experiment! In an improperly aligned experiment this action causes a bright–dark gradient of the illumination intensity on the screen. We will see in the following descriptions why it is reasonable to only partially block the image of the slit with the razor blade.

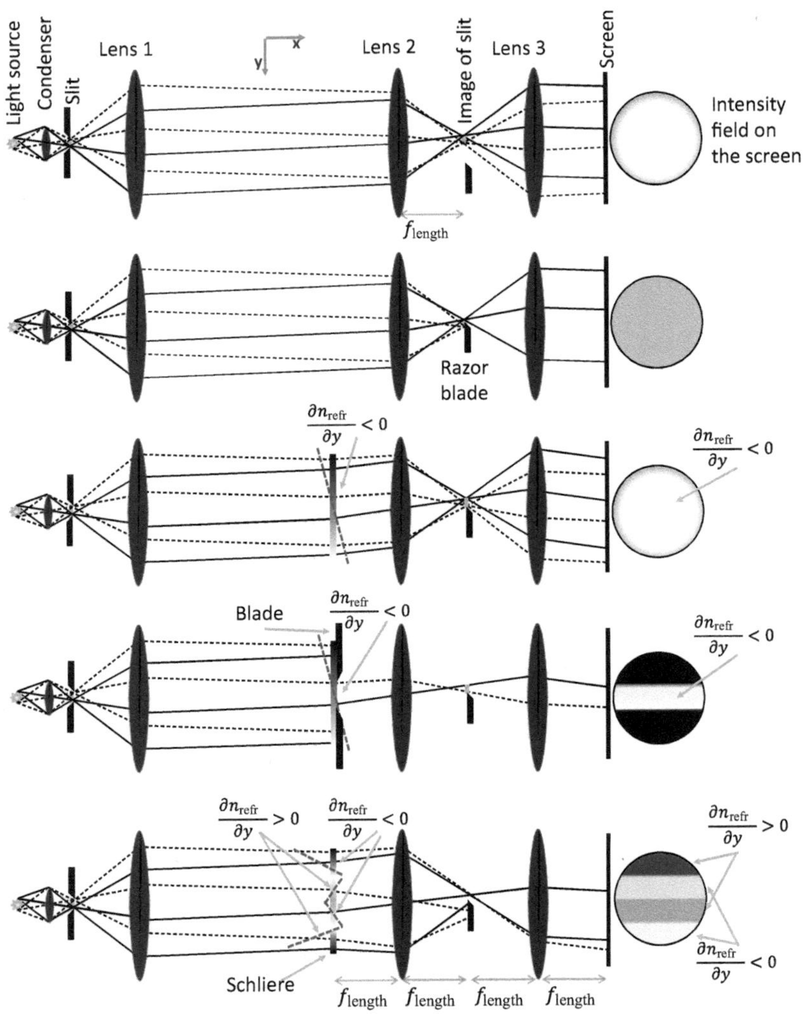

FIGURE 4.4 Sketch of the light paths in a schlieren experiment using an extended light source (slit) instead of a pinhole. (Top row) Razor blade does not cut any parts of the image of the slit so that all the light passing the slit can hit the screen (bright signal on the screen). (Second row) Razor blade blocks only half of the image of the slit so that 50% of the light passing the slit hits the screen (medium signal on the screen). (Third row) Schliere of constant index of refraction tilts the light paths and thus lifts the image of the slit above the razor blade so that all the light passing the slit hits the screen (bright signal on the screen). (Fourth row) Two blades cut the light passing the schliere so that only the effectively illuminated region of the schliere is imaged on the screen (this region is imaged bright in the screen). (Fifth row) Schliere of different gradients of the index of refraction tilts light passing different schlieren regions differently onto or above the razor blade so that light passing certain regions of the schliere hits the screen with different intensities (varying brightness on the screen).

The sketch shown in the third row of Fig. 4.4 contains a schliere featuring a constant gradient of the index of refraction, which is represented by the straight dashed purple line. The highest index of refraction is at the top of the schliere and the smallest at the bottom. The light rays passing the schliere are bent towards the regions of higher index of refraction upward. Due to the angular ray deflection ε_y upwards the image of the slit is also moved upward. In Fig. 4.4 the angular ray deflection ε_y is sufficiently large to lift the entire image of the slit above the razor blade. Therefore the screen is illuminated as bright as with no razor blade being present. Let us assume that the gradient of the index of refraction was smaller so that the image of the slit was lifted, but not completely lifted above the razor blade. Some portion of the image of the slit is then still blocked by the razor blade. Consequently the screen would appear brighter than the background illumination level but not as bright as with no razor blade installed. Let us now assume that the gradient of the index of refraction was positive. Then the light rays would be bent downward to the bottom and the image of the slit would be lowered with respect to the edge of the razor blade. Thus the screen would be illuminated less than the background illumination level. Summarising, the use of a slit instead of a pinhole renders two essential advantages over the pinhole based schlieren set-up. Firstly, it makes possible the measurement of positive and negative gradients of the index of refraction. Secondly, it makes possible the measurement of the magnitude of the gradients of the index of refraction.

I prepared the fourth sketch in Fig. 4.4 in order to show that the two lenses – lens 2 and lens 3 – image the schliere onto the screen, similarly to how the two lenses – lens 1 and lens 2 – image the slit between lens 2 and lens 3. This is possible if the distances between the lens 2 and the schliere and the distance between the lens 3 and the screen are the focal lengths f_{length} of the lenses. Light rays can only pass through a limited region of the schliere, as the upper and lower region of the schliere are blocked with two blades. I want to highlight here that according to Fig. 4.4 light rays originating from the upper and the lower edge of the slit pass the free schlieren region. If this is the case for light rays coming from the two extreme positions within the slit we can conclude that light rays originating from everywhere from the slit pass the free schlieren region. The same would be true if the free region was further confined or shifted up or down. Through each region of the schliere light rays originating from each location within the slit pass. Thus the two lenses – lens 1 and lens 2 – still produce a complete image of the slit, but using for this purpose less light rays. In the fourth sketch in Fig. 4.4 the negative gradient of the index of refraction is strong enough to lift the image of the slit completely above the edge of the razor blade. As I previously mentioned that the schliere is imaged onto the screen. Consequently only the free schlieren region appears brighter than the background illumination level on the screen. The blocked regions are completely dark, as from these regions no light rays reach the screen. Here I want to mention that the schliere can also be imaged without lens 3, if the distance between the schliere and lens 2 and the distance between lens 2 and the screen are chosen as sketched in Fig. 4.3.

From the sketches already discussed, we know that the schliere is imaged onto the screen and that not only the sign of the gradient of the index of refraction but also its magnitude can be visualised, if a slit instead of a pinhole is used. Therefore the purpose of the fifth sketch in Fig. 4.4 is to summarise the advantages of the slit-based schlieren experiment. I inserted a schliere with varying gradients of the index of refraction. The indices of refraction are represented by the purple dashed lines. Schlieren regions featuring a negative gradient of the index of refraction appear brighter on the screen than the background illumination level. They appear the brighter the larger the value of the respective gradient is. Schlieren regions featuring a positive gradient of the index of refraction appear darker on the screen than the background illumination level. They appear the darker the larger the value of the respective gradient is.

It should be mentioned that the unblocked height and the entire height of the image of the slit play an important role in the schlieren technique. Firstly, the entire height of the slit determines the measurement range covered by the schlieren experiment. The larger the height of the slit is the larger is also the maximum gradient of the index of refraction the set-up can still detect below the detection limit. Larger gradients would lift or lower the image of the slit entirely above or below the edge of the razor blade. If the latter occurs, the measurement range of the schlieren system is exceeded and the respective schlieren regions appear either white or black on the screen. Secondly, the height of the unblocked slit image determines the sensitivity of the schlieren experiment. The smaller the unblocked height of the schlieren image is, the more clearly an already small change of the angular deflection changes the illumination level on the screen. Thirdly, the unblocked height in relation to the entire slit height determines the level of background illumination and with this the measurable values of gradients of the indices of refraction. If the background illumination level is set to 50%, positive as well as negative gradients of the index of refraction can be measured to the same extend. If the background illumination level is set above or below 50%, the range of gradients measurable is shifted towards the positive or the negative ones, respectively. For a more comprehensive description of the influence of the shape of the slit onto the measurement range and measurement sensitivity I refer to Settles's book [3].

Figure 4.5 shows a Z-type schlieren experiments in which the lenses (not the condenser lens) are replaced by parabolic mirrors. Replacing these lenses by mirrors has two advantages. Firstly, mirrors in contrast to lenses do not suffer from chromatic aberrations. Secondly, achromatic lenses, which have to be utilised in order to minimise the chromatic aberrations, are expensive compared to parabolic mirrors, especially if a large free diameter is required. We will see in the 'case studies' that Z-type schlieren set-ups are frequently used. For a detailed description of the Z-type schlieren system, its 'pros' and 'cons' and

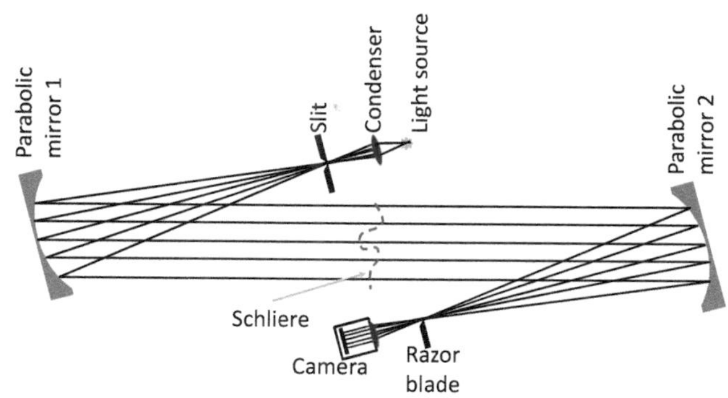

FIGURE 4.5 Sketch of a Z-arranged schlieren experiment.

its noteworthiness I again refer to Settles's book [3], which also includes a construction guidance of such type of set-up.

4.1.4 Distinction Between the Shadowgraph and the Schlieren Technique

In the previous descriptions I already mentioned the specialities of the shadowgraph and the schlieren technique. Here I want to summarise them briefly.

Figure 4.6 shows a collimated shadowgraph set-up. The one shown in Fig. 4.2 is a diverging set-up. Irrespective of whether or not a schliere is present in the collimated part of the set-up, the screen is illuminated with the same signal level. Also in Fig. 4.2 we saw that regions featuring different gradients of the index of refraction appeared with the same brightness on the screen. On the contrary we saw previously that in a schlieren experiment the brightness on the screen is a function of the magnitude of the gradient of the index of refraction. This brings us to the first difference between the shadowgraph and the schlieren technique.

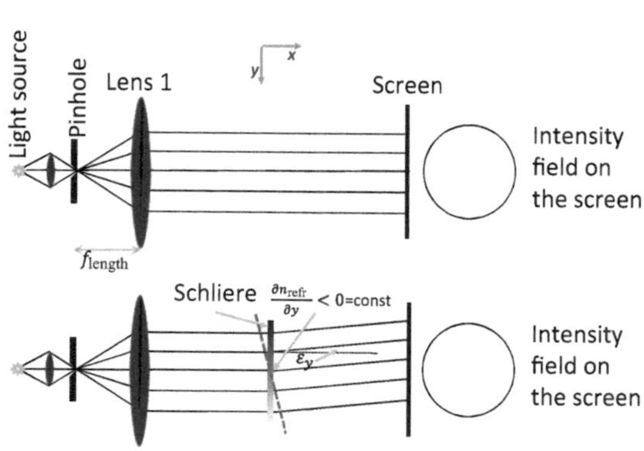

FIGURE 4.6 Sketch of the light path in a collimated shadowgraph experiment without a schliere (top) and with a schliere of constant gradient of index of refraction being present.

The schlieren technique detects gradients of the index of refraction $\partial n_{refr}/\partial y$ and converts it into illumination intensities on the screen, whereas the shadowgraph technique detects a curvature of the field of the index of refraction $\partial^2 n_{refr}/\partial y^2$ and detects it as ray displacement on the screen.

The shadowgraph experiments shown in Figs 4.2 and 4.6 do not contain an optical element which images the schliere on the screen. Therefore the structures visible on the screen of a shadowgraph experiments may be artefacts. Well-separated regions in the schliere may overlap on the screen if for example one lower region bends the rays up and a higher region bends the rays down. This brings us to the second difference between schlieren and shadowgraph experiments. **The schlieren experiment images the schliere on the screen and features a depth of focus, whereas the shadowgraph experiment does not image the schliere.**

The third difference between the schlieren and the shadowgraph technique is their complexity. **A shadowgraph technique is straightforward and also can be easily applied in large fields, whereas the schlieren technique requires more experimental efforts, especially for measurements of large schlieren objects.**

4.2 BALLISTIC SPRAY IMAGING: A SPECIAL SHADOWGRAPH TECHNIQUE

In this section I describe briefly the working principle of a special shadowgraph technique, which is known as ballistic spray imaging, and show one example application. For a more detailed description I refer to Refs. [8–14]. As there are several optical effects cleverly combined in a ballistic spray imaging experiment, I decided to present this technique in an extra section on a prominent position and not to shift it to the following section of case studies.

Ballistic spray imaging has been developed in order to analyse dense sprays in the vicinity of the nozzle exit. In this context 'dense sprays' means that many droplets obstruct the view on the core structure of the liquid stream. Before 2004, when Paciaroni and Linne [8] introduced the ballistic imaging technique for dense spray analysis at ambient pressure, no optical measurement technique was able to provide these insights which were urgently required in order to better understand the atomisation behaviour of the streams coming out of the injector. Applications of this technique followed for spray analysis at elevated pressure [12–15].

Ballistic spray imaging is a shadowgraph technique which in contrast to the shadowgraph techniques described above exploits elastic light scattering. Elastic light scattering of photons from drops is described in Section 2.3.4. Figure 4.7 shows that when collimated light is incident to a spray, different photons travel different paths through the spray, depending on their scattering progression.

Photons passing through the drop cloud without any interaction with drops are called ballistic photons. The group of quasi-ballistic photons represent the

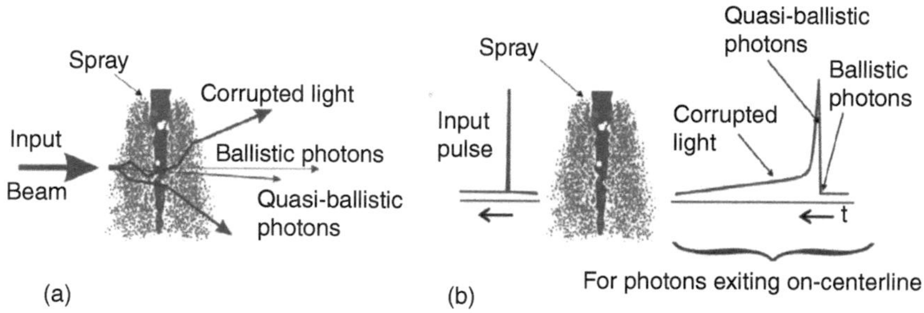

FIGURE 4.7 (a) Geometric and (b) time dependence of various classes of photons as they exit a spray. *(Source: Ref. [16].)*

forward scattered photons (see Figure 2.29 bottom right and bottom centre). The group of corrupted light comprises the photons which due to multiple scattering mechanisms leave the spray off-axis. Figure 4.7b shows that if the incident photons arrive at the spray in an infinitely short light pulse, the different groups of photons exit the spray on-centreline at different times, depending on the different path lengths they travelled through the spray.

It is mainly the group of corrupted photons which obscures a common shadowgram acquired from a dense spray. Therefore ballistic imaging techniques in general try to detect the ballistic photons. In ballistic spray imaging a time-gating method, which is described further, makes possible the temporal discrimination of the different groups of photons. Unfortunately the number of ballistic photons passing a spray is insufficient to detect a clear shadowgram. Thus next to the ballistic photons also the quasi-ballistic photons are detected in ballistic spray imaging, or in other words, mainly the corrupted photons are temporally rejected.

Figure 4.8 shows a sketch of a ballistic spray imaging experiment [16]. Typically a Ti:Sapphire regenerative amplifier with a repetition rate of 1 kHz is used, which is seeded with a mode-locked Ti:Sapphire laser. The amplifier typically emits 50–150 fs (femtoseconds) light pulses with a single-shot pulse energy of 4–10 mJ and a centre wavelength of 800 nm. A beam splitter splits the pulse in an imaging and in a switching beam. The imaging beam passes a time delay arm in order to exactly synchronise temporally the imaging and the switching beam in the OKE gate. A polarisation analyser, which is not contained in Fig. 4.8, assures linear polarisation of the imaging beam. The linear polarisation of the imaging beam is then rotated to 45 linear polarisation. A telescope extents the laser beam to the size of the spray region to be analysed. Behind the spray the laser pulse passes a lens which focuses the laser beam through a second polarisation analyser into a cell filled with CS_2. The second polarisation (left of the CS_2 cell) analyser lets pass only light which is polarised linear at 45° like the light before it interacted with the spray. The third polarisation analyser (right of the CS_2 cell) is aligned normal to the polarisation of the second polarisation analyser. Thus – if the switching beam is not activated – light passing the second

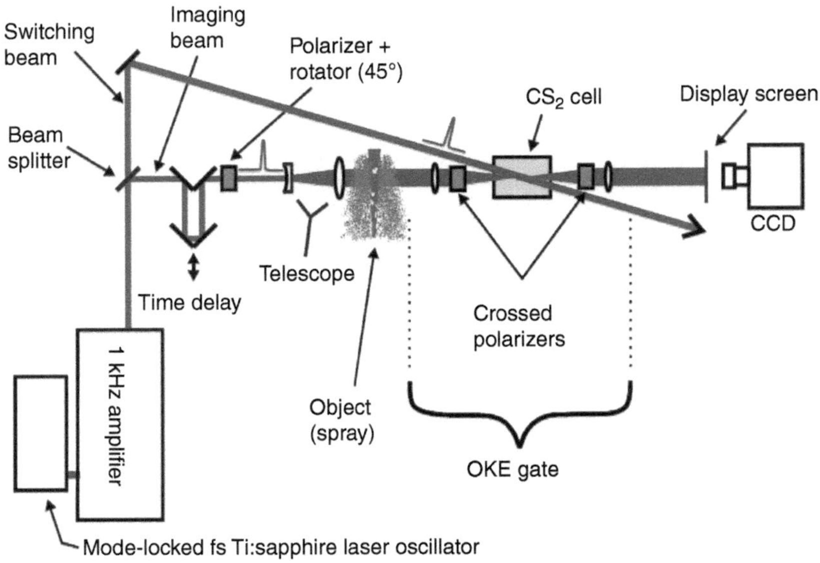

FIGURE 4.8 Schematic of a typical single-shot, single-image implementation of OKE time gated ballistic imaging. *(Source: Ref. [16].)*

polariser will be rejected at the third polariser. When the switching beam passes through the CS_2 cell, the CS_2 molecules in the beam path align according to the polarisation of the switching beam and make the CS_2 birefringent.

The birefringence is maintained according to the molecular relaxation time of CS_2 for 1.5–1.8 ps, which is longer than the pulse duration of the switching beam. If the imaging beam passes the CS_2 cell while it maintains the birefringence, the polarisation of the imaging beam is rotated and as a consequence can pass the third polariser. The combination of the two crossed polarisers and the CS_2 cell is called optical Kerr effect (OKE) gate. The OKE gate here is utilised as an ultrafast shutter. Such short shutter times cannot be realised with ICCD detectors. The last lens produces the shadowgram on a screen, which can be observed by a low-light-sensitive camera, or directly on the chip of a camera. Now it becomes clear what the time delay arm is useful for. The delay time between the switching beam and the imaging beam is adjusted here. During the time when the ballistic and the quasi-ballistic photons of the imaging beam pass the CS_2 chamber, the CS_2 has still to be birefringent in order to rotate the polarisation of the ballistic and quasi-ballistic light so that it can pass to the detector. When the corrupted photons appear in the CS_2 cell the birefringence of the CS_2 must not be maintained anymore. Then the polarisation of the corrupted light is not changed and thus the corrupted light cannot pass the third polarisation analyser.

In Fig. 4.9 Linne [16] compares for the same spray a common shadowgram with a ballistic shadowgram.

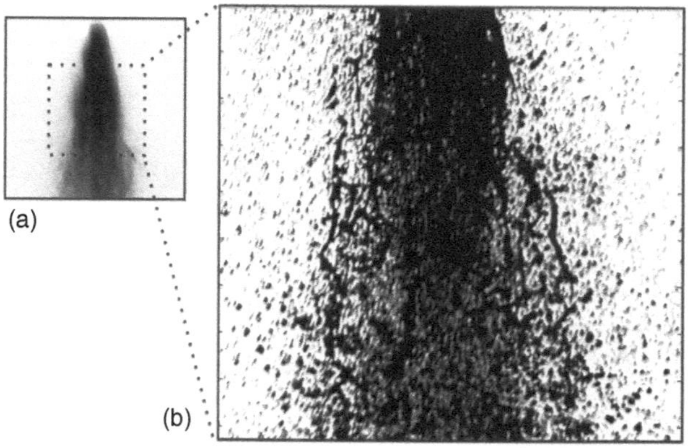

FIGURE 4.9 **Ballistic images of an effervescent spray discussed by Ref. [11].** (a) A laser shadowgram, and (b) a ballistic image of the region enclosed in the square in (a). *(Source: Ref. [16].)*

It must be clarified that for the common shadowgram the camera employed was not able to temporally resolve (freeze) the spray motion. Therefore the comparison of the not temporally resolved common shadowgram and the temporally resolved ballistic shadowgram is not fair, according to Linne's statements. But one can see that the ballistic shadowgraph method provides a clear image of the core of the spray. The dense drop region surrounding the core of the spray does not seem to obscure the shadowgram.

Finally I want to highlight that the experimental set-up shown in Fig. 4.8 combines several measures in order to filter corrupted photons. The most obvious one is the temporal filtering for which the OKE gate is installed. The second measure in order to filter the corrupted photons is based on the lens behind the spray. The farther this lens is away from the spray, the smaller is the probability that corrupted photons, which exit the spray off-axis are collected by this lens. The ballistic and the quasi-ballistic photons are on-axis or nearly on axis, so that their collection efficiency does not really depend on the distance between this lens and the spray. The third measure to filter the corrupted photons is put into practice by installing a polarisation analyser behind the spray which only lets pass light polarised like the imaging beam before interaction with the spray. The more often light is scattered from drops, the more probable the polarisation of the light is changed. Thus the probability that the corrupted light features a polarisation other than the ballistic or the quasi-ballistic light is high. In other words, the polarisation analyser lets pass preferentially not corrupted light.

4.3 CASE STUDIES

In the following subsections I describe some case studies covering sprays and jets, drying experiments, pendant, sessile and levitated drop experiments and applications of shadowgraphy in micro capillaries or in microfluidic devices.

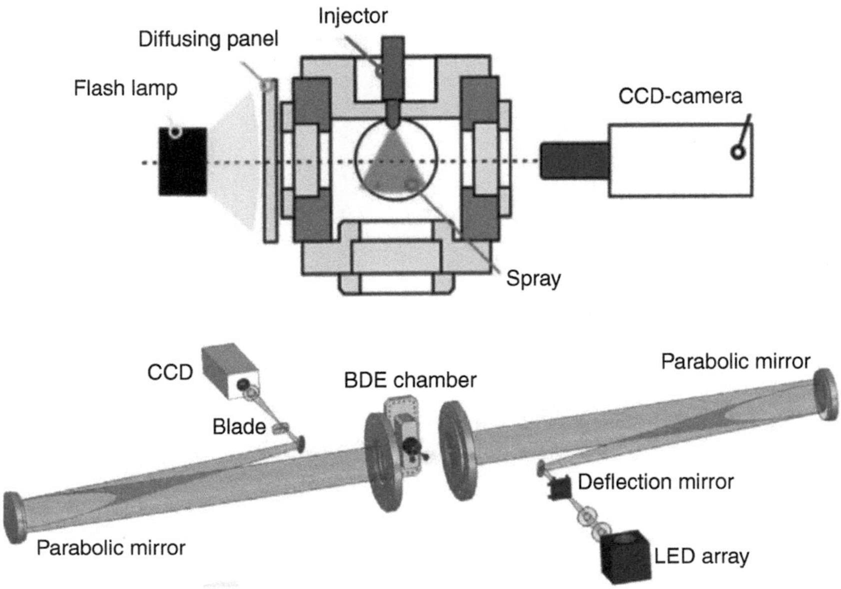

FIGURE 4.10 (Top) Shadowgraph set-up build around a high-pressure high-temperature injection chamber and (bottom) schlieren Z-type configuration build around the same injection chamber. *(Source: Ref. [38].)*

From the previous chapters we know that shadowgraph and schlieren techniques can always be applied when the gradients of the index of refraction deflect the incident light. So potential fields of application of these two measurement techniques are unlimited in high-pressure process technology and in supercritical fluid science and technology.

In the following subchapters you will stumble several times over the expression 'pseudo-shadowgraph'. In the context of Fig. 4.10 I put a lot of emphasise in order to explain why many shadowgraph techniques reported in literature do not really produce shadowgrams but images of the schlieren object. This is due to the fact that often a camera lens is used between the schlieren object and the detector-chip. Nevertheless in literature it is well accepted to also call this measurement technique a shadowgraph technique.

4.3.1 Characterisation of Sprays, Jets

Sprays or jets play an important role in by far more high-pressure disciplines than are obvious at first glance. Schlieren and shadowgraph techniques have been applied successfully in order to analyse

- the penetration behaviour, the evaporation behaviour and the mixture generation of fuel sprays and jets with high-pressure environment [17–21],

- the extension of hydrothermal flames [22–24] for the oxidation of waste or as heat source in a supercritical water reactor [25],
- the extension and the penetration of supercritical water jets for the fragmentation of rocks [26],
- the cutting performance of supercritical and liquid CO_2 streams [27,28],
- the mixing behaviour of organic solution sprays/jets with a supercritical environment for the generation of particles [29–35],
- the atomisation and expansion of pressurised high viscous fluids also for particle generation [36,37],
- and for sure for many more.

I decided to report in more detail on two shadowgraph/schlieren case studies related to sprays and jets, which are interesting from several points of view. The following fuel-spray-evaporation case study [17] exposes nicely the differences of a pseudo-shadowgram and a schlieren image acquired from the same spray. It will become clear below why I decorated this shadowgraph technique with the attribute 'pseudo'. The second CO_2-jet-cutting case study [27,28] is also a pseudo-shadowgraph technique and reveals nicely that thermodynamics is not always able to explain the observed phenomena.

With respect to the fuel-spray-evaporation study Knorsch et al. [17] analysed the evaporation behaviour of different pure fuel compounds and mixtures of them. They were especially interested in analysing the evaporation behaviour of sprays of azeotropes, such as the mixture of isooctane and ethanol. For this purpose they analysed the evaporation behaviour of fuel sprays with a pseudo-shadowgraph and a Z-type schlieren technique, which both are shown in Fig. 4.10. While the Z-type schlieren system is very similar to the one described in Fig. 4.5, the shadowgraph technique is very different to the ones described previously. The main difference is that in the previous shadowgraph set-ups – shown in Fig. 4.2 as diverging or in Fig. 4.6 as collimated technique – the light rays did not intersect with each other before they interacted with a schliere and furthermore – and this is the decisive part – no imaging element was used behind the schliere. On the contrary, in Fig. 4.10 a diffusion panel and a camera lens are utilised. The transparent diffusion panel is illuminated from the backside with a flash lamp and thus it illuminates the spray homogeneously. But light rays originating from different locations on the diffusion panel intersect with each other on their way through the chamber, even without schlieren existing. This also means that through each point of the spray plane, light rays pass that originate from different locations from the diffusive plate and thus feature different angles of incidence. The camera lens is utilised in order to image the spray plane onto the CCD chip of the camera. This changes the working principle of the common and the pseudo-shadowgraph experiment essentially and here makes possible the visual separation of regions in the spray where still liquid fuel is present (pseudo-shadowgram left in Fig. 4.11) and other regions where the spay has already evaporated (difference of the spray extension of the schlieren image right in Fig. 4.11 and the shadowgram).

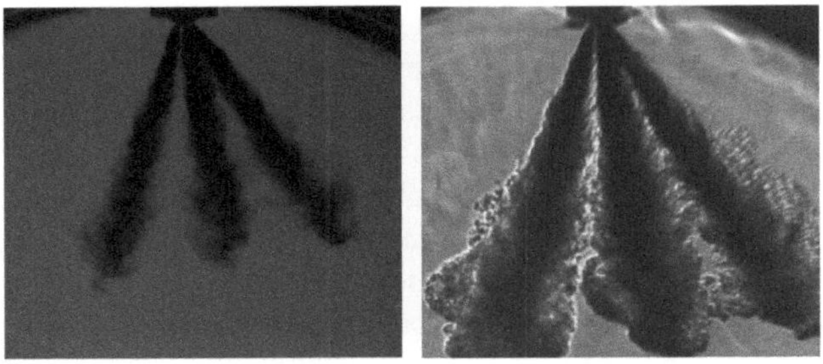

FIGURE 4.11 (Left) Pseudo shadowgram taken from a fuel spray and (right) corresponding schlieren image acquired from the same spray. *(Source: Ref. [17].)*

Due to the imaging of the spray-plane onto the CCD chip the recorded image does no more contain information about the displacement of light rays – as it was the case in Figs 4.2 and 4.6 for common shadowgraph experiments – but about the light intensity passing different regions within the imaged spray plane and still arriving on the CCD chip. The more light is scattered off-axis due to the drop–light interactions, the less light can be collected on axis from these regions with the camera lens and the less light is detected from these regions on the CCD chip. Thus, spray regions attenuating – due to off-axis light scattering – the incident light on axis more than others appears darker than others in the acquired image. Considering this, the described shadowgraph technique is more or less an extinction method and mainly makes visible the liquid (droplet) phase distribution in the spray. This is why I added the attribute 'pseudo' to 'shadowgraph'. One also could compare the working principle of the pseudo-shadowgraph technique with that one of a schlieren set-up working without a razor blade.

The dark regions in the pseudo-shadowgram in the left part of Fig. 4.11 indicate the extension of the liquid phase of the spray. Regions in which the injected fuel has already evaporated and which definitely feature gradients of the index of refraction are not visible in the pseudo-shadowgram. In these regions the incident light rays are deflected according to Fig. 4.1, but only marginally so that the deflected rays still pass the camera lens and thus can be detected as bright regions on the CCD chip.

The schlieren image right in Fig. 4.11 represents the distribution of the gradients of the index of refraction within the imaged spray plane. As already mentioned, there are gradients of the index of refraction in regions where already all of the injected liquid fuel has evaporated. These regions surround the regions in which still liquid fuel is present.

The images acquired with the pseudo-shadowgraph technique and the schlieren experiment can be compared with respect to the spatial extension of the spray, as both techniques image the spray plane onto the CCD chip. Therefore a comparison of both reveals regions in which the injected fuel is either liquid or/and gaseous (schlieren image), or regions in which the injected fuel is still

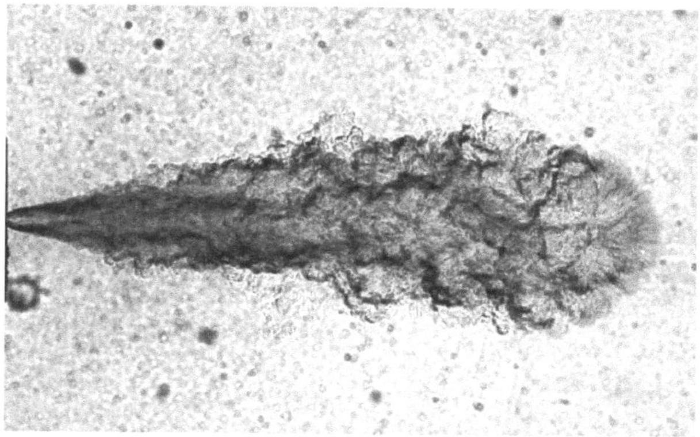

FIGURE 4.12 **Schlieren image of a hydrogen jet injected at 298 K with 10 MPa into a chamber purged with nitrogen at 1 MPa.** Injector opens for 1000 ms. Image acquired 300 ms after visible start of injection. *(Source: Image provided by courtesy of the group of engine combustion from the Lehrstuhl für Technische Thermodynamik from the Friedrich-Alexander-Universität Erlangen-Nürnberg, Germany; Group headed by Michael Wensing.)*

liquid (pseudo-shadowgram). Regions which are dark in the pseudo-shadowgram are also dark in the schlieren image, due to the fact that the light incident to these regions in the spray plane has been scattered away off axis and could not be detected. Regions which are dark in the schlieren image are not dark in the pseudo-shadowgram if the light passing the respective regions in the spray plane was only refracted but not too far off axis so that it still can be detected.

Figure 4.12 shows a schlieren image of a hydrogen jet injected into pressurised nitrogen and is supposed to show the differences between a gas-like single-phase and a liquid–gas two-phase mixing process shown before in Fig. 4.11. No droplets which potentially could scatter the incident light far off axis are formed during the mixing process of the two non-liquid fluids hydrogen and nitrogen. Thus the majority of the light rays detectable without the jet can also be detected with the jet being present. Now the dark regions mainly result from the fact that one portion of the refracted light is blocked by the razor edge.

With respect to the CO_2-jet-cutting case study, Engelmeier et al. [27,28] investigated into the suitability of CO_2 jets for cutting purposes, where especially no residual cutting fluid is supposed to contaminate the processed product. Applying also a pseudo-shadowgraph technique (they imaged the jet-plane onto the CCD chip) they were able to demonstrate the existence of liquid CO_2 jets at ambient pressure (0.1 MPa). This observation was surprising due to the CO_2 triple point pressure being at 0.518 MPa, below which – in thermodynamic equilibrium – CO_2 can either exist as solid or as gas, but not as a liquid. Figure 4.13 shows pseudo-shadowgrams of some of these liquid jets, which were generated when injecting CO_2 pre-cooled to 259 K with injection pressures between 20 MPa and 100 MPa into ambient air.

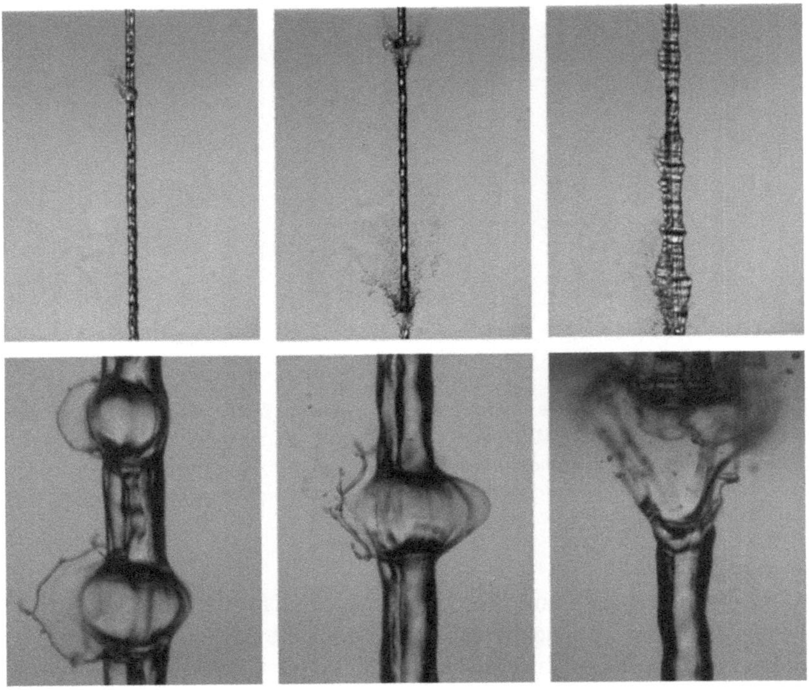

FIGURE 4.13 Emergence of CO_2 gas bubbles in a coherent liquid CO_2 jet at a pressure of 0.1 MPa at which CO_2 cannot exist in the liquid state in thermodynamic equilibrium. *(Sources: Provided by courtesy of the Lehrstuhl für Feststoffverfahrenstechnik of the Ruhr Universität Bochum, Germany. Shadowgrams acquired by Lena Engelmeier.)*

The shadowgrams reveal the emergence of gas bubbles popping up within the CO_2 jets. The bubbles break the coherent liquid CO_2 jet into parts. The lifetime of the liquid jets and with this their length was long enough to cut different material.

4.3.2 Pendant, Sessile and Levitated Drops at High Pressure

The pendant and sessile drop methods are frequently applied at elevated pressure in order to measure the interfacial tension or in order to measure mass transport phenomena [39–48]. Usually a drop of fluid A is positioned either as pendant drop on a capillary or on a rod or as sessile drop on a plate, while the drop is surrounded by a high-pressure environment. The environment is composed of fluid B. The shape of the drop is a function of the mutual interfacial tensions between fluid A fluid B and the solid, which is either the capillary or the rod in the case of the pendant drop or the baseplate in the case of the sessile drop. From the shape of the drop the mutual interfacial tensions can be deduced.

Mass will be transported between the drop and its environment if fluids A and B are partially miscible. Depending on which fluid features a higher solubility in the other fluid, the net mass flow can be directed into or out of the drop. The net mass flow approaches zero when the drop and its surrounding are in equilibrium. If the mass transport across the drop's surface is slow compared to the temporal resolution of the shadowgraph experiment (usually pseudo-shadowgram) and if the time required to generate the drop is short compared to the equilibration time, the evolution of the drop's shape can reveal mass transport quantities as well as the evolution of the interfacial tension during the equilibration process.

In the following I decided to describe one sessile-drop experiment carried out by Mattea et al. [49] and one levitated drop experiment carried out by Borosa et al. [42]. I chose the first one as in contrast to common sessile or pendant drop experiments Mattea et al. analysed the mass transport in a system containing three and not two fluids. This study aimed to better understand the supercritical extraction of emulsion process for the generation of particles. I chose the second one as the construction of an acoustic drop levitator working at both, high pressure and high temperature, is an engineering masterstroke.

Figure 4.14 shows in the top row the evolution of a drop of organic solvent which is surrounded by water. The drop is saturated with water. What is not visible in this figure is that the drop and the surrounding water are both contained in a cuvette which itself is positioned in a chamber pressurised with CO_2. Fig. 4.14a represents an instant before pressurisation with CO_2 and Fig. 4.14b–d represents instants after pressurisation to 10 MPa. The drop changes its shape as the pressurisation induces mass transfer mainly between the CO_2-phase, the water-phase and the drop. At 10 MPa the organic solvent and the CO_2 are

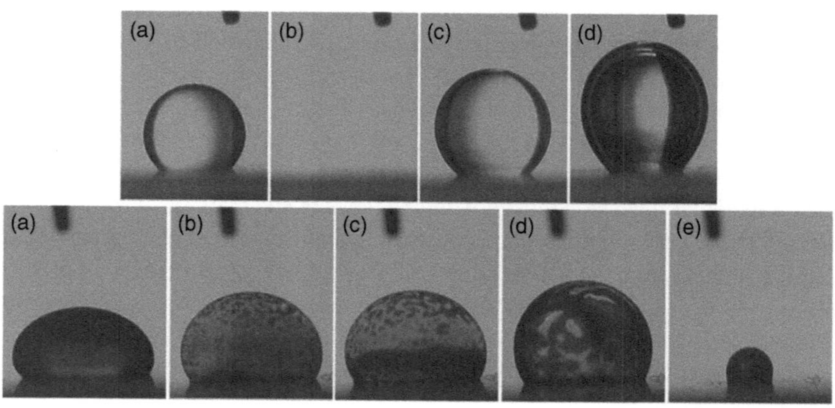

FIGURE 4.14 (Top) Drop evolution in an experiment at a pressure of 10 MPa and temperature of 308 K and (bottom) β-carotene particle precipitation during an experiment at pressure of 5 MPa and temperature of 308 K: (a) initial condition, (b) beginning of particle formation, (c, d) particle agglomeration, and (e) final condition after the drop detachment from the cell surface. *(Source: Ref. [49].)*

completely miscible with each other, but the water phase separates them from each other. Thus the mass transferred between the drop and the CO_2-phase outside the cuvette has to pass through the water phase, which limits the mass transport. Due to the mass transport of CO_2 through the water-phase into the drop and due to the mass-transport of the organic solvent through the water-phase into the CO_2-phase the composition, the mass and the surface tension of the sessile drop changes. The droplet grows until it lifts from the bottom of the cuvette after the instant indicated by the Fig. 4.14d. At the instant indicated by the Fig. 4.14b the drop has not disappeared but is just invisible because its index of refraction is equal to the one of its surrounding water-phase. The bottom row of Fig. 4.14 shows the evolution of an organic drop containing some β-carotene. Again the drop is positioned in a cuvette filled with water and this cuvette again is positioned in a high-pressure apparatus which now is pressurised to 5 MPa.

These conditions represent the circumstances prevailing during the extraction of emulsions with compressed CO_2. In the shadowgram (Fig. 4.14a) one can see that there are no β-carotene particles before the pressurisation with CO_2. After the pressurisation with CO_2 β-carotene particles are formed which are visible on the surface of the drop. This is due to the anti-solvent effect of the CO_2 migrating into the drop and due to the simultaneous reduction of organic solvent from the drop. The formed particles agglomerate and when the drop lifts from the bottom of the cuvette a particle resides at the bottom of the cuvette.

The pendant drop and the sessile drop experiment have in common that the drop is in contact with a solid surface and not completely surrounded by another fluid. This of course influences the processes the drop is involved in. As examples, mass and heat transfer is hindered or influenced through the solid–fluid interface between the drop and the solid part and furthermore the solid–fluid interface can act as source for heterogeneous nucleation if particle precipitation studies should be performed. Thus under some circumstances the contact-less levitation of drops brings some advantageous. One method of drop levitation is the so-called acoustic levitation, which at ambient pressure is a frequently applied tool. But the adoption of an acoustic levitator to a high-pressure view cell bears some challenges, as the tight combination of the levitator components and the pressure cell on the one hand side has to be sealed against high pressure but on the other hand side must not detune the acoustic features of the levitator. Therefore the construction of an acoustical levitator working at pressures up to 20 MPa and at temperatures between 253 K and 453 K by Borosa et al. [42,50] for sure was not easy but already has provided new insight to high-pressure research and science.

Figure 4.15 shows in the left part a pseudo-shadowgram of a drop levitated in a high-pressure environment within the Borosa acoustic levitator. Next to the fact that information of mass transport can be deduced from the evolution of the drop's shape, also particle formation processes taking place inside the drop can be observed, as it is shown in the right part of Fig. 4.15 for different stadia of the crystallisation progress.

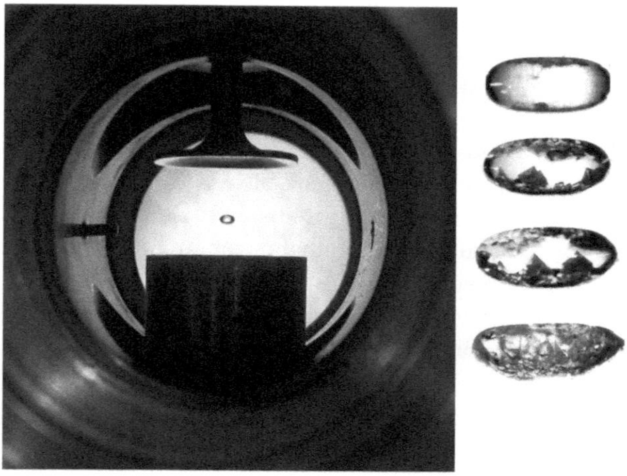

FIGURE 4.15 (Left) Pseudo-shadowgram of a drop levitated acoustically in a high-pressure environment and (right) pseudo-shadowgrams of the crystallisation during the CO_2 hydrate formation in a drop. *(Source: Provided by courtesy of Borosa Acoustic Levitation, Germany.)*

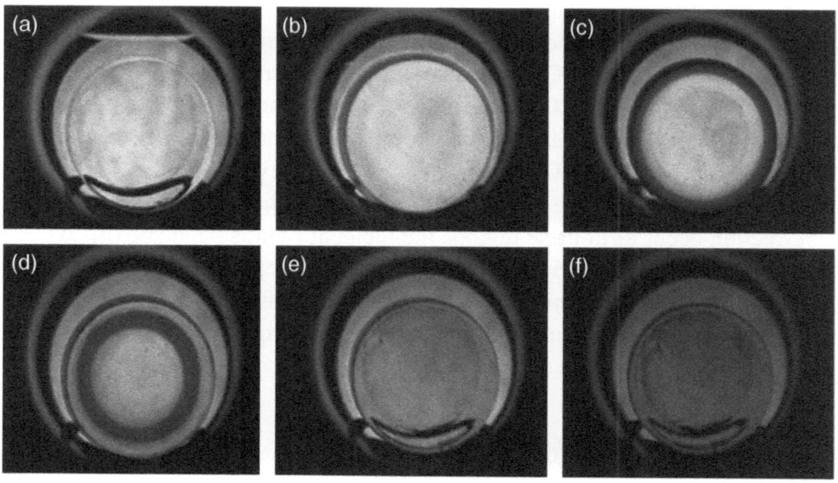

FIGURE 4.16 Captures of the aerogels during supercritical drying (axial view): (a) before pressurising (immersed in methanol); (b) after pressurising; (c) after connection of the recirculation pump; (d) during supercritical washing (crown formation); (e) after supercritical drying and (f) after depressurisation. *(Source: Ref. [55].)*

4.3.3 Supercritical Drying of Gels

Figure 4.16 shows pseudo-shadowgrams of a silica-based cylindrical aerogel during the supercritical drying process [51–53]. The aerogel cylinder is positioned laying on its curved surface while the camera takes the pseudo-shadowgrams looking towards one plane end of the cylinder. In Fig. 4.16a one can see the meniscus of organic solvent – here it is methanol – above the aerogel indicating that the high-pressure view cell is first filled with methanol. Also the pores of the aerogel are still filled with methanol. Therefore at this stadium it is correct to call it an alcogel instead of an aerogel. In Fig. 4.16b the chamber is pressurised and purged with supercritical CO_2 which washes the excess of methanol out of the view cell. The meniscus has disappeared. The methanol captured inside the pores is also extracted slowly out of pores. The drying is accomplished due to the unlimited mutual miscibility of methanol and CO_2 at the set pressure and temperature. Thus, while the CO_2 replaces the methanol inside the pores the methanol is transported out of the gel structure. During the replacement of methanol with CO_2 inside the gel, the view cell is continuously washed with fresh CO_2 at isobaric and isothermal conditions. In Fig. 4.16c one can see a dark ring at the inner boundary of the curved gel surface. In Fig. 4.16d this ring already decreased radially towards the axial centre of the cylinder. The ring appears dark in the pseudo-shadowgram, as light rays incident on the respective regions from the back side are deflected by gradients of the index of refraction so much off-axis that these deflected light rays no more pass through the camera lens and thus cannot be detected on the CCD chip. The gradients of the index of refraction emerge as during the exchange of methanol with CO_2 one fluid with a higher index of refraction, which is methanol, is replaced by a fluid with a lower index of refraction, which is compressed CO_2. The index of refraction of mixtures of CO_2 and methanol are not known for these operational conditions but in general can be in between or even smaller or larger than the indices of refraction of pure CO_2 and methanol [54]. Nevertheless, in that regions which are indicated by the dark ring, the highest radial gradients of the index of refraction occur as in these regions the incident light is deflected off-axis the most. Sanz-Moral et al. [55] used this dark ring as reference for the propagation of a proceeding gradient through the aerogel and – based on the velocity of the ring shrinkage – deduced information about the drying kinetics and the role of the CO_2 washing flow rate on the drying time.

4.3.4 Microfluidics for High Pressure

Microfluidic devices have been not only used for fundamental analysis in high-pressure science but also for application-oriented studies and for process realisations in high-pressure process technology. Some unique features are inherent to microfluidics. Short heat- and mass-transport paths have to be overcome in order to equilibrate a system inside the capillaries. Thus the radial concentration and temperature profiles often are flat. The large contact area between the fluid and the material of the microfluidic device makes possible an efficient in- and out-coupling of heat into and out of the fluids flowing inside the capillary what enables the realisation of isothermal process routes of endo- or exothermic mechanisms. The large contact area between non-miscible fluids arranged in alternating drop/plug-flows promotes a fast mass transport between the drop and the plug phase which promotes a fast equilibration. The well-defined flow structure as well as the well-defined residence time of a fluid element flowing through the capillaries assures a high reproducibility of experiments and very homogeneous product characteristics. The small fluid volumes contained in the capillaries promotes the handling of toxic or dangerous fluids. Due to the small kinematic viscosity of fluids in their supercritical state compared to their liquid state, the realisation of turbulent microfluidic flows is a unique feature of supercritical microfluidics.

Although there have been many optical investigations into processes taking place in microfluidics at ambient pressure, only few are relevant for high-pressure process technology [56–66]. Especially because of the short optical path length through the micro-capillaries they are attractive for optical measurements. A common view cell filled with an emulsion usually is not transparent. Thus the realisation of any optical measurement technique for the analysis of any hypothetic process taking place within the emulsion contained within the common view cell is more than challenging. In contrast, the same emulsion would still be transparent if contained within a micro capillary, simply because of the short path length the light has to pass through the capillary [64]. The same is true for particles-from-spray process. In common high-pressure vessels the path lengths through the optically dense particle loaded jets is too far so that the respective regions appear dark in the shadowgram. In contrast, if particles are formed in a micro capillary, the respective regions are still transparent.

Figure 4.17 shows a sketch of a micro capillary experiment for investigations into the mass transfer relevant for the supercritical extraction of emulsions process.

I want to remind that the experiments described in the context of Fig. 4.14 have been developed in order to investigate into the same direction. In Fig. 4.17

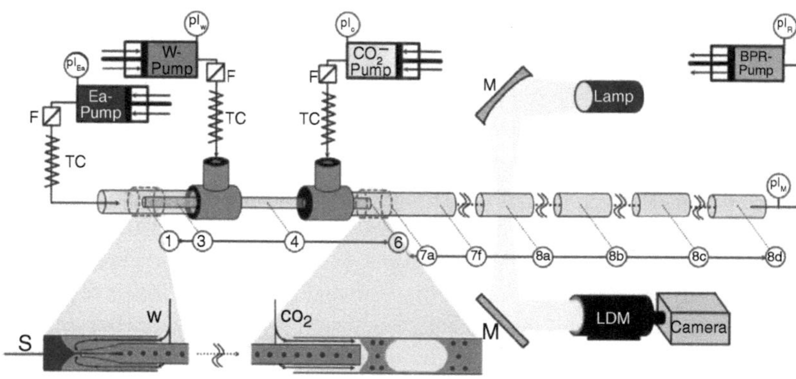

FIGURE 4.17 Schematic of the experimental micro capillary system (MCS) indicating all positions 1–8d accessible with the camera. Ea, ethyl acetate; W, water; C, CO_2; F, sinter inline filter; pI, pressure indicator; TC, temperature control; M, mirror; LDM, long distance microscope; BPR-pump, back pressure regulation pump. *(Source: Ref. [64].)*

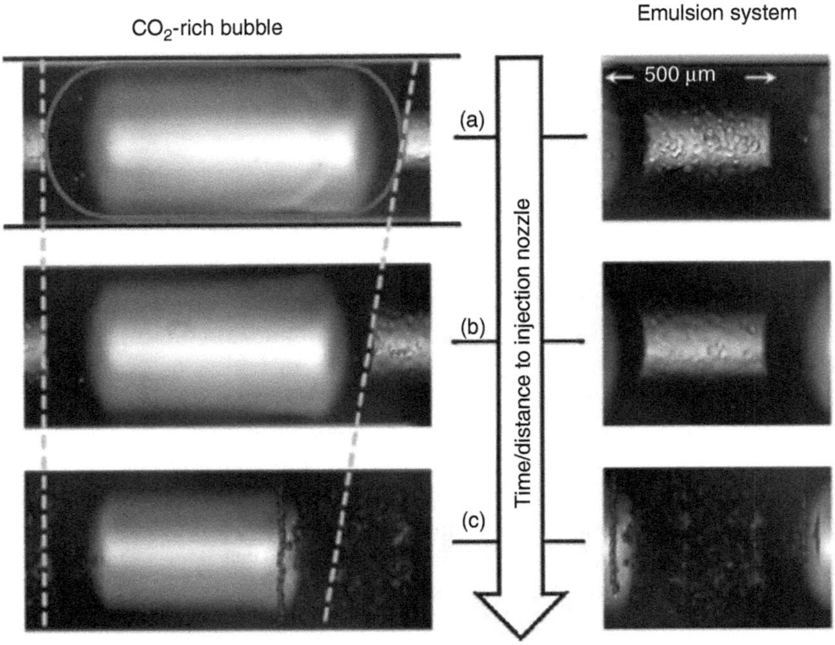

FIGURE 4.18 Visualisation of the reduction of the bubble volume (left column) and the changing appearance of the emulsion system (right column) with increasing residence time in the measuring capillary (a–c). *(Source: Ref. [64].)*

at location 1 an emulsion of ethyl acetate in water is formed by means of a focussing flow capillary arrangement. At point 6 the formed emulsion, which more or less is a chain of ethyl acetate drops saturated with water surrounded by water saturated with ethyl acetate, gets in contact with compressed CO_2 in a co-flow capillary arrangement. Downstream of the mixing point 6, CO_2 bubbles form which are separated from each other by plugs of the ethyl acetate–water emulsion. The pseudo-shadowgraph experiment composed of a light source, two mirrors and a detection unit can be moved along the capillary in order to take snapshots of the CO_2 bubbles and the emulsion plugs at different locations downstream of the mixing point. Figure 4.18 shows example pseudo-shadowgrams of the CO_2 bubbles on the left side and of the emulsion plugs on the right side acquired at the locations referenced in Fig. 4.17 as (8a), (8b) and (8c). One can see that the bubbles shrink the further they move downstream. This is due to the transfer of CO_2 from the bubble through the water into the ethyl acetate and due to the transfer of ethyl acetate from the drops through the water into the CO_2 bubbles. The latter processes increases the density of the bubbles significantly so that they shrink even though mass is transported into them. The shadowgrams taken from the emulsions show that the drops become nearly invisible at the measurement point (8b). This means that at this point the indices of refraction of the drops and the water phase surrounding the drops are similar. Luther et al. [64] evaluated the change of the volume of the CO_2 bubbles with respect to measurement location at which the drops inside the plugs became invisible in order to deduce how much mass of which compound (water, CO_2 and ethyl acetate) has been transported between the different phases.

REFERENCES

[1] Tropea C, Yarin AL, Foss JF. Springer handbook of experimental fluid mechanics. Berlin: Springer Science & Business Media; 2007.

[2] Feldmann O, Mayinger F. Optical measurements: techniques and applications. Berlin: Springer Science & Business Media; 2012.

[3] Settles G. Schlieren and shadowgraph techniques: visualizing phenomena in transparent media. Berlin: Springer Science & Business Media; 2001.

[4] Wang Y, Liu X, Im K-S, Lee W-K, Wang J, Fezzaa K, Hung DLS, Winkelman JR. Ultrafast X-ray study of dense-liquid-jet flow dynamics using structure-tracking velocimetry. Nat Phys 2008;4:305–9.

[5] Willers A,-I, Schardin H, Das B. Toeplersche Schlierenverfahren, Grundlagen für seine Anwendung und quantitative Auswertung, ZAMM. J Appl Math Mech/Zeitschrift für Angewandte Mathematik und Mechanik 1935;15:388–388.

[6] Vasilév LA. Schlieren methods, Israel Program; (distributed by Keter Inc.), New York; 1971.

[7] Goldstein R. Optical systems for flow measurement. Fluid Mechanics Measurements. Washington, DC: Hemisphere; 1983. p. 377–417.

[8] Paciaroni M, Linne M. Single-shot, two-dimensional ballistic imaging through scattering media. Appl Optics 2004;43:5100–9.

[9] Linne MA, Paciaroni M, Berrocal E, Sedarsky D. Ballistic imaging of liquid breakup processes in dense sprays, Proceedings of the combustion institute, 2009;32:2147–2161.

[10] Sedarsky D, Gord J, Carter C, Meyer T, Linne M. Fast-framing ballistic imaging of velocity in an aerated spray. Optics Lett 2009;34:2748–50.

[11] Linne M, Sedarsky D, Meyer T, Gord J, Carter C. Ballistic imaging in the near-field of an effervescent spray. Exp Fluids 2010;49:911–23.

[12] Duran S, Porter J, Parker T. Ballistic imaging of sprays at diesel relevant conditions. In: Proceedings of the Twelvth International Conference on Liquid Atomisation and Spray Systems (ICLASS), Heidelberg, Germany, 2012.

[13] Falgout Z, Rahm M, Wang Z, Linne M. Ballistic imaging in the spray formation region for the ECN spray. In: ILASS Europe September 2013.

[14] Duran SP, Porter JM, Parker TE. Picosecond ballistic imaging of diesel injection in high-temperature and high-pressure air. Exp Fluids 2015;56:1–12.

[15] Linne M, Rahm M, Falgout Z, Wang Z, Paciaroni M, Matlok S, Hult J. Correlation of internal flow and spray breakup for a fuel injector used in ship engines. In: 8th US National Combustion Meeting 2013.

[16] Linne M. Imaging in the optically dense regions of a spray: a review of developing techniques. Prog Energ Combust 2013;39:403–40.

[17] Knorsch T, Heldmann M, Zigan L, Wensing M, Leipertz A. On the role of physiochemical properties on evaporation behavior of DISI biofuel sprays. Exp Fluids 2013;54:1–14.

[18] Erfan I, Chitsaz I, Ziabasharhagh M, Hajialimohammadi A, Fleck B. Injection characteristics of gaseous jet injected by a single-hole nozzle direct injector. Fuel 2015;160:24–34.

[19] Decker M, Schik A, Meier UE, Stricker W. Quantitative Raman imaging investigations of mixing phenomena in high-pressure cryogenic jets. Appl Optics 1998;37:5620–7.

[20] Smith JJ, Schneider G, Suslov D, Oschwald M, Haidn O. Steady-state high pressure LOx/H2 rocket engine combustion. Aerosp Sci Technol 2007;11:39–47.

[21] Mayer W, Schik A, Sch-aring M, ffler H, Tamura. Injection and mixing processes in high-pressure liquid oxygen/gaseous hydrogen rocket combustors. J Propul Power 2000;16:823–8.

[22] Schilling W, Franck EU. Combustion and diffusion flames at high pressures to 2000 bar. Berichte der Bunsengesellschaft für physikalische Chemie 1988;92:631–6.

[23] Steeper RR, Rice SF, Brown MS, Johnston SC. Methane and methanol diffusion flames in supercritical water. J Supercrit Fluids 1992;5:262–8.

[24] Serikawa RM, Usui T, Nishimura T, Sato H, Hamada S, Sekino H. Hydrothermal flames in supercritical water oxidation: investigation in a pilot scale continuous reactor. Fuel 2002;81:1147–59.

[25] Wellig B, Weber M, Lieball K, Prikopsky K, von Rohr PR. Hydrothermal methanol diffusion flame as internal heat source in a SCWO reactor. J Supercrit Fluids 2009;49:59–70.

[26] Rothenfluh T, Schuler MJ, von Rohr PR. Penetration length studies of supercritical water jets submerged in a subcritical water environment using a novel optical Schlieren method. J Supercrit Fluids 2011;57:175–82.

[27] Engelmeier L, Pollak S, Kilzer A, Weidner E. Liquid carbon dioxide jets for cutting applications. J Supercrit Fluids 2012;69:29–33.

[28] Lena E, Michael K, Stefan P, Eckhard W. Schneiden und Bohren mit flüssigem Kohlendioxid, Chemie Ingenieur Technik. Avaialble from: http://onlinelibrary.wiley.com/doi/10.1002/cite.201500075/abstract

[29] Carretier E, Badens E, Guichardon P, Boutin O, Charbit G. Hydrodynamics of supercritical antisolvent precipitation: characterization and influence on particle morphology. Ind Eng Chem Res 2003;42:331–8.

[30] Badens E, Boutin O, Charbit G. Laminar jet dispersion and jet atomization in pressurized carbon dioxide. J Supercrit Fluids 2005;36:81–90.

[31] Lesoin L, Crampon C, Boutin O, Badens E. Preparation of liposomes using the supercritical anti-solvent (SAS) process and comparison with a conventional method. J Supercrit Fluids 2011;57:162–74.

[32] Reverchon E, Torino E, Dowy S, Braeuer A, Leipertz A. Interactions of phase equilibria, jet fluid dynamics and mass transfer during supercritical antisolvent micronization. Chem Eng J 2010;156:446–58.

[33] De Marco I, Knauer O, Cice F, Braeuer A, Reverchon E. Interactions of phase equilibria, jet fluid dynamics and mass transfer during supercritical antisolvent micronization: The influence of solvents. Chem Eng J 2012;203:71–80.

[34] De Marco I, Rossmann M, Prosapio V, Reverchon E, Braeuer A. Control of particle size, at micrometric and nanometric range, using supercritical antisolvent precipitation from solvent mixtures: application to PVP. Chem Eng J 2015;273:344–52.

[35] Lengsfeld CS, Delplanque JP, Barocas VH, Randolph TW. Mechanism governing microparticle morphology during precipitation by a compressed antisolvent: atomization vs. nucleation and growth. J Phys Chem B 2000;104:2725–35.

[36] Jan-Martin L, Sabine K, Andreas K, Marcus P. Particle formation from gas enriched polymeric melts and polymeric solutions. In: Udo F, editor. DFG Priority Program 1423 Process Spray. Berlin, Heidelberg: Springer; 2015/2016.

[37] Pavlinka I, Andreas K, Eckhard W. Pulverisation of emulsions with supercritical CO_2, 2015/2016.

[38] Keller P, Knorsch T, Wensing M, Hasse C. Experimental and numerical analysis of iso-octane/ethanol sprays under gasoline engine conditions. Int J Heat Mass Transfer 2015;84:497–510.
[39] Seifried B, Temelli F. Interfacial tension of marine lipids in contact with high pressure carbon dioxide. J Supercrit Fluids 2010;52:203–14.
[40] Chen X, Adkins SS, Nguyen QP, Sanders AW, Johnston KP. Interfacial tension and the behavior of microemulsions and macroemulsions of water and carbon dioxide with a branched hydrocarbon nonionic surfactant. J Supercrit Fluids 2010;55:712–23.
[41] Sutjiadi-Sia Y, Jaeger P, Eggers R. Interfacial phenomena of aqueous systems in dense carbon dioxide. J Supercrit Fluids 2008;46:272–9.
[42] Borosa D, Brinkmann J, Kareth S, Kilzer A, Petermann M. Vergleich des Stofftransports von hängenden und akustisch levitierten Wassertropfen in CO_2. Chemie Ingenieur Technik 2014;86:666–74.
[43] Yang C, Gu Y. A new method for measuring solvent diffusivity in heavy oil by dynamic pendant drop shape analysis (DPDSA). SPE J 2006;11:48–57.
[44] Chiquet P, Daridon J-L, Broseta D, Thibeau S. CO2/water interfacial tensions under pressure and temperature conditions of CO_2 geological storage. Energ Convers Manage 2007;48: 736–44.
[45] Jaeger PT, Eggers R, Baumgartl H. Interfacial properties of high viscous liquids in a supercritical carbon dioxide atmosphere. J Supercrit Fluids 2002;24:203–17.
[46] Pastore Carbone MG, Di Maio E, Scherillo G, Mensitieri G, Iannace S. Solubility, mutual diffusivity, specific volume and interfacial tension of molten PCL/CO_2 solutions by a fully experimental procedure: effect of pressure and temperature. J Supercrit Fluids 2012;67:131–8.
[47] Pastore Carbone MG, Di Maio E, Musto P, Braeuer A, Mensitieri G. On the unexpected nonmonotonic profile of specific volume observed in PCL/CO_2 solutions. Polymer 2015;56:252–5.
[48] Carbone MGP, Di Maio E, Iannace S, Mensitieri G. Simultaneous experimental evaluation of solubility, diffusivity, interfacial tension and specific volume of polymer/gas solutions. Polym Test 2011;30:303–9.
[49] Mattea F, Martín Á, Schulz C, Jaeger P, Eggers R, Cocero MJ. Behavior of an organic solvent drop during the supercritical extraction of emulsions. AIChE J 2009;56:1184–95.
[50] Borosa D, Kareth S, Petermann M. Apparatur zur Untersuchung der Stofftransportmechanismen an schwebenden Tropfen unter erhöhten Drücken. Chemie Ingenieur Technik, 2012;84:145–148.
[51] Vanbommel MJ, Dehaan AB. Drying of silica aerogel with supercritical carbon-dioxide. J Non-Cryst Solids 1995;186:78–82.
[52] Wawrzyniak P, Rogacki G, Pruba J, Bartczak Z. Diffusion of ethanol–carbon dioxide in silica gel. J Non-Cryst Solids 1998;225:86–90.
[53] García-González CA, Camino-Rey MC, Alnaief M, Zetzl C, Smirnova I. Supercritical drying of aerogels using CO_2: effect of extraction time on the end material textural properties. J Supercrit Fluids 2012;66:297–306.
[54] Sun Y, Shekunov BY, York P. Refractive Index of supercritical CO_2-ethanol solvents. Chem Eng Comm 2003;190:1–14.
[55] Sanz-Moral LM, Rueda M, Mato R, Martin A. View cell investigation of silica aerogels during supercritical drying: analysis of size variation and mass transfer mechanisms. J Supercrit Fluids 2014;92:24–30.
[56] Assmann N, Kaiser S, Rudolf von Rohr P. Supercritical extraction of vanillin in a microfluidic device. J Supercrit Fluids 2012;67:149–54.
[57] Assmann N, Werhan H, Ładosz A, von Rohr PR. Supercritical extraction of lignin oxidation products in a microfluidic device. Chem Eng Sci 2013;99:177–83.

[58] Marre S, Jensen KF. Synthesis of micro and nanostructures in microfluidic systems. Chem Soc Rev 2009;39:1183–202.

[59] Kuhn S, Hartman RL, Sultana M, Nagy KD, Marre S, Jensen KF. Teflon-coated silicon microreactors: impact on segmented liquid-liquid multiphase flows. Langmuir 2011;27:6519–27.

[60] Lorber N, Sarrazin F, Guillot P, Panizza P, Colin A, Pavageau B, Hany C, Maestro P, Marre S, Delclos T. Some recent advances in the design and the use of miniaturized droplet-based continuous process: applications in chemistry and high-pressure microflows. Lab Chip 2011;11:779–87.

[61] Marre S, Roig Y, Aymonier C. Supercritical microfluidics: opportunities in flow-through chemistry and materials science. J Supercrit Fluids 2012;66:251–64.

[62] Pinho B, Girardon S, Bazer-Bachi F, Bergeot G, Marre S, Aymonier C. Simultaneous measurement of fluids density and viscosity using HP/HT capillary devices. J Supercrit Fluids 2015;105:186–92.

[63] Luther SK, Braeuer A. High-pressure microfluidics for the investigation into multi-phase systems using the supercritical fluid extraction of emulsions (SFEE). J Supercrit Fluids 2012;65:78–86.

[64] Luther SK, Schuster JJ, Leipertz A, Braeuer A. Microfluidic investigation into mass transfer in compressible multi-phase systems composed of oil, water and carbon dioxide at elevated pressure. J Supercrit Fluids 2013;84:121–31.

[65] Luther SK, Will S, Braeuer A. Phase-specific Raman spectroscopy for fast segmented microfluidic flows. Lab Chip 2014;14:2910–3.

[66] Couto R, Chambon S, Aymonier C, Mignard E, Pavageau B, Erriguible A, Marre S. Microfluidic supercritical antisolvent continuous processing and direct spray-coating of poly (3-hexylthiophene) nanoparticles for OFET devices. Chem Commun 2015;51:1008–11.

Chapter 5

Laser-Induced Fluorescence (LIF) and Phosphorescence (LIP) Techniques

The laser-induced fluorescence (LIF) and laser-induced phosphorescence (LIP) techniques both are emission processes that follow after the absorption of light. In this context Section 2.4.1.6 provides the fundamentals of both measurement techniques. There it is also described under which circumstances LIF can be considered to be a quantitative composition or concentration indicating measurement technique. This description is built up onto the fundamentals of the photophysics of fluorescing molecules and thus has been presented already in the fundamentals-related Chapter 2 of this book. On the contrary, the more empirical semi-physical approaches for temperature measurements and/or the spectral discrimination between vapour and liquid fuel fraction in sprays are presented here in the Sections 5.1 and 5.2. Section 5.3 finally reports some case studies of LIF and LIP measurements at elevated pressure. I chose some examples which are not very typical for LIF and LIP measurements at high pressure, but which are very much related to measurements in supercritical environments. A detailed justification about the selection of these case studies is provided at the beginning of Section 5.3.

Of course it is not possible to cover all potential applications and already completed studies employing LIF and LIP at high pressure here in this chapter. I just want to mention few other LIF applications relevant for high-pressure process technology which I am aware of, but which I did not describe. One example is the fluorescence of ruby. In many different studies the fluorescence of ruby has been analysed in order to measure the temperature of the ruby or the pressure to which the ruby is exposed [1–7]. I did not cover this measurement technique as the respective signals do not originate from the fluid treated in a process, but from a ruby which can be positioned in a high-pressure vessel and remains there passively. Another example is the analysis of the pressure-induced protein folding and unfolding, for which also LIF has frequently been applied [8–12]. The related LIF techniques are based on the circumstance that in the folded and unfolded state of a protein the chromophore

Supercritical Fluid Science and Technology, Vol 7. http://dx.doi.org/10.1016/B978-0-444-63422-1.00005-5

groups (they naturally exist in proteins) of the protein are either surrounded by the large folded protein molecule or in the case of the unfolded protein they are exposed to the solvent. We already saw in Section 2.4.1.6 that the fluorescence is influenced by its environment, meaning that the fluorescence of proteins can be utilised to differentiate between unfolded and folded proteins. Similarly, LIF tracers whose fluorescence spectrum indicates the pH-value rely on the effect the environment has on the fluorescence [13]. They also have been applied at elevated pressures in order to investigate into the microorganism deactivation potential of compressed CO_2 [14] or for monitoring the effect of pressure on the pH value [15]. These protein and pH-value-related LIF techniques are also not covered in this book, as firstly a person with a profound background in biotechnology was definitely a more suitable author than me and secondly most of these studies have been carried out not really *in situ*.

5.1 LIF AND LIP-THERMOMETRY

With respect to LIF and LIP a variety of methods suitable for probing temperature exist. They are based on either the temperature sensitive absorption characteristics, or the temperature sensitive fluorescence/phosphorescence quantum yield of molecules, or on the temperature sensitive combination of both effects together. In this context the fundamentals are described in Section 2.4.1.6 of this book. Next to the LIF and LIP-thermometry methods described further, certainly other emission-based thermometry methods exist, which technically are applicable also at high pressures, such as for example the two-line atomic fluorescence thermometry method [16]. But, mainly due to their seldom application in high-pressure environment or due to their debatable suitability for high-pressure environments they are not contained in the following descriptions.

5.1.1 Laser-Induced Fluorescence (LIF) for Liquid-Phase Thermometry

The temperature dependency of the fluorescence signal intensity of dye molecules has been exploited for measuring temperature in liquid systems [17–21]. Usually the LIF signal intensity emitted from a constant number of dye molecules, such as rhodamine or fluorescein, decreases with increasing temperature. On this effect trace amounts of dye molecules are dissolved in the liquid systems in which the temperature is to be measured. The dissolved amount should be small in order to keep the absorption of the excitation laser beam low but high enough to detect the LIF signals with sufficient signal-to-noise ratios. If the dye molecules are distributed homogeneously in the liquid system and if the attenuation of the excitation laser can be neglected, the detected fluorescence signal intensities directly – on the basis of a beforehand carried-out calibration – reveal the temperature. The corresponding technique is referred to as single-dye/one-colour LIF thermometry [22]. Of course this technique fails if the single-shot energy of the excitation laser fluctuates, if minor absorption effects attenuate the excitation laser within the probe volume and if the dye molecules are not homogeneously distributed throughout the system. In all these cases the detected fluorescence signal intensity would no more be solely a function of the temperature. On this account the single-dye/two-colour [23–27] and the two-dyes/two-colour [26–35] LIF thermometers have been developed. The single-dye/two-colour LIF thermometry concept makes use of the fact that not only the intensity of the fluorescence signal but also its spectral shape is a function of temperature. Thus it envisages the detection of two spectrally separated regions within the fluorescence band of the dye molecule, whose intensities feature different temperature sensitivities. The ratio of the signal intensities of both regions is temperature sensitive, but not affected by shot-to-shot fluctuations of the excitation laser, by absorption-based attenuation effects or by an inhomogeneous distribution of the dye molecules themselves. All these effects cancel out due to the consideration of the ratio of signal intensities of two separate spectral regions of the same molecule. The same is true for the two-dyes/two-colours concept, which is based on the fact that two different dyes feature different temperature sensitivities. Again, this concept is based on the consideration of signal intensity ratios, but this time one detected signal originates from the first dye species and the other one from the second dye species. Ideally both dyes can be excited at the same wavelength but their fluorescence spectra are spectrally shifted to each other.

It should be underlined here that the fluorescence spectrum of many dye molecules next to temperature also very much depends on the chemical composition of the system. A change of the composition or the pH-value would change the fluorescence signal intensity and the spectral shape of the fluorescence signal. In fact, there are some studies reporting on the influence of varying conditions such as pressure or oxygen (LIF quencher) on the photophysics of dye molecules and other temperature indicating LIF tracers [36–38], but only very few real applications of these methods for temperature measurements in the liquid phase but at high pressure are available in fact [39].

5.1.2 Laser-Induced Phosphorescence (LIP) for Liquid-Phase Thermometry

Next to the LIF the LIP can also act as temperature sensing method [40–51]. Although after the excitation with a short laser pulse the lifetime of fluorescence emissions is in the order of nanoseconds, the lifetime of phosphorescence emissions spreads over micro to milliseconds. This makes possible the temporal discrimination of the phosphorescence signal from interfering elastically scattered stray light or from interfering fluorescence emissions. But – due to their long lifetime – phosphorescence can also be very sensitive to quenching [52]. We will see in the further progress of this section that phosphorescing molecules can be either dissolved in the fluid under investigation or isolated as dispersed

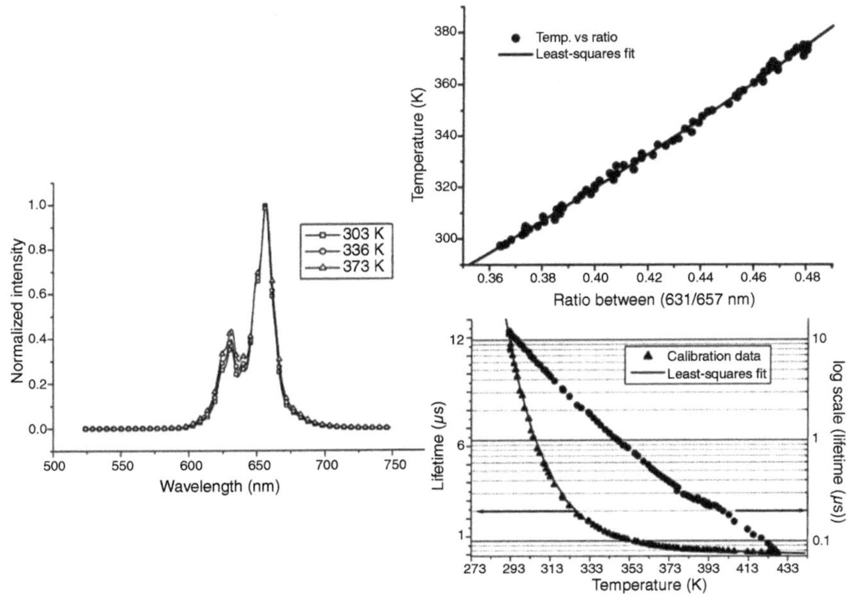

FIGURE 5.1 (Left) Spectra of Mg_4FGeO_6:Mn in distilled water at three temperatures. The 631-nm line was seen to increase with temperature relative to the 657-nm line. (Right top) Results from calibration measurements made in a liquid tank. The ratio between the 631-nm and the 657-nm lines is plotted against cell temperatures). (Right bottom) La_2O_2S:Eu lifetime measured at several temperatures. The lifetime decreases exponentially as the temperature increases. The lifetime is plotted on a logarithmic scale to illustrate the sensitivity of the technique up to 433 K. The sensitivity is directly related to the slope of the logarithm of the lifetime versus temperature. *(Source: Ref. [54].)*

particles. In the latter case, where the phosphorescing molecules are more or less encapsulated in solid dispersed particles, they do not really get in contact with the surrounding fluid. Then, – this is not true for all phosphors but for most of them [52] – the phosphorescence behaviour is not influenced by the chemical composition of their environment, but solely by temperature [53].

Figure 5.1 shows in the left part phosphorescence spectra of Mg_4FGeO_6:Mn particles dispersed in liquid water at different temperatures.

In the top right part of Fig. 5.1 the temperature sensitivity of the phosphorescence ratio between the two peaks at 631 nm and 657 nm is given. In the bottom right part of Fig. 5.1 the phosphorescence lifetime is shown as a function of temperature for La_2O_2S:Eu dispersed in water. Thus according to Fig. 5.1 either the ratio of two spectral regions of the phosphorescence spectrum [55] or the lifetime of the phosphorescence emission [56] can be utilised for the measurement of temperature in liquids. The temperature sensitivity of both approaches is based on the non-radiative de-excitation rate of the triplet state (starting point of phosphorescence compare Section 2.4.1.6), which grows with temperature, but not uniformly over the range of phosphorescence emissions.

As already mentioned – due to the long lifetime of the phosphorescence – the phosphorescence of molecules suffers enormously from quenching, if they are in contact with quenchers. If there are potential quenchers such as oxygen around, the phosphorescence can be easily quenched completely. Already the small amount of oxygen dissolved in liquid water significantly quenches the phosphorescence of molecules such as acetone. Therefore the application of MOLECULAR phosphors (not of phosphor particles) as temperature indicating tracers relied on the availability of novel molecular phosphors which are not affected by quenching [57–59]. Koochesfahani et al. [60] and Gendrich et al. [61] applied the quenching resistant phosphorescence tracer 1-BrNp_Gβ-CD_ROH firstly solely for molecular tagging velocimetry, a technique for the measurement of velocity fields in flows that is not described in the context of the book further. 1-BrNp_Gβ-CD_ROH is a triplex formed by mixing the lumophore (1-BrNp), certain alcohols (indicated collectively by ROH), and an aqueous solution of glucosyl-β-cyclodextrin, Gβ-CD [60,61]. Due to the temperature sensitivity of the phosphorescence, Thomson and Maynes [62] extended the molecular tagging velocimetry measurement technique also for measurement of temperature fields. Figure 5.2 shows in the top part the absorption, the fluorescence and the phosphorescence spectrum of a similar triplex molecular phosphorescence tracer and in the bottom part how sensitively the intensity of the phosphorescence signal is affected by a small variation of temperature. Comparing the influence of temperature onto the phosphorescence intensity exhibited in the left part of Fig. 5.1 and in the bottom part of Fig. 5.2 it becomes obvious that different phosphors show different temperature sensitivities. Therefore the identification of a suitable phosphor is crucial for high sensitivity temperature measurements. The temperature sensitivity of the 1-BrNp_Gb-CD_ROH triplex for example is also superior to the temperature sensitivity of dye-fluorescence tracers.

Finally I want to mention that neither LIF nor LIP-tracer molecules nor particles can be arbitrarily seeded to liquid systems. The presence of particles in liquid systems significantly influences the viscosity of the system [63,64]. The presence of dissolved tracer molecules influences the equilibria and thus may impact heat- as well as mass-transport processes. With respect to evaporating droplets, one has to be aware that the large tracer molecules or tracer particles do not co-evaporate with the solvent (water or an organic solvent). The tracers concentrate in the droplets which also affects the emission characteristics of the tracer molecules and the evaporation behaviour of drops.

5.1.3 Laser-Induced Fluorescence (LIF) for Vapour-Phase Thermometry

For LIF thermometry applications in the gas phase, temperature indicating fluorescence tracers must be present in the gas phase. They must be available in the gas phase at concentration high enough to make a detection of the LIF signals with sufficient signal-to-noise ratios possible. Thus, compounds

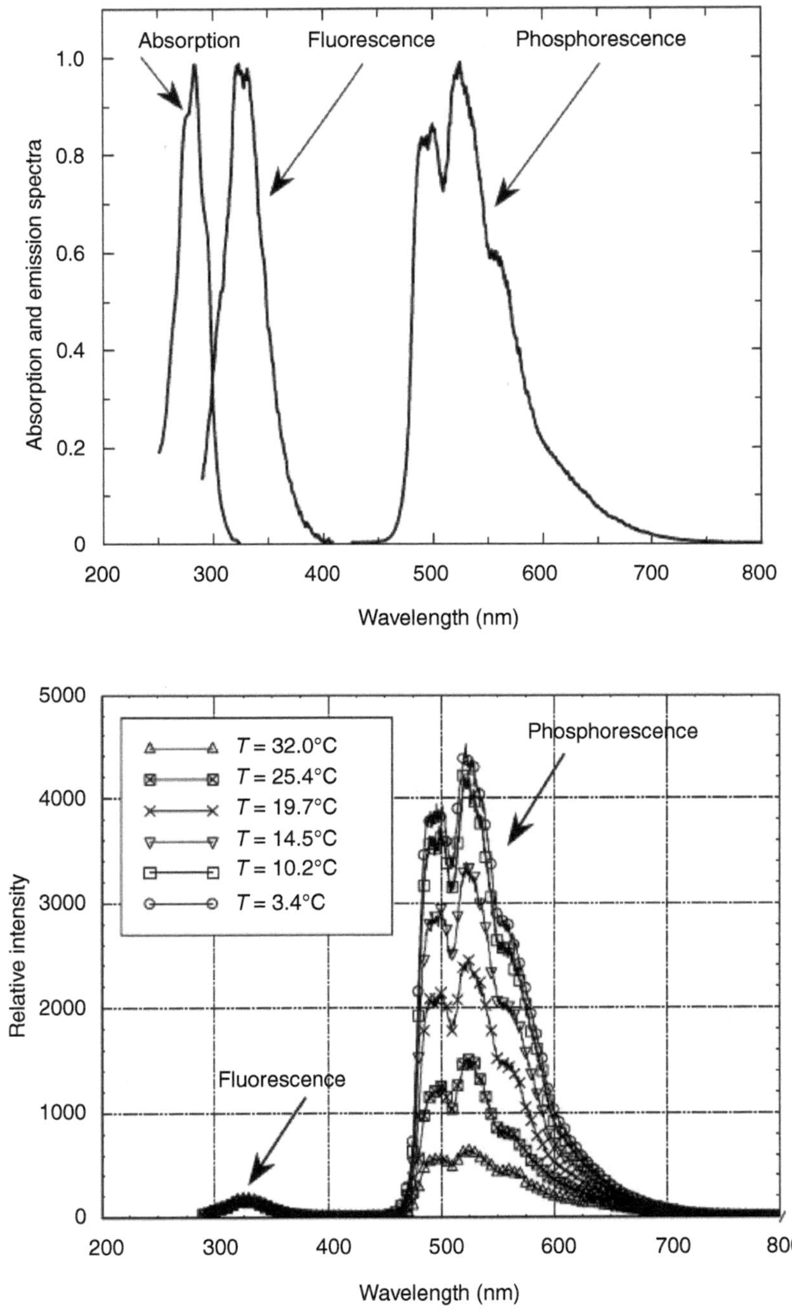

FIGURE 5.2 Absorption and emission spectra of the 1-BrNp_Mβ-CD_ROH complex. (Top) Normalized absorption and emission spectra at room temperature. (Bottom) Emission spectra at different temperatures excited at 280 nm. *(Source: Ref. [65].)*

featuring a rather high vapour pressure, such as acetone, 3-pentanone, triethylamine or toluene are frequently applied LIF tracers for temperature measurements in the gas phase. Alternatively, molecules which are naturally present in the system, such as NO molecules or OH radicals in combustion systems, can act as tracers for LIF thermometry. Common dye molecules are no choice because of their by far too small vapour pressure and their decomposition at high temperatures. From Section 2.4.1.6 we know already that – mainly due to their very different lifetimes in the excited state – the fluorescence of ketones (acetone and 3-pentanone) is rather insensitive to a change of the chemical composition of its environment, whereas the fluorescence of aromatic molecules (toluene) is sensitively influenced by its chemical environment. Therefore the thermometry strategies for the gas phase using acetone and 3-pentanone or aromatic molecules as temperature indicating LIF tracers are described in Sections 5.1.3.1 and 5.1.3.2. A comprehensive review on the photophysics of different LIF tracers is provided by Schulz and Sick [66], also including acetone, 3-pantone and toluene.

5.1.3.1 Ketone LIF Thermometry (Acetone and 3-Pentanone)

Due to extensive studies of several research groups, the photo-physical behaviour of acetone and 3-pentanone is well described for a wide range of temperatures and pressures [67–79]. It is understood to such an extent that it has been possible to reflect the photo-physical behaviours of acetone and 3-pentanone in semi-physical models [72,75]. These models comprise the different inter- and intramolecular processes discussed in the context of Fig. 2.44. In the following I will describe only the suitability of acetone as temperature indicating ketone LIF tracer for the gas phase at elevated pressure. For more detailed information on 3-pentanone I refer to the refs. [66,79,80].

Thurber et al. [72] quantified for ambient pressure conditions the influence of temperature onto the absorption cross-section of acetone vapour in a nitrogen bath-gas as it is shown in Fig. 5.3. The absorption cross-section at 248 nm remains nearly constant in the analysed range of temperatures, but with increasing wavelength the absorption cross-sections increase with temperature.

After the absorption cross-sections Thurber et al. [72] also quantified the LIF signal intensities per acetone molecule for different excitation wavelength and also at ambient pressure conditions, as it is shown in the left part of Fig. 5.4. Depending on the excitation wavelength the temperature sensitivity of the LIF signal per acetone molecule is more or less pronounced and can be utilised for LIF thermometry in gas systems whenever the acetone concentration in nitrogen (bath gas) is known. For the case that the acetone concentration is not known, Thurber et al. [72] recommend the consideration of ratios of acetone LIF intensities excited consecutively at two different wavelengths, as it is shown in the right part of Fig. 5.4.

In a subsequent study Thurber and Hanson [73] investigated the influence of pressure and gas composition at ambient temperature conditions and then,

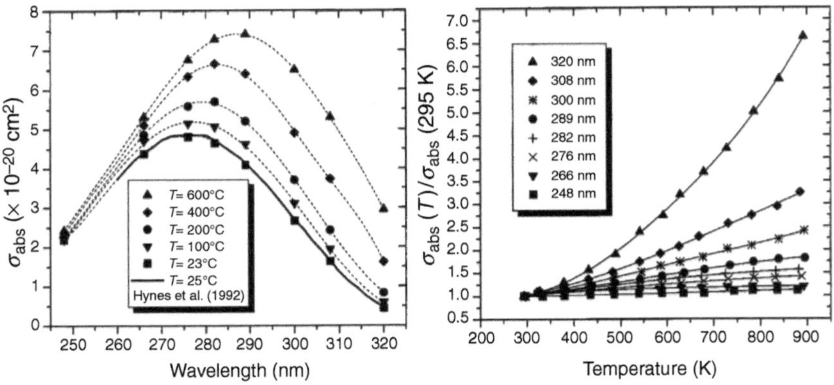

FIGURE 5.3 (Left) Absorption spectra of acetone for different temperatures. (Right) The normalized absorption cross-section of acetone shows a steady rise with temperature, with the effect increasing with wavelength. *(Source: Ref. [72].)*

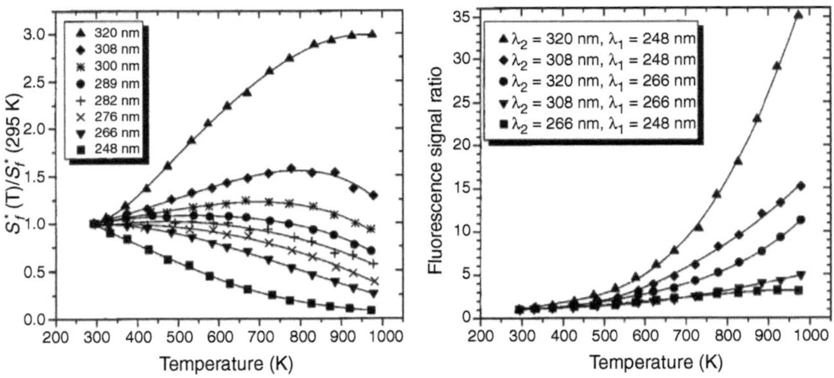

FIGURE 5.4 (Left) Fluorescence per molecule per unit laser fluence at atmospheric pressure, normalised to the room-temperature value for each wavelength and plotted as a function of temperature for all eight excitation wavelengths. For this figure and subsequent figures in this section, symbols represent experimental data points and lines are fits to the data. Error is estimated to be ±2% at low temperatures and from ±3 to ±4% at higher temperatures. (Right) Fluorescence signal ratios ($P=1$ atm) produced by five different wavelength combinations, plotted as a function of temperature. Each data point represents the ratio of the longer-wavelength fluorescence signal at a given temperature to the shorter-wavelength signal at the same temperature. *(Source: Ref. [72].)*

considering the previous study in combination with the latter one, summarised their findings in a semi-physical LIF model for acetone vapour, covering engine related pressure and temperature fields. Other researchers also investigated into the influence of pressure and temperature onto the LIF signal intensity of acetone [68,70,71,81].

Löffler et al. [76,78] calibrated the LIF signal intensity simultaneously as a function of temperature and pressure and additionally as a function of the

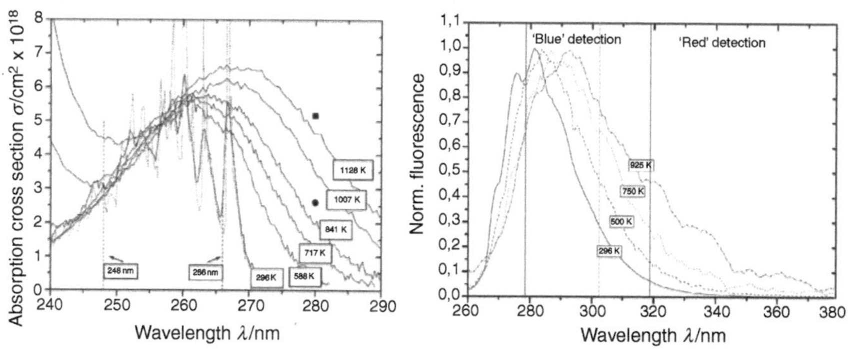

FIGURE 5.5 (Left) Solid lines, absorption spectra from a shock tube experiment; dotted line, room temperature spectrum [82]; ● 850 K, ■ 1200 K [83]; vertical lines illustrate the laser wavelengths used in the flow-cell experiments (Source: Ref. [84]). (Right) Red shift of the emissions spectrum with increasing temperature (Source: Ref. [85]).

bath-gas composition in order to being able to simultaneously measure the gas composition, and the gas temperature in internal combustion engines with exhaust gas re-circulation.

It should be made clear that for the thermometry concept yielding the results as shown in the right part of Fig. 5.4, two lasers with different emission wavelengths are required. The fluorescence spectrum of acetone vapour is not affected by the excitation wavelength. Thus the two LIF signals have to be excited one after the other and detected separately. If temperature fields have to be resolved temporally in non-stationary flows, the two signals also have to be generated and detected shortly one after the other. The time delay between the two LIF excitations and detections must be as short as effectively the flow does not move from one to the subsequent excitation. Therefore either two separate cameras are needed, which are each synchronised to one excitation laser, or one camera with double shutter option is required, where the second image can be acquired during the read-out event of the previously acquired image. The features of different CCD detector architectures are described in Section 3.5.2.

5.1.3.2 Toluene-LIF Thermometry

The left part of Figure 5.5 shows the temperature dependency of the absorption spectrum of toluene. The fine structure of the absorption spectrum gets lost with increasing temperature. Furthermore the absorption cross-sections increase especially in the red tail of the spectrum with increasing temperature. The dominant absorption in the blue tail of the spectrum observable for the two highest temperatures of 1128 K and 1007 K is due to the excitation of the second electronic excited state. Fluorescence spectra of toluene normalised to their peak maximum are shown for different temperatures in the right part of Fig. 5.5. Due to the normalisation of the fluorescence spectra to their maximum it cannot

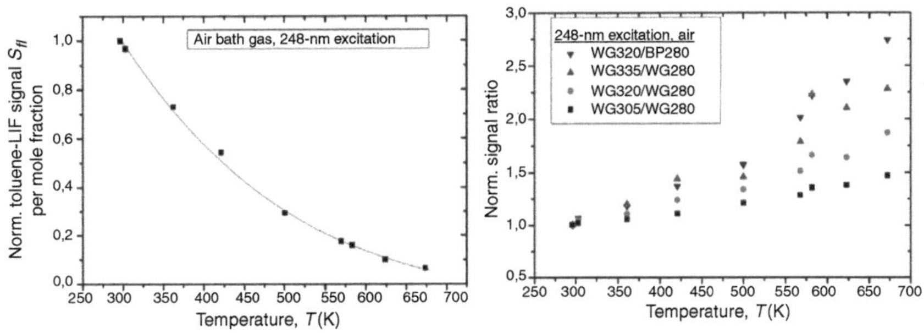

FIGURE 5.6 (Left) Calibration curves for the determination of temperature from the toluene-LIF signal in a homogeneously seeded air flow. (Right) Plots of the fluorescence signal ratio as a function of temperature for various filter combinations in air. *(Source: Ref. [85].)*

be seen that the entire fluorescence intensity drops with increasing temperature, as it is shown for toluene dissolved in an air bath gas in the left part of Fig. 5.6. Referring back to the right part of Fig. 5.5, we can see that with increasing temperature the fluorescence intensity drops in the bluish part of the spectrum while it rises in the reddish part of the fluorescence spectrum. On that effect the right part in Fig. 5.6 shows the temperature dependency of the intensity ratio of the reddish to the bluish spectral parts of the fluorescence spectra of toluene in air bath-gas. The different curves represent different filter combinations utilised for the detection of the bluish and the reddish spectral parts.

Of course the considerations of intensity ratios (right part in Fig. 5.6) instead of absolute intensities (left part in Fig. 5.6) makes the experiment independent of shot-to-shot fluctuations of the laser and laser attenuation due to absorption.

So far it is not considered that toluene belongs to the tracer class of aromatic tracers and therefore its fluorescence is quenched in the presence of oxygen. Figure 5.7 shows fluorescence spectra of toluene in bath-gases containing different concentrations of the quencher oxygen.

A comparison of the spectra on the basis of absolute values was not reasonable, as already small amounts of oxygen quench the fluorescence so heavily that the fluorescence spectra in oxygen containing bath-gases would more or less appear as flat horizontal lines. Therefore, spectra normalised to their peak intensity are compared in Fig. 5.7. While there are significant differences in the shapes of the fluorescence spectra corresponding to 0 mbar and 130 mbar oxygen, the differences in the shape of the spectra corresponding to 130 mbar and 300 mbar are marginal. This suggests that the shape of the fluorescence spectrum of toluene in the presence of more than 200 mbar oxygen is mainly affected by temperature and not by the quencher [87]. Toluene-LIF temperature measurements in internal combustion engines had been based on this assumption [88]. A combined study of the influence of temperature and pressure in nitrogen and in air followed [89].

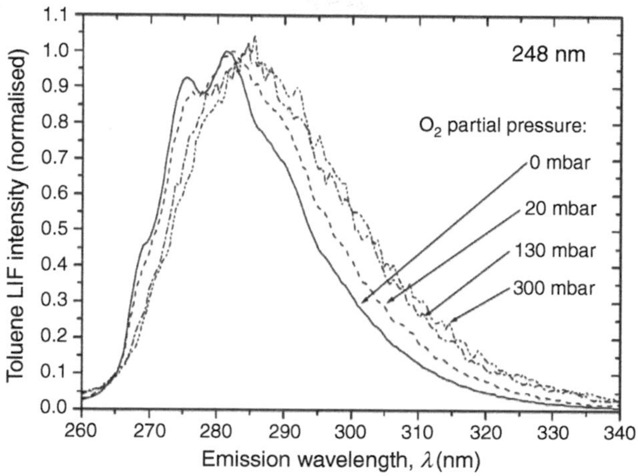

FIGURE 5.7 Toluene-LIF spectra with 248-nm excitation at room temperature for different oxygen partial pressures. Spectra are corrected for instrumental spectral response. *(Source: Ref. [86].)*

5.1.3.3 Thermometry Using LIF of Small Molecules (NO and OH)

In this section the NO-LIF thermometry concept is described briefly. With respect to the similar OH-LIF thermometry concept mainly references are provided for interested readers. All the LIF and LIP-based thermometers described previously have in common that the fluorescence spectrum or its intensity emitted after the excitation at one wavelength is analysed. In the case of NO-thermometry this is different. For the measurement of temperature here a tunable and narrow-band excitation laser is scanned stepwise in small increments over a certain range of the absorption spectrum of NO. At each scanned excitation wavelength the intensity of the fluorescence emission is recorded, which results in an excitation scan. Excitation scans have already been mentioned in the context of Fig. 2.47.

Figure 5.8 shows in the left part a simulated excitation scan for NO and O_2 at 2000 K and for 1 bar and 20 bar. A spectral region can be identified between 225.944 nm and 226.112 nm in which the excitation san of NO is not interfered by oxygen fluorescence. This is of special importance for combustion environments.

With increasing pressure the fine structure of the excitation scan spectrum gets lost. Nevertheless the excitation scan spectrum remains a function of temperature. Therefore Lee et al. [90] compared, as it is shown in the right part of Fig. 5.8, experimentally measured excitation scan spectra of NO with simulated ones in order to identify the temperature of the experimentally measured excitation scan spectrum. The experimental spectra were measured in NO-containing regions of a stabilised flame. In this context I want to mention that LIFSim accounts for inter and intra-molecular processes affecting the excitation scan spectrum. As the excitation wavelength has to be scanned over a certain range of

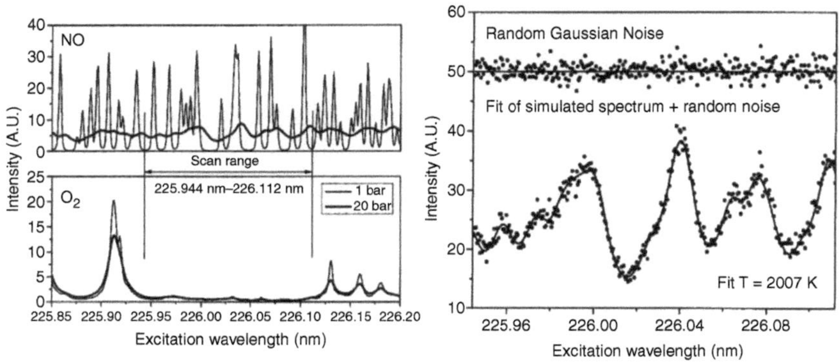

FIGURE 5.8 (Left) Numerical simulation (LIFSim) of NO and O_2 LIF excitation spectra in the range of 225.8–226.2 nm at 2000 K. The scan range (225.944–226.112 nm) is located in the region of minimum O_2 LIF interference. As spectra overlap caused by collisional broadening at 20 bars can be observed in the NO spectra, whereas O_2 is less affected because of predissociation. The intensities of NO and O_2 are on the same scale. (Right) Example of sensitivity analysis by using simulated spectra with added noise (Gaussian) and fitted to extract temperature at 20 bars (225.944–226.112 nm region). Random Gaussian noise is added to simulated spectra for temperature fitting to simulate experimental conditions. *(Source: Ref. [90].)*

the absorption spectrum of NO, this technique is applicable to stationary flows or to strictly repetitive flows. This kind of NO-LIF thermometry has even been applied in order to measure the gas phase temperature in sprays [91,92]. For these studies NO had to be seeded as temperature indicating tracer to the gas into which the spray was injected.

With respect to the thermometry of the OH radical I refer interested readers to refs [93–97].

5.1.4 Laser-Induced Phosphorescence (LIP) for Vapour Phase Thermometry

What has been described in Section 5.1.2 on LIP for liquid phase thermometry can be transferred to the vapour phase thermometry, as long as phosphorescence particles are used as temperature indicating tracers. In gas phase systems temperatures exceeding those of liquid phase systems can be realised, wherefore a variety of phosphors exist. Again, different phosphors feature different temperature sensitivities over different temperature intervals. Thus a suitable phosphor has to be chosen carefully.

In Fig. 5.9 Alden summarises which phosphors are applicable in which temperature range, depending on whether the temperature is derived from lifetime studies of the phosphorescence or from the intensity ratio of the phosphorescence (see Section 5.1.2).

An example calibration of the temperature sensitivity of the phosphorescence spectrum of the compound BAM:Eu after excitation at 355 nm is shown in the

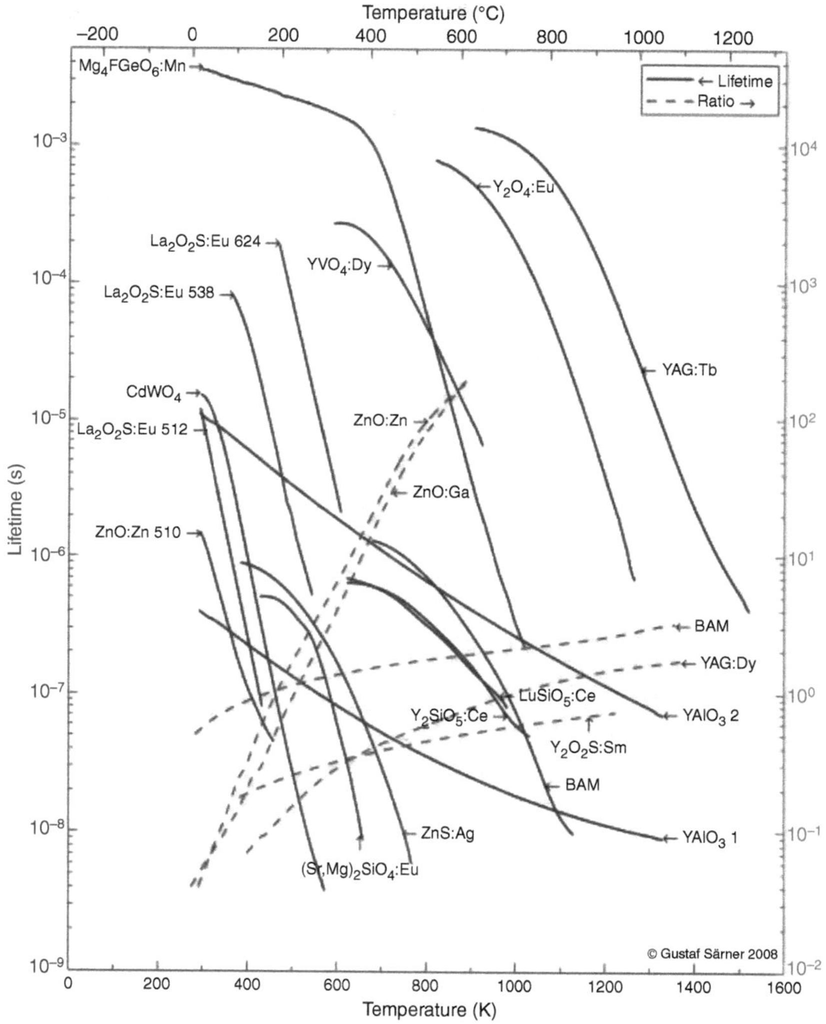

FIGURE 5.9 **Temperature sensitivity for some phosphors.** Ratio calibrations are dashed and can be read on the **y**-axis at the right *(Source: Ref. [53].)*

left part of Fig. 5.10. When considering the intensity ratios of the two spectral filter regions (filter A and filter B) also indicated in the left part of Fig. 5.10, the temperature sensitivity shown in the right part of Fig. 5.10 results [98].

Brübach et al. [52] analysed the influence of the bath-gas composition and pressure (up to 3 MPa) on the phosphorescence behaviour. They found that the temperature-sensitive behaviour of some phosphors is neither affected by the bath-gas composition nor by pressure. Thus, after a calibration at ambient pressure in air they can be applied as universal temperature sensors in other

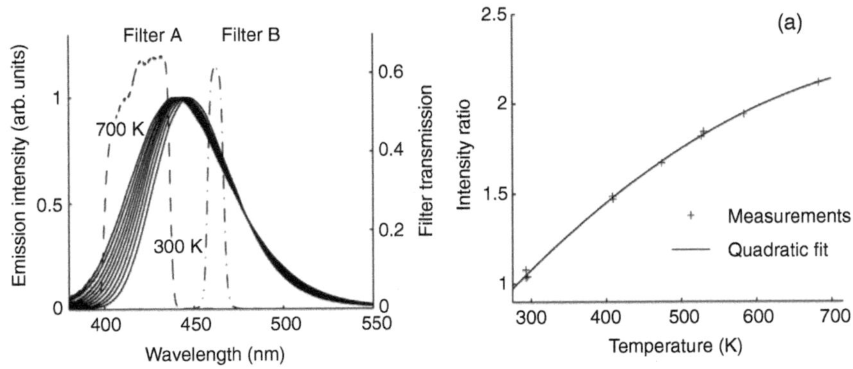

FIGURE 5.10 (Left) BAM:Eu emission spectrum following 355 nm excitation, normalised to the emission band peak and recorded at 50 K intervals (1 nm spectral resolution). The measured transmission curves of the filters used in this study are superimposed on the spectrum. (Right) Intensity ratio response measured in the gas phase. *(Source: Ref. [98].)*

gases and at other pressures, at least in the analysed pressure range. But for the phosphor Y_2O_3:Eu they found a strong oxygen-pressure dependency and even irreversible effects [52]. The irreversible behaviour of the phosphorescence of phosphors once exposed to high temperatures in air and therefore oxidised, today is exploited in order to provide thermal history sensors for harsh environments [99].

5.2 LASER-INDUCED EXCITED-COMPLEX FLUORESCENCE (LIEF)

The laser-induced excited-complex fluorescence (LIEF) is a variant of LIF using two tracer compounds which in the excited state form an excited-complex. The excited-complex is also referred to as exciplex, for which reason LIEF is also called laser-induced exciplex fluorescence.

With respect to engine-combustion-research at elevated pressure, Melton [100] first exploited LIEF for the spectral separation of the LIF signal originating from the vapour phase of the already evaporated fuel and the LIEF-signal originating solely from the liquid phase of the injected fuel.

The energy scheme shown in Fig. 5.11 illustrates the working principle of the LIEF technique, when using naphthalene (N) and dimethylaniline (DMA) as exciplex forming monomers. Naphthalene as well as dimethylaniline can be excited at the same wavelength in the ultraviolet spectral region. After the excitation of the monomers, the monomers can fluoresce back to the ground state. In the case of the excited naphthalene monomer the fluorescence spectra would peak at 340 nm. The fluorescence of the excited dimethylaniline-monomer would peak at 335 nm. If one excited naphthalene-monomer forms an exciplex with one non-excited dimethylaniline monomer, or, if the other way round one excited dimethylaniline monomer forms an exciplex with one non-excited

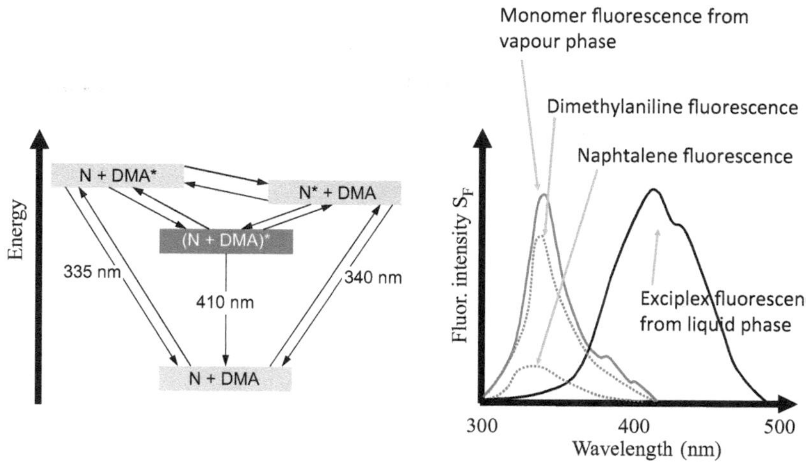

FIGURE 5.11 (Left) Schematic summary of photophysics naphthalene/dimethylaniline. (Right) Sketch of the fluorescence spectra of the monomers, originating from the vapour phase of the solution where no exciplex is formed, and of the exciplex originating from the liquid phase of the solution. *(Source: Graphs provided on courtesy of Andreas Paul Fröba from the Erlangen Graduate School in Advanced Optical Technologies of the Friedrich-Alexander-Universität Erlangen-Nürnberg, Germany.)*

naphthalene-monomer, the energy of the excited complex is less than that one of the excited monomer and thus the fluorescence of the exciplex is red-shifted with respect to the fluorescence of the monomers. In this case the fluorescence spectrum of the exciplex would peak at 410nm.

The probability of the formation of an exciplex very much depends on the average distance between the monomers forming the exciplex. As the intermolecular distance in the liquid phase is much shorter than in the vapour phase, the probability of exciplex formation in the liquid phase is high and in the vapour phase is low. If the concentration of the two monomers naphthalene and dimethylaniline within the non-fluorescing model fuel are chosen properly, the formation of exciplexes in the vapour phase can be excluded completely, while they still can form in the liquid phase. This makes possible the spectral separation of the LIEF signal originating from the liquid phase only, and the LIF signal originating from the vapour and the liquid phase. LIEF has been applied frequently in order to analyse the evaporation of fuel sprays at elevated pressure [101–107].

Melton used as exciplex formers the tracers naphthalene and tetramethyl-*p*-phenylene diamine [100]. Fröba et al. [108] suggested as LIEF tracers trimethylamine and benzene. Kim et al. [109] proposed trimethylnaphthalene and dimethylaniline. For the couple fluorobenzene and diethyl-methyl-amine the co-evaporation with the model fuel *n*-hexane has been proven [110,111].

All these afore mentioned LIEF systems suffer from fluorescence quenching in the presence of oxygen as a quencher. When the LIEF-tracer doped fuel is injected into air, the vapour phase LIF signal of the monomers is quenched

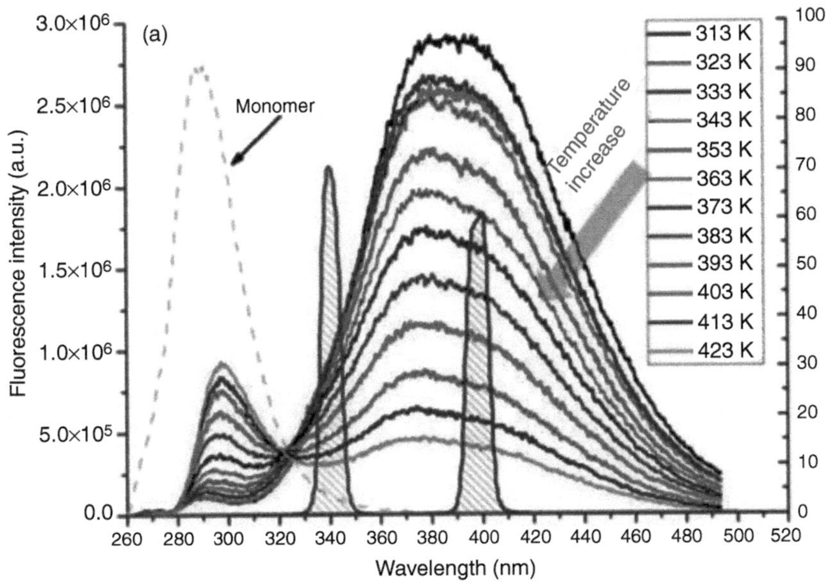

FIGURE 5.12 The exciplex fluorescence spectra at various temperatures (2% FB, 9% DEMA, 89% *n*-hexane using 266-nm excitation). The hatched red and blue area indicates the transmittance of filters. *(Source: Ref. [112].)*

significantly, which makes a quantitative interpretation of the vapour phase signals challenging. A qualitative visualisation of the vapour phase regions of the spray is still possible. But not only the fluorescence of the vapour phase is quenched while it mixes with air, also the exciplex fluorescence of the liquid phase can be quenched. At elevated pressures the absorption of the gas surrounding liquid droplets into these droplets is rather fast. Therefore the oxygen concentration inside the liquid phase also can influence the shape as well as the intensity of the exciplex fluorescence spectrum. This also makes the quantitative interpretation of liquid phase LIEF signals challenging, for which quantitative studies are mainly carried out in nitrogen environment.

It has been mentioned already that the LIEF tracer couple fluorobenzene (FB) and diethyl-methyl-amine (DEMA) co-evaporates together with the model fuel *n*-hexane. Thus it can be assumed that the chemical composition of an evaporating liquid *n*-hexane drop doped with this tracer couple remains constant during the evaporation process. Zhang et al. [112] calibrated the temperature sensitivity of the exciplex fluorescence of this tracer couple in the liquid phase of *n*-hexane, as it is shown in Fig. 5.12. During the calibration experiments the liquid and with tracer doped *n*-hexane was kept in a nitrogen environment in order to exclude any quenching from oxygen. With increasing temperature the exciplex fluorescence drops compared to the monomer fluorescence, which itself rises with increasing temperature. Therefore the ratio of the fluorescence signals of the monomer, which increases with increasing temperature, and the excimer, which decreases with increasing temperature, would provide a highly sensitive thermometry technique. But, unfortunately, this is not an option for spraying processes. In spraying processes the liquid droplets are surrounded by the vapour phase whose fluorescence signal is dominated by monomer fluorescence. Thus in practical applications the monomer fluorescence never originates solely from the liquid phase, but also from the vapour phase, or in other words the monomer fluorescence signal of the liquid phase is interfered with the monomer fluorescence signal from the vapour phase, or the other way round. This is the reason why the thermometry method for temperature measurements of the liquid phase has to rely on signals whose origin can be assigned to clearly the liquid phase.

This is why Zhang et al. [112] considered for their temperature measuring method the two spectral regions indicated in Fig. 5.12 as two narrow peaks centred at 340 and 400 nm. The temperature dependency of the exciplex fluorescence in both spectral regions is different so that also the ratio of the fluorescence intensities recorded from these respective regions is a function of temperature. Zhang et al. [112] also found that in a nitrogen environment up to 2.2 MPa, pressure does not influence their LIEF thermometry method.

5.3 CASE STUDIES

In the sections that follow four case studies are described. Three of them have a direct relation to supercritical fluids. LIF and LIP applications in supercritical environments are not typical and the reason why I decided to pick them here as case studies. On the contrary, third case study is in fact related to high pressure but has no relation to supercritical fluids. Nevertheless I decided to discuss this study here as it according to my opinion nicely illustrates how and that LIF, LIP and LIEF measurement techniques are affected by their environment.

5.3.1 Drops on Demand at Near Critical Conditions

This case study describes research that has been carried out within the still ongoing Collaborative Research Centre Transregio 75 of the German Research Foundation (DFG). This research centre is on ‘Droplet dynamics under extreme ambient conditions’. Therefore some sub-projects also focus on the ‘mass transfer’ between drops and their environment at conditions relevant for example for diesel or rocket engines. In these combustion devices liquid or supercritical fuel is injected into an environment which itself is near-critical or supercritical. Though combustion and many other high-pressure mixing processes – for example for particle generation, extraction or drying – at first glance have nothing in common except the high pressure, the circumstances at which mass or heat is transferred is very similar. So these at first glance very different fields, have a common feature in that the mass- and heat transport processes which dominate the final product of the process all take place at near-critical or supercritical conditions.

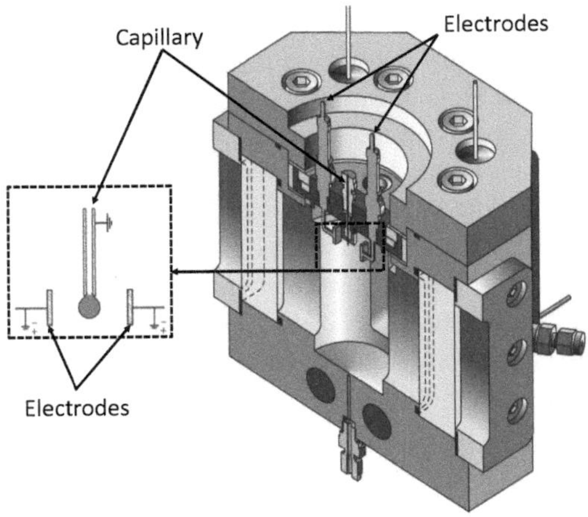

FIGURE 5.13 (Left) Schematic of the electric droplet detachment system. The steel capillary is grounded, the electrodes can be charged with up to 10kV, during time intervals ranging from milliseconds to several seconds. (Right) Experimental setup: a three-dimensional view of the interior of the drop on demand high-pressure chamber. *(Source: Figures provided on courtesy of the Institute of Aerospace Thermodynamics of the Universität Stuttgart, Germany, and of the Institute Reactive Flows and Diagnostics of the Technische Universität Darmstadt, Germany; Collaborative Research Centre Transregio 75 [113].)*

Within this Collaborative Research Centre Transregio 75 Oldenhof et al. [113] built up a drop on demand cell, which is optically accessible and can be operated at high pressures and high temperatures. The respective chamber is shown in Fig. 5.13 on the right side.

The drop-on-demand generator is mounted in the centre of the chamber's top and shown in a magnified view on the left side of Fig. 5.13. Through a capillary liquid can be fed into the chamber which accumulates as a drop at the bottom of the capillary. The electric field, generated when a voltage is applied between the grounded capillary and the side electrodes, triggers the detachment of the drop from the capillary. Oldenhof et al. [113] demonstrated the reproducibility of the drops and the drops' detachment from the capillary using a shadowgraph technique.

Bork et al. [114] has used the LIF and LIP optical set-up shown on the left side in Fig. 5.14 for the characterisation of the falling drops.

They used the 320nm excitation wavelength of a dye laser for the excitation of the acetone molecules. On the right side in Fig. 5.14 it is shown that the fourth harmonic wavelength emitted by a Nd:YAG laser (266nm) is absorbed too strongly by the acetone, while the dye laser excitation wavelength of 320nm provides a rather homogeneous illumination of the drop.

The LIF measurements brought some light into a debate which came up because of the shadowgram of the near-critical falling acetone drop shown on the left side in Fig. 5.15. Before I describe the debate in more detail I want to

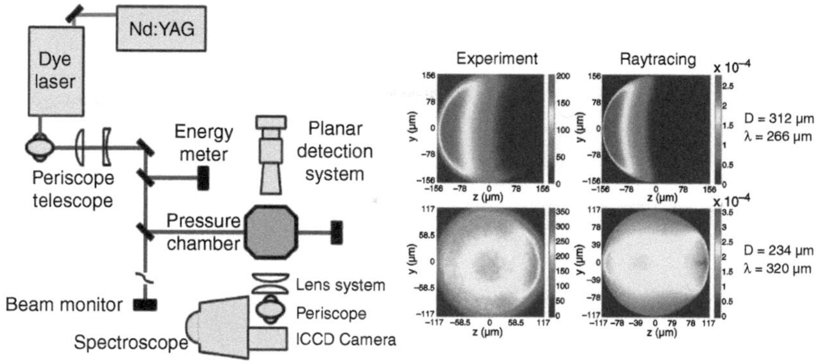

FIGURE 5.14 (Left) Sketch of the experimental setup for imaging and spectroscopic investigations into the fluorescence and the phosphorescence of acetone in nitrogen and air at high pressures and elevated temperatures. (Right) Computed and experimentally measured penetration of a laser beam into/through drops of different radii and using two different excitation wavelengths. *(Source: Figures provided on courtesy of the Institute Reactive Flows and Diagnostics of the Technische Universität Darmstadt, Germany; Drop light penetration images acquired by Benjamin Bork and Andreas Dreizler within the Collaborative Research Centre 75 project B2.)*

provide some details regarding the operational conditions of this experiment. The acetone drop is falling in a nitrogen environment. The temperature and the pressure inside the chamber are 483 K and 5.8 MPa, respectively. With respect to acetone the pressure exceeds and the temperature is below its critical parameters (4.7 MPa and 508 K). Nevertheless, the operational conditions are chosen near the critical parameters of acetone. So the debate was about whether the droplet's horn, which is visible in the shadowgram and labelled as horn, is liquid or gaseous. In Section 4.1.2 we saw that the shadowgraph technique detects the second derivative of the field of the index of refraction, which can take extreme values easily exceeding the first derivative. Thus the horn visible in the shadowgram could be either because of a liquid acetone ligament detaching from the drop or because of a gas phase schliere of already evaporated acetone. One cannot be excluded from the other solely based on the shadowgram.

But the fluorescence image shown in the right part of Fig. 5.15 was able to answer this question. The LIF image of the acetone drop shows a rather spherical drop. The LIF signal of the liquid drop is much more intense than that one of the acetone vapour in the wake of the falling drop, as the density of acetone in the liquid drop is much higher than in the vapour phase, where acetone is additionally diluted. If there were liquid acetone ligaments detaching from the falling drop, they would be visible in the LIF image, as they also would be as dense as the liquid acetone in the main body of the drop. As no liquid ligaments could be identified at these operational conditions in any LIF image, it became clear that the horn visible in the shadowgram was due to a gas phase schliere. This observation corroborates the findings by a previous study of Weckenmann et al. [115], which combined shadowgrams and diffuse light images

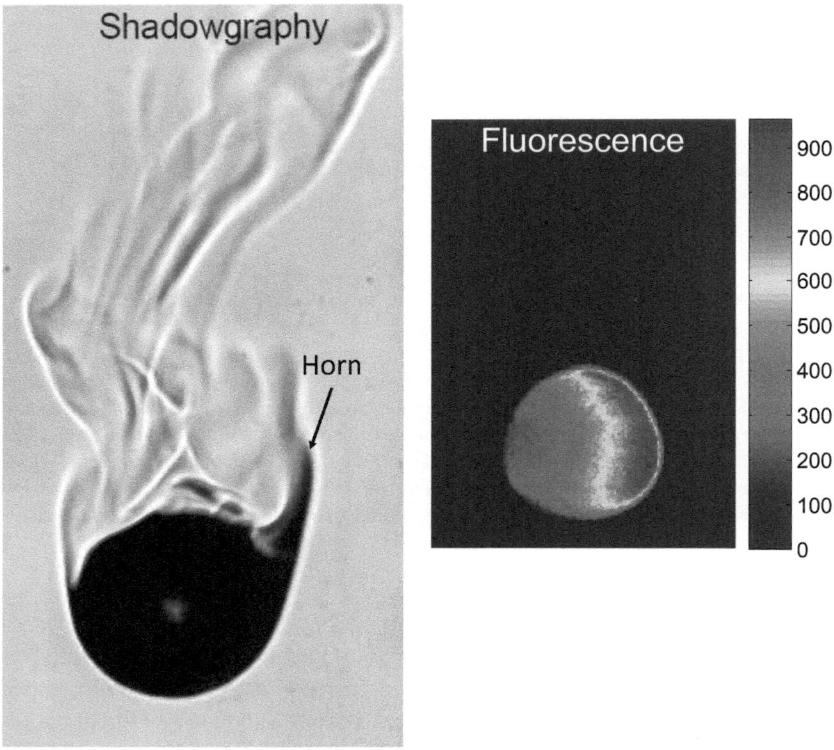

FIGURE 5.15 (Left) Shadowgram of an acetone drop falling through nitrogen at 5.8 MPa and 520 K. (Right) LIF image of acetone drop at similar conditions. *(Source: (Left) Figure provided on courtesy of the Institute of Aerospace Thermodynamics of the Universität Stuttgart, Germany, Shadowgram acquired by Florian Weckenmann and Grazia Lamanna within the Collaborative Research Centre Transregio 75 project B2. (Right) Figure provided on courtesy of the Institute Reactive Flows and Diagnostics of the Technische Universität Darmstadt, Germany; Drop images acquired by Benjamin Bork and Andreas Dreizler within the Collaborative Research Centre Transregio 75 project B2.)*

with a thermodynamic analysis. The inclusion of binary mixtures effects in the analysis of experimental data demonstrated that the appearance of horns detaching from the drop's surface cannot be interpreted as a transition to supercritical state, as commonly assumed in literature.

5.3.2 Jet Mixing in the Vicinity of the Critical Point

Roy et al. [116] carried out some LIF measurements in order to analyse the mixing behaviour between an injected jet and its environment at near and supercritical conditions. With their study they intended to complement the work I already described in Section 3.4.6, which cared mainly about the mixing of sub-critical

TABLE 5.1 Selected Test Cases [116].

Case	$T_{r,ch}$	$T_{r,inj}$	$P_{r,ch}$	$P_{r,inj}$	m (g s^{-1})	U_{inj} (m s^{-1})
1	0.69	1.13	1.38	1.51	19.31	20.98
2	0.72	1.21	1.86	1.97	18.33	17.95
3	0.76	1.29	1.88	1.98	17.24	20.31
4	0.80	1.31	1.38	1.51	16.99	30.00
5	1.04	1.00	1.26	1.34	17.82	7.07
6	1.06	1.08	1.37	1.47	16.78	14.56
7	1.08	1.08	1.41	1.50	17.35	14.47
8	1.09	1.18	1.47	1.67	18.64	20.57

jets with a supercritical environment. On the contrary Roy et al. [116] focus in their study on supercritical jets injected into sub-critical environment. They injected FK-5-1-12 [$CF_3CF_2C(O)CF(CF_3)_2$] – also known as fluoroketone – into nitrogen. Fluoroketone was chosen because of its mild critical parameters of 1.86 MPa and 441 K and its spectroscopic characteristics. It can be excited at a wavelength of 355 nm (third harmonic of a Nd:YAG laser) to fluoresce and its LIF signal is rather insensitive to temperature and pressure at the here relevant conditions [117]. They recorded the LIF signals using an ICCD-camera (see Section 3.5.2.6). The operational conditions of the experiments they report in [116] are summarised in Table 5.1 as reduced temperatures T_r and reduced pressures P_r for the chamber (index: ch) and the injected fluid before injection (index: inj). The reduced parameters are the ratio of the set pressure P or temperature T and the critical pressure P_{crit} or the critical temperature T_{crit} of the fluoroketone, not that of the nitrogen.

Cases 1–4 thus represent the injection of a supercritical fluid into a sub-critical environment and cases 5–8 represent the injection of a supercritical fluid into a supercritical environment, referring to the critical parameters of the fluoroketone.

Example density fields of the fluoroketone, which were derived from the LIF images, are shown for the cases 1–4 in Fig. 5.16. In the magnified view of the mixing zone between the fluoroketone and the nitrogen, droplets can be clearly identified. Thus, though the fluoroketone is injected as a supercritical fluid, the mass transfer is dominated by the evaporation of liquid fluoroketone into a gaseous atmosphere.

Example density fields for the injection of supercritical fluoroketone into supercritical environment according to the cases 5–8 are shown in Fig. 5.17. The magnified sections clearly show that no droplets are formed. Therefore the mass is transferred like in a single-phase mixing process. Mass has not to be transferred across interfaces.

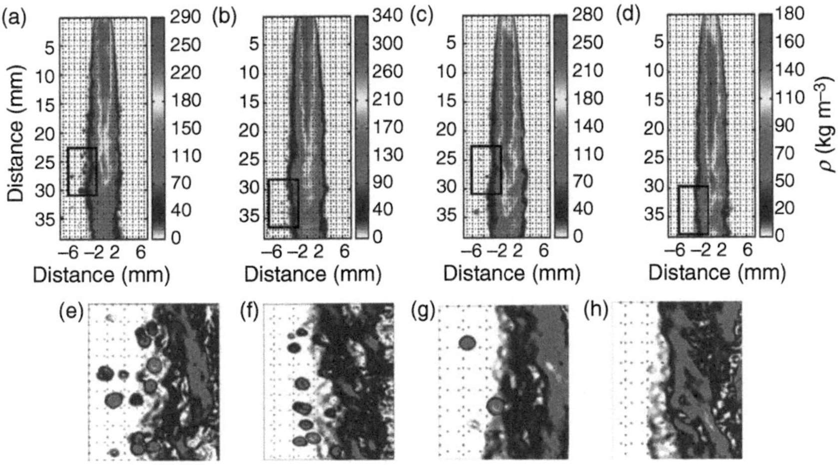

FIGURE 5.16 Scaled images of supercritical jets injected into subcritical (with respect to the fluoroketone) chamber conditions. Test conditions correspond to cases 1–4 in Table 5.1: (a–d) density images; (e–h) zoomed-in density gradient images. *(Source: Ref. [116].)*

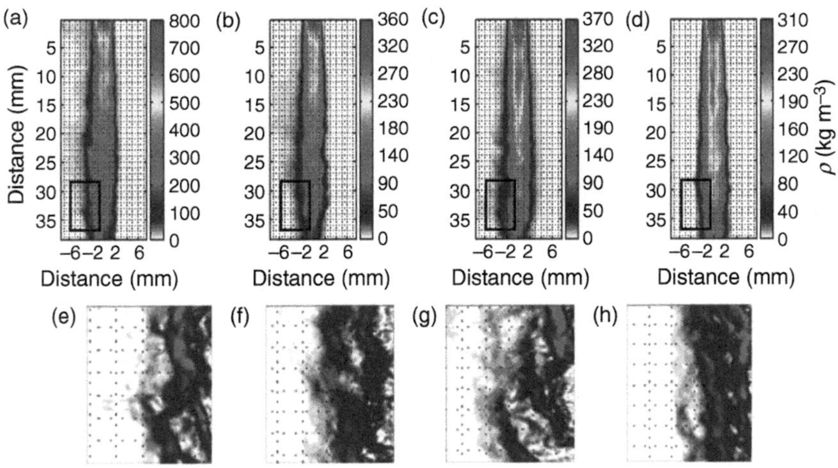

FIGURE 5.17 Scaled images of a supercritical jet injected into supercritical (with respect to the fluoroketone) chamber conditions. Test conditions correspond to cases 5–8 in Table 5.1: (a–d) density images; (e–h) zoomed-in density gradient images. *(Source: Ref. [116].)*

It should be mentioned here that neither the critical parameters of the injected fluid nor that one of the fluid in the environment alone determine whether a mixing process takes place in the single-phase mixing region without the formation of droplets or in the two-phase mixing regime with formation of droplets and mass transfer across interfaces. Only the critical point of the mixture (not that ones of the pure compounds) can be considered as reference point in order to distinguish between single-phase and two-phase mixing [118].

5.3.3 Comprehensive Multi-Parameter Measurements in Transparent Internal Combustion Engines

This case study is supposed to show that a comprehensive multi-parameter study of a high-pressure mixing process – here in an internal combustion engine – by means of fluorescence-based techniques demands a detailed knowledge of the photo-physical behaviour of the fluorescing compounds at various conditions. Within the respective LIEF case study of Wieske et al. [102] they reported very thoroughly about which measures they took in order to quantify the detected signals.

For their LIEF study they used the LIEF-tracer couple fluorobenzene (FB) and diethyl-methyl-amine (DEMA) in the model fuel *n*-hexane. It already is mentioned in Section 5.2 that for this system the co-evaporation of the tracers and the fuel has been proven [110,111]. Methyl-*tert*-butylether (MTBE) was added in order to assure a knocking free engine operation. The LIEF process was excited with a frequency quadrupled Nd:YAG laser at 266 nm. Figure 5.18 shows how the laser light sheet was positioned inside the internal combustion chamber of the optically accessible engine. The injected spray, the spark plug position and the inlet and outlet vales are indicated or labelled. Wieske et al. [102] positioned a 4-way optical mirror system in front of the camera lens in order to image simultaneously four separate images of the same object on the image intensifier of the ICCD-camera and to detect them on the CCD chip of the camera. The working principle of an ICCD-camera is described in Section 3.5.2.6.

In Section 5.2 the temperature sensitivity of the exciplex couple FB and DEMA in the liquid phase is discussed in the context of Fig. 5.12. With increasing temperature the reddish part of the exciplex fluorescence decreases while the bluish part increases. The bluish part of the exciplex fluorescence represents the fluorescence of the monomers in the liquid phase. The entire LIEF concept is based on probing the monomer fluorescence from the vapour and the liquid fuel fraction and the exciplex fluorescence solely from the liquid fuel fraction. For qualitative analysis this concept works. But Wieske et al. [102] aimed at measuring the quantity of the vapour fuel fraction. So they had to correct for the contribution of the monomer signal coming from the liquid phase in order to get the monomer signal solely coming from the vapour fuel fraction. This interference of the liquid phase signal to the vapour phase signal is in literature often referred to as 'cross-talk'. On this account Wieske et al. [102] occupied two of their optical detection channels for the detection of LIEF-signals in the spectral ranges 335-345 nm and 435-445 nm. From the ratio of the intensities detected in these spectral regions and based on a beforehand carried out calibration they were able to determine the temperature of the liquid

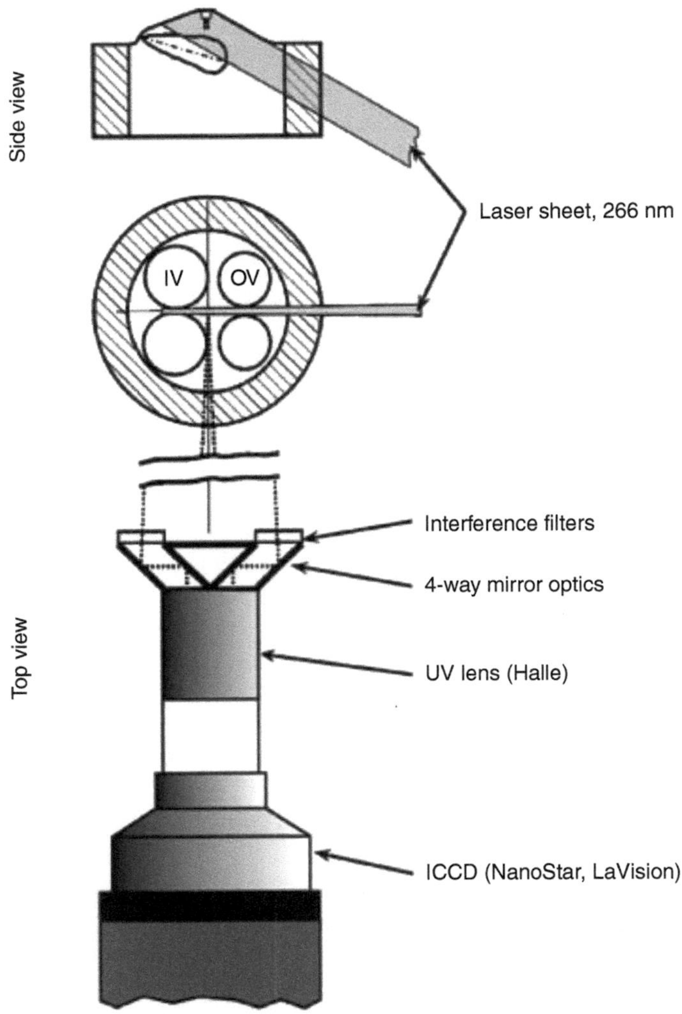

FIGURE 5.18 Optical setup of multi-parameter imaging experiment in an internal combustion engine. *(Source: Ref. [102].)*

fuel fraction. In general this temperature evaluation strategy is identical to the one described in the context of Fig. 5.12, except that other spectral regions were considered. Once the temperature is known, the temperature information can be utilised in combination with the exciplex signal detected between 435 nm and 445 nm in order to compute the monomer signal from the liquid phase at 295 nm (cross-talk), which is the wavelength range at which the monomer fluorescence signal coming from the vapour fuel fraction is detected. Thus the third optical channel of the detection system is occupied for the detection of the monomer signal at 290 nm. After cross-talk correction and assuming a homogeneous quenching everywhere inside the combustion chamber, the monomer signal reveals the fuel/air-ratio, from which the lambda-value is computed. The fourth channel of the detection system is occupied for the detection of elastically scattered light. This channel makes visible the distribution of the liquid phase inside the light sheet. Figure 5.19 summarises the parameter fields obtainable simultaneously with this measurement technique.

5.3.4 Brief Comparison Between Elastic Light Scattering and LIF for Transcritical Jet Mixing Analysis

I chose the final case study in order to once more show that each optical measurement technique can measure something, but that the measurement results have to be evaluated carefully in order to get new and real insights into processes. Wensing et al. [118] demonstrated this clearly by comparing images made from the same spray/jet but with different measurement techniques.

Sprays generated by injectors are commonly analysed with elastic light scattering techniques. In this case the spray is illuminated integrally with for example a flash lamp. In case of pulsed fuel sprays the penetration of the spray into the combustion chamber can be analysed as a function of the time after the visible start of injection (vSOI). It can be checked for example whether or not all spray streams injected by a multi-hole injector are identical, whether or not the injected spray impinges on the wall of the combustion chamber or on the engine's piston before it has evaporated and starts to burn and whether the injection starts and stops abruptly or there might be bouncing of the opening and closing injector valve.

Elastic light scattering from droplets is described in Section 2.3.4 of this book. From the previous case studies we saw that droplets only form during mixing of the injected fuel and the environment, if a surface tension exists between a liquid and a vapour phase. At diesel-combustion-related conditions the critical point of the mixture of the diesel fuel and the air can be below the operation pressure and temperature. Thus, even if the fuel is injected at moderate temperature as a liquid into the combustion chamber filled with hot and compressed air, it heats up while it penetrates through the hot environment so that the mixing process undergoes a transition from a two-phase mixing process near the nozzle exit to a single-phase mixing process farther away from the nozzle exit. This transition from a two-phase to a single-phase mixing process is accompanied by the disappearance of the interfacial tension, which only can exist as long as there are a liquid and a gaseous phase in a two-phase mixing process. Consequently also the phase boundaries, from which the incident light is scattered elastically, disappear.

In Fig. 5.20 this scenario is shown and its implications for elastic light scattering and LIF methods become obvious.

The spray images shown in the left and the right part of Fig. 5.20 correspond to the same operational conditions. While in the left part the

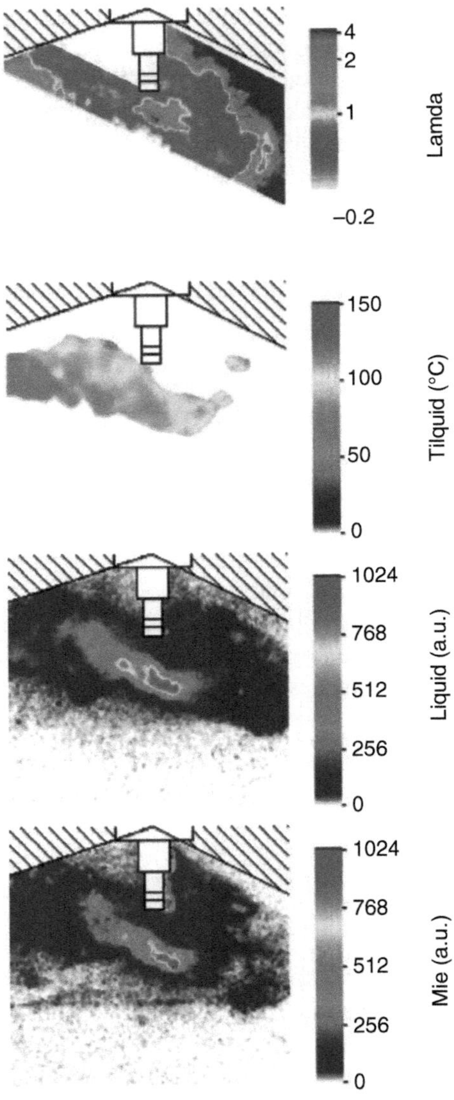

FIGURE 5.19 Simultaneous measurement of: λ-(vapour), liquid temperature and liquid fuel distribution (exciplex fluorescence at 340 nm), and Mie scattering (240–270 nm), 50° BTDC. *(Source: Ref. [102].)*

entire spray is represented in the bottom view based on an elastic light scattering method, only one stream of the spray is represented in the right part, which has been acquired using the LIF technique. In this case Wensing et al. [118] did not use a non fluorescing model fuel but instead used the standard fuel which can be excited directly to fluoresce with the third harmonic of a Nd:YAG laser. Suitable LIF tracers for diesel spray analysis also with a

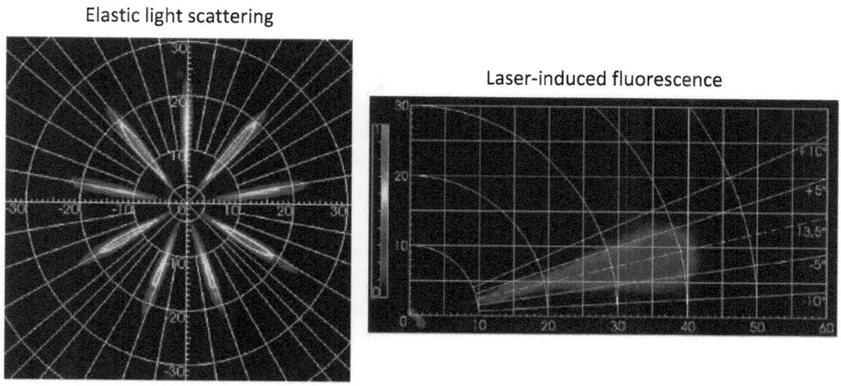

FIGURE 5.20 (Left) Spray image of a whole fuel spray. (Right) LIF image of a fuel cone 1000 ms after vSOI. $P_{chamber}$, 7 MPa; $T_{chamber}$, 900 K; p_{fuel}, 140 MPa; T_{fuel}, 298 K; $m_{injected}$, 50 mg. *(Source: Ref. [118].)*

prospect of measuring temperature are reported in ref. [119]. The operational conditions are chosen in favour of achieving a transition from a two-phase to a single-phase mixing regime. 1000 ms after visible start of injection (vSOI) the elastic light scattering image shows a penetration of the spray streams of approximately 20 mm into the chamber. On the contrary, at the same time the LIF image shows a penetration of more than 40 mm into the environment. Thus it becomes obvious that at these operational conditions elastic light scattering method cannot provide correct information about the mixing performances of injectors. Elastic light scattering methods can only indicate the regions in which the mixing process is still a two-phase one. In contrast, the LIF technique can visualise the penetration behaviour also in single-phase mixing regimes further downstream of the injector after the transition from the two-phase to the single-phase mixing process.

On this effect elastic light scattering is also compared with the shadowgraph technique in ref. [118], but not further described here. The shadowgraph technique can make visible the real penetration of the injected fluid – such as LIF – while the shortcomings of the elastic light scattering technique are still the same.

REFERENCES

[1] Piermarini GJ, Block S, Barnett J, Forman R. Calibration of the pressure dependence of the R1 ruby fluorescence line to 195 kbar. J Appl Phys 1975;46:2774–80.

[2] Mao H, Bell P, Shaner Jt, Steinberg D. Specific volume measurements of Cu, Mo, Pd, and Ag and calibration of the ruby R1 fluorescence pressure gauge from 0.06 to 1 Mbar. J Appl Phys 1978;49:3276–83.

[3] Xu J, Mao H, Bell P. High-pressure ruby and diamond fluorescence: observations at 0.21 to 0.55 terapascal. Science 1986;232:1404–6.

[4] Grasset O. Calibration of the R ruby fluorescence lines in the pressure range [0–1 GPa] and the temperature range [250–300 K]. High Pressure Res 2001;21:139–57.

[5] Raju SV, Zaug JM, Chen B, Yan J, Knight JW, Jeanloz R, Clark SM. Determination of the variation of the fluorescence line positions of ruby, strontium tetraborate, alexandrite, and samarium-doped yttrium aluminum garnet with pressure and temperature. J Appl Phys 2011;110:023521.

[6] Bauer JD, Bayarjargal L, Winkler B. Ruby fluorescence lifetime measurements for temperature determinations at high (p, T). High Pressure Res 2012;32:155–62.

[7] Bush B, Klavins P, Lawson M, Shirer K, Curro N. Development and calibration of a ruby fluorescence pressure measurement system. Bull Am Phys Soc 2013;58, http://meetings.aps.org/Meeting/CAL13/Event/206435

[8] Weber G, Drickamer HG. The effect of high pressure upon proteins and other biomolecules. Quart Rev Biophys 1983;16:89–112.

[9] Mozhaev VV, Heremans K, Frank J, Masson P, Balny C. High pressure effects on protein structure and function. Protein Struct Funct Genet 1996;24:81–91.

[10] Lakowicz JR, Weber G. Quenching of fluorescence by oxygen. Probe for structural fluctuations in macromolecules. Biochemistry 1973;12:4161–70.

[11] He X-H, Liu H-Z, Liu L, Zhao G-L, Wang Q, Chen Q-L. Effects of high pressure on the physicochemical and functional properties of peanut protein isolates. Food Hydrocoll 2014;36:123–9.

[12] Dellarole M, Royer CA. High-pressure fluorescence applications. Fluorescence Spectroscopy and Microscopy, Springer, 2014, pp. 53–74.

[13] Martin MM, Lindqvist L. The pH dependence of fluorescein fluorescence. J Lumin 1975;10:381–90.

[14] Spilimbergo S, Quaranta A, Garcia-Gonzalez L, Contrini C, Cinquemani C, Van Ginneken L. Intracellular pH measurement during high-pressure CO_2 pasteurization evaluated by cell fluorescent staining. J Supercrit Fluids 2010;53:185–91.

[15] Molina-Gutierrez A, Stippl V, Delgado A, Gänzle MG, Vogel RF. In situ determination of the intracellular pH of Lactococcus lactis and Lactobacillus plantarum during pressure treatment. Appl Environ Microbiol 2002;68:4399–406.

[16] Kaminski C, Engström J, Aldén M. Quasi-instantaneous two-dimensional temperature measurements in a spark ignition engine using 2-line atomic fluorescence. In: Symposium (International) on Combustion, Elsevier, 1998, pp. 85–93.

[17] Sakakibara J, Hishida K, Maeda M. Vortex structure and heat transfer in the stagnation region of an impinging plane jet (simultaneous measurements of velocity and temperature fields by digital particle image velocimetry and laser-induced fluorescence). Int J Heat Mass Transfer 1997;40:3163–76.

[18] Pringsheim P, Schulman JH. Fluorescence and phosphorescence. Phys Today 1950;3:35.

[19] Coolen M, Kieft R, Rindt C, Van Steenhoven A. Application of 2-D LIF temperature measurements in water using a Nd: YAG laser. Exp Fluids 1999;27:420–6.

[20] Seuntiëns H, Kieft R, Rindt C, Van Steenhoven A. 2D temperature measurements in the wake of a heated cylinder using LIF. Exp Fluids 2001;31:588–95.

[21] Ross D, Gaitan M, Locascio LE. Temperature measurement in microfluidic systems using a temperature-dependent fluorescent dye. Anal Chem 2001;73:4117–23.

[22] Estrada-Pérez CE, Hassan YA, Tan S. Experimental characterization of temperature sensitive dyes for laser induced fluorescence thermometry. Rev Sci Instrum 2011;82:074901.

[23] Bruchhausen M, Guillard F, Lemoine F. Instantaneous measurement of two-dimensional temperature distributions by means of two-color planar laser induced fluorescence (PLIF). Exp Fluids 2005;38:123–31.

[24] Castanet G, Lavieille P, Lebouché M, Lemoine F. Measurement of the temperature distribution within monodisperse combusting droplets in linear streams using two-color laser-induced fluorescence. Exp Fluids 2003;35:563–71.

[25] Lavieille P, Lemoine F, Lavergne G, Lebouché M. Evaporating and combusting droplet temperature measurements using two-color laser-induced fluorescence. Exp Fluids 2001;31:45–55.
[26] Shim J, Hwang H-Y, Lee M-Y, Lee D-Y. Two color Laser induced confocal fluorescent thermometry: design and experiments. Int J Precis Eng Man 2015;16:567–71.
[27] Labergue A, Delconte A, Castanet G, Lemoine F. Study of the droplet size effect coupled with the laser light scattering in sprays for two-color LIF thermometry measurements. Exp Fluids 2012;52:1121–32.
[28] Coppeta J, Rogers C. Dual emission laser induced fluorescence for direct planar scalar behavior measurements. Exp Fluids 1998;25:1–15.
[29] Funatani S, Fujisawa N, Ikeda H. Simultaneous measurement of temperature and velocity using two-colour LIF combined with PIV with a colour CCD camera and its application to the turbulent buoyant plume. Meas Sci Technol 2004;15:983.
[30] Hishida K, Sakakibara J. Combined Planar laser-induced fluorescence–particle image velocimetry technique for velocity and temperature fields. Exp Fluids 2000;29:S129–40.
[31] Kim H, Kihm K, Allen J. Examination of ratiometric laser induced fluorescence thermometry for microscale spatial measurement resolution. Int J Heat Mass Transfer 2003;46:3967–74.
[32] Natrajan V, Christensen K. Two-color laser-induced fluorescent thermometry for microfluidic systems. Meas Sci Technol 2009;20:015401.
[33] Sakakibara J, Adrian R. Whole field measurement of temperature in water using two-color laser induced fluorescence. Exp Fluids 1999;26:7–15.
[34] Sakakibara J, Adrian R. Measurement of temperature field of a Rayleigh-Bénard convection using two-color laser-induced fluorescence. Exp Fluids 2004;37:331–40.
[35] Kosseifi N, Yoshikawa H, Biwole P, Mathis C, Rousseaux G, Boyer S, Coupez T. Application of two-color LIF thermometry to nucleate boiling. J Mater Sci Eng B 2013;3:281–90.
[36] Parigger C, Plemmons DH, Litchford RJ, Jeng S-M. Exciplex liquid-phase thermometer using time-resolved laser-induced fluorescence. Optics Lett 1998;23:76–8.
[37] Bai F, Melton LA. High-temperature, oxygen-resistant molecular fluorescence thermometers. Appl Spectrosc 1997;51:1276–80.
[38] Kadota T, Miyoshi K, Tsue M. Exciplex method for remote probing of fuel droplet temperature. Effects of ambient conditions on fluorescence. JSME Int J Ser B 1993;36:371–7.
[39] Yoshizaki T, Takemura Y, Hisaeda T, Nishida K, Hiroyasu H. Planar measurements of the liquid phase temperature in diesel sprays injected into high-pressure and high-temperature environments. SAE Technical Paper; 1996.
[40] Wickersheim K, Sun M. Phosphors and fiber optics remove doubt from difficult temperature-measurements. Res Dev 1985;27:114–9.
[41] Goss LP, Smith AA, Post ME. Surface thermometry by laser-induced fluorescence. Rev Sci Instrum 1989;60:3702–6.
[42] Allison SW, Gillies GT. Remote thermometry with thermographic phosphors: Instrumentation and applications. Rev Sci Instrum 1997;68:2615–50.
[43] Marino R, Westring B, Laufer G, Krauss RH, Whitehurst R. Digital strain and temperature imaging technique. AIAA J 1999;37:1097–101.
[44] Omrane A, Ossler F, Aldén M. Two-dimensional surface temperature measurements of burning materials. In: Proceedings of the combustion institute, 2002;29:2653–2659.
[45] Knappe C, Andersson P, Algotsson M, Richter M, Linden J, Alden M, Tuner M, Johansson B. Laser-induced phosphorescence and the impact of phosphor coating thickness on crank-angle resolved cylinder wall temperatures. SAE Int J Eng 2011;4:1689–98.

[46] Jaber A, Zigan L, Sakhrieh A, Leipertz A. Surface temperature measurements in a porous media burner using a new laser-induced phosphorescence intensity ratio technique. Meas Sci Technol 2013;24:075202.

[47] Jovicic G, Zigan L, Will S, Leipertz A. Phosphor thermometry in turbulent hot gas flows applying Dy: YAG and Dy: Er: YAG particles. Meas Sci Technol 2015;26:015204.

[48] Feist J, Heyes A, Seefelt S. Thermographic phosphor thermometry for film cooling studies in gas turbine combustors. In: Proceedings of the Institution of Mechanical Engineers. J Power Energy 2003;217: 193–200.

[49] Särner G, Richter M, Aldén M. Investigations of blue emitting phosphors for thermometry. Meas Sci Technol 2008;19:125304.

[50] Venturini F, Bürgi R, Borisov SM, Klimant I. Optical temperature sensing using a new thermographic phosphor. Sensor Actuat A-Phys 2015;233:324–9.

[51] van Lipzig J, Yu M, Dam N, Luijten C, de Goey L. Gas-phase thermometry in a high-pressure cell using BaMgAl10O17: Eu as a thermographic phosphor. Appl Phys B 2013;111: 469–81.

[52] Brübach J, Dreizler A, Janicka J. Gas compositional and pressure effects on thermographic phosphor thermometry. Meas Sci Technol 2007;18:764.

[53] Aldén M, Omrane A, Richter M, Särner G. Thermographic phosphors for thermometry: a survey of combustion applications. Prog Energy Combust Sci 2011;37:422–61.

[54] Omrane A, Juhlin G, Ossler F, Aldén M. Temperature measurements of single droplets by use of laser-induced phosphorescence. Appl Optics 2004;43:3523–9.

[55] Brübach J, Patt A, Dreizler A. Spray thermometry using thermographic phosphors. Appl Phys B 2006;83:499–502.

[56] Omrane A, Särner G, Aldén M. 2D-temperature imaging of single droplets and sprays using thermographic phosphors. Appl Phys B 2004;79:431–4.

[57] Ponce A, Wong PA, Way JJ, Nocera DG. Intense phosphorescence triggered by alcohols upon formation of a cyclodextrin ternary complex. J Phys Chem 1993;97:11137–42.

[58] Mortellaro M, Nocera D. A turn-on for optical sensing. Chemtech 1996;26:17–23.

[59] Hartmann WK, Gray MH, Ponce A, Nocera DG, Wong PA. Substrate induced phosphorescence from cyclodextrin· lumophore host-guest complexes. Inorganica Chimica Acta 1996;243:239–48.

[60] Koochesfahani M, Cohn R, Gendrich C, Nocera D. Molecular tagging diagnostics for the study of kinematics and mixing in liquid phase flows. In: Invited Plenary Paper, Proceedings of the eighth international symposium on applications of laser techniques to fluids mechanics, Lisbon, Portugal, 1996, pp. 1.2.

[61] Gendrich CP, Koochesfahani MM, Nocera DG. Molecular tagging velocimetry and other novel applications of a new phosphorescent supramolecule. Exp Fluids 1997;23:361–72.

[62] Thomson S, Maynes D. Spatially resolved temperature measurements in a liquid using laser induced phosphorescence. J Fluids Eng 2001;123:293–302.

[63] Mooney M. The viscosity of a concentrated suspension of spherical particles. J Colloid Sci 1951;6:162–70.

[64] Brinkman H. The viscosity of concentrated suspensions and solutions. J Chem Phys 1952;20. 571-571.

[65] Hu H, Koochesfahani M, Lum C. Molecular tagging thermometry with adjustable temperature sensitivity. Exp Fluids 2006;40:753–63.

[66] Schulz C, Sick V. Tracer-LIF diagnostics: quantitative measurement of fuel concentration, temperature and fuel/air ratio in practical combustion systems. Prog Energy Combust Sci 2005;31:75–121.

[67] Lozano A, Yip B, Hanson RK. Acetone: a tracer for concentration measurements in gaseous flows by planar laser-induced fluorescence. Exp Fluids 1992;13:369–76.

[68] Grossmann F, Monkhouse PB, Ridder M, Sick V, Wolfrum J. Temperature and pressure dependences of the laser-induced fluorescence of gas-phase acetone and 3-pentanone. Appl Phys B 1996;62:249–53.

[69] Thurber MC, Grisch F, Hanson RK. Temperature imaging with single- and dual-wavelength acetone planar laser-induced fluorescence. Optics Lett 1997;22:251–3.

[70] Yuen L, Peters J, Lucht R. Pressure dependence of laser-induced fluorescence from acetone. Appl Optics 1997;36:3271–7.

[71] Ossler F, Alden M. Measurements of picosecond laser induced fluorescence from gas phase 3-pentanone and acetone: Implications to combustion diagnostics. Appl Phys B-Lasers O 1998;64:493–502.

[72] Thurber MC, Grisch F, Kirby BJ, Votsmeier M, Hanson RK. Measurements and modeling of acetone laser-induced fluorescence with implications for temperature-imaging diagnostics. Appl Optics 1998;37:4963–78.

[73] Thurber MC, Hanson RK. Pressure and composition dependences of acetone laser-induced fluorescence with excitation at 248, 266, and 308 nm. Appl Phys B 1999;69:229–40.

[74] Einecke S, Schulz C, Sick V. Measurement of temperature, fuel concentration and equivalence ratio fields using tracer LIF in IC engine combustion. Appl Phys B 2000;71:717–23.

[75] Koban W, Koch JD, Sick V, Wermuth N, Hanson RK, Schulz C. Predicting LIF signal strength for toluene and 3-pentanone under engine-related temperature and pressure conditions. In: International Symposium on Combustion, Combustion Institute, 2005, pp. 1545–1553.

[76] Löffler M, Beyrau F, Leipertz A. Acetone laser-induced fluorescence behavior for the simultaneous quantification of temperature and residual gas distribution in fired spark-ignition engines. Appl Optics 2010;49:37–49.

[77] Trost J, Löffler M, Zigan L, Leipertz A. Simultaneous quantitative acetone-PLIF measurements for determination of temperature and gas composition fields in an IC-engine. Phys Procedia 2010;5:689–96.

[78] Löffler M, Kröckel K, Koch P, Beyrau F, Leipertz A, Grasreiner S, Heinisch A. Simultaneous quantitative measurements of temperature and residual gas fields inside a fired SI-engine using acetone laser-induced fluorescence. In: SAE Technical Paper; 2009.

[79] Trost J, Zigan L, Leipertz A. Quantitative vapor temperature imaging in DISI-sprays at elevated pressures and temperatures using two-line excitation laser-induced fluorescence. P Combust Inst 2013;34:3645–3652.

[80] Koch JD, Hanson RK. Temperature and excitation wavelength dependencies of 3-pentanone absorption and fluorescence for PLIF applications. Appl Phys B 2003;76:319–24.

[81] Heicklen J. The fluorescence and phosphorescence of biacetyl vapor and acetone Vapor1. J Am Chem Soc 1959;81:3863–6.

[82] Burton CS, Noyes WA. Electronic energy relaxation in toluene vapor. J Chem Phys 1968;49:1705–14.

[83] Hippler H, Troe J, Wendelken HJ. Collisional deactivation of vibrationally highly excited polyatomic molecules. II. Direct observations for excited toluene. J Chem Phys 1983;78:6709–17.

[84] Koban W, Koch JD, Hanson RK, Schulz C. Absorption and fluorescence of toluene vapor at elevated temperatures. Phys Chem Chem Phys 2004;6:2940–5.

[85] Luong M, Koban W, Schulz C. Novel strategies for imaging temperature distribution using toluene LIF. J Phys Conf Ser, 2006;45:133.

[86] Koban W, Koch JD, Hanson RK, Schulz C. Oxygen quenching of toluene fluorescence at elevated temperatures. Appl Physics B 2005;80:777–84.

[87] Koban W, Koch JD, Hanson RK, Schulz C. Toluene LIF at elevated temperatures: implications for fuel-air ratio measurements. Appl Physics B 2005;80:147–50.

[88] Luong M, Zhang R, Schulz C, Sick V. Toluene laser-induced fluorescence for in-cylinder temperature imaging in internal combustion engines. Appl Physics B 2008;91:669–75.

[89] Devillers R, Bruneaux G, Schulz C. Investigation of toluene LIF at high pressure and high temperature in an optical engine. Appl Physics B 2009;96:735–9.

[90] Lee T, Bessler WG, Kronemayer H, Schulz C, Jeffries JB. Quantitative temperature measurements in high-pressure flames with multiline NO-LIF thermometry. Appl Optics 2005;44:6718–28.

[91] Kronemayer H, Bessler WG, Schulz C. Gas-phase temperature imaging in spray systems using multi-line NO-LIF thermometry. Appl Physics B 2005;81:1071–4.

[92] Kronemayer H, Omerbegovic K, Schulz C. Quantification of the evaporative cooling in an ethanol spray created by a gasoline direct-injection system measured by multiline NO-LIF gas-temperature imaging. Appl Optics 2007;46:8322–7.

[93] Devillers R, Bruneaux G, Schulz C. Development of a two-line OH-laser-induced fluorescence thermometry diagnostics strategy for gas-phase temperature measurements in engines. Appl Optics 2008;47:5871–85.

[94] Seitzman JM, Hanson R, DeBarber P, Hess C. Application of quantitative two-line OH planar laser-induced fluorescence for temporally resolved planar thermometry in reacting flows. Appl Optics 1994;33:4000–12.

[95] Malmqvist E, Jonsson M, Aldén M, Bood J. OH-thermometry with photofragmentation laser-induced fluorescence. In: Laser Applications to Chemical, Security and Environmental Analysis, Optical Society of America, 2014, pp. LM2D. 4.

[96] Eckbreth AC. Laser diagnostics for combustion temperature and species. 2nd ed. Amsterdam, The Netherlands: Gordon and Breach; 1996.

[97] Vaidyanathan A, Gustavsson JP, Segal C. Oxygen/hydrogen-planar-laser-induced fluorescence measurements and accuracy investigation in high-pressure combustion. J Propul Power 2009;25:864–74.

[98] Fond B, Abram C, Heyes AL, Kempf AM, Beyrau F. Simultaneous temperature, mixture fraction and velocity imaging in turbulent flows using thermographic phosphor tracer particles. Opt Express 2012;20:22118–33.

[99] Yáñez-González Á, Ruiz-Trejo E, van Wachem B, Skinner S, Beyrau F, Heyes A. A detailed characterization of BaMgAl 10 O 17: Eu phosphor as a thermal history sensor for harsh environments. Sensor Actuat A-Phys 2015;234:339–45.

[100] Melton LA. Spectrally separated fluorescence emissions for diesel fuel droplets and vapor. Appl Optics 1983;22:2224–6.

[101] Zhao H, Ladommatos N. Optical diagnostics for in-cylinder mixture formation measurements in IC engines. Prog Energy Combust Sci 1998;24:297–336.

[102] Wieske P, Wissel S, Grünefeld G, Pischinger S. Improvement of LIEF by wavelength-resolved acquisition of multiple images using a single CCD detector- Simultaneous 2D measurement of air/fuel ratio, temperatrure distribution of the liquid phase and qualitative distribution of the liquid phase with the Multi-2D technique. Appl Phys B 2006;83:323–9.

[103] Kashdan JT, Docquier N, BruneauxG. Mixture preparation and combustion via LIEF and LIF of combustion radicals in a direct-injection, HCCI diesel engine. SAE Technical Paper; 2004.

[104] Fujimoto H, Choi D, Shima Y, Senda J. Two-dimensional imaging of fuel-vapour concentration by use of LIEF technique during mixture formation process in a DI diesel engine. Meas Sci Technol 2002;13:391.

[105] Choi D, Iwamuro M, Shima Y, Senda J, Fujimoto H. The effect of fuel-vapor concentration on the process of initial combustion and soot formation in a DI diesel engine using LII and LIEF. SAE Technical Paper; 2001.

[106] Ipp W, Wagner V, Krämer H, Wensing M, Leipertz A, Arndt S, Jain AK. Spray formation of high pressure swirl gasoline injectors investigated by two-dimensional Mie and LIEF techniques. SAE Technical Paper; 1999.

[107] Fröba A, Rabenstein F, Münch K, Leipertz A. Mixture of TEA and benzene as a new seeding material for the quantitative two-dimensional LIEF imaging of vapor and liquid fuel inside SI engines. Combus Flame 1998;112:199–209.

[108] Fröba AP, Rabenstein F, Münch KU, Leipertz A. Mixture of Triethylamine (TEA) and Benzene as a new seeding material for the quantitative two-dimensional laser-induced exciplex Fluorescence imaging of vapor and liquid fuel inside SI engines. Combust Flame 1998;112:199–209.

[109] Kim JU, Golding B, Schock HJ, Keller P, Nocera DG. Exciplex fluorescence visualization systems for pre-combustion diagnosis of an automotive gasoline engine. SAE Technical Paper; 1996.

[110] Ghandhi J, Felton P, Gajdeczko B, Bracco F. Investigation of the fuel distribution in a two-stroke engine with an air-assisted injector. SAE Technical Paper; 1994.

[111] Melton LA. Exciplex-based vapor/liquid visualization system appropriate for automotive gasolines. Appl Spectrosc 1993;47:782–6.

[112] Zhang Y, Zhang G, Xu M, Wang J. Droplet temperature measurement based on 2-color laser-induced exciplex fluorescence. Exp Fluids 2013;54:1–10.

[113] Oldenhof E, Weckenmann F, Lamanna G, Weigand B, Bork B, Dreizler A. Experimental investigation of isolated acetone droplets at ambient and near-critical conditions, injected in a nitrogen atmosphere. Progress in Propulsion Physics, EDP Sciences; 2013, pp. 257–270.

[114] Bork B, König T, Lamanna G, Weckenmann F, Dreizler A. Thermometry of evaporating acetone droplets in near-critical conditions by combined phosphorescence/fluorescence measurements. In: 10th international conference on laser-light and interactions with particles; 2014.

[115] Weckenmann F, Bork B, Lamanna G, Dreizler A, Weigand B. Experimental investigation of droplet injections in the vicinity of the critical point. In: 14th European meetin on supercritical fluids, Marseille, France; 2014.

[116] Roy A, Joly C, Segal C. Disintegrating supercritical jets in a subcritical environment. J Fluid Mech 2013;717:193–202.

[117] Gustavsson JP, Segal C. Fluorescence Spectrum of 2-Trifluoromethyl-1, 1, 1, 2, 4, 4, 5, 5, 5-nonafluoro-3-pentanone. Appl Spectrosc 2007;61:903–7.

[118] Wensing M, Vogel T, Götz G. Transition of diesel spray to a supercritical state under engine conditions. Int J Eng Res 2015;doi: 10.1177/1468087415604281.

[119] Trost J, Zigan L, Leipertz A, Sahoo D, Miles PC. Characterization of four potential laser-induced fluorescence tracers for diesel engine applications. Appl Optics 2013;52:8001–7.

Chapter 6

Absorption Spectroscopy

This chapter mainly describes in Section 6.1 how absorption spectroscopy is put into practise in general, and in Section 6.2 how absorption measurements are realised at high pressure. The fundamentals of absorption in the infrared (IR), the near infrared (NIR) and the ultraviolet and visible (UV/VIS) spectral range are provided in Section 2.4.1 of this book. Therein it is already mentioned that the absorption cross-sections of many compounds are very large in the IR and especially in the UV/VIS spectral range. Therefore the absorbing species have to be attenuated before reliable absorption spectra can be measured. The process of attenuation (sample preparation) clearly conflicts the realisation of *in situ* measurements in real technically relevant systems. Thus in Section 6.2 I do not present any *in situ* UV/VIS absorption study related to high pressure but one out of many in literature available *in situ* IR absorption studies in high-pressure systems. In this case – though the absorption cross-sections in the IR spectral range are large – the overall absorbance is kept moderate by making the path length along which the incident radiation is attenuated very short, shorter than the wavelength of the corresponding incident IR radiation. Such short path lengths can be realised in an evanescent field emerging when the incident light is totally reflected inside the so-called attenuated total internal reflexion (ATR) crystal. The Section 6.1.3 and the final Case Study 6.2.3 describe the working principle of this technique, even related to imaging applications.

In Section 6.1.2 I describe in detail the working principle of Fourier-transform spectrometers, which have not been discussed so far in one of the previous chapters of this book. In Chapter 3 I mentioned already that Fourier-transform Raman spectroscopy is a powerful measurement technique, but did not describe how it works. The descriptions provided here in Section 6.1.2 can directly be transferred to understand how Fourier-transform Raman spectroscopy works. Most of the commercially available absorption spectrometers for the (N)IR spectral range are Fourier-transform spectrometers.

Sections 6.1.4 and 6.2.2 describe time-domain absorption spectroscopic applications, which are very interesting more from an optical engineer's view than from a process engineer's view. Here, due to the clever combination of different optical devices, some optical effects are exploited which produce from a spectrally narrow short laser pulse a wavelength scan, which can be sent through the sample cuvette and afterwards be detected only using a fast photo detector. Neither a Fourier-transform nor a grating-based spectral

Supercritical Fluid Science and Technology, Vol 7. http://dx.doi.org/10.1016/B978-0-444-63422-1.00006-7

analyser is required for this measurement technique. Furthermore, as this absorption technique works in the NIR spectral range, where the absorption cross-sections of compounds are small, it can be utilised as *in situ* measurement tool.

The spectral range in which absorption phenomena are discussed in this book spreads from the UV to the IR. In Section 3.5.2 we already saw that detectors which are suitable for the detection of VIS radiation are not automatically suitable for the detection of UV or NIR radiation. Thus it is clear that in order to cover the spectral range between the UV and the IR, different detectors as well as different light sources have to be utilised. These experimental details are not covered in this book. Throughout this chapter 'light source' and 'detector' are used irrespectively for which spectral range the absorption spectrometer should be applied.

6.1 WORKING PRINCIPLES OF ABSORPTION SPECTROMETERS

Four different set-ups of absorption spectrometers are described in the four sections that follow. It will become clear that the four described set-ups are variants of the usage of different apparatuses for one and the same purpose, which is the generation of an absorbance spectrum. In general typically used absorption spectrometers can be grouped in three groups. The first group comprises all spectrometers which use a grating-based spectrometer or monochromator for the direct detection of an intensity-over-wavelength spectrum. Such a device is described in Section 6.1.1. The second group comprises spectrometers which use a Michelson interferometer with a moveable mirror in order to detect an intensity-over-delay signal, which is afterwards converted via the discrete Fourier-transform into an intensity-over-wavelength spectrum. The respective spectrometers are called Fourier-transform spectrometers. The third group comprises spectrometers which use fast wavelength scans for the generation of the absorbance spectrum. In this case, waves of different wavelength pass the sample volume temporally delayed to each other. Thus the intensity-over-time signal is detected in the time-domain and afterwards converted into an intensity-over-wavelength spectrum. The respective spectrometers are called time-of-flight or time-domain absorption spectrometers.

Irrespective of whether a grating-based, a Fourier-transform or a time-domain spectrometer is utilised, the to-be-absorbed radiation can be sent through the sample either in transmission or in an evanescent field. Only Section 6.1.3 describes the absorption in an evanescent field.

6.1.1 Absorption in Transmission

Figure 6.1 shows a sketch of a conventional absorption spectrometer working in transmission.

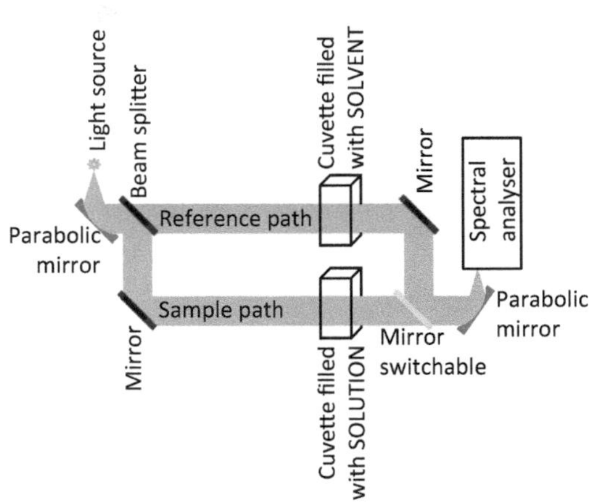

FIGURE 6.1 Sketch of a conventional absorption spectroscopic set-up working in transmission and using a spectral analyser.

Electromagnetic radiation originating from a light source is first collimated. Here the term 'light source' is used irrespective of whether it is a source for IR, NIR, VIS or UV radiation. The collimated beam is split into a reference and a sample path. The sample path passes through the cuvette (or reactor) containing the substance that should be analysed. Let this be a solute dissolved in a solvent. Behind the cuvette the sample beam is guided into a spectral analyser, which can be a spectrometer, or a monochromator. In case of Fourier-transform absorption spectroscopy it also can be a Michelson interferometer, as it will be described later in the context of Section 6.1.2. The working principle of a common spectrometer is described in Section 2.3.3.

In order to measure absorption spectra according to the general Eqn (2.80) (here slightly modified)

$$A(\lambda) = -\log_{10}(T(\lambda)) = \log_{10}\left(\frac{I_{\text{ref}}(\lambda)}{I_{\text{sample}}(\lambda)}\right). \tag{6.1}$$

of solely the solute and not of the solvent, the acquisition of an absorption spectrum requires a measurement $I_{\text{sample}}(\lambda)$ in a cuvette containing the solution and a reference measurement $I_{\text{ref}}(\lambda)$ in a cuvette only containing the solvent. When the sample beam $I_{\text{sample}}(\lambda)$ is analysed spectrally, the switchable mirror in Fig. 6.1 is turned out of the beam path so that only the sample beam is guided to the spectral analyser. If the switchable mirror is turned back into the beam path, the sample beam is blocked or reflected from the mirror, while then the reference beam $I_{\text{ref}}(\lambda)$ is guided to the spectral analyser and spectrally analysed. The reference cuvette is usually filled solely with the solvent also contained in the other cuvette. In Section 2.4.1 we saw that some solvents can absorb light in different spectral regions. As the solvent is contained in both cuvettes, the absorption differences of the reference and in the sample cuvette is only due to the solute.

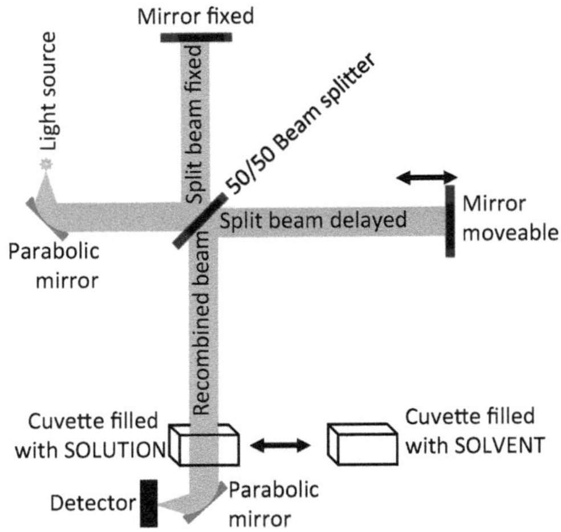

FIGURE 6.2 Sketch of a Fourier-transform absorption spectrometer working in transmission.

6.1.2 Fourier-Transform Absorption Spectroscopy

Figure 6.2 shows a Fourier-transform absorption spectrometer.

The set-up works like a Michelson interferometer with the special feature that one of the two reflecting mirrors is moveable back and forth. The Fourier-transform absorption spectrometer does not contain a monochromator or a spectrometer. The spectral information is computed on the basis the Fourier-transform from the intensity detected with the detector as a function of the optical path difference (OPD) between the fixed split beam and the delayed split beam. A more detailed and illustrative description will follow after the description of the set-up.

Again the term 'light source' here covers sources of IR, NIR, VIS and UV radiation, depending on which kind of absorption spectroscopy is to be applied. The broadband radiation originating from the light source is collimated and split – ideally in equal portions – into two paths. One path between the beam splitter and the fixed mirror is referred to as 'split beam fixed', whereas the other one between the beam splitter and the moveable mirror is referred to as 'split beam delayed'. The mirrors reflect the collimated radiation back to the 50/50 beam splitter where the radiation from both paths partially recombines in the 'recombined beam'. Of the radiation coming from the fixed mirror, 50% does not propagate into the recombined path but is reflected back to the light source. The same is true for the radiation coming from the moveable mirror. Here also 50% pass the beam splitter to the light source, while the other 50% are reflected into the recombined beam path. Thus in the recombined beam, equal portions of radiation coming from the fixed beam path and from the delayed beam path recombine. Consequently the electromagnetic waves interfere and modulate the intensity of the recombined beam as a function of the 'delay' between the light coming from the fixed beam path and the delayed beam path. The 'delay' or the OPD can be varied by moving the moveable mirror back and forth. I prepared the four figures that follow in order to illustrate the interference, when the radiation coming from the fixed beam path and the delayed beam path recombine in the recombined beam path.

Figure 6.3 shows snapshots of three different waves – red, green and blue – in the recombined beam path. The top illustration in Fig. 6.3 is labelled with 'split beam fixed'. This does not mean that the snapshots show the three waves in the split beam path. It shows the waves in the recombined beam, which come from the split beam path. The same is true for the middle illustration which is labelled 'split beam delayed'. It shows snapshots of the waves in the recombined beam path, which come from the delayed beam path. The red wave represents a long, the green wave a medium, and the blue wave a short wavelength wave. The black line represents the summation of these three waves.

If the OPD between the fixed beam path and the delayed beam path is zero, the radiation coming from the delayed beam path is said to be not delayed with respect to the radiation coming from the fixed beam path. In this case the distance between the beam splitter and the fixed mirror is exactly the distance between the beam splitter and the moveable mirror. Then there is no phase shift between the waves in the top and the middle illustrations in Fig. 6.3. Consequently the red, the green and the blue waves coming from the fixed and

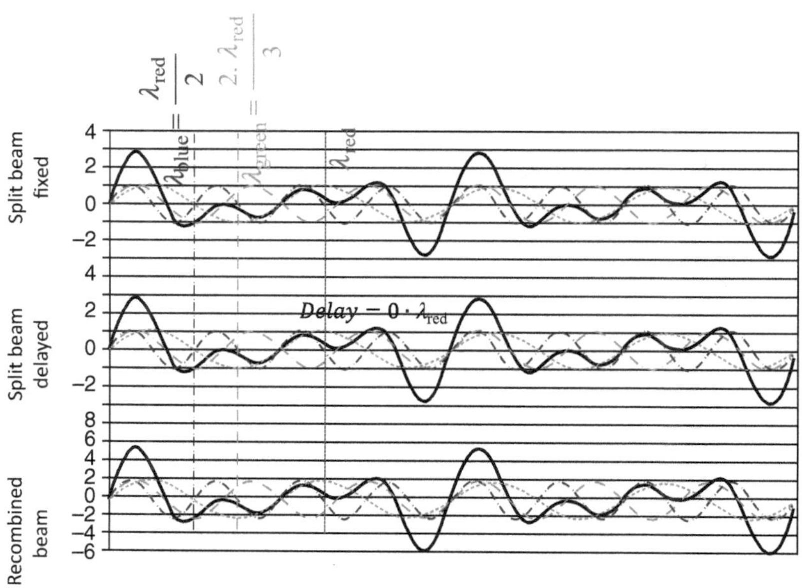

FIGURE 6.3 (Top) Representation of three example waves (dotted, dashed and dotted-dashed (red, green and blue in the web version, respectively)) and their summation (solid black) coming from the fixed beam path. (Middle) Representation of the same waves coming from the delayed beam path with a delay of $0 \cdot \lambda_{red}$. (Bottom) Representation of the recombination (summation) of the three waves coming from the fixed and from the delayed beam paths.

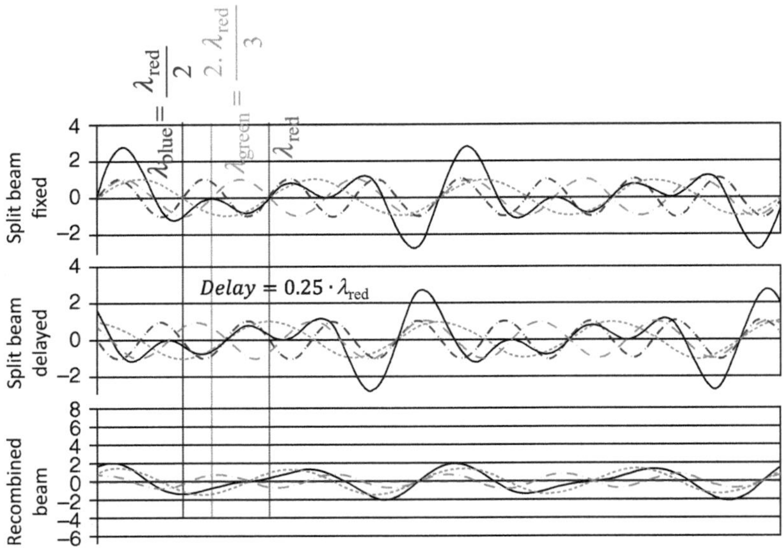

FIGURE 6.4 (Top) Representation of three example waves (dotted, dashed and dotted-dashed (red, green and blue in the web version, respectively)) and their summation (solid black) coming from the fixed beam path. (Middle) Representation of the same waves coming from the delayed beam path with a delay of $0.25 \cdot \lambda_{red}$. (Bottom) Representation of the recombination (summation) of the three waves coming from the fixed and from the delayed beam paths.

the delayed beam paths interfere in the recombined beam path positively with each other and thus the amplitudes of the 'recombined waves' in the bottom illustration in Fig. 6.3 is simply two times the amplitude of the waves shown in the middle and the top illustration. In this case all waves contained in the beam paths interfere positively with each other and consequently the intensity of the recombined beam is maximum.

In Fig. 6.4 the delay between the waves coming from the fixed and the delayed beam path is $(0.25 \cdot \lambda_{red})$ in the recombined beam path. This delay can be adjusted if the moveable mirror is moved back by $(0.125 \cdot \lambda_{red})$. The delay or the OPD is two times the distance the moveable mirror is moved, as the radiation travels two times between the beam splitter and the mirror, once before it is reflected and once after its reflection. As the blue wave's wavelength is half of the red wave's wavelength the delay of $(0.25 \cdot \lambda_{red})$ corresponds to a delay of $(0.5 \cdot \lambda_{blue})$.

The blue waves coming from the delayed beam are shifted by half of their wavelength with respect to the blue waves coming from the fixed beam path. Under this circumstance the blue waves interfere destructively and extinguish each other. Also the other waves interfere with each other but do not extinguish each other. Nevertheless – due to the interference – the intensity in the recombined beam is decreased and for this mirror position the detector detects less intensity than before where there was no delay between the paths.

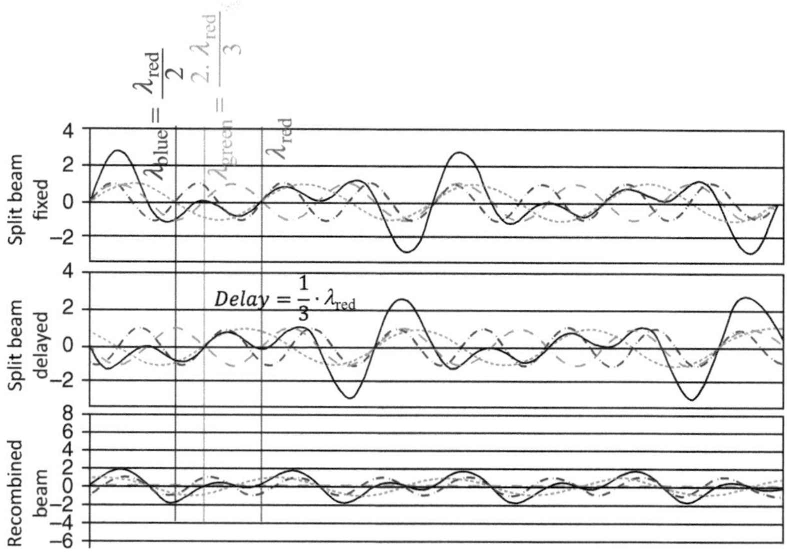

FIGURE 6.5 (Top) Representation of three example waves (dotted, dashed and dotted-dashed (red, green and blue in the web version, respectively)) and their summation (solid black) coming from the fixed beam path. (Middle) Representation of the same waves coming from the delayed beam path with a delay of $1/3 \cdot \lambda_{red}$. (Bottom) Representation of the recombination (summation) of the three waves coming from the fixed and from the delayed beam paths.

If we move the moveable mirror farther back we could reach a situation shown in Fig. 6.5.

The delay corresponds to $(1/3 \cdot \lambda_{red})$. Now the green waves coming from the delayed beam are shifted by half of their wavelength with respect to the green waves coming from the fixed beam path. Thus the green waves interfere destructively and extinguish each other. Also the other waves interfere as it is shown in Fig. 6.5. This again modulates the intensity of the recombined beam.

Figure 6.6 shows that for a delay of $(0.5 \cdot \lambda_{red})$ the red waves would interfere completely destructively with each other, while the other waves also interfere. Again the intensity of the recombined beam is modulated.

In a Fourier absorption spectroscopy experiment the moveable mirror is moved back and forth to many different delay positions and the corresponding intensities are recorded. A corresponding interferogram is shown in Fig. 6.7.

Using the discrete Fourier-transform

$$I(\bar{v}) = 4 \int_{\mathrm{OPD}=-\infty}^{\mathrm{OPD}=+\infty} \left\{ \left[I(\mathrm{OPD}) - 0.5 \cdot I(\mathrm{OPD}=0) \right] \cdot \cos(2\pi\bar{v} \cdot \mathrm{OPD}) \right\} d(\mathrm{OPD}) \quad (6.2)$$

the spectrum of the light source can be computed, if there is no cuvette in the recombined beam path.

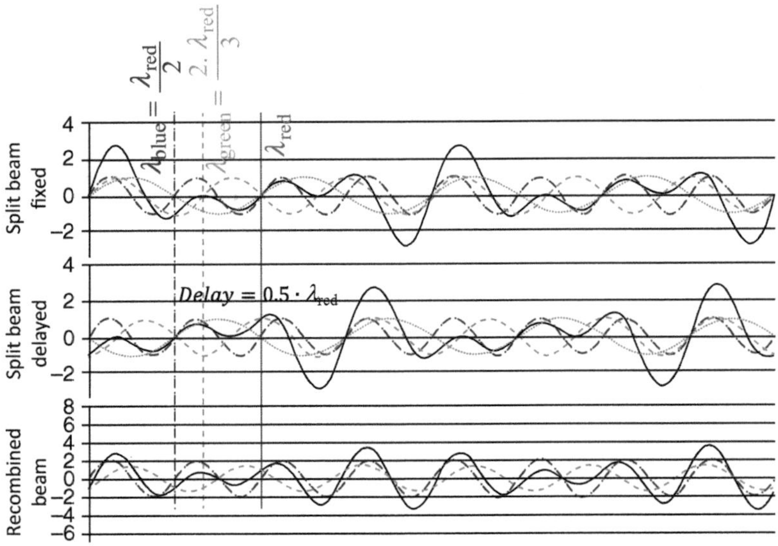

FIGURE 6.6 (Top) Representation of three example waves (dotted, dashed and dotted-dashed (red, green and blue in the web version, respectively)) and their summation (solid black) coming from the fixed beam path. (Middle) Representation of the same waves coming from the delayed beam path with a delay of $0.5 \cdot \lambda_{red}$. (Bottom) Representation of the recombination (summation) of the three waves coming from the fixed and from the delayed beam paths.

This is why Fourier-transform spectrometers are also utilised for the characterisation of the emission spectra of light sources. $I(\text{OPD})$ is the detected intensity as a function of the OPD. $I(\text{OPD}=0)$ is the maximum signal which is detected for no OPD between the two paths. The waves contained in the emitted spectrum of the light source are characterised by their wave number $\bar{\nu} = 1/\lambda$. Having recorded the interferogram shown in Fig. 6.7, all quantities contained in Eqn (6.1) are known, except of $\bar{\nu}$, for which different values can be scanned in order to get the spectral intensity distribution of the light source $I(\bar{\nu})$, which can easily be converted into a wavelength scale $I(\lambda = 1/\bar{\nu})$.

If a cuvette containing an absorbing solution is inserted into the recombined beam path, the spectrum of the light source is modulated according to the absorption of the absorbing medium. Of course, the cuvette can also be positioned in the collimated beam path between the light source and the beam splitter, which might seem to be more logic. Nevertheless, the absorbing medium is usually placed in the recombined beam path. In order to measure absorption spectra according to the general Eqn (2.80) (here slightly modified)

$$A(\lambda) = -\log_{10}(T(\lambda)) = \log_{10}\left(\frac{I_{\text{ref}}(\lambda)}{I_{\text{sample}}(\lambda)}\right).$$

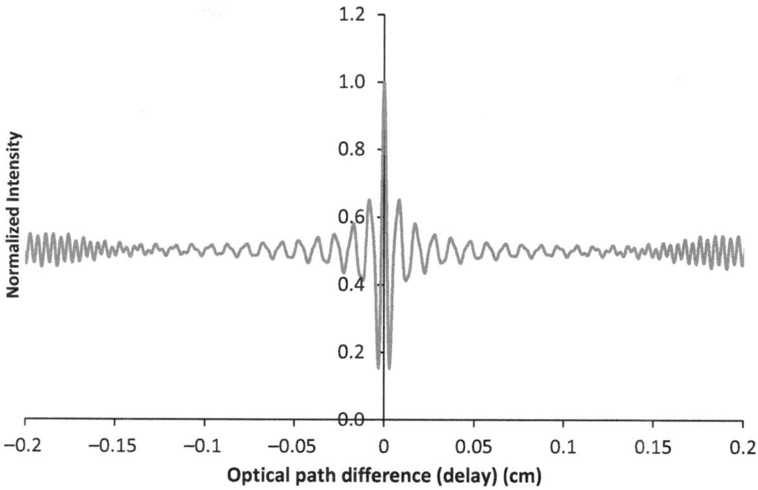

FIGURE 6.7 Fourier absorption spectroscopy interferogram. The detected intensity is given as a function of the delay between the fixed and the delayed beam path, or as a function of the mirror position of the moveable mirror. For the computation of the interferogram 60 waves in steps of 100 nm between 4000 nm and 10000 nm are considered and the moveable mirror is moved in increments of 500 nm to different OPDs. *(Source: Interferogram computed by Julian Schuster from the Lehrstuhl für Technische Thermodynamik of the Friedrich-Alexander-Universität Erlangen-Nürnberg, Germany.)*

of solely the solute and not of the solvent, the acquisition of an absorption spectrum requires a measurement $I_{\text{sample}}(\lambda)$ in a cuvette containing the solution and a reference measurement $I_{\text{ref}}(\lambda)$ in a cuvette only containing the solvent. The reference and the sample cuvette can be exchanged in the recombined beam path.

Finally I want to mention which settings in a Fourier-transform spectrometer influence the absorption spectrum in which way. The spectral resolution of the Fourier-transform spectrometer is determined by the maximum OPD. The larger the OPD (the farther the distance along which the mirror is moved) the better is the resolution of the obtained spectrum. The number of measurements taken in between OPD = 0 and OPD = max does not influence the spectral resolution! It mainly determines how accurately the real spectral intensity profile of the broadband light source can be reflected. The smaller the step width of the moving mirror, the more accurately the real spectral intensity profile of the broadband light source can be reproduced with the discrete Fourier-transform.

6.1.3 Absorption in the Evanescent Field

In Sections 2.4.1.3 and 2.4.1.5 on UV/VIS and IR absorption it is mentioned that the compounds from which an absorption spectrum is to be measured have to be diluted for the measurements in non-absorbing solvents. This is called the

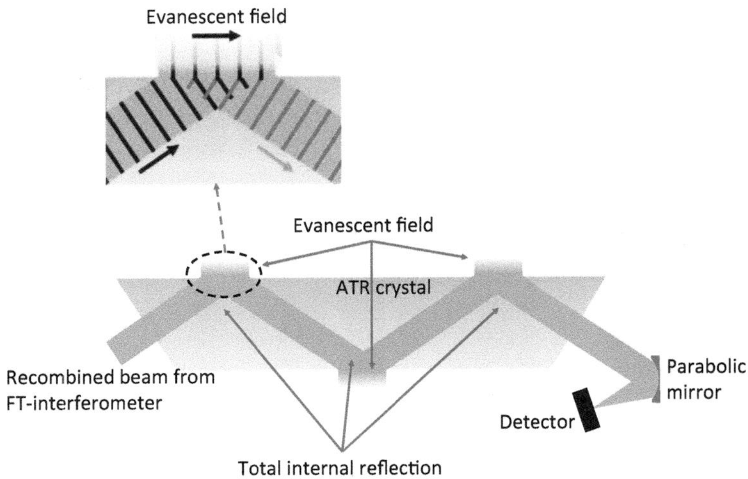

FIGURE 6.8 Sketch of the propagation of a collimated beam through an ATR crystal utilised in ATR absorption devices. Magnified excerpt shows the plane wave fronts in the incident and the refracted beam as well as in the evanescent field.

preparation of the sample. The sample preparation is mainly required due to the large absorption cross-sections of many compounds in the respective spectral regions. If they were not diluted, the incident UV/VIS or IR radiation would be extinguished completely inside the cuvette so that no light – at least at the absorbed wavelengths λ – would be transmitted through the cuvette $I_{sample}(\lambda) = 0$. Under this circumstance the quantification of the absorbance according to Eqn (2.80) is not possible, simply because the denominator $I_{sample}(\lambda)$ would be zero. Under many circumstances the large absorption cross-sections make an *in situ* absorption measurement in transmission in real systems impossible. On this effect the attenuated total reflexion (ATR) absorption spectroscopy offers one way out. In an ATR absorption device the incident radiation is absorbed in the evanescent field that is extended into the absorbing medium only for few micrometres. Due to the very short path lengths along which the incident radiation is in contact with the absorbing medium, *in situ* measurements very close to the surface of absorbing media or objects are possible without any need for sample preparation.

Figure 6.8 shows the propagation of a collimated beam through a crystal which is utilised for ATR absorption devices. In Fig. 6.8 the incident beam is labelled as coming from the Fourier-transform spectrometer as many ATR absorption spectrometers use the Fourier-transform approach for the determination of the spectrum. But in general the collimated incident beam can also originate from a broadband light source which is afterwards send to the Fourier-transform device or to a spectrometer or monochromator. The incident beam is coupled into the ATR crystal under a suitable angle assuring the total internal reflexion at the top and bottom plane surface of the ATR crystal. In Fig. 6.8 the beam is three times totally reflected before it leaves the ATR crystal on the right side. The number of total internal reflexions can be more or less, depending on the length of the crystal.

The evanescent field exists outside the ATR crystal in the vicinity of the interface where the beam is totally reflected. The intensity of the evanescent field is maximum at the interface and decays exponentially with the distance from the interface. For a more illustrative description I want to recall the contents of Section 2.3.2. There we saw that if light rays travel from a medium of large index of refraction through an interface to a medium of small index of refraction, the light rays are in general partially reflected and refracted. When we increased the angle of incidence we realised that the angle of the refracted light rays increased faster. We reached the angle of total internal reflexion where the refracted light rays were travelling parallel to the interface between the two media of different indices of refraction. Such a scenario is sketched in the magnified excerpt in Fig. 6.8. The tilt black lines represent the plane wave fronts of a monochromatic beam. In case of a real absorption experiment there would be many different wavelengths contained in the beam. The tilt grey lines represent the plane wave fronts in the reflected beam. The vertical black lines represent the plane wave fronts in the refracted beam, which would travel parallel to the interface. This also represents the evanescent field, which only exists where the beam is totally reflected internally. In Fig. 6.8 one also can see that the amplitude of the wave front in the evanescent field decreases with increasing distance from the interface. The intensity of the evanescent field decays exponentially with increasing distance from the interface. The penetration depth of the evanescent field shrinks with increasing ratio of the indices of refraction of the ATR crystal and the medium in the evanescent field, and with increasing angle of incidence. With respect to ATR absorption spectrometers the intensity of the evanescent field typically decays to e^{-1} within 1/3 of one wavelength from the ATR crystal's surface. If there were absorbing compounds present in the evanescent field, they would attenuate the radiation and consequently the absorption spectrum can be measured. Colloquially describing, all the light intensity of the incident beam passes the evanescent field. Thus theoretically, the incident beam can be extinguished completely, if there are extreme absorbers accumulating on the surface of the ATR crystal. In the case of ATR absorption spectroscopy one always has to be aware that the absorption spectrum represents a thin layer of the absorbing medium. If ATR absorption spectra have to be acquired from weakly attenuating media, long ATR crystals with many evanescent fields sites can be employed. Effectively the interaction length between the beam and the medium is extended. For measurements in strongly absorbing media ATR crystals with less evanescent field sites can be used. Then effectively the interaction length between the beam and the absorbing medium is decreased.

The circumstance that in an evanescent field the absorption takes place in a thin layer can also be exploited for 'chemical imaging'. In this case the beam coming from the Fourier-transform spectrometer set-up is not analysed with a zero-dimensional detector, but – as it is sketched in Fig. 6.9 – with a

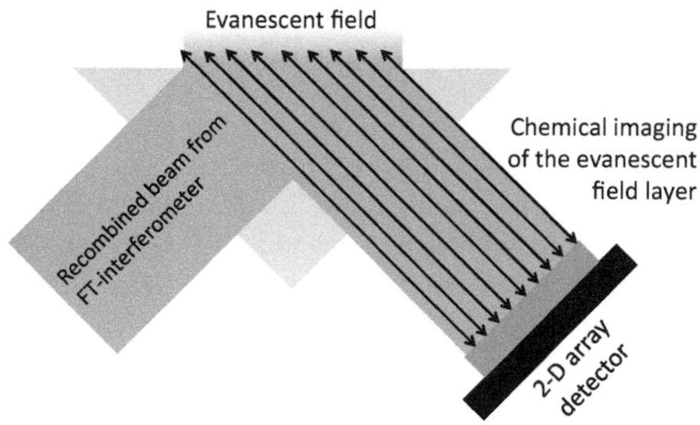

FIGURE 6.9 Sketch of an imaging ATR absorption spectrometer concept. The two-dimensional detector array images the thin layer of the evanescent field. The corresponding measurement technique is also referred to as 'chemical imaging'.

two-dimensional detector array. Then the different pixels of the two-dimensional array detector measure the absorption spectra corresponding to different locations within the surface at which the beam is totally reflected. In fact, to make this clear another time, they do not directly measure the absorption spectra. They measure the intensity as a function of the delay between the fixed and the delayed beam path. Then for each pixel the discrete Fourier-transform has to convert the intensity-over-delay measurement into an intensity-over-wavelength spectrum. From the different absorption spectra, which are assigned to different locations within the thin layer of the evanescent field (represented by the black arrows in Fig. 6.9), the chemical composition of the absorbing medium can be computed spatially resolved, wherefore the technique is also called 'chemical imaging'.

6.1.4 Supercontinuum Time-of-Flight Absorption Spectroscopy

Next to the conventional absorption spectroscopy that uses a common grating spectrometer or monochromator for spectral analysis and next to the Fourier-transform absorption spectroscopy that uses a Michelson interferometer and a Fourier-transform for spectral analysis, the supercontinuum time-of-flight absorption spectroscopy offers a third variant of absorption spectroscopy. Here the spectral information is decoded in a temporal information, meaning that waves with different wavelengths pass the sample cuvette one after the other and therefore also arrive successively on the detector [1,2]. If the temporal delay of the waves of different wavelengths is known, the detector can detect the 'intensity-over-time', what then can be transferred into an 'intensity-over-wavelength' spectrum.

Figure 6.10 shows a sketch of an example supercontinuum time-of-flight NIR absorption set-up [3].

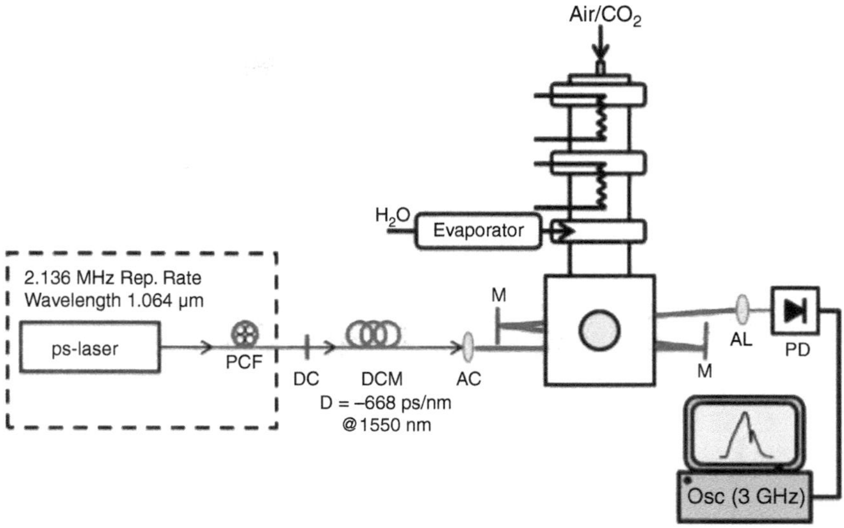

FIGURE 6.10 Supercontinuum absorption spectroscopy set-up for temperature and multi-species measurements. PCF, photonic crystal fibre; DC, dichroic mirror; DCM, dispersion compensation module; AC, achromatic collimator; M, mirror; AL, achromatic lens; PD, photodiode; Osc, oscilloscope. *(Source: Ref. [3].)*

The original light source is a quasi-monochromatic short pulse laser. Here it is a picosecond laser emitting at 1064 nm. Of course for an absorption experiment a broadband light source is required but not a quasi-monochromatic laser. Therefore the laser pulse which also is referred to as pump pulse is launched into a glass fibre – here it is a photonic crystal fibre – in order to transfer the ingoing spectrally narrow and short pump pulse into a spectrally broad and still short outgoing laser pulse, which is called a supercontinuum (SC) pulse. The initially narrow band pump pulse gets spectrally broadened due to effects such as four-wave mixing, stimulated Raman scattering and Brillouin scattering, which take place inside the photonic crystal fibre [4]. As the outgoing supercontinuum pulse can cover continuously the spectral range from the visible to the mid-IR, it carries the attribute supercontinuum in its name. At this point there is already a spectrally broadband radiation which can be utilised without any further measures for conventional absorption spectroscopy using common spectrometers. But for the time-of-flight absorption spectroscopy the waves of different wavelengths still have to be separated in the time-domain. Therefore the spectrally broad but still short SC-pulse is sent through a strongly dispersive fibre, as they are for example mounted in dispersion compensation modules (DCM) for telecommunication. In telecommunication they are used for exactly the opposite purpose they are used here. Here the DCM fibre is utilised to separate waves of different wavelengths in time. At this point I refer to Fig. 2.21 which shows that waves of different wavelength travel through the dispersive fibre with different velocities. Consequently waves of different wavelength – if launched quasi instantaneous into the DCM fibre as a short pulse – exit the DCM fibre temporally delayed to each other. It depends on the dispersion, which can be normal (positive) or anomalous (negative) whether waves with longer or the shorter wavelengths travel faster. The length of dispersive fibres can be several kilometres and broaden the initial picosecond pulse into a so-called wavelength-scan of several 100s of nanoseconds. Now a fast detector (photodiode) can be utilised to record the temporal shape of the wavelength scan. If the dispersion of the dispersive fibre is characterised, the experimenter can convert the recorded 'intensity-over-time' signal into an 'intensity-over-wavelength' spectrum. Of course, in order to compute the absorbance according to Eqn (2.80) two measurements are required, one in a reference and another one in the sampling cuvette (reactor).

6.2 CASE STUDIES

Three case studies are reported in the sections that follow. The first one reports the usage of a Fourier-transform NIR spectrometer working in transmission in order to analyse a high-pressure extraction process. I chose this case study as the absorption sensor has not been applied under idealised laboratory conditions but in a rather large and technically relevant extraction plant.

The second case study reports time-domain NIR absorption measurements in mixtures of dry air and water vapour at high temperatures and pressures up to 3 MPa. I chose this one in order to demonstrate that a lot of different information can be extracted from absorption spectra, not only mixture compositions but also temperatures and most probably also pressures.

The third case study reports 'chemical imaging' using an ATR Fourier-transform infrared (FTIR) apparatus equipped with a two-dimensional detector. I chose this case study in order to give a demonstration how ATR FTIR chemical imaging, which has been described briefly in the context of Fig. 6.9, works in fact. Here IR radiation is used instead of NIR radiation that was used in the case studies one and two.

Absorption measurements for the analysis of polymerisation at elevated pressures are not reported in this section, but can be found in literature [5–7].

6.2.1 NIR Absorption Spectroscopy for Inline Monitoring of Extraction Processes

With respect to rising energy costs it might be useful to run extraction processes only while the extraction yield is high and to stop them already before the long lasting and thus energy intensive extraction of the remaining residuals. The adoption of the operational conditions of the extraction process to the yield – of course – anticipates that the extraction yield is permanently known. To this end Böhm et al. [8] implemented an inline NIR absorption high-pressure and high-temperature cell into a rather large technical laboratory plant for the extraction of hop. Compressed CO_2 flows through a bed of hop pellets, which are filled into

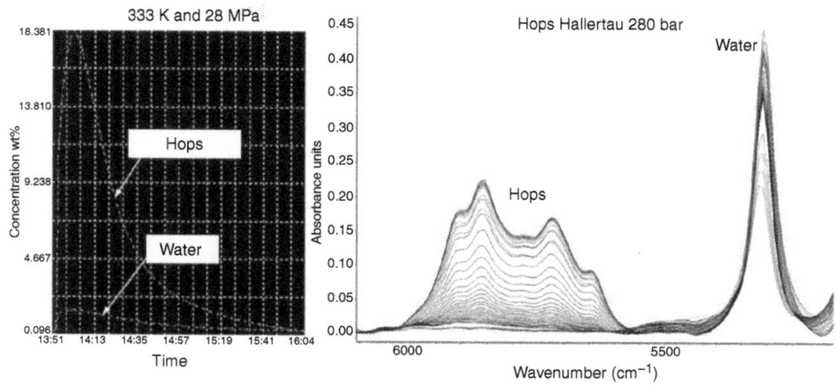

FIGURE 6.11 **Result of inline monitoring for hops extraction at 333 K and 28 MPa.** *(Source: Ref. [8].)*

a high-pressure vessel, and extracts the 'hop-acid' as well as the water from the hop pellets. Before the 'hop-acid' and the water are separated from the CO_2 flow in a separator, the mixture passes through the inline absorption cell. The inline absorption cell is combined via two glass fibres with a Fourier-transform absorption spectrometer. One fibre guides the light from the recombined beam of the Michelson interferometer (compare Fig. 6.2) to the inline absorption cell. The light then in transmission passes through the measurement cell and is guided in the second fibre to the detector of the Fourier-transform absorption spectrometer.

Figure 6.11 (right) shows that at 333 K and 28 MPa the absorption features of hop and water can be separated from each other. Consequently the two absorbances of both compounds can be regarded as not influenced by each other and – after a calibration – the conversion of absorbance into concentration is straightforward. The concentration profiles of the 'hop-acid' and water as a function of extraction time are shown in the left part of Fig. 6.11.

The authors of this case study report that calibration measurements are required for each pressure and temperature at which the extraction process is operated, at least when the pressure is varied a lot between two measurements. This could be circumvented if a spectral region was found in which next to water and the 'hop-acid' also CO_2 absorbs. Then the absorption of CO_2 can act as a reference-signal and the amount of extracted 'hop-acid' and water can be quantified – no more as concentration – but as ratio normalised to the amount of CO_2.

6.2.2 Supercontinuum Absorption Spectroscopy for Temperature Sensing at Elevated Pressure

Werblinski et al. [9] analysed the suitability of time-domain supercontinuum absorption spectroscopy for temperature and composition measurements at 3 MPa. This technique is described in Section 6.1.4, where it is called time-of-flight absorption spectroscopy. The experimental set-up used in this study was very

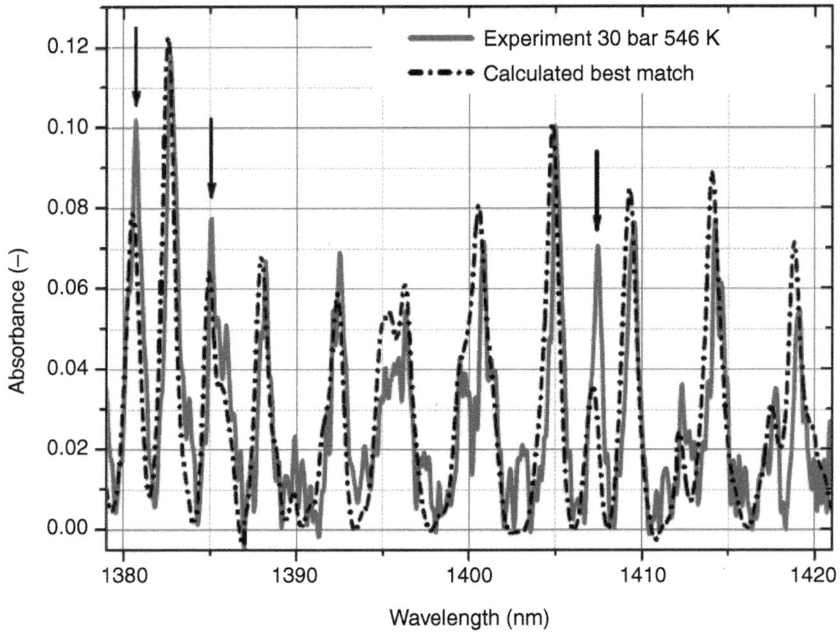

FIGURE 6.12 Comparison of the experimental spectra at 3 MPa and 546 K (x_{water} = 0.0463, L = 25 mm) and the calculated best match at 3 MPa and 569 K (x_{water} = 0.05, L = 25 mm). *(Source: Figure provided on courtesy by Thomas Werblinski from the Lehrstuhl für Technische Thermodynamik of the Friedrich-Alexander-Universität Erlangen-Nürnberg, Germany; from [9].)*

similar to the one shown in Fig. 6.10, except that only one transmission path was realised through the high-pressure high-temperature cell. Due to the nearly ubiquitous presence of water in the environment and in many technical systems their measurement technique was based on the detection of absorption lines of water vapour, which was added in defined amounts to dry air. Therefore they analysed the NIR spectral range from 1330 nm to 1450 nm, which they could easily cover with their supercontinuum light source and which exhibits many overtone and combinative absorption bands of water. Figure 6.12 shows how many different absorption bands exist in the small spectral region between 1380 nm and 1420 nm. Most of the bands Werblinski et al. [9] measured experimentally are also present in the absorption spectrum calculated using the BT2 [10] database (http://www.tampa.phys.ucl.ac.uk/ftp/astrodata/water/BT2/) for the localisation of the band positions and HITRAN08 (see Section 2.4.1.3) for the consideration of broadening effects, as it has been presented before by Voronin et al. [11]. In Fig. 6.12 peaks showing a large discrepancy between the experiment and the computation are decorated with black arrows. These peaks were not considered for the temperature evaluation. The remaining peaks were considered for the evaluation of temperature. On this account the temperature of the calculated spectrum was varied until the peak intensity distribution of the computed and the experimentally measured spectrum showed the best fit.

For the evaluation of temperature the absolute peak intensities are not relevant, only the distribution of the peak intensities in the considered spectral range. Here I want to refer to Section 2.2.7 where the temperature sensitive distribution of molecules over their energy states is described (Boltzmann distribution). Thus, the temperature measurements described in this case study are – in a simplified view – based on the comparison of experimentally measured and computed Boltzmann distributions of water molecules. Once the temperature is identified, also the absolute intensity of the computed spectrum was fitted to the experimental one by varying the water content in the computations. Consequently next to the temperature also the water content in the dry-air/water-vapour mixture was quantified.

With the time-domain supercontinuum absorption spectroscopy method Werblinski et al. [9] measured a temperature of 569 K and a water vapour content of $x_{water} = 0.05$ when the chamber conditions were experimentally set to 546 K and a water vapour content of $x_{water} = 0.0463$ at 3 MPa.

6.2.3 *In Situ* Imaging of Diffusion and Dissolution Phenomena at High Pressures

Fleming et al. [12] applied an ATR FTIR absorption spectroscopic method for imaging the interdiffusion of different miscible polymers in the presence of compressed CO_2. The respective imaging method is described in the context of Fig. 6.9 in Section 6.1.3. This research aimed at contributing to the understanding of the diffusion and separation of different polymers in polymer formulations, which are frequently applied in pharmaceutics [13–15].

For this purpose a high-pressure cell is assembled around the ATR crystal, which in the left part of Fig. 6.13 is labelled as diamond [16]. Inside the high pressure cell two polymers, here polyvinylpyrrolidone (PVP) and poly(ethylene

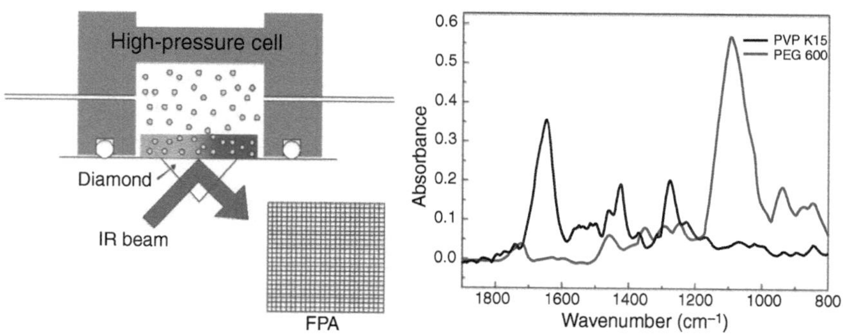

FIGURE 6.13 (Left) Schematic diagram of high-pressure cell showing the interface of PVP (light grey (green in the web version)) and PEG (dark grey (blue in the web version)) in the presence of CO_2 molecules (white (yellow in the web version)). The interfacial region of the polymer system covers the diamond ATR crystal. (Right) ATR-FTIR spectra of PVP K15 (dotted (black dotted in the web version)) and PEG 600 (dashed (red dashed in the web version)) measured with 8 cm^{-1} spectral resolution. *(Source: Ref. [12].)*

glycol) (PEG), are positioned next to each other on top of the diamond, so that the evanescent field of the IR beam extends into the bottom layer of the polymers. The IR beam coming as recombined beam from the Michelson-interferometer-part of the Fourier-transform device (compare Fig. 6.2) is totally internally reflected at the diamond and guided to the focal plane array (FPA) detector. The evanescent field emerges outside the diamond above the surface where the IR beam is totally reflected. The FPA detects an image of the evanescent-field layer. For each pixel on the FPA the intensity is recorded as a function of the delay between the fixed and the delayed-split-beam (compare Fig. 6.2). From this 'intensity-over-delay' function the 'intensity-over-wavelength' spectrum can be computed for each pixel applying the discrete Fourier-transform as described in Section 6.1.2. Thus, having made a reference and a sample measurement, the absorbance spectrum can be computed according to Eqn (2.80) for each pixel. The right part of Fig. 6.13 shows two example absorbance spectra of PVP and PEG. The absorbance spectrum of PVP was acquired with one pixel of the FPA detector located in the image region of PVP whereas the absorbance spectrum of PEG was acquired with another pixel which is located in the image region of PEG. It is important to underline that for each pixel of the FPA an absorbance spectrum is measured, meaning that each pixel records information of the chemical composition from a specific location in the evanescent-field-layer.

The right part of Fig. 6.13 shows that the absorbance spectra of PVP and PEG show spectral features which can be considered not-interfered with each other. Therefore Fleming et al. [12] chose the spectral range between $1600\,cm^{-1}$ and $1720\,cm^{-1}$ for the integration of the PVP absorbance and the spectral range between $1000\,cm^{-1}$ and $1175\,cm^{-1}$ for the integration of the PEG absorbance.

Figure 6.14 shows the spatial distribution of the integrated absorbances of PVP for different contact times between the miscible polymers without and with CO_2 being present at different pressures and 308 K. The left column represents the starting conditions when the two polymers are still separated by a sharp interface. As time proceeds, the sharp step in the spatial distribution of PVP degrades in each row. With increasing pressure, more CO_2 can be dissolved in the polymers [17]. The sorption of CO_2 into the polymer is accompanied with swelling of the polymer and a decrease in the chain–chain interactions which increases the mobility of the molecules. Thus with increasing pressure the initially sharp boundary between PVP and PEG degrades faster into a smooth transition from PVP-rich regions to PVP-poor regions.

The swelling of PVP can also be observed in Fig. 6.14 when comparing the absorbance in the PVP-rich regions far away from the PVP/PEG boundary at 0 min and later. A slight decrease in the integrated absorbance can be seen. This is due to the fact that in a swollen polymer – compared to a not swollen one – the PVP density is decreased and therefore simply less PVP molecules can absorb the IR radiation in the thin evanescent-field layer.

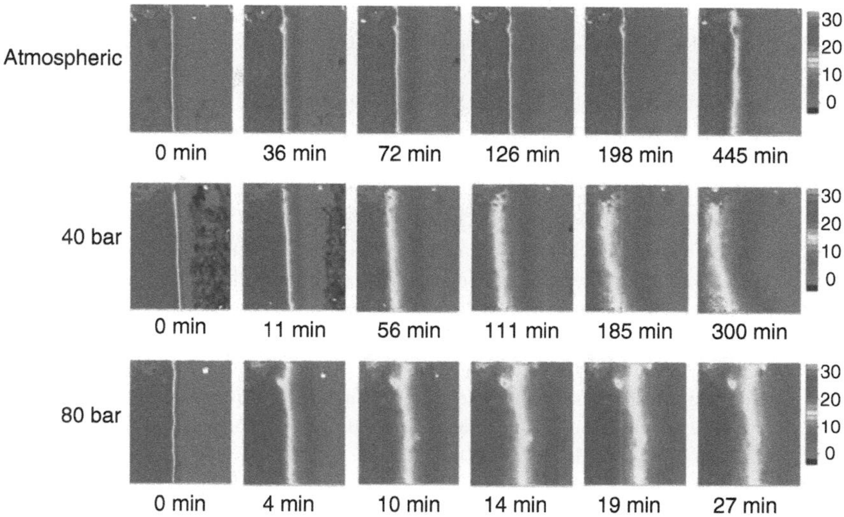

FIGURE 6.14 FTIR-ATR images corresponding to PVP in contact with PEG 600 as a function of pressure at 308 K. The images are based on the distribution of the integrated absorbance of the carbonyl band of PVP. *(Source: Ref. [12].)*

Next to PEG-specific and PVP-specific integral absorbances Fleming et al. [12] also evaluated the integrated absorbance of CO_2 in the spectral range between $2300\,cm^{-1}$ and $2350\,cm^{-1}$. The temporal and spatial correlation of the PVP, the PEG and the CO_2 integrated absorbances made possible a thorough analysis how the interplay of these three compounds influences the interdiffusion process between PVP and PEG in the presence of compressed CO_2. Further details are very much related to mass transport for which reason I refer to Ref. [12] for interested readers.

Finally I want to mention that the pixels located in the upper-left corner of the FPA are simply damaged and cannot provide reasonable signals. This explains the artefacts visible in the upper-left corner of the images in Fig. 6.14.

REFERENCES

[1] Werblinski T, Mittmann F, Altenhoff M, Seeger T, Zigan L, Will S. Temperature and water mole fraction measurements by time-domain-based supercontinuum absorption spectroscopy in a flame. Appl Phys B 2015;118:153–8.

[2] Hult J, Watt RS, Kaminski CF. High bandwidth absorption spectroscopy with a dispersed supercontinuum source. Opt Express 2007;15:11385–95.

[3] Werblinski T, Engel SR, Engelbrecht R, Zigan L, Will S. Temperature and multi-species measurements by supercontinuum absorption spectroscopy for IC engine applications. Opt Express 2013;21:13656–67.

[4] Schenkel B. Supercontinuum generation and compression, in, Diss., Naturwissenschaften. Eidgenössische Technische Hochschule ETH Zürich, Nr 2004;15570:2004.

[5] Buback M, Vögele HP, Winkels HJ. High-pressure absorption spectroscopy in polymerizing systems, in: Makromolekulare Chemie. Macromolecular Symposia, Wiley Online Library; 1986, pp. 69–74.

[6] Buback M. Spectroscopy of fluid phases – the study of chemical reactions and equilibria up to high pressures. Angewandte Chemie 1991;30:641–53. International Edition in English.

[7] Yamawaki H, Yoshida M, Kakudate Y, Usuba S, Yokoi H, Fujiwara S, Aoki K, Ruoff R, Malhotra R, Lorents D. Infrared study of vibrational property and polymerization of fullerene C60 and C70 under pressure. J Phys Chem 1993;97:11161–3.

[8] Böhm D, Grau T, Igl-Schmid N, Johnsen S, Kaczowka E, Klotz A, Schulmeyr J, Türk M, Wiegand G, Wuzik A, Zehnder B. Demonstration of NIR inline monitoring for hops extraction and micronization of benzoic acid in supercritical CO_2. J Supercrit Fluids 2013;79:330–6.

[9] Werblinski T, Mittmann F, Zigan L, Will S. Temperature measurements in gas flows at elevated pressures using supercontinuum absorption spectroscopy. In: 17th International symposium on applications of laser techniques to fluid mechanics, Lisbon, Portugal; 2014.

[10] Barber R, Tennyson J, Harris G, Tolchenov R. A high-accuracy computed water line list. Mon Not R Astron Soc 2006;368:1087–94.

[11] Voronin BA, Lavrentieva NN, Mishina TP, Chesnokova TY, Barber MJ, Tennyson J. Estimate of the J′ J ″dependence of water vapor line broadening parameters. J Quant Spectrosc Radiat Transfer 2010;111:2308–14.

[12] Fleming OS, Chan KA, Kazarian SG. High-pressure CO_2-enhanced polymer interdiffusion and dissolution studied with in situ ATR-FTIR spectroscopic imaging. Polymer 2006;47:4649–58.

[13] Bairamov D, Chalykh A, Feldstein M, Siegel R, Platé N. Dissolution and mutual diffusion of poly (*N*-vinyl pyrrolidone) in short-chain poly (ethylene glycol) as observed by optical wedge microinterferometry. J Appl Polym Sci 2002;85:1128–36.

[14] Bairamov DF, Chalykh AE, Feldstein MM, Siegel RA. Impact of molecular weight on miscibility and interdiffusion between poly (*N*-vinyl pyrrolidone) and poly (ethylene glycol). Macromol Chem Physic 2002;203:2674–85.

[15] Klein J. Interdiffusion of Polymers. Science 1990;250:640–6.

[16] Kazarian SG, Chan KA. FTIR imaging of polymeric materials under high-pressure carbon dioxide. Macromolecules 2004;37:579–84.

[17] Pastore Carbone MG, Di Maio E, Scherillo G, Mensitieri G, Iannace S. Solubility, mutual diffusivity, specific volume and interfacial tension of molten PCL/CO_2 solutions by a fully experimental procedure: effect of pressure and temperature. J Supercrit Fluids 2012;67:131–8.

Subject Index

D

H

I

K

L

M

N

O

P

Q

R

S

FSC
www.fsc.org
MIX
Papier | Fördert
gute Waldnutzung
FSC® C083411

Kolibri 360 GmbH
Ferdinand-Jühlke-Straße 7
99095 Erfurt, Deutschland
produktsicherheit@kolibri360.de